UNIVERSAL WELL CONTROL

UNIVERSAL WELL CONTROL

GERALD RAABE
SCOTT JORTNER

ELSEVIER

Gulf Professional Publishing
An imprint of Elsevier

Gulf Professional Publishing is an imprint of Elsevier
50 Hampshire Street, 5th Floor, Cambridge, MA 02139, United States
The Boulevard, Langford Lane, Kidlington, Oxford, OX5 1GB, United Kingdom

Library of Congress Cataloging-in-Publication Data
A catalog record for this book is available from the Library of Congress

British Library Cataloguing-in-Publication Data
A catalogue record for this book is available from the British Library

ISBN: 978-0-323-90584-8

For information on all Gulf Professional publications
visit our website at https://www.elsevier.com/books-and-journals

Cover Image Credit: Hung Tran, all rights reserved.

Publisher: Charlotte Cockle
Senior Acquisitions Editor: Katie Hammon
Editorial Project Manager: Aleksandra Packowska
Production Project Manager: Poulouse Joseph
Cover Designer: Matthew Limbert

Typeset by STRAIVE, India

We dedicate this book to oil and gas professionals everywhere who have dedicated their lives to producing the energy which affords the world the highest standard of living in the history of mankind; and to rig crews and support personnel whose essential careers and diligent efforts are such an integral part of the success of our industry.

Finally, we dedicate this book to our wives, Peggy and Sheila, and our families. Your unending love and support illuminated a clear path paved with creative energy.

Contents

For additional information on the topics covered in the book, visit the companion site: https://www.elsevier.com/books-and-journals/book-companion/9780323905848

Authors biography

Gerald "Gerry" Raabe

Gerald Raabe is a Sr. Drilling and Well Control Engineer with a BS in Petroleum Engineering from Texas A&M University. His career spans over 42 years where, in his early years, he began as a floor hand, progressing through each rig position including tool pusher and drilling representative. Over the years, Gerry has advanced to management and executive roles within a major oil company, an oil well service company, and the world's leading well control company. His wealth of experience has come from many challenging assignments including resident assignments in Indonesia, Nigeria, and Kuwait overseeing operations within rank-wildcatting, high pressure/high temperature, steam, inland waterways, offshore, deepwater, arctic, tropic, and desert environments. These assignments afforded him the opportunity to work closely with colleagues in multicultural environments. As a 3D illustrator/animation hobbyist, Gerry has dedicated himself to sharing of many of the lessons learned to improve safety for rig crews as coauthor of *Universal Well Control.*

Charles "Scott" Jortner

Scott received his BS Degree in Petroleum Engineering from Louisiana State University in 1973 and is a licensed professional engineer in the State of Texas. Scott has over 45 years of experience in drilling, workover, production, and well control operations. He has worked as a Drilling Engineer, Production Superintendent, Drilling Superintendent, Drilling Manager, Well Control Engineer, and Well Control Manager for both domestic US and international operations—onshore and offshore (including deepwater). Well control operations include pressure control, blowouts, contingency planning, kick modeling, and dynamic kills. Extensive time in support of these operations was spent in the field. He has also participated in API and AADE special committees reviewing certain drilling practices. Scott has been a member of the Society of Petroleum Engineers (SPE) and the American Association of Drilling Engineers (AADE). Scott is also an avid pickleball player.

Foreword

I felt very honored when the authors asked me to write the foreword for this book. Then I became very humbled when the thought hit me "I have never written a Preface to a book before, what do I write?" So, I decided that this would be a great opportunity to express my philosophy of well control that I have developed during my 40+ years as a Drilling Engineer, most of which has focused on well control. I have been involved in well control during several phases of my career.

- First as an undergrad student at Texas A&M learning the basics of drilling;
- Second as a young engineer sitting through basic and refresher well control courses taught by the industry;
- Third as a well control practitioner with both Pennzoil and HNG/Enron Oil and Gas;
- Fourth as a well control instructor at the University of Houston–Victoria, Petroleum Training Institute. This is where I actually began to learn well control procedures.
- Fifth as a graduate student back at Texas A&M. Here, I continued as a trainer, but began to understand how fluids actually behave with changing temperature, pressure, and composition.
- Finally, as an educator (not to be confused with trainer) and researcher on the faculty at Texas A&M.

The last phase is where I began to realize the more you learn about a topic or process, the more you realize how little you actually know! As an industry, we know very little how fluids interact within the wellbore during a well control situation. What we think we know, which may not be correct, has gotten us into some deadly situations. Unfortunately, it is usually the only time the industry attempts to gain an understanding of what really happened after one of these disasters. There are research projects undertaken, and when the projects are complete, the industry pretty much goes back to doing the same thing they did before. After all, kick detection and kick circulation should not be difficult **IF** the kick progresses as we learn in our Basic Well Control courses. The vast majority of kicks can be handled with the skills gained from these courses.

I may be exaggerating some. There have been significant improvements in equipment, techniques, and procedures, such as Managed Pressure Drilling, riser gas handling equipment, and so on. These improvements aid in drilling more complex wells (horizontal, expended reach, and ultradeep water) where we need to rethink well control. Continuous improvements in drilling technology necessitate continuous improvement in well control.

The authors of this manual "Universal Well Control," Gerald Raabe and Scott Jortner, both have years of experience and extensive knowledge in drilling, workover and completion operations. They both have worked as well control professionals and have conducted well control training classes. It is my opinion that they have used this extensive experience and knowledge to produce the most complete and up-to-date Well Control Manual available to the industry. Their intended purpose for this work is to provide guidance to the question "We have taken a kick, what do I do next?" and I believe that the information and procedures explained in this Manual are sufficient to safely handle and kill the majority of kicks that will be taken.

Even with following what is written in this Manual, there will still be blowouts. People make mistakes because they do not understand what is happening down hole. The industry, government, academia, research labs, etc. have to work together for research programs (not just projects) to gain a better understanding of what is actually going on in the well and at the surface during well control situations.

Jerome J. Schubert
The Harold Vance Department of Petroleum Engineering,
Texas A&M University, College Station, TX, United States

Quick tips: What to do now?

Gerald Raabe and Scott Jortner

A kick has been taken, the well shut in, and it's been many months since your last well control school. What do you do now? The following criteria should be used as a guideline for developing a well kill plan and discussed with both field and office personnel before initiating well kill circulation. Operational policy or bridging document may dictate type of well control method to be employed.

Worst-case analysis—Assume the influx is gas

1. **For water-based muds**, gas will migrate at rates up to 4000 feet (ft) per hour. To offset impact of migration rates, the **Driller's Well Control Method** is suggested. The first Circulation of Driller's Method will be used to circulate out the influx providing adequate time to thoroughly mix KMW (see *Driller's Method Recap*).
2. **For oil-based muds,** gas migration is less of a concern as the gas will remain in solution minimizing migrations rates.
   a. The **Driller's Well Control Method** should be considered first to rapidly remove the influx (see *Driller's Method Recap*).
   b. If rig contains appropriate mud mixing equipment or well contains close kick tolerances, the **Wait and Weight Well Control Method** is an acceptable alternative (see *Wait and Weight Recap*).
3. If the influx may contain **hazardous gases** (i.e., H_2S hydrocarbons) which may exceed 10 ppm H_2S or 8% LEL for hydrocarbons and may poise an eminent threat to regional area population and infrastructure, and crew safety, or if the rig has not been outfitted with H_2S emergency systems, the **Bullheading Method** should be considered (see *Bullheading Recap*).

You are OK

1. If the well has been successfully shut-in and the kick has been contained without jeopardizing the casing seat, **you are ok.**
2. Using any constant bottom-hole pressure well control method, high probability exists in which the influx can be safely mitigated.

Quick tips

Bumping the float

- With the well shut-in, notify supervisor(s) accordingly (see Fig. 1).
- Accurately record stabilized shut-in casing pressure and pit gain.
- Bump float by appropriate means and record stabilized shut-in drillpipe pressure.
- Bumping float procedure
  - Record $SIDPP_{initial}$, $SICP_{initial}$, and Pit Gain.
  - With choke closed, engage pumps (10 spm).
  - DP will start increasing over time, followed by gradual increase in CP.
    - For light-weight muds, a pressure "lull" or multiple readings of the same pressure may be observed.
    - For heavier-weight muds, a pressure "lull" may not occur.
  - Stop pumps after CP increases (allow no more than 100 psi rise).
  - Subtract the $SICP_{inital}$ from $SICP_{final}$. = Trapped Pressure (Δ P).
  - Subtract Trapped Pressure (Δ P) from $SIDPP_{final}$ for SIDPP.
- The amount of time for stabilization will vary depending on field conditions.
- Record active pit volumes, barite, and chemical stocks.

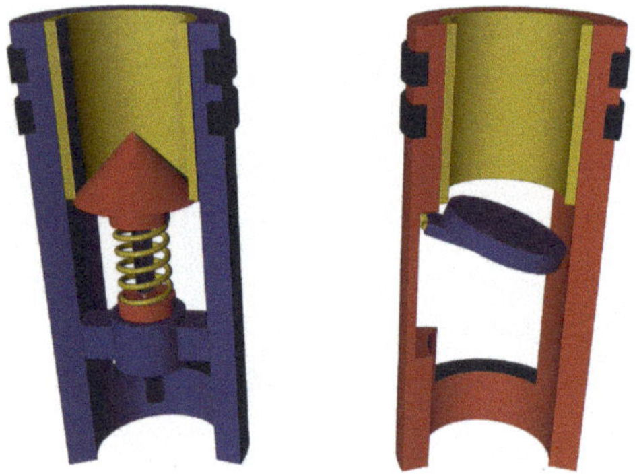

Fig. 1 Plunger and flapper type float valves.

PreJob safety meeting

Team

- For the most part, do not be in a hurry (unless abnormal conditions exist).
- Team effort is needed—ensure importance of good communication for each team member.
- Everyone is important—from Roustabout to Company Man.
- Speak the truth—do not embellish, do not boiler house (repeat what you believe superiors want to hear), do not guess.
- Hold question and answer period, ensure all questions are answered to complete satisfaction.
- Review instructions or directions.
  - Ensure responders repeat instructions in their own words to ensure instructions are clearly understood. Be patient.

- Review well control method to be employed (see appropriate well control method recap).
- Variances in mud density are expected when mixing mud while pumping. State correct mud densities.
- Notify supervisor of any change or unusual occurrence.
- Wells can be killed without knowing SPR (slow pump rates). Can be estimated during circulating.
- Additional back pressure (keeping circulating pressure higher than calculated) is not needed and can be detrimental if too much is applied. The safety factors described below show how additional pressure is applied to the bottom of the hole.
  - Safety Factor 1—Circulating friction pressure is added to HP.
  - While circulating at the SPR, some slight additional back pressure on the formation is exerted by the friction loss in the annulus. There is not a lot of friction loss due to the slow pump rate, but it does give a slight overbalance to the formation.
  - Safety Factor 2—KMW is rounded up adding additional HP.
  - The calculated KMW is rounded up to the nearest 0.1 ppg and never down. This slight increase in MW adds some additional hydrostatic pressure to the bottom of the hole. For example, in a 12,000 ft. straight hole, the rounded-up mud weight of 10.3 ppg versus the calculated KMW of 10.22 ppg adds an additional 50 psi of hydrostatic pressure.
  - Safety Factor 3—Trapped pressure after bumping the float (if not bled off).
  - This trapped pressure on the annulus can be used as a safety factor while bringing the pumps up to speed.
  - Safety Factor 4—Leak-Off Test/Formation Integrity Test is rounded down.
  - LOT/FIT is rounded down, meaning the MAASP may be slightly higher than calculated, allowing the shoe to withstand additional pressure before fracturing.
- Execute notification plans for surrounding infrastructure (i.e., rigs).
- BOP back-up procedures should be discussed if primary BOP malfunctions during kick.
- Know where the tooljoint is positioned within BOP.
- Review H_2S procedures in case alarms sound.
- Review hand-off procedures for crew changes and bathroom and smoking breaks.

Driller

- Re-zeroes pit level gain/losses and strokes.
- Records total strokes, pit levels, DP, and Casing pressures.
- Monitors Active System Volume and records every 10 min. If applicable, notifies Choke Operator by loudspeaker.

Assistant driller

- Oversees continuous operations by visiting each station.
- Reports back to Driller.

Pit staff

- Continually monitors returns.
- Immediately notifies Driller of any changes.
- Reports back any problems during mud mixing operations.
- Does not transfer fluids without notifying Driller first. Driller to relate information to Choke Operator.

Mud company representative

- Provides MW_{in}/MW_{out} every 15 min or until asked to change.
- Provides Barite + Mud and/or Base Oil addition rate (bbl/h).
- Uses Active pit system, if possible.
- Any and all mud transfers to be announced. Driller to record all changes.

Choke operator

- See Kill Circulation (Choke Operator)

Choke manifold and BOPs

- Isolate and operate all chokes before initiating kill circulation.
- Open appropriate isolation valves upstream of choke manifold. Open all appropriate valves downstream of choke manifold. Direct returns to MGS.
- Assign continuous visual inspection for BOP and choke manifold for quick leak detection during circulation. BOP testing is performed with fluid, and influx may be gaseous and could leak more easily than fluid.
- Place markers on all valves showing opening and closing position.

MGS, flare, pits, and pumps

- Fill mud leg of MGS with Kill Mud Weight (KMW) before initiating circulation. Ensure MGS mud leg is clear of debris.
- Flare stack ignition source should be verified as operable before initiating kill activities.
- MGS butterfly valve flanges are weakest link and are known to leak during well control. Have back-up gaskets available in case of a leak.
- Verify H_2S and Gas monitors are functioning and engaged during circulation. Limits below are for exposure of 8-h period and may vary due to operational requirements. Please see Supervisor for exact limits (see Fig. 2).
    - **10 ppm** H_2S is highest threshold for activities.
    - **8% LEL** levels are highest threshold for activities.

Fig. 2 Verify limits.

- Fluid volumes handling procedures should be discussed during prejob safety meeting.
  - Mud volume increases due to barite and chemical additions.
  - Mud volume increases due to gas expansion during well kill.
- Degasser to remain on throughout circulation.
- Ensure back-up mud pump is readied for service.

Kill circulation

Choke operator

- Understand weather forecast for duration of well kill operation (see Fig. 3).
- Fatigue during long well control circulations can be expected. Plan for it.
  - Appropriate Well Control Kill Sheet.
  - High seat stool chair (bar stool).
  - Hand-held calculator.
  - Weather-proof (Wind-resistant) clip board.
  - Flashlight.
  - Hard candies (to keep mouth most).
  - DO NOT USE PHONES—not intrinsically safe and can cause spark.

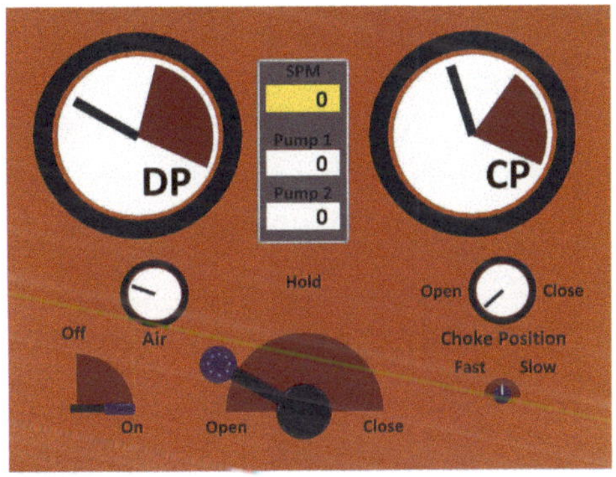

Fig. 3 Choke panel.

- Commit to Memory
  - ICP calculated.
  - FCP.
  - Strokes to Bit.
  - Total Strokes.
- Review common problems which may be encountered during circulation (see *Common Problems*).
- Use only ONE set of gauges to perform kill activities (normally choke panel gauges).
- Monitor pit levels throughout circulation. Driller to notify choke operator every 10 min of pit level throughout circulation.
- Record the following:
  - Time Interval
  - Strokes
  - SPM
  - Barrels pumped
  - DP Actual (vs. DP calculated)
  - CP Actual
  - Choke Position
  - Pit Volume

  - MW_{in}
  - MW_{out}
- Remember lag time between choke manipulation and response on drillpipe pressure gauge (2 s per 1000 ft).
- When at SPR, choke operating position should be approximately half open (50% open).
  - If at any time during DP step-down circulation, the choke position is at or near full open at SPR, shut-down and shut-in (probable partially plugged bit or choke).
- When in doubt, shut-in. All pump variances must be done by holding casing pressure constant.
  - DO NOT OPEN THE CHOKE when gas reaches the surface.
  - If drillpipe becomes plugged during circulation, well control can be maintained using the Volumetric Method.

On bottom kill and well shut-in

Water-based fluids

If the well contains Water-Based fluids, gas influxes may rise fast without circulation. Rising gas in a shut-in well will greatly increase the pressure at the casing shoe and greatly increase the chances of breaking down the shoe.

(a) Gas Migration Distance

$$\text{Gas Migration Distance}_{\text{TVD}_{\text{ft}}} = \left(\frac{\text{Rise in SICP}}{\left(\text{MW}_{\text{ppg}}\,\text{x}\,0.052\right)}\right) \tag{1}$$

(b) Gas Migration Rate

$$\text{Gas Migration Rate}_{\text{TVD}_{\text{ft}}/\text{min}} = \frac{\text{Gas Migration Distance}_{\text{TVD}_{\text{ft}}}}{\text{Time of Rise}_{\textbf{min}}} \tag{2}$$

After organizational meetings, begin circulation as soon as possible. Remember, this is a two-circulation well kill method.

Driller's method recap

First circulation

$$\overline{\text{HP}}_{\text{Drillpipe}} + \overline{\text{SIDPP}} + \text{Friction}\uparrow = \overline{\text{BHP}} = \overline{\text{HP}}_{\text{Annulus}} + \overline{\text{SICP}} + \text{Friction}^{\nearrow\emptyset}$$

Fig. 4 First Circ—Anytime pumps are varied, hold CP constant.

$$\overline{\text{HP}}_{\text{Drillpipe}} + \overline{\text{Friction}}_{\text{Circulating}} = \overline{\text{BHP}}_{\text{Circ Until Influx is Out}}$$

Fig. 5 First Circ—Hold DP constant until influx has be circulated out.

1. Stage pumps up to SPR, while holding CP constant (see Fig. 4).
2. Read and record ICP on DP (see Fig. 5).
3. Maintain constant ICP at SPR until Influx is circulated out.
4. Stage pumps down to off while holding CP constant.
5. If SICP > SIDPP, repeat steps 1–4.
6. Using SIDPP, determine KMW.
7. Monitor well while mixing KMW.

Second circulation

$$\overline{HP}_{Annulus} + \overline{SICP} + \text{Friction}^{\uparrow\emptyset} = \overline{BHP}_{KMW\ to\ Bit}$$

Fig. 6 Second Circ—Hold CP constant until KMW reaches bit.

$$\overline{HP}_{Drillpipe} + \overline{\text{Friction}}_{Circulating} = \overline{BHP}_{KMW\ to\ Surface}$$

Fig. 7 Second Circ—After KMW reaches bit, Hold DP constant until KMW to surface.

1. Mix and route KMW for Kill Operations.
2. Stage pumps up to SPR, while holding CP constant.
3. Hold CP constant until KMW reaches bit (see Fig. 6).
4. Hold DP constant until KMW reaches surface (see Fig. 7).
5. $MW_{in} = MW_{out}$ for three consecutive readings over 15 min or 1.5 times total circulation volume to ensure the well is dead.

Notes

1. Monitor shut-in casing pressure during preparations to initiate the Driller's Method. If SICP increases significantly, suggest using the Volumetric Control method to allow influx to expand under controlled environment.
2. If problems occur during first circulation of Driller's Method well kill operations, be prepared to cease pumping operations and switch to Volumetric Control.
3. Initial circulating pressure (ICP) should be the sum of the shut-in drill pipe pressure (SIDPP) plus the slow pump pressure (SPP).
4. Keep the pump rate constant at the SPR, even if the ICP is not exactly what was calculated.
5. Choke operating position should be approximately half open (or somewhat less) at SPR. If at any time circulation, the choke position is at or near full open at SPR, shut-down and shut-in (probable partially plugged bit or choke).
6. During first Circulation, the resultant ICP should be held constant until the kick is circulated out of the hole.

- Suggest overdisplacing (~1–1/2 hole volumes) to ensure entire kick has been circulated out.

7. While holding casing pressure constant, slow the pumps down and shut-in the well trapping the pressure on the casing. This pressure should be the same value as the initial SIDPP.
   - If SICP > SIDPP, an influx is probably in the annulus and need to repeat first circulation of Driller's Method.
   - If SIDPP = SICP, proceed to second circulation of the Driller's Method.
8. During second Circulation, after staging pumps up to speed holding casing pressure constant, casing pressure is to be maintained until KMW reaches the bit.
9. During second Circulation, after KMW reaches bit, the FCP is to be maintained until $MW_{in} = MW_{out}$ has three consecutive readings over 15 min or 1.5 times total circulation volume.
10. Open choke to check for flow. Do not open BOP before opening choke to check for pressure.
11. Ensure floor is clear of personnel when BOPs are opened as trapped pressure below preventer elements may exist.

Oil-based drilling fluids

If the well contains Oil-Based Fluids, generally speaking, gas influxes will migrate extremely slow.

1. Wait and Weight can be used, as time is not of the essence.
   a. The Driller's Method may still be the best consideration to use as it takes less time to initiate circulation, and there are fewer calculations to make and generally fewer choke manipulations required.
   b. Sufficient time is available to prepare proper kill weight mud and allow chemicals time to properly yield for more consistent mud, prior to initiating the second circulation.
2. Gas will break out suddenly at the bubble point, and rapidly expanding gas may blow mud out of the MGS mud leg followed by gas. Gas escaping from the mud leg will be directed to the mud pit area and will be dangerous for personnel. Shut-in the well if this occurs.
3. May want to slow down pump rate as gas nears 80% of annular volume.

Varying pumps, estimate new pump pressure

$$\text{New Pump Pressure}_{\mathbf{psi}} = \text{Current Pump Pressure}_{\mathbf{psi}} \times \left(\frac{\text{New SPM}}{\text{Old SPM}}\right)^2 \quad (3)$$

Wait and weight well control method recap

$$\overline{HP}_{Drillpipe} + \overline{SIDPP} + Friction \uparrow = \overline{BHP} = \overline{HP}_{Annulus} + \overline{SICP} + Friction^{\uparrow 0}$$

Fig. 8 Anytime pumps are varied, hold CP constant.

$$HP_{Drillpipe} \uparrow + DP \downarrow = \overline{BHP}$$

Fig. 9 Step down chart compensates for increase in HPdrillpipe by reducing circulating drillpipe pressure.

$$\overline{HP}_{Drillpipe} + \overline{Friction}_{Circulating} = \overline{BHP}_{KMW\ to\ Surface}$$

Fig. 10 Circ KMW to surface.

1. Slowly stage pumps up to speed hold casing pressure constant (see Fig. 8)
   a. Maintain constant SPR for duration of circulation.
2. Once Surface Volume has been pumped, reset stroke counters.
3. If $ICP_{actual} = ICP_{calculated}$, follow Drillpipe Step-Down Chart (see Fig. 9).
   - If $ICP_{actual} > ICP_{calculated}$, use formula below to determine SPR $Pressure_{actual}$.
   - Recalculate Step-Down Chart.
4. Once the KMW is at the bit, the circulating DP will be at FCP. Maintain constant FCP on DP until influx has been circulated out (see Fig. 10).
5. Suggest circulating 1.5 total hole volume or when $KMW_{in} = KMW_{out}$ and verified by three consecutive and equal MW measurements over 15 min to ensure the kick has been circulated out of the hole.
6. Open choke to check for flow. Do not open BOP before opening choke to check for pressure.
7. Ensure floor is clear of personnel when BOPs are opened as trapped pressure below preventer elements may exist. With no flow, well is dead.

SPR $Pressure_{actual}$

1. SPR $Pressure_{actual} = ICP_{actual} - SIDPP$
2. Substitute SPR $Pressure_{actual}$ in formula Below and Recalculate FCP_{actual}
3. FCP actual = SPR Pressure $_{actual} \times (KMW/OMW)$
4. Recalculate step down chart using ICP_{actual} and FCP_{actual}

Notes

1. If problems occur during Wait and Weight Method well kill operations, be prepared to cease pumping operations. May consider using Volumetric Control depending on circumstances.
2. Choke operating position should be approximately half open (or somewhat less) at SPR. If at any time circulation, the choke position is at or near full open at SPR, shut-down and shut-in (probable partially plugged bit or choke).
3. Perform step-down chart.
4. FCP should be held constant until the kick is circulated out of the hole.
   - Suggest overdisplacing (~1–1/2 hole volumes) to ensure entire kick has been circulated out.
5. While holding casing pressure constant, slow the pumps down and shut-in the well trapping the pressure on the casing. This pressure should be the same value as the initial SIDPP.
   - If SICP > SIDPP, an influx is probably in the annulus and need to repeat first circulation of Driller's Method.
   - If SIDPP = SICP, well should be dead and well may contain trapped pressure. Consider raising KMW density prior to tripping.
6. Open choke to check for flow. Do not open BOP before opening choke to check for pressure.
7. Ensure floor is clear of personnel when BOPs are opened as trapped pressure below preventer elements may exist.

Off-bottom kill (mud cap)

If the bit is off bottom when an influx occurs and the well is shut-in, several calculations will need to be performed to determine if the well can be killed with the bit off bottom or if the bit need to be "stripped" back to bottom to perform the well kill. Use Variable Bit Depth Kill sheet or determine as follows.

1. Determine if the density of KMW needed for an off-bottom kill will exceed the casing/liner shoe leak off pressure or formation integrity test pressure.
   a. The reservoir pressure calculated by adding the hydrostatic pressure of the original mud in the hole to the SIDPP.
   b. To equal reservoir pressure, the combined hydrostatic pressures in the annulus/open hole (the hydrostatic pressure of the length of the kick plus the hydrostatic pressure of mud from the top of the kick to the bit plus the hydrostatic pressure of KMW from the bit to the surface) must equal the reservoir pressure. Only the mud from the bit to surface can be weighted up (mud cap kill).

c. Divide current SIDPP by 0.052 and the true vertical depth of the bit off bottom. This calculation estimates the additional mud weight needed to be added to current mud weight for KMW off bottom.

$$\Delta MW_{\text{Bit Off Bottom TVD}} = \frac{\text{SIDPP}}{(0.052) \times \text{Bit TVD}} \tag{4}$$

$$KMW_{\text{Bit off Bottom}} = OMW_{\text{ppg}} + \Delta MW_{\text{Bit Off Bottom TVD}} \tag{5}$$

d. Compare the hydrostatic pressure of calculated $KMW_{\text{bit off bottom}}$ at the casing/liner shoe to the LOT/FIT. If it is greater, then stripping is required.

2. The Volumetric Method can be employed in water-based fluids to let a gas bubble rise above the bit, and then it may be possible to circulate influx out using the Driller's Method.
   a. Once complete, stripping will then be necessary to go back to bottom and circulate out the light mud using the second circulation of the Driller's Method.
3. For stripping operations, suggested equipment options include:
   a. Use calibrated tripping (strip) tank. Stripping tank is preferable due to more accurate readings for small volumes of mud, but a tripping tank can be used if needed. Trip tank (or stripping tank) to be connected to the choke manifold at point located downstream from the operating shock.
   b. Annular Preventer without Surge Bottles
      i. Ensure means of effective communication between Driller and person at remote BOP closing unit (Accumulator).
      ii. Liberally coat each tooljoint with grease and prepare for stripping. Grease is preferable to Pipe Dope as pipe dope contains solids which may be abrasive to the annular element.
      iii. Ensure a fluid level exists on top of annular preventer (observe from rig floor). If not, top off annular preventer with water.
      iv. Position remote BOP closing unit operator and be ready to adjust annular regulator. Ensure effective communications can occur throughout stripping operation. Record annular closing pressure for full effective seal on drillpipe body ($\text{Pressure}_{\text{reference}}$).
      v. Adjust Annular Preventer closing pressure to allow small seepage of mud. Record primary annular closing pressure for seepage ($\text{Pressure}_{\text{DP body}}$).
      vi. Slowly strip tooljoint into element and reduce Annular Closing Pressure to allow seepage when tooljoint has entered into the annular element and forces element to partially open. Record secondary annular closing pressure for seepage ($\text{Pressure}_{\text{Tooljoint}}$).

vii. Operating annular closing pressure "stripping range" will be between $\text{Pressure}_{\text{Tooljoint}}$ and $\text{Pressure}_{\text{DP body}}$. When the drillpipe body is being stripped, the Annular Closing Pressure to be adjusted to $\text{Pressure}_{\text{body}}$. When the tooljoint enters annular and until it passes complete through the element, the Annular Closing Pressure will be decreased to the $\text{Pressure}_{\text{Tooljoint}}$. Periodic adjustments in pressure ranges may be necessary due to annular element wear.

viii. Strip into hole until the bottom of the hole is reached.

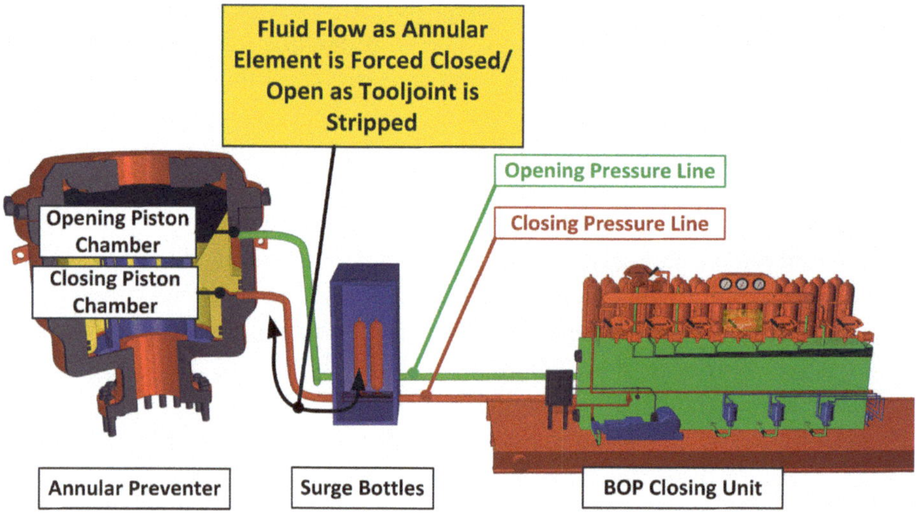

Fig. 11 Surge bottle diagram.

c. Annular Preventer with Surge Bottles. Surge bottles are used to automatically reduce and increase annular preventer closing pressures as tooljoints pass through the annular preventer. Two (2) ten-gallon accumulator bottles along with a simple manifold can be installed. Determine operating Stripping Pressure range as follows (see Fig. 11):

<u>For initial installation or during BOP Test, Perform Stripping Test:</u>

i. Install Annular Surge System (Surge bottles and manifold) on the closing line of the Annular Pressure. Precharge Annular Surge system must equal the BOP Closing Unit's precharge pressure of the accumulator bottles.

ii. Perform pressure test of Annular Surge System at rated operating range of BOP Closing Unit.

iii. Surge bottle bladder should be precharge to 50% of the minimum required closing pressure for the preventer.

iv. Prepare for Stripping test by ensuring a fluid level exists on top of annular preventer (observe from rig floor). If not, top off annular preventer with water.

v. Position operator at remote BOP closing unit for functioning Annular Pressure regulator. Ensure effective communications can occur throughout stripping operation.

vi. Lower test joint and locate tooljoint above Annular Preventer.

vii. Close Annular Preventer onto Test Joint's drillpipe body with full operating pressure according to manufacturer specifications. Record Annular Closing Pressure for full effective seal on drillpipe body ($Pressure_{Reference}$).

viii. Adjust Annular Preventer closing pressure to allow small seepage of mud (fluid level drops). Record annular closing pressure for seepage ($Pressure_{DP\ body}$)

ix. Slowly strip tooljoint into element. BOP Closing Unit operator will reduce the Annular Closing Pressure to allow seepage when tooljoint has entered into the annular element. Record $Pressure_{Tooljoint}$.

x. Open annular preventer and Operator at BOP Closing Unit to rest Annular Regulator pressure back to the original manufacture specifications ($Pressure_{Reference}$)

xi. Record Pressure $_{Reference}$, Stripping Range ($Pressure_{Tooljoint}$ $Pressure_{body}$) on Tour Sheet. The stripping range will be those pressure observed between $Pressure_{Tooljoint}$ and $Pressure_{DP\ body}$. to form an effective seal while allowing small seepage. When the drillpipe body is being stripped, the Annular Closing Pressure to be adjusted to $Pressure_{body}$. When the tooljoint enters annular and until it passes complete through the element, the Annular Closing Pressure will be decreased to the $Pressure_{Tooljoint}$. Periodic adjustments in pressure ranges may be necessary due to annular element wear.

After Installation—Ensure precharge pressures are verified during every Accumulator test and functional operability during BOP Tests.

d. Circulating and calibrated strip tank. Alternately, circulating and calibrate trip tank may be used.

e. Surge bottle affixed to the Annular Preventer closing port (or line). Surge bottles are suggested as to automatically reduce and increase annular preventer closing pressures as tooljoints pass through the annular preventer.

i. If surge bottles are unavailable, the annular opening chamber can be vented and pressures reduced/increased on the BOP Closing system annular regulator. This operation cannot be normally accomplished remotely, therefore adequate communication between rig floor and BOP Closing unit must be assured.

High H_2S and/or close proximity to the public

Most governments and companies do not want to circulate out a kick to the surface which may contain a high percentage of H_2S gas. H_2S gas can be a danger to the rig crews and to any nearby populations of people or livestock. If a kick has occurred in a formation with high H_2S gas, bullheading the kick back into the formation is generally an acceptable method of dealing with this type of kick. Care must be taken to not exceed the leak-off pressure at the casing shoe.

Bullheading

In most types of kick situations while drilling, bullheading as a means of primary well control is not advised. If bullheading is considered for well control, review the following:

1. In order to effectively kill the well by bullheading, the kick must be pumped back into the formation from which the kick originated. For the most part, bullheading can be effectively accomplished in carbonate formations.
2. Normally, but not always, a well has the weakest formation strength at the shoe, while each formation penetrated has higher formation strength. Bullheading will force fluids into the weakest formations, which may be uphole (shoe) from the original kick zone. Therefore, bullheading may breakdown the shoe resulting in a worse situation or fluids may enter another zone and not kill the well.
3. Bullheading is often used to pump kill fluids down the drillpipe/tubing to kill a producing well.
4. Ensure isolation exists between drillpipe and annulus.

Bullheading kill considerations

The following is a quick review for Bullheading operations.

1. Determine annular and drillpipe volumes from surface to bottom of hole.
2. Determine maximum bullheading pressure and rate.
   - **a.** Maximum Bullheading pressure will normally be calculated from 80% of casing burst at the shoe. Variances between 70% and 90% of burst may occur due to condition of casing and cement.
   - **b.** Stage pumps slowly until breakover occurs. Monitor casing pressure and do not exceed 80% of casing burst. Note breakover pressure.
   - **c.** Once breakover occurs, stage pumps to achieve maximum SPR while keeping pump pressures below 80% of casing burst.
   - **d.** Maximum pump pressures may also be limited to the pressure rating of the drilling pumps. If a cementing pump is used, the rate is usually the limiting factor.

   i. The practical limit for pumping rates through a 200-ft-long, 2 in nominal (ID of 1.128″) kill line is about 7 to 8 bpm. With light mud, a slightly higher rate can be obtained and a slightly slower rate with heavy mud. The pump pressure limit used for these observations is 5000 psi.
   ii. With light and heavy mud, the practical limit for pumping rates through a 200-ft long, 3 in nominal (ID of 1.8″) kill line increases up to 16 bpm with pressures below 3000 psi.
   iii. Please note that the maximum rates for pumping through a kill line do not represent the actual bullheading rates and pressure for the subject well. These maximum rates and pressures are only examples of kill line and pump limits.
3. Bullheading half of the annular volume first, followed by half of the drillpipe volume. The next step will be to continue bullheading the remaining half of the annular volume, followed by the remaining half of the drillpipe volume **+open hole volume +10%**.
   a. By bullheading the DP at the conclusion of bullheading operations, this will ensure the drill bit jets are not plugged and minimized any U tube effects.
   b. Ensure open hole volume is included in the annular bullheading calculations.
4. For most bullheading operations, the drill sting will have been pulled above the casing seat, in order to minimize drill string differential sticking potential (see Fig. 12).

DP Volume		bbls		strokes
Annular Volume		bbls		strokes
Burst		psi		
PopOff		psi		

Start Time	Stks	Bbls	DP	CP	SPM	Ck Pos	Pit Vol	MW In	Comments
	0	0			0				
					10				
					20				
					30				Note Breakover Pressure
					40				
					50				

Fig. 12 Example of tally book bullheading records.

Gas expansion (allow casing pressure to increase)

$$\overline{BHP} = HP_{Mud}\downarrow + HP_{Influx}\downarrow + CP\uparrow$$

Fig. 13 Influx pushes mud out as it expands, casing pressure increases.

One of the common misunderstandings which may occur during well control is the need to allow the casing pressure to rise in a controlled manner through choke manipulations while circulating. During the circulation, the influx must be allowed to expand in order to keep bottom-hole pressure constant (see Fig. 13).

If during the circulation, casing pressure does not increase using constant bottom-hole pressure methods, more than likely the influx only contains traces of gas and formation water/oil. Since these fluids will not expand rapidly, minimal gas expansion will take place resulting in minimal changes to casing pressure.

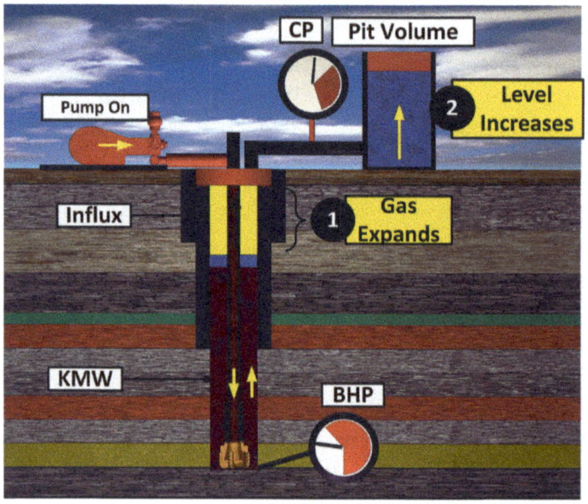

Fig. 14 Gas at surface pushing fluid from wellbore.

If gas expansion occurs, the gas will "push" mud from the wellbore. As gas begins to escape from the well, the choke is normally pinched by several closing choke manipulations in order to keep DP constant (allowing for lag time). The pressure of the gas must be maintained to offset the reduction of hydrostatic pressure of the mud "pushed" from the well. To maintain pressure within the gas influx, casing pressure MUST be allowed to increase (see Fig. 14).

Note: If casing pressure is held constant (by opening the choke), the pressure contained within the gas will lower and the bottom-hole pressure will be greatly reduced. If this mistake occurs, valuable time will be needed in order to repressurize the gas influx in order to achieve desired bottom-hole pressure.

Common problems during well control

If problems occur during well kill operations, be prepared to cease pumping operations. If influx is in annulus and SICP continues to rise, suggest switching to Volumetric Control Method, until circulation can be re-established.

1. **Gas blows out the mud leg on the MGS:**
   a. Cease pumping and shut-in well.
   b. Recalculate pressures using a slower pump rate and attempt to circulate after filling the MGS mud leg with mud and attempt to circulate out kick.
   c. If the same problem occurs, use the Lube and Bleed Method to remove gas from the wellbore. Circulation can then be resumed.
2. **If the choke line freezes downstream of the choke**
   a. Switch to back-up choke.
   b. Cease pumping and shut-in the well. Use Volumetric Method to safely bring gas to the surface (bleed off mud using the manual choke) while thawing line.

3. **Partial or Total lost Circulation**
    - **a.** Slow pump rate to see if it has any effect on circulation.
    - **b.** Shut-in well and mix LCM pill to pump through BHA and bit. Use as heavy a concentration as possible for the BHA in use.
    - **c.** If LCM and a slower pump rate do help, a different type of LCM may be needed such as a reactant plug.
    - **d.** Consideration should be given to contacting a well control specialist with well control company and mud specialist with mud company.

4. **Plugged choke**—It can occur at any time, and therefore, choke opening values and DP and CP must be continually recorded. If the casing pressure starts rising followed by rise in drillpipe pressure, plugging of the choke may be occurring. This may happen quickly or gradually (see Fig. 15).

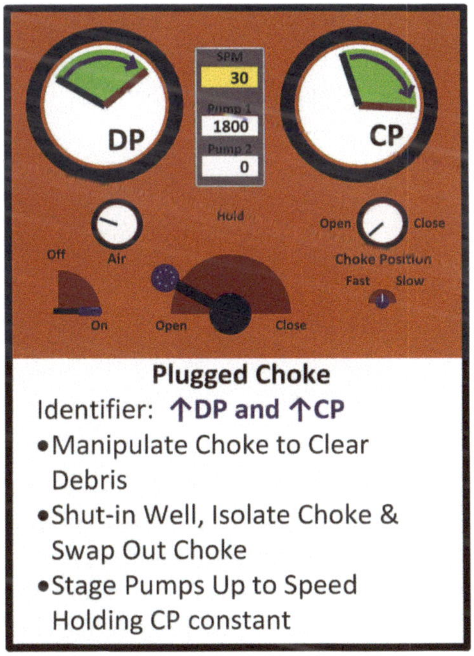

Fig. 15 Plugged choke.

<u>Gradual Plugging:</u> For gradual plugging while circulating, small opening manipulations are made to keep BHP constant. If pressures continue to rise as choke is opened, the well is to be secured by stopping pumps and shutting the well in. The current choke is to be isolated by closing isolation valves, while a secondary choke is lined up and readied for service. The pumps are staged up to speed holding previously recorded CP before plugging occurred.

<u>Complete Plugging:</u> For complete plugging while circulating, large opening choke manipulations will be unable to keep BHP constant. If pressures suddenly rise with the choke continually opened, the well is to be secured by stopping pumps and shutting the well in. The current choke is to be isolated by closing isolation valves, while a secondary choke is lined up and readied for service. The pumps are staged up to speed holding previously recorded CP before plugging occurred.

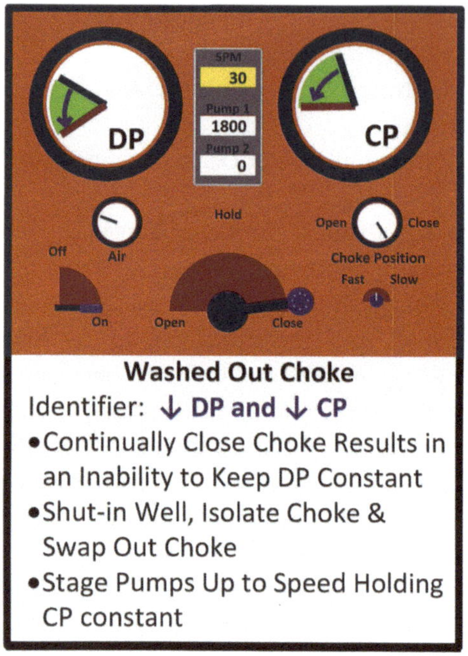

Fig. 16 Washed out choke.

5. **Washed out choke**—Can occur at any time. During circulation, small closing manipulations are made to keep BHP constant. If pressures continue to drop as choke is closed, the well is to be secured by stopping pumps and shutting the well in. The current choke is to be isolated by closing isolation valves, while a secondary choke is lined up and readied for service. The pumps are staged up to speed holding current casing pressure constant (see Fig. 16).

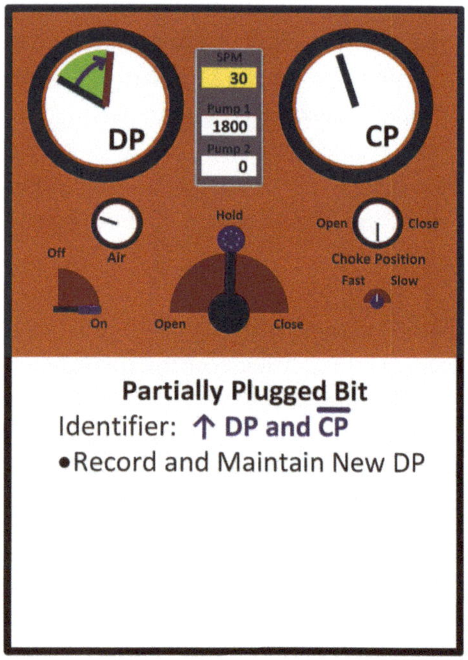

Fig. 17 Partially plugged bit.

6. **Partially Plugged Bit**—Can occur at any time and can be expected anytime pump is shut-off. Drillpipe pressure will suddenly increase without any increase in casing pressure (see Fig. 17).

   At ICP (KMW has not reached the bit). If pumps are staged up to speed holding casing pressure constant, record actual ICP. Recalculate step-down chart as needed. Do not open choke to achieve desired calculated ICP, as this action will cause BHP to become low enough to initiate additional kicks.

   At FCP (KMW has reached the bit). If kill mud weight has reached the bit, record new drillpipe pressure and maintain this pressure throughout circulation.

Fig. 18 Washed out bit.

7. **Bit wash out**—Can occur at any time. Drillpipe pressure will suddenly decrease without any increase in casing pressure (see Fig. 18)

At ICP (KMW has not reached the bit). If pumps are staged up to speed holding casing pressure constant, record actual ICP. Recalculate step down chart as needed. Do not close choke to achieve desired calculated ICP, as this action will cause BHP to increase and may fracture casing seat.

At FCP (KMW has reached the bit). If kill mud weight has reached the bit, record new drillpipe pressure and maintain this pressure throughout circulation.

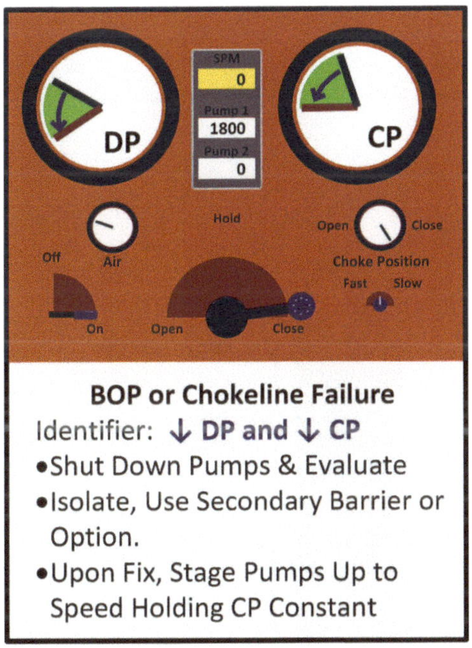

Fig. 19 BOP/choke line failure.

8. **BOP or choke line failure**—Can occur at any time. Drillpipe pressure and casing will decrease without changes in SPR

Shut down pumps immediately and shut-in well. Investigate equipment failure, isolate, and repair BOP/Choke Line. Use secondary barrier (i.e., BOP) as needed (see Fig. 19).

Upon completing repairs, stage pumps up to speed holding CP constant.

Fig. 20 Pump failure.

9. **Pump failure**—Can occur at any time. Pump pressure cavitation will cause Kelly/top drive hose to vibrate. Drillpipe pressure and casing will decrease without changes in SPR. If severe, Kelly hose will vibrate as pump(s) start failing (see Fig. 20)

Shut down pumps immediately and shut-in well. Isolate defective pump and line-up replacement pump.

Stage pumps up to speed holding casing pressure constant.

Quick tips (before kick)

Rig-up and testing

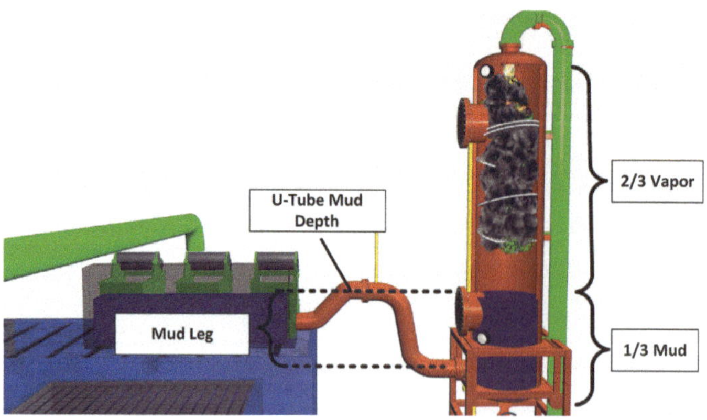

Fig. 21 MGS volumes and gauges.

- Verify discrepancies between Driller's Console gauges, choke panel, and choke manifold gauges. Repair or replace as needed. Record finding per daily record for reference (see Fig. 21).
- Use only one set of gauges of well control (choke panel)
  - o Have spare set of gauges available in case of needed replacement.
- Complex multifunction standpipe manifolds can leak (by-pass from standpipe to annulus side). Thoroughly test to ensure no by-pass exists.
- MGS should be installed to ensure at least one-third volume of mud leg.
  - o Weakest connection is mud leg flange gasket.
  - o For optimum operation, suggest two pressure gauges should be installed with output on Driller' Console. Upper gauge measure vapor pressure near top of vessel. Secondary gauge measure hydrostatic pressure of mud leg and located immediately above mud leg entry point.
- Choke line installation on BOP stack normally consists of one remote-controlled HCR valve and one manual valve.
  - o Suggest installing additional HCR valve for first exit point from drilling spool. Followed by manual gate valve and HCR#2. HCR #2 would be used for closing/opening during well control operations. HCR#1 would be used as an isolation valve to prevent solids from settling and plugging choke line. Manual gate valve would be used for emergency service isolation to repair HCR#2 (see Fig. 22).

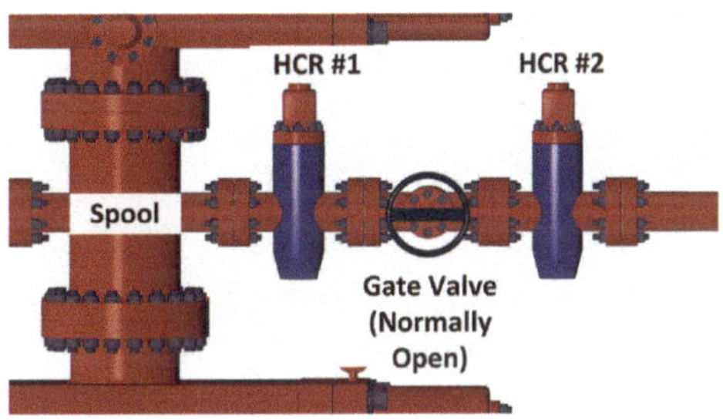

Fig. 22 Two HCR's and one gate valve.

  - o No manual intervention is necessary during well control operations.
- Choke panels should be outfitted with sufficient lights to enhance visibility of panel during nighttime operations.
- Choke manifolds should be outfitted with both Drillpipe Pressure and Casing Pressure gauges. Gauges should be mounted on portable stand which may be repositioned throughout choke manifold during operations.
- Outfit choke manifold with Open and Close metal placards which can be read from rig floor. Install and test high-pressure flowline from choke manifold to Trip Tank. Entry point downstream of chokes with high-pressure isolation valve rated at same pressure rating of choke manifold. This line may be needed for Volumetric Control, Lubricate and Bleed, and Stripping.
- Install and test high-pressure pump-in sub to choke manifold from alternate high-pressure pump. Entry point upstream of chokes with high-pressure isolation valve

rated at same pressure rating as valves in the choke manifold. The pump in line is designed to provide secondary pump-in source besides kill line pump-in source.

- Flare ignition sources
  - Automatic flare ignitors are preferred method for ignition.
  - Manual flare ignitors which use propane source should be ignited before circulation is initiated.
  - Flare pistols (flare gun) or flare rifle should be made available as secondary ignitor.
  - Manual flare buckets (diesel, kerosene, etc.) are discouraged. If the fire were to extinguish, individual interceding may be necessary. This will require appropriate use of SCBA and controlled egress.
- Flare lines, emergency by-pass lines should be installed with minimum bends using only targeted "Tees" for 90-degree bends and 30- or 45-degree welding bends for more gradual radial bends.
- Flare lines and emergency by-pass lines should be outfitted with removable end-plugs for hydraulic testing. Test plugs must be removed before operations commence.
- Flare lines should be securely anchored (barite bags, cement blocks, etc.)
- Verify H_2S and Gas meters are functioning and engaged during circulation.

After cementing casing

- Verify accuracy between Driller's Console gauges, choke panel, and choke manifold gauges. Adjust as needed (see Fig. 23).
- Perform various SPR and record pressures and MW on daily log.
  - Record SPR using each choke.
  - Close choke to halfway position, record increase in pressure. *Recorded pressure with the choke partially closed can be used to estimate back pressure while circulating.*
- Perform fingerprinting of equipment as needed.

Fig. 23 Choke panel choke opening gauge at 50% open.

CHAPTER ONE

Well control discussion and theories

Introduction

This manual was conceived for Oil and Gas industry personnel to improve their understanding of correct procedures and methods that can be used to safely maintain control of a well. This manual incorporates basic theory, discussions, practical applications, suggestions, and generalized procedures for well control. We offer brief explanations of the forces that can cause well control situations to arise. Practical applications and their mathematical representations of physical events and solutions are offered for review. Throughout this manual, the special "highlighted" sections have been included to highlight important procedural information, equipment suggestions, and general tips based on authors' experiences.

By eliminating unfamiliarity and confusion associated with well control procedures and equipment, our expectation will be to greatly reduce or eliminate irreversible damage to the individual and/or environment during well control events. Proper training along with targeted experience will result in a heightened level of confidence to organize and successfully direct a well killing operation.

Since well control situations are normally resolved within a team environment, effective communication between all participants must be established. An experienced and responsible individual will seek out the best advice on unusual problems. Since experience and knowledge are key factors in well control techniques, the Senior Person-in-Charge will be able to assist in formulating a unique solution to a particular well control problem.

Commentary and suggestions

In an effort to separate basis of theory from particular theory, our authors have attempted to showcase real-life experiences in each section beginning with "Commentary." Every commentary can be viewed as a quick-reference for each title where we merge both theoretical and practical advice, observations, and solutions based upon authors' own experiences. Our goal with each commentary will be to interject considerations and potential solutions for the most common problems facing rig crews during well control events.

Well control event is a planned event. The basis of proper planning will be to consider all problems and mitigating solutions before operations begin. Planning begins with matching and selecting the best rig, which meets all the criteria for drilling or working over a well. Within this planning, equipment is matched, selected, and tested

Universal Well Control
https://doi.org/10.1016/B978-0-323-90584-8.00001-0

for the rigors of drilling or workover for expected formation depth, pressures, and temperatures as well as providing suitable barriers to minimize chances of complex well control situations. Unfortunately, due to nonhomogeneous formations drilled (pressure, temperature, and environment) and with equipment wear, even the most well-planned well may become difficult to control.

WELL CONTROL IS A PLANNED EVENT

1. **WITH PROPER PLANNING AND ROUTINE TESTING RIG CREWS CAN BE ASSURED EQUIPMENT WILL PERFORM AS DESIGNED**
2. **RIG CREWS MUST BE DILIGENT IN THEIR WELL CONTROL PREPARATIONS AND PERFORM INSTALLATION AND TESTING DUTIES WITH ALL SERIOUSNESS, AS THEIR LIVES DEPEND ON EACH TEST**
3. **DO NOT ATTEMPT TO SHORTCHANGE, FALSIFY OR SKIRT ANY PRESSURE TEST**

For the most part, well control operations occur sporadically, meaning even seasoned, well-trained crews may not perform operations correctly. As well control operations are nonroutine, inexperienced rig crews may suffer confusion in operational procedures. To aid both experienced and inexperienced rig crews to a successful mitigation of any well control operation, safety equipment installation and testing procedures should be rigidly followed.

Our first commentary is offered to suggest the following suggestions and additions be made to normal rig equipment complement in order to facilitate a wider range of implemented well control techniques.

Location of choke panels

Fig. 1.1 Exterior choke panel example.

Location of the choke panel differs on each rig. In today's world, about half of the rigs worldwide have choke panels located outdoors or near the rig floor. Normally positioned to view both the choke manifold and MGS, these panels represent a robust means for manipulating chokes during well control operations. Although suitable for service, outdoor choke panels may increase fatigue of crew members due to ambient temperatures and inclement weather conditions. Newer rigs with positive pressure Driller's cabins offer significant benefits if choke panels are relocated within the cabin. Within the quiet Driller's Cabin environment with proper lighting, remote choke panels can offer distinct advantages when compared to rig floor mounted panels especially with enhanced rig crew communication (Fig. 1.1).

Dual HCRs on choke line

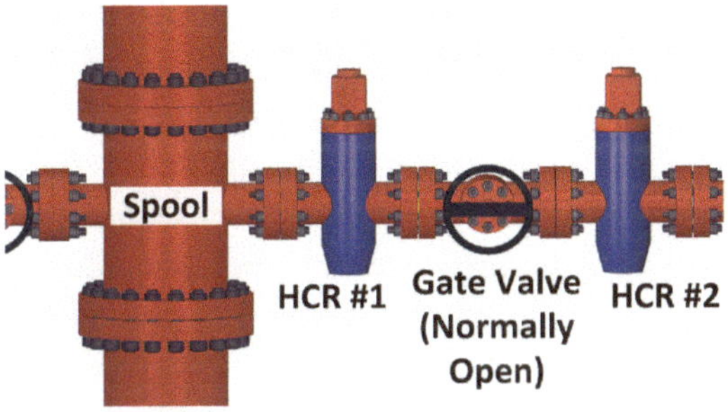

Fig. 1.2 Two HCR's and one gate valve.

Installation of one additional HCR valve within the choke line from drilling spool would be beneficial. Currently, most companies utilize an HCR and Manual gate valve, with placement of the manual gate valve (either to the inside or outside of the HCR) determined by company preference. Unfortunately, the current two-valve arrangement translates to manual interaction during any well control event. To alleviate personnel from manually interfacing with pressurized valves, we suggest adding one additional HCR valve to form an HCR/Manual/HCR configuration, whereupon the manual valve remains open. With manual valve open and relegated to isolation duties, HCRs can be operated remotely to direct fluids through the choke line (Fig. 1.2).

Mud gas separators with pressure sensors (gauges)

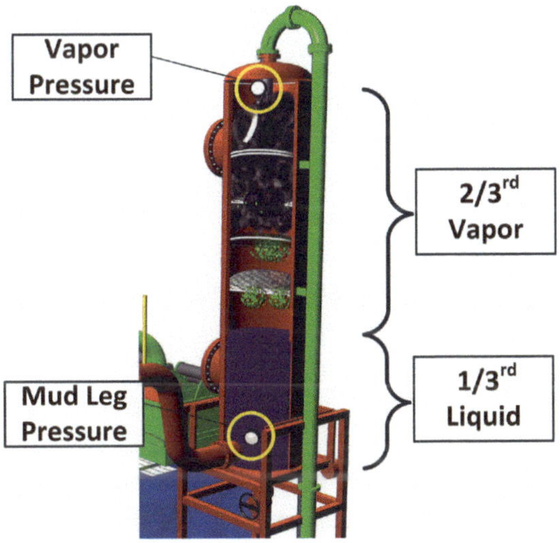

Fig. 1.3 MGS volumes and gauges.

Most MGS do not employ a complement of low-pressure gauges. Without gauges, MGS pressure thresholds cannot be monitored, and blow through the mud leg can occur with little or no foresight. For anyone who has experienced blow through, it is obvious that the sudden release of formation gases into the active pit region can easily become a life-threatening event. Since MGS are an integral part of the well control equipment complement, employing specialized pressures gauges to both vapor and liquid regions will provide critical information to determine likelihood of potential blow through. To further enhance crew safety, we suggest installing two pressure sensors as depicted and routing their output to the Driller's Console. This will allow the Driller to monitor vessel pressures during well control operations (Fig. 1.3).

Portable DP/CP panel stand

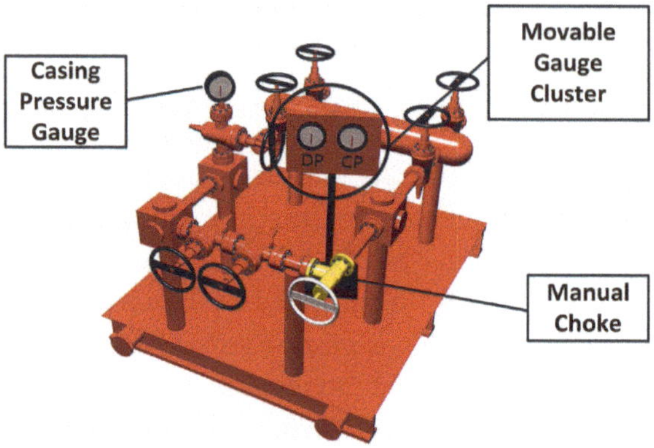

Fig. 1.4 Manual choke and gauge.

Most choke manifolds include at least one hydraulic and one manual choke for use during well control operations. It is imperative these chokes be routinely serviced to ensure smooth operation during emergency services. In order to facilitate improved understanding of pressures during well control operations, most choke manifolds are outfitted with a casing pressure gauge. Unfortunately, these gauges are not always visible from the manual choke. Our team suggests outfitting a small portable panel that can be outfitted with both casing pressure and pump pressure gauges within an independent panel which may be repositioned as needed for any choke on the choke manifold. If a manual choke is to be used, the ability to ascertain critical casing and pumping pressures is paramount for success (Fig. 1.4).

Surge rack

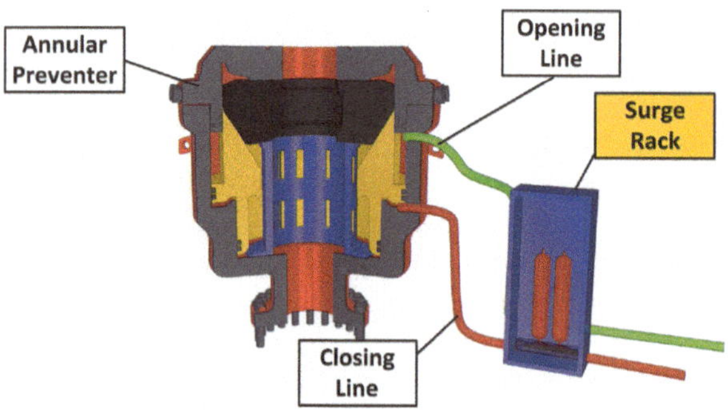

Fig. 1.5 Surge rack.

Since most well control events occur when the bit is off bottom, the rig crews must decide on whether to strip the bit back to bottom or kill the well at current depth. For those wells with compromised or questionable casing shoe fracture pressures, the need for stripping increases. For normal stripping operations, crews would have to manually adjust annular closing pressures each time a tooljoint passes through the preventer. With a simple installation of a surge rack or bottles near or on the annular preventer, as the tooljoint passes through the preventer, the closing fluid is automatically vented to/from the surge rack. Using a surge rack will allow tooljoints to pass through the annular preventer automatically and alleviate manual control interfacing (Fig. 1.5).

Easy to read analog gauges or digital gauges

The following suggestion is offered for those rigs using analog pressure gauges. **Digital gauges are preferred since they are easy to read.**

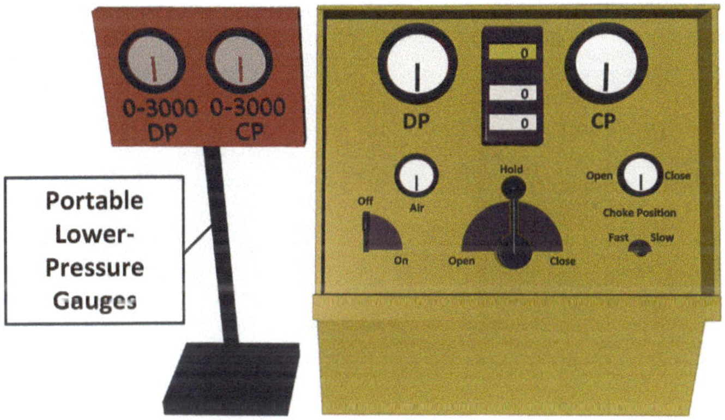

Fig. 1.6 Accurate gauges.

One of the largest problems in well control is utilizing high-pressure gauges to perform low-pressure well control operations. For proficient rig crews, the ability to recognize and react to kick indicators will minimize the overall size of the kick. For these small volume kicks, increases of kill mud weights will be minor in nature along with displacement pressures. In order to keep bottom hole constant during displacement of kill weight muds while allowing the kick to expand in the annulus, rig-supplied pressure gauges are used (Choke Panel and Driller's Console/Manifold). These gauges are normally matched to the operational pressures of the BOPs for testing purposes. Testing pressures may exceed thousands of psi. However, displacement pressures may be several magnitudes lower than maximum test pressures. Therefore, incremental pressure changes during displacement circulations at slow circulating rates with minimal increases in kill weight muds may not be easily observed on a high-pressure gauge. If the well control action plan (kill sheet) estimates that small pressures will be encountered during the circulation, the use of a calibrated smaller pressure range gauge may be used in lieu of the standard high-pressure gauge (Fig. 1.6).

Calibrated trip and strip tanks

Today's rig equipment complements normally consist of a calibrated trip tank and occasionally outfitted with a strip tank. The calibration of these tanks has been developed from construction drawings which may not always reflect actual build dimensions. It is advisable to perform a routine check of the internal dimensions of the actual build of both trip and strip tanks. The trip and strip tanks should be outfitted with flowlines to the active pit system as well as feed lines from the flowline and downstream of the choke baffle chamber. Trip and strip tanks plumbed in this fashion allow the greatest flexibility during well control operations (Fig. 1.7).

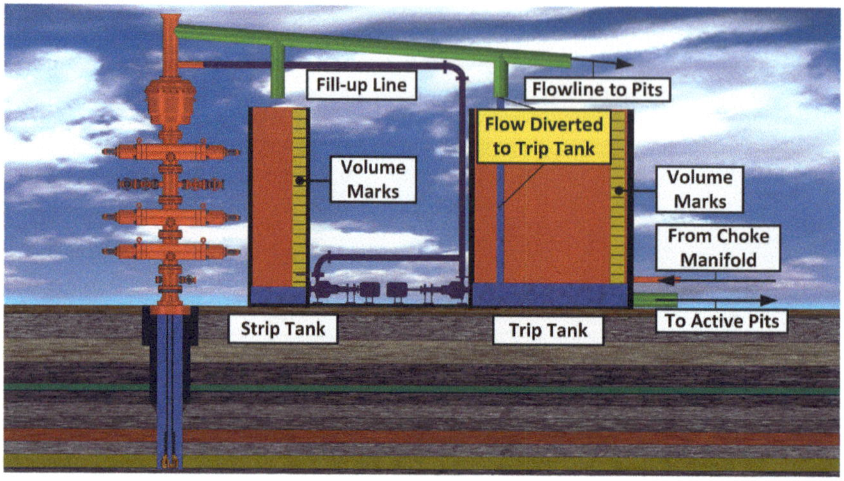

Fig. 1.7 Strip and trip tanks.

Pit level call-out

Most exterior choke panels are remotely located away from the Driller's console, and therefore, pump and pit volume information may not easily be ascertained. For these rigs, it is important the Driller corresponds with the choke operator throughout the well kill circulation, especially during any time the pumps are brought up to speed or slowed. Using the intercom system, the Driller should communicate changes in pump rate and speed along with active pit volumes.

One of the most difficult problems to identify during well control circulations will be the partial plugging of the bit while circulating. During the initial kill rate, pumps are slowly increased while the choke is manipulated to keep casing pressure constant. If done correctly, when the pumps have achieved appropriate slow pump rate, the initial circulating pressure should match estimated values. If not, the step-down pressure chart should be recalculated.

Unfortunately, this common solution is sometimes overlooked, and the choke is opened to achieve desired ICP. If this occurs, the BHP may be reduced, resulting in additional kick(s) occurring. If pits gains are observed during initial portion of well control, more than likely, the BHP has been decreased. This will be extremely difficult to view at the exterior choke panel.

If pit volumes are continually communicated via the intercom during the circulation, the choke operator should be able to identify problems more readily.

Off-bottom well control action plans (formerly entitled kill sheets)

Since most well control problems occur with the bit off bottom, the introduction and use of enhanced Off-Bottom Well Control Action Plans are suggested. These action plans allow team members to evaluate a variety of information, prepare for the well kill circulation, perform job safety analysis, and execute kill operations safely. These enhanced action plans allow for defining many outstanding characteristics necessary for safe mitigation. First and foremost, these enhanced action plans will help determine if the well can be safely killed at current bit depth. With more information readily available within the Action Plans, rig crews can easily remind themselves of any problems before proceeding forward.

Local and governmental requirements

Governmental requirements include a broad background in various methods and techniques in well control. Where applicable, these methods and techniques will be presented in "*quoted and italic text.*" Our well control guidelines meet and exceed governmental requirements but will <u>limit</u> the methods and techniques to the most accurate and field applicable well control methods.

Team and individual responsibilities

As crews must operate as a team, efficient, effective, and knowledgeable communications must be continuous during all activities. These teams must ensure effective two-way communication in order to achieve a higher awareness in well control and safety. As a team member, each individual will be responsible to provide diligent efforts toward ensuring safe operations at all time. Each crew and team member must follow appropriate governmental rules and regulations at all times.

Blowout prevention equipment (BOP or BOPE)

<u>All</u> Blowout Prevention Equipment (BOP or BOPE) must be correctly installed and tested to the manufacturer and governmental specifications. Thorough equipment testing must be performed and verified as if one's own life may depend on it. By thoroughly and completely testing each piece of primary blowout prevention equipment (even auxiliary equipment), a high level of confidence for proper functioning of BOPs at the time of a critical event is assured.

The selection of the appropriate Blowout Preventer Equipment and pressure rating must exceed and contain any anticipated surface pressure. Equipment rating and Maximum Allowable Surface Pressures must be established before permits to drill can be issued. From these pressure limits, an appropriate BOP stack arrangement can be developed for a specific well classification.

Once installed and tested, crews must ensure pipe and/or variable pipe rams are outfitted to fit the actual tubular in use. If pressure seals are broken to replace rams, these rams must again be pressure tested. Any previously tested pressure seal must be retested if it is broken for any reason. Pressure rating of any BOP must exceed the anticipated surface pressure. The only exception to this limit will be the pressure limit of the annular blowout preventer.

For an annular preventer when anticipated surface pressure exceeds rated working pressure, crews will have to develop a well control procedure, which indicates how the annular preventer will be utilized, and the pressure limitations, which will be applied during each mode of pressure control.

Rams-type BOPs and related control equipment including the choke manifold shall be tested at the anticipated surface pressure or at 70% of the minimum internal yield pressure of casing, whichever is less.

- *"Annular-type BOP shall be tested at 70% of its rated working pressure or 70% of the minimum internal working pressure of casing whichever is less."*
- *"Frequency of testing shall not exceed once per three weeks."*
- *"Recommend using a cup tester positioned 15–30 feet below the casing head to test stacks which are physically impossible to be tested by a test plug."*

Well control responses

Each team individual is responsible for the proper detection and executing proper well control procedures. To support this responsibility, each team member must ensure proper maintenance and testing of equipment. Diligent efforts are to be extended to maintaining drilling or workover fluid properties and densities. Every individual is empowered with stop work authority, employing any warning sign of a kick and initiating securing the well.

Prerecorded data sheet

All pertinent information for operations must be recorded and verified within the Prerecorded Data Sheet. This includes all relevant well data, tubular data, testing data, fluid volumes as well as pump data. This data sheet is to be updated every time the fluid weight is changed by 0.2 ppg to provide the most up-to-date information or change in bottom-hole assembly configurations. The Prerecorded Data Sheet should be employed in the day-to-day operations supplying needed information before the well kicks.

Slow pump rate data

The following information must be updated and maintained within the Pre-recorded information sheet for use in emergency situations. Pump rate data are to be recorded on each well at following intervals:

1. *On initial rig up*
2. *After fluid densities change more than 0.2 ppg*
3. *After pump repairs and/or changes in Liner/Swab size*
4. *Before drilling out casing shoes*
5. *Anytime there are indications of the wellbore being close to underbalance*

Selected rates will vary depending upon pump sizes, hole parameters, and fluid densities. Generally speaking, rates of 2 bpm, 3 bpm, and 5 bpm are normally selected as an optimum rate of mixing mud solids with mud hopper. Faster rates can be achieved with larger hopper. Slower rates down to as low as 1/2 bpm may be needed for deepwater to reduce back pressure.

Posting information

Communication is an effective key to unlocking the confusion surrounding well control. Each team member should strive to communicate effectively with all personnel. One easy method of communication is to affix the following information in a highly visible place on the Rig floor. This information should include, but is not limited to, the following:

1. *Maximum Wellhead and Casing Burst Pressure.*
2. *Maximum Pump Pressure Before Pump Pop Off*
3. *Maximum Number of Stands Pulled Prior to Filling the Hole.*
   a. *While Pulling Drillpipe or Tubing.*
   b. *While Pulling Heavy-weight Drillpipe*
   c. *While Pulling Drill Collars*
4. *Shut-In Procedures and Crew Responsibilities for Current Operations.*

Each of these responsibilities will be discussed throughout this manual. Once again, if you have questions concerning any of these topics, please discuss them with the Person-in-Charge.

Risk management and risk assessments

Risk management is the process of determining the maximum acceptable level of overall risk from a proposed activity and then developing and executing activities to increase control and reduce risk to an acceptable level. In any modern day-to-day oilfield activity, risk management and mitigation are an integral part of safety.

In support of the risk management process, a thorough well control hazard risk assessment may be employed. The simplest approach of a detailed risk assessment is to hold a communication meeting between participating companies and service providers before operations are to begin or before any critical operation is to occur. Each company assists in developing an itemized listing of controllable risks and mitigation strategies needed to reduce probable risks. This combined list will define clear goals and expectations.

Bridging document

A Bridging document consists of a detailed and documented operational plan that defines individual operations and terminology developed by two or more companies. These plans outline alignment and agreement between organizations to safely manage individual operational elements which may not be defined within their respective contract. This document may include provisions to overcome misunderstandings, conflicts, or alleviate misappropriation of resources between companies, before initiating projects.

Barriers

Once the set of Blowout Prevention Equipment has been installed on a well and tested, there are four potential flow paths of charged formation fluids and gases to reach the surface (Fig. 1.8).

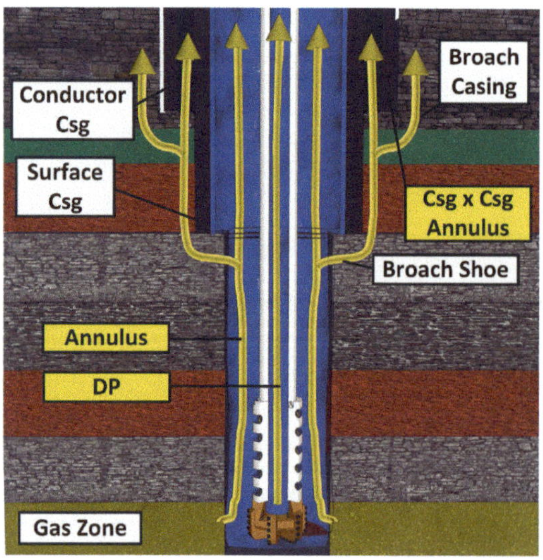

Fig. 1.8 Four flow paths.

- **a.** Casing × casing annulus
- **b.** Workstring tubular × casing annulus
- **c.** Workstring tubular
- **d.** Broach from shoe to surface outside all casing strings.

In order to maintain control of the charged formation fluids, a physical pressure isolation fluid, solidifying fluid or solid, is used. These pressure isolation devices are called barriers.

Isolation Fluid Barrier: This occurs when a stable fluid column of sufficient density exerts sufficient hydrostatic pressure to cease encroachment of charged formation fluids/gases into the wellbore.

$$HP_{fluid} > BHP$$

Note: Over time, fluid densities can vary due to particle settling, thermal expansion, fluid loss, and hole deviation.

Solidifying fluid barrier

The use of solidifying fluid, such as cement, resins, or solids settling plugs, may be considered to isolate charged formations fluids/gas by forming a solid isolation plug. Normally, cements are used to provide both tubular isolation (i.e., cemented casing) and fluid isolation (open hole balanced cement plugs). Each type of solidifying fluid must be designed, tested, mixed, and displaced according to specification standards employed by the individual service provider. Each plug design must include:

- Ability to be mixed and pumped on location
- Properties which are resistant to degradation due to wellbore gases, formation fluids, and corrosive environments
- Properties which remain liquid throughout entire displacement
- Provide resistance to premature setting due to high temperatures
- Short transition time from fluid to solid state after displacement
- Minimum free water to minimize chances of flow after displacement
- Ensure sufficient long-term curing strength properties

Solid barrier

Normally, a mechanical sealing device installed in the well provides an elastometric or metal seal, such as an isolation plug, full opening safety valve, inside blowout preventer, and BOP equipment.

Pressure seal verification

Pressure seal verification can be performed in a variety of ways, depending upon the type of barrier. Any pressure testing of barriers must include a review of the accuracy of the pressure monitoring devices in order to observe and detect small volume bleed-offs.

- Isolation Fluid Barrier may require a simple visual verification of no flow, followed by constant monitoring.
- Solidifying Fluid Barriers may require tagging solid after sufficient curing time has elapsed. Depending on where this plug is place, further testing may require lowering densities of the fluids above plug to perform a "negative pressure test" as well as a "positive test."
- Solid barriers may require simple verification of engagement by noting number of turns necessary to set the device or sufficient weight applied as a downward force to ensure deployment.

Barrier procedures

To ensure proper pressure isolation, installation and testing of barriers must be contained within a common set of barrier procedures. These procedures must include detailed installation, testing, and verification criteria. Each procedure should include the primary installation and testing procedure as well as an alternate solution in case the primary procedure fails.

Number and Type of Barriers: Depending on governmental regulation or client policies, the number and type of barriers may be included within the permit to drill. At all times, crews must follow and not deviate from these directives. Any questions regarding barriers should be routed to the Person-in-Charge.

Barrier policy

A barrier policy is a subset of the barrier philosophy, governing the type and number of barriers required for any given well operation. A barrier policy can be established either by government regulation or, in the absence of regulation, by the company-specific policy of the operator.

Autonomous Barrier: A barrier, which fails, will not lead to failure of any other barrier within the well.

Pressurized Formation Fluids/Gas Wells—Minimum of two barriers needed for each flow path.

Nonpressured Formation Fluid/Gas Wells—Minimum of one barrier is needed.

Initial barrier: The initial barrier is the first barrier to provide isolation from pressurized formation fluids/gases.

Back-up barrier: The back-up barrier, sometimes called the secondary barrier, will be independent of the initial barrier. The back-up barrier serves to isolate pressurized formation fluids/gases, if the initial barrier were to fail.

Critical well control skills

As every well control situation occurs on a customized rig within a unique operating environment, each team will be confronted with a unique set of challenges. Each team should have a balance of skills, experience, and knowledge in order to safety control a well as listed below:

- Obtain Accurate Information: Each well control situation begins after closing and isolating the wellbore. Obtaining timely and accurate information will prevent future problems from occurring.
- Inquire: Before killing operations commence, ensure all questions are reviewed. Everyone must understand their individual role and responsibility.
- Flexible Thinking: If alterations in pressures, volumes, and equipment functioning are observed before killing operations commence, ensure these observations are brought to the Person-in-Charge.
- Communication: During the kill, ensure communication between all rig crew members is established and ongoing.
- Collaboration: If problems arise during the well kill, ensure the team collaborates on appropriate solution.

Management of change

Management of Change (MOC) is a method to allow changes to be made to approved plans should the need arises. There is a specific process, and changes have to be approved by specific individuals. It may utilize lessons learned and best practices to minimize safety, health, and environmental risks when deviations to approved well plan are needed. Examples of this include changes to or substitution of alternate equipment due to failure or lack of availability or altering an approved well plan due to unforeseen hole problems such as sidetracking, etc.

Emergency response plan (ERP)

Emergency Response Plans organize a set of actions to be performed during an emergency. These procedures may include immediate or temporary action necessary to prevent personnel or environmental harm. Used as a planning tool, these procedures are normally developed in cooperation with all affected parties, allowing for maximum communication in a variety of occupational problems. ERPs are updated as needed, due to changing conditions, and are seen as dynamic.

Incident command system (ICS)

The Incident Command System (ICS) is defined as a standardized management of command, control, and coordination of emergency response. The ICS provides a normalized and common hierarchy within which individual emergency responders from multiple departments/sections/locations including governmental agencies can be effectively coordinated and administered. ICS was initially developed to address problems of interagency responses to wildfires and is now a component of the US National Incident Management System (NIMS). ICS allows its users to adopt an integrated organizational structure to match the complexities and demands of single or multiple incidents without being hindered by jurisdictional boundaries.

The ICS provides guidance for asset organization and outlines processes necessary to manage responses throughout entire life of an emergency. These responses can be separated into five functional areas:

1. Incident Command
2. Emergency Operations
3. Planning
4. Logistics
5. Finance and Administration

The ICS is based on the following characteristics which contribute to the successful application of this system.

- Common Terminology and Clear Text—use of similar terms and definitions including organizational structure, function, facilities, and resources across disciplines.
- Modular Organization—defining and assigning response resources according to their responsibilities.
- Management by Objectives—establishing team objectives and overseeing execution of processes necessary in safely obtaining objectives.
- Reliance on an Incident Action Plan—define a cohesive single-point document outlining incident goals, objectives, and strategies within a unified incident command. IAP can be used for "practice" events.
- Manageable Span of Control—response organization is structured so each supervisory level oversees an appropriate number of assets (varies based on size and complexity of the event so it can maintain effective supervision).
- Predesignated Incident Locations and Facilities—defining locations where expected critical incident-related functions will occur. Used as part of the IAP and can be used for "practice" events.
- Resource Management—customized system to organize, maintain, identify, request, and track resources. May include computer programs or simple white boards.
- Integrated Communications—establish electronic broadcast and/or electronic text systems capable of sending/receiving/distributing critical communications throughout ICS.
- Chain of Command and Unity of Command—establish and populate chain of command by structural definition and assigning individuals. Updated as needed due to personnel changes.
- Unified Command—establish a common set of objectives throughout organization. Define strategies to prevent rework and conflicts of duplication of efforts.
- Transfer of Command—within ICS, define hand-off procedures to provide relief and 24-h coverage.
- Accountability—assign accountabilities for each position throughout ICS structure by written description.
- Mobilization—define process for mobilization and needed timeframe. To be used within an IAP practice session to ensure efficiency. Apply lessons learned as needed.
- Information and Intelligence Management—defined procedure necessary to produce and archive information within defined executing timetable.

Standardized units of measure

In order to alleviate confusion surrounding chosen fields of measurement, this Universal Well Control Manual has standardized on the following units of measurement:

Depth	=	Feet (ft)
Density	=	Pounds per gallon (ppg)
Pressure	=	Pounds per square inch (psi)
Volume	=	Barrels (bbls)
Hydrostatic Pressure (HP)	=	Pounds per square inch (psi)

Density and weight relationship

Formations encountered in drilling and workover operations consist of a rock matrix with voids between rock particles filled with fluid. Under pressure, gases and oils may become trapped within the formation fluid. Forces act upon the matrix and fluids including gravity, heat, and pressure. These forces act on the trapped fluids within the rock matrix consisting of saltwater, oil, and entrained gas.

Density: The density of a fluid is defined as its mass per unit volume. The mathematical expression of this is:

$$D=\frac{m}{V} \tag{1.1}$$

where

$D =$ Density

$M =$ Mass

$V =$ Unit Volume

From a scientific definition, mass and weight are completely different properties. Mass is the amount of matter calculated from atomic level. Mass of an object cannot be changed, unless mass is added to or subtracted from the original object.

Weight is the measurement of gravitational pull on an object. Weight of an object can vary depending upon where on the earth this measurement is obtained due to changes in gravitational pull at different geological locations. Fortunately, these gravitational differences are extremely small and can be disregarded.

The mass of an object involves complex calculations. For practical simplicity, mass and weight are considered to be the same. And since weights of an object can be easily obtained, this is the preferred method for determining the density of a fluid and expressed as pounds per gallon (ppg).

MASS = WEIGHT (PPG).

Hydrostatic pressure

ALWAYS USE TRUE VERTICAL DEPTHS FOR PRESSURE CALCULATIONS.

Hydrostatic pressure is the force per unit area exerted by a vertical column of fluid of known density. This primary equation is used to calculate pressures along the vertical axis of wellbore.

$$HP = 0.052 \times MW \times TVD \tag{1.2}$$

The equation is dependent on two variables, true vertical depth and mud weight. True vertical depth, not measured depth, must be used when calculating hydrostatic pressure. Measured depths, especially on deviated wellbores, will be greater than calculated true vertical depths. Therefore, if measured depths are used, calculated hydrostatic pressures will be greater than actual hydrostatic pressures.

Fluid density is the second dependent variable within the equation. It must be understood that both liquids and gasses can exert hydrostatic pressure. The two common units of measure used to describe fluid weight are:

Pounds per gallon (ppg)

Pounds per cubic foot (pcf)

A conversion factor is used to calculate hydrostatic pressure. The conversion factor is calculated by dividing the laboratory measured gradient for fresh water by the density for fresh water.

$$\frac{0.433\ \text{psi/ft}}{8.33\ \text{ppg}} = \frac{0.052\ \text{psi} - \text{gal}}{\text{ft} - \text{lbs}} \tag{1.3}$$

$$\frac{0.433\ \text{psi/ft}}{62.31\ \text{pcf}} = \frac{.007\text{psi} - \text{ft}^3}{\text{ft} - \text{lbs}} \tag{1.4}$$

Surface pressure

Under well control situations, surface pressure is any pressure which is exerted at the top of a fluid column, recorded in psi. Under well control situations with static shut-in well conditions, surface pressure is represented as either SICP (Shut-in Casing Pressure) or SIDPP (Shut-in Drillpipe Pressure).

Bottom-hole pressure

Bottom-hole pressure for a static fluid column is the summation of all pressures acting in the well. Bottom hole pressure is equal to the hydrostatic pressure of the well plus surface pressure. It is expressed mathematically as

$$BHP = HP + SP \tag{1.5}$$

Choke pressure

Choke pressure is simply the back pressure associated by circulating through a restricted orifice or throttling devices (i.e., choke). This pressure can easily be read by the casing gauge when circulating through the choke.

$$BHP = HP + SP + Friction_{choke} \tag{1.6}$$

Capacity (volume)

ALWAYS USE MEASURED DEPTHS WHEN CALCULATING VOLUMES AND CAPACITIES.

Capacity is the volume of a fluid contained within a specified vessel or well. The terms capacity and volume can be used interchangeably. Capacity is calculated by multiplying the end area of a vessel by its length. The equations below represent both volume and capacity using our standardized units of measure.

$$\text{Volume} = \text{Capacity Factor} \times \text{Length}_{\text{Measured Depth}} \tag{1.7}$$

where
  Volume = barrels (bbl)
  Capacity Factor = barrels/foot (bbl/ft)
  Length = Measured Depth (ft)

$$\text{Capacity} = \text{Capacity} \times \text{Length}_{\text{Measured Depth}} \tag{1.8}$$

where
  Capacity = barrels (bbl)
  Capacity Factor = barrels/foot (bbl/ft)
  Length = Measured Depth (ft)

Capacity may be represented as:

1. Internal Capacity or the internal volume of tubulars.
2. Annular Capacity or the external volume of between open hole (casing/liner) and tubular.

Capacity conversion constant

If capacity is the volume for a vessel, then the capacity factor is simply the volume per foot of the vessel. Therefore, capacity factor and area may be used interchangeably. The equation below represents the derivation for converting area in square inches to capacity factor of bbl/ft.

Capacity Factor is defined as:

$$\text{Capacity Factor} = \frac{\pi(d)^2}{4} \times 1 \text{ ft} \tag{1.9}$$

where: Capacity factor (bbl/ft)
  d = Diameter (inches)
  l = Unit length of 1 ft

$$\text{Capacity Factor} = \left(\frac{(\text{diameter})^2}{\left(\frac{4}{\pi} x \frac{144 \text{ in}^2}{\text{ft}^2} x \frac{5.6146 \text{ ft}^3}{\text{bbl}} \right)} \right) \times \text{Length} \tag{1.10}$$

$$\text{Capacity Factor} = \left(\frac{(\text{diameter})^2}{(1029.4)} \right) \times \text{Length} \tag{1.11}$$

$$\text{Capacity} = \text{Capacity Factor} \times \text{Length} \tag{1.12}$$

Internal capacity factor

$$\text{Capacity Factor}_{\text{Internal}} = \left(\frac{(\text{Internal Diameter})^2}{(1029.4)}\right) \tag{1.13}$$

where

Capacity $\text{Factor}_{\text{Internal}}$ = Internal Capacity Factor (bbls/ft)
Internal Diameter = Internal Diameter of Tubular (in)
1029.4 = Capacity Factor Conversion Constant (in^2-ft/bbl)

Annular capacity factor

$$\text{Capacity Factor}_{\text{Annulus}} = \left(\frac{(\text{Outside Diameter})^2 - (\text{Internal Diameter})^2}{(1029.4)}\right) \tag{1.14}$$

where

Capacity $\text{Factor}_{\text{Annulus}}$ = Annular Capacity Factor (bbls/ft)
Outside Diameter = Outside Diameter of Tubulars (in)
Internal Diameter = Internal Diameter of Tubular (in)
1029.4 = Capacity Factor Conversion Constant (in^2-ft/bbl)

Pipe displacement

Pipe displacement is defined as the volume of steel either removed or added to the well as pipe is run into or pulled from the hole. In order to maintain a safe hydrostatic pressure in the well, a volume of mud equal to the pipe displacement (volume of steel) must be pumped into the annulus as pipe is removed. Conversely as pipe is run into the hole, a volume of mud equal to the pipe displacement should be recovered in the mud pits/trip tank.

$$\text{Displacement} = \text{Displacement Factor} \times \text{Length}_{\text{Measured Depth}}$$

where

Displacement = barrels (bbl)
Displacement Factor = barrels/foot (bbl/ft)
Length = Measured Depth (ft)

Pipe displacement factor

$$\text{Displacement Factor}_{\text{Pipe}} = \left[(\text{Outside Diameter pipe})^2 - (\text{Internal diameter})^2\right]/1029.4$$

where

Displacement $\text{Factor}_{\text{pipe}}$ = Displacement Factor (bbls/ft)
Outside Diameter = Outside Diameter of pipe (in)
Inside Diameter = Inside Diameter of same pipe (in)
1029.4 = Conversion Constant (in^2-ft/bbl)

Removing tubulars from the well

The process of removing tubulars from a well is called tripping. If the pipe pulls dry, it means the mud inside the drillstring falls out into the well. As the drillstring is pulled from the hole, the height of the mud drops inside the well because of the volume of steel (displacement) which has been removed. This volume is converted into a length and then to a pressure. If the drillstring is pulled wet, this means the mud inside the pipe does not fall out and will not remain within the well. Thus, the volume of mud which drops is equal to the volume of steel removed plus the capacity of the pipe removed.

The method of calculation for determining the maximum number of stands which can be pulled prior to filling the hole is the same no matter what type of tubular is being pulled. What does matter is whether the tubular is able to empty as it is pulled. Therefore, two methods of calculation are shown: (1) Pulling Pipe Dry and (2) Pulling Pipe Wet.

MAXIMUM ALLOWABLE REDUCTION IN HYDROSTATIC PRESSURE = 75 PSI
WHILE REMOVING PIPE FROM THE WELL

FILL-UP FREQUENCY IS AS SIMPLE AT 5, 3,1

EVERY 5 STANDS OF DRILLPIPE OR LOSS OF 75 PSI, WHICHEVER COMES FIRST
EVERY 3 STANDS OF HEAVIWATE OR LOSS OF 75 PSI, WHICHEVER COMES FIRST
EVERY 1 STAND OF DRILLCOLLARS OR LOSS OF 75 PSI, WHICHEVER COMES FIRST.

IN LIEU OF FOLLOWING THE ABOVE GENERAL RULE OF THUMB, A CIRCULATING TRIP TANK MAY BE USED WHEREUPON THE HOLE IS CONTINUALLY FILLED WHILE PULLING THE DRILLSTRING FROM THE WELL.

Maximum feet or number of stands pulled prior to filling the hole

THE BELOW CAPACITIES CALCULATIONS DO NOT ACCOUNT FOR COUPLINGS.
THEREFORE, SPECIFIC DRILLSTRING (HEAVIWATE DRILLPIPE, TUBING, ETC. ARE IMPORTANT) SHOULD BE RESEARCHED IN SERVICE COMPANY PROVIDED CAPACITY AND DISPLACEMENT TABLES

Trip sheet calculations for tripping dry

See Fig. 1.9.

Trip Sheet Calculations for Tripping Dry

Barrel of Steel Out = Barrel of Fluid In

$$\text{Displacement} = \left(\frac{(\text{OD})^2 - (\text{ID})^2}{(1029.4)}\right) \qquad \text{Ann Cap} = \left(\frac{(\text{ID}_{\text{casing}})^2 - (\text{OD}_{\text{tubular}})^2}{(1029.4)}\right) + \left(\frac{(\text{ID}_{\text{tubular}})^2}{(1029.4)}\right)$$

1. Calculate Displacement of Steel Removed from Well
To be calculated for each type of tubular in well (i.e., drillpipe, heaviwate, drill collar)

Displacement Factor = [(□ OD (in) × □ OD (in)) − (□ ID (in) × □ ID (in))] ÷ 1029.4 = □ bbls/ft (Step #1)

Displacement = (□ Disp Factor Step #1 (bbls/ft) × □ Length of Pipe (1 ft)) = □ bbls (Step #2)

2. Calculate Annular Fluid Capacity Factor

Casing Capacity Factor = [(□ Csg ID (in) × □ Csg ID (in)) − (□ DP OD (in) × □ DP OD (in))] ÷ 1029.4 = □ bbls/ft (Step #3)

Internal Capacity Factor = (□ DP ID (in) × □ DP ID (in)) ÷ 1029.4 = □ bbls/ft (Step #4)

3. Calculate Total Fluid Capacity Factor

Total Fluid Capacity Factor = (□ Step #3 (bbls/ft) + □ Step #4 (bbls/ft)) = □ bbls/ft (Step #5)

4. Length Fluid Falls in Annulus when Removing DRY Pipe

Fall Length = (□ Step #2 (bbl) ÷ □ Step #5 (bbl/ft)) = □ ft (Step #6)

5. Reduction in Wellbore Hydrostatic Pressure per Foot (Maximum Limit = 75 psi)

Press Reduction = (□ Step #6 (ft) × □ OMW (ppg) × 0.052 Coversion Factor) = □ psi/ft (Step #7)

6. Maximum Number of Feet which can be tripped BEFORE fill-up

Max Length (ft) = (75 Pr Reduction (psi) ÷ □ Step #7 (psi/ft)) = □ ft (Step #8)

7. Maximum Number of Stands which can be tripped BEFORE fill-up

Max Length (Std) = (□ Step #8 (ft) ÷ 90 Max Std Length (ft/std)) = □ stds (Step #9)

If length of stand is unknown, use 30 ft for single, 60 ft for double or 90 ft for triple

Fig. 1.9 Tripping DRY calculation sheet.

Example tripping DRY fill-up worksheet

See Fig. 1.10.

Universal Well Control DRY Tripping Worksheet

Input: UWC DRY Tripping Worksheet (Displacement)											Total
Descriptions	Units	Wrksht	DC1	DC2	Other	HWDP1	HWDP2	DP1	DP2	DP3	
No. of Stands											
OMW	ppg										
Tubular OD	in										
Tubular ID	in										
Csg ID/Hole Size	in										
Avg. Stand Length	ft										
Displacement Factor	bbls/ft	Step #1&2									
Csg Capacity Factor	bbls/ft	Step #3									
Internal Cap Factor	bbls/ft	Step #4									
Total Fluid Cap Factor	bbls/ft	Step #5									
Length Fluid Falls Dry Trip	ft	Step #6									
Pressure Reduction	psi/ft	Step #7									
Max Tripped Length Before Fillup	ft	Step #8									
Max Number of Stands for 75 psi	stds	Step #9									
Fill-up Interval Check	stds										
Est Fill-up Volume per Interval	stds										
Est Displacement Volume	bbls										

Output: UWC DRY Tripping Worksheet (Displacement)

Selected Tubular	Length		Selected Tubular Displacement		Calculated Fill-Up (Trip Sheet)	Calculated Cumulative (Trip Sheet)	Actual Cumulative (Rig Sensor)	Differences Calc and Rig Sensor	Actual Cumulative (Mud Logger)	Differences Calc and Mud Logger
	ft		bbls/ft		bbls	bbls	bbls	bbs	bbls	bbs
#N/A		x	#N/A	=	#N/A	#N/A				
#N/A		x	#N/A	=	#N/A	#N/A				
#N/A		x	#N/A	=	#N/A	#N/A				
#N/A		x	#N/A	=	#N/A	#N/A				
#N/A		x	#N/A	=	#N/A	#N/A				
#N/A		x	#N/A	=	#N/A	#N/A				
#N/A		x	#N/A	=	#N/A	#N/A				
#N/A		x	#N/A	=	#N/A	#N/A				
#N/A		x	#N/A	=	#N/A	#N/A				
#N/A		x	#N/A	=	#N/A	#N/A				
#N/A		x	#N/A	=	#N/A	#N/A				
#N/A		x	#N/A	=	#N/A	#N/A				
#N/A		x	#N/A	=	#N/A	#N/A				
#N/A		x	#N/A	=	#N/A	#N/A				
#N/A		x	#N/A	=	#N/A	#N/A				
#N/A		x	#N/A	=	#N/A	#N/A				
#N/A		x	#N/A	=	#N/A	#N/A				
#N/A		x	#N/A	=	#N/A	#N/A				
#N/A		x	#N/A	=	#N/A	#N/A				
#N/A		x	#N/A	=	#N/A	#N/A				
#N/A		x	#N/A	=	#N/A	#N/A				
#N/A		x	#N/A	=	#N/A	#N/A				

Fig. 1.10 Example tripping DRY Fill-up worksheet.

Trip sheet calculations for tripping WET

See Fig. 1.11.

Trip Sheet Calculations for Tripping Wet

Barrel of Steel Out = Barrel of Fluid In

$$\textbf{Displacement} = \left(\frac{OD^2}{(1029.4)}\right) \qquad \textbf{Annular Capacity} = \left(\frac{(ID_{casing})^2 - (OD_{tubular})^2}{(1029.4)}\right)$$

1. Calculate Displacement of Steel Removed from Well PLUS Internal Capacity
To be calculated for each type of tubular in well (i.e., drillpipe, heaviwate, drill collar)

Displacement Factor = ([ ] OD (in) × [ ] OD (in)) ÷ 1029.4 = [ ] bbls/ft Step #1

Displacement = ([ ] Disp Factor Step #1 (bbls/ft) × [ ] Length of Pipe (1 ft)) = [ ] bbls Step #2

2. Calculate Total Fluid Capacity Factor

Total Fluid Capacity Factor = [([ ] Csg ID (in) × [ ] Csg ID (in)) − ([ ] DP OD (in) × [ ] DP OD (in))] ÷ 1029.4 = [ ] bbls/ft Step #3

3. Length Fluid Falls in Annulus when Removing WET Pipe

Fall Length = ([ ] Step #2 (bbls) ÷ [ ] Step #3 (bbl/ft)) = [ ] ft Step #4

4. Reduction in Wellbore Hydrostatic Pressure per Stand (Maximum Limit = 75 psi)

Press Reduction = ([ ] Step #4 (ft) × [ ] OMW (ppg) × [0.052] Coversion Factor) = [ ] psi/ft Step #5

5. Maximum Number of Feet which can be tripped BEFORE fill-up

Max No of Stands = ([75] Max Pr Reduction (psi) ÷ [ ] Step #5 (psi/ft)) = [ ] ft Step #6

6. Maximum Number of Stands which can be tripped BEFORE fill-up

Max Length (Std) = ([ ] Step #6 (ft) ÷ [90] Max Std Length (ft/std)) = [ ] stds Step #7

If length of stand is unknown, use 30 ft for single, 60 ft for double or 90 ft for triple

Fig. 1.11 Tripping WET calculation sheet.

Example tripping WET fill-up worksheet

See Fig. 1.12.

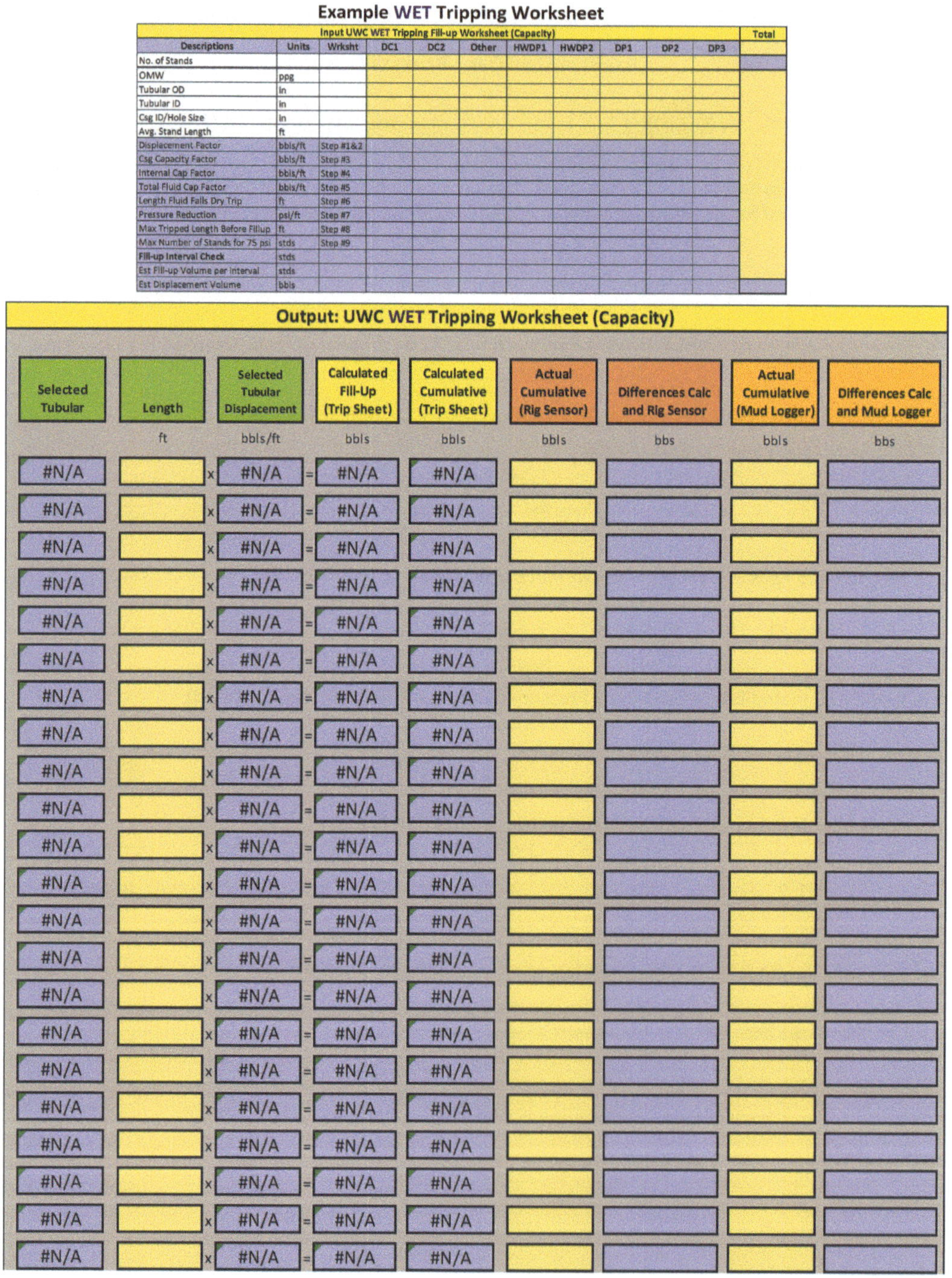

Example WET Tripping Worksheet

Input UWC WET Tripping Fill-up Worksheet (Capacity)											Total
Descriptions	Units	Wrksht	DC1	DC2	Other	HWDP1	HWDP2	DP1	DP2	DP3	
No. of Stands											
OMW	ppg										
Tubular OD	in										
Tubular ID	in										
Csg ID/Hole Size	in										
Avg. Stand Length	ft										
Displacement Factor	bbls/ft	Step #1&2									
Csg Capacity Factor	bbls/ft	Step #3									
Internal Cap Factor	bbls/ft	Step #4									
Total Fluid Cap Factor	bbls/ft	Step #5									
Length Fluid Falls Dry Trip	ft	Step #6									
Pressure Reduction	psi/ft	Step #7									
Max Tripped Length Before Fillup	ft	Step #8									
Max Number of Stands for 75 psi	stds	Step #9									
Fill-up Interval Check	stds										
Est Fill-up Volume per Interval	stds										
Est Displacement Volume	bbls										

Output: UWC WET Tripping Worksheet (Capacity)

Selected Tubular	Length		Selected Tubular Displacement		Calculated Fill-Up (Trip Sheet)	Calculated Cumulative (Trip Sheet)	Actual Cumulative (Rig Sensor)	Differences Calc and Rig Sensor	Actual Cumulative (Mud Logger)	Differences Calc and Mud Logger
	ft		bbls/ft		bbls	bbls	bbls	bbs	bbls	bbs
#N/A		x	#N/A	=	#N/A	#N/A				
#N/A		x	#N/A	=	#N/A	#N/A				
#N/A		x	#N/A	=	#N/A	#N/A				
#N/A		x	#N/A	=	#N/A	#N/A				
#N/A		x	#N/A	=	#N/A	#N/A				
#N/A		x	#N/A	=	#N/A	#N/A				
#N/A		x	#N/A	=	#N/A	#N/A				
#N/A		x	#N/A	=	#N/A	#N/A				
#N/A		x	#N/A	=	#N/A	#N/A				
#N/A		x	#N/A	=	#N/A	#N/A				
#N/A		x	#N/A	=	#N/A	#N/A				
#N/A		x	#N/A	=	#N/A	#N/A				
#N/A		x	#N/A	=	#N/A	#N/A				
#N/A		x	#N/A	=	#N/A	#N/A				
#N/A		x	#N/A	=	#N/A	#N/A				
#N/A		x	#N/A	=	#N/A	#N/A				
#N/A		x	#N/A	=	#N/A	#N/A				
#N/A		x	#N/A	=	#N/A	#N/A				
#N/A		x	#N/A	=	#N/A	#N/A				
#N/A		x	#N/A	=	#N/A	#N/A				
#N/A		x	#N/A	=	#N/A	#N/A				
#N/A		x	#N/A	=	#N/A	#N/A				

Fig. 1.12 Example tripping WET fill-up worksheet.

Charged formations (abnormal pressure)

The formation fluids (saltwater/oil/gas) are contained within a pressure envelope developed by gravity, true vertical depth, pressure from layers of rock above, and movement of the matrix due to geological events, etc. Formations deposited in a salt water environment (sedimentary), which have the ability for formation waters to migrate out of the sediments as additional sediments are deposited on top, have a typical formation pressure gradient of 0.465 psi/ft.

If a significant geological event occurs, i.e., upward movement with fluids unable to migrate (trapped), the resident formation pressure may become overpressured when compared to surrounding or uphole formations. When overpressure occurs, the formations will deviate from 0.465 psi/ft. and are considered abnormal or charged formations.

RANGE OF FLUID GRADIENTS FRESH WATER TO SALT WATER = 0.433 to.465 PSI/FT

What causes the well to flow?

When formation pressure exceeds the hydrostatic pressure of the fluid column in the well and the formation contains sufficient interconnected voids (porosity and permeability), formation fluids may enter the well. Mathematically represented below, this condition is known as "differential pressure." Not every formation with this condition will flow, as each formation layer has individual characteristics which may prevent flow combined with the properties of the fluid used to penetrate this formation (i.e., wall cake, gel strengths, etc.)

$$\Delta DP = FP - HP \tag{1.15}$$

where

ΔDP = Differential Pressure (psi)
FP = Formation Pressure (psi)
HP = Hydrostatic Pressure (psi)

Fracture pressure

Fracture pressure is the pressure in the wellbore at which a formation will structurally fail and allow fluids to enter the formation. Depending on the formation type, this failure may be permanent whereupon the strength of the rock has forever been reduced (i.e., hard rock formations such as limestone, Chert, etc.). Other formations share stresses between the rock and the fluid within the structure. Induced pressures may lead to a temporary failure, but the rock may return to original strength when applied pressures are stopped. These temporarily fractured formations are sometimes called "plastic" and are represented by sedimentary clays and some siltstones.

If upper zones are too weak to withstand rigors of higher pressures downhole, the weaker zones are normally cased off and cemented. The lowest point of the casing/liner is called the casing/liner shoe or seat. Once the casing isolates and bonds to the upper formations by displacement of cement, the cement is allowed to cure. Once hardened, the cement left inside the casing is drilled out, and the formation immediately below the casing seat is pressure tested.

WEAK FORMATIONS ARE CASED OFF AND CEMENTED TO ALLOW FOR HEAVIER FLUIDS DOWNHOLE.

MASP—Maximum allowable surface pressure at static conditions

The Maximum Allowable Surface Pressure before a formation fracture occurs is converted to an equivalent mud weight (EMW or MWE). Once determined, the equivalent mud weight is used as the limit which cannot be exceeded before reaching the next casing point.

Normally, well construction designs incorporate overdesigned high strength casing and BOPs at surface. This shifts any failure point below surface in the case of fracture-induced well control. The formation just below the casing shoe will fracture before the surface pressure systems fail (i.e., BOPs and casing). It is important that the fracture point of the casing seat is measured to ensure this limit will not be exceeded. This surface casing pressure limit corresponding to this casing seat formation failure pressure is called the Maximum Allowable Surface Pressure (MASP).

MASP IS A STATIC FLUID TEST, WHERE PRESSURE IS APPLIED TO A STATIC FLUID COLUMN UNTIL THE FORMATION FRACTURE LIMIT IS REACHED.

IF STATIC MASP LIMIT IS REACHED DURING INITIAL SHUT-IN OF THE KICK, NO ATTEMPT SHOULD BE MADE TO LOWER PRESSURE BY OPENING THE CHOKE. IF THE CHOKE IS OPENED DURING STATIC CONDITIONS, A LARGER INFLUX WILL BE ALLOWED TO ENTER WELL.

UNDER NORMAL CIRCUMSTANCES, IF A WELL IS SECURED AND THE MASP IS NOT EXCEEDED, THE WELL CAN BE KILLED THROUGH CONVENTIONAL WELL KILL OPERATIONS.

Determining MASP from LOT/FIT

Maximum Allowable Surface Pressure is defined as the maximum allowable pressure exerted in the annulus at which casing shoe formations will fracture during initial closure or shut-in conditions. MASP can be determined by testing the casing shoe formation (LOT/FIT), normally performed after setting and drilling out casing.

To accurately determine this value, the well must contain a (1) homogenous (same) mud weight throughout the entire well, and (2) precise Leak-Off Test pressure or accurate formation integrity test information is known. The pressure required to fracture the formation decreases, as the fluid density increases.

To calculate MASP, four conditions must be met.

(1) True Vertical Depth to the Casing Shoe (ft) must be known.

(2) Current Mud Weight (ppg)—Must be same density through well. Therefore, circulate and condition mud before performing test.

(3) Maximum Mud Weight (ppg)—The equivalent mud weight from Leak-Off Test or Formation Integrity Test.

(4) Kill Mud Weight Density

MASP with original mud weight

$$\text{MASP} = \left(\text{Max MW}_{\text{LOT or FIT}} - \text{Current MW}_{\text{ppg}}\right) \times 0.052 \times \text{Shoe TVD}_{\text{ft}} \quad (1.16)$$

where

MASP = Maximum Allowable Surface Pressure (psi)

Max $\text{MW}_{\text{LOT or FIT}}$ = Maximum Mud Weight from Leak-Off or Formation Integrity Test (ppg)

Current MW_{ppg} = Current Mud Weight (ppg)

0.052 = Conversion Constant (psi/ft/ppg)

Shoe TVD_{ft} = True Vertical Depth to Casing Shoe (ft)

Example of MASP with original mud weight

Example well with OMW

TD: 11,400 ft. MD/TVD

OMW = 11.4 ppg

LOT = 16.2 ppg

Casing Shoe = 9–5/8″ at 5000 ft (MD/TVD)

$$\text{MASP} = (16.2 - 11.4) \times 0.052 \times 5000 = 1248\ \text{psi} \quad (1.17)$$

MASP with kill mud weight

With an influx in the well, the original mud weight will be replaced with kill mud weight. This increase in kill mud weight will cause the Maximum Allowable Surface

Pressure to be reduced. The Maximum Allowable Surface Pressure can be calculated using the following formula.

$$MASP = \left(Max\ MW_{LOT\ or\ FIT} - KMW_{ppg}\right) \times 0.052 \times Shoe\ TVD_{ft} \tag{1.18}$$

where

MASP = Maximum Allowable Surface Pressure (psi)
Max $MW_{LOT\ or\ FIT}$ = Maximum Mud Weight from Leak-Off or Formation Integrity Test (ppg)
KMWppg = Kill Mud Weight (ppg)
0.052 = Conversion Constant (psi/ft/ppg)
Shoe TVD_{ft} = True Vertical Depth to Casing Shoe (ft)

Example of MASP with kill mud weight

Example well with KMW:

TD: 11,400 ft MD/TVD
KMW = 12.5 ppg
LOT = 16.2 ppg
Casing Shoe = 9–5/8″ at 5000 ft (MD/TVD)

$$MASP = (16.2 - 12.5) \times 0.052 \times 5000 = 962\ psi \tag{1.19}$$

Transition of static MASP to dynamic MASP

Once a well kill operation begins, the wellbore fluids transition from static conditions to flowing conditions through a throttling device (i.e., choke). Once well kill operations begin, the static MASP is no longer valid as the fluids are not static. Due to the complexities of calculations of flowing fluids with unknown influx types, the ability to recalculate dynamic MASP is not possible. As a result, once the well kill has begun, MASP is no longer considered to be the pressure limit.

Casing pressure during well control kill operations

Under dynamic conditions, the dynamic MASP will not be exceeded during well control operations, as long as the influx isn't of sufficient size and volume to displace open hole annulus from bottom of the hole to casing shoe. Multiple safety factors exist within the MASP, such as:

- Leak-Off test pressure/Formation Integrity Pressures are lower than the formation breakdown pressure. LOT EMW determinations are rounded down, and FIT is selected to be less than fracture pressures.
- If constant BHP pressure kill method is used, circulating friction pressures should not be added as the choke is opened to offset pumping friction pressures.
- Kill Mud Weight determinations are rounded-up ensuring their densities will be greater than shut-in conditions.

- The influx is contained in the annulus between bottom of hole and casing shoe, and shoe is not fractured.
- Casing shoe pressures will reach maximum when the influx reaches shoe. Once the shoe has been reached, the influx in the open hole section will be replaced by kill weight fluid.
- Casing pressures will increase if the influx contains gas. Gas is allowed to expand under controlled conditions.
- Shoe tests are normally performed in a freshly drilled hole. Over time, oilfield cements continue to cure along with the placement of mud cake over the formations and generally ensure test pressures will increase over time.

DURING WELL KILL OPERATIONS, CASING PRESSURES MAY EXCEED FORMATION TESTS DUE TO GAS EXPANSION IN THE ANNULUS. MASP LIMITS ARE ONLY RELEVANT IN STATIC CONDITIONS.

Equivalent circulating density (ECD)

Equivalent circulating density is the effective density of a circulating fluid in the wellbore resulting from the sum of the hydrostatic pressure imposed by the static fluid column and the friction pressure.

$$ECD = \frac{(MW + APL)}{(0.052 \times TVD)} \tag{1.20}$$

where

ECD = Equivalent Circulating Density (ppg)
MW = Current Mud Weight (ppg)
APL = Annular Pressure Loss (psi)
TVD = True Vertical Depth (ft)

For changes in mud weights or pump rates, the APL can be effectively determined by:

$$APL_{New} = APL_{Old} \times \frac{(MW_{New})}{(MW_{Old})} \tag{1.21}$$

where

APL_{New} = New Mud Weight Annular Pressure Loss (psi)
APL_{Old} = Old MW Annular Pressure Loss (psi)
MW_{New} = New Mud Weight (ppg)
MW_{Old} = Old Mud Weight (ppg)

Or

$$APL_{New} = APL_{Old} \times \frac{(\text{Pump Speed}_{New})^2}{(\text{Pump Speed}_{Old})^2} \tag{1.22}$$

where

APL_{New} = New Mud Weight Annular Pressure Loss (psi)
APL_{Old} = Old MW Annular Pressure Loss (psi)
Pump $Speed_{New}$ = New Mud Pump Speed)
Pump $Speed_{Old}$ = Old Mud Pump Speed)

Ballooning theory

Well ballooning is a term used to identify a type of "false indication" of a kick. The phenomenon involved is the acceptance of drilling fluid into a fracture in the formation at normal circulation rates and pressures and the return of the fluid back into the wellbore when pumps are stopped and pressures are reduced. This flow back has the surface appearance of a kick with increase in pit levels and flow with pumps off.

With the pumps engaged, the Equivalent Circulating Density (ECD) exceeds formation fracture pressure, microfractures are created within the formation, and drilling mud will be deposited into these small induced formation fractures. The microfractures can be propagated causing temporary storage (under pressure) of large mud volumes downhole. Microfractures normally do not cause severe losses or total losses (Figs. 1.13 and 1.14).

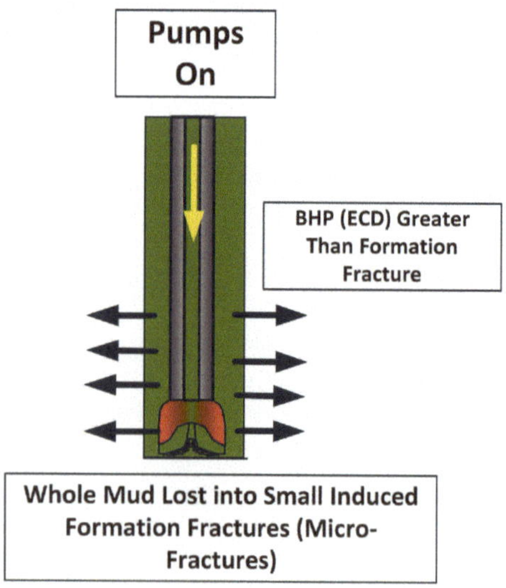

Fig. 1.13 Temporary storage.

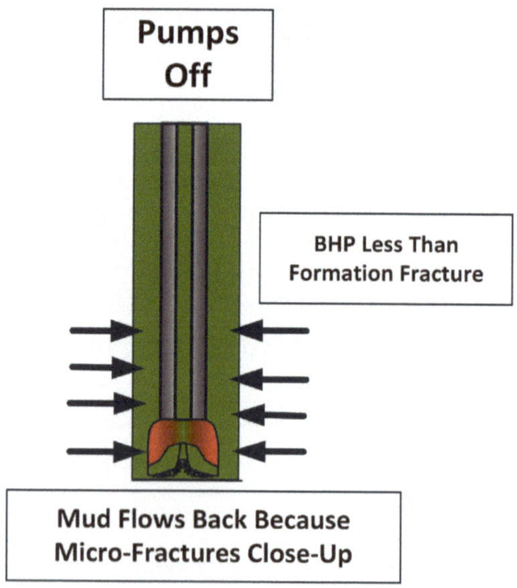

Fig. 1.14 Flow back of mud.

When pumps are disengaged, the ECD is reduced as the annular pressure losses approach zero. Since the microfractures have been storing fluid under pressure, the wellbore pressure is reduced and therefore, the microfractures will close causing drilling mud to flow back into the wellbore. Mud flow back can also bring unwanted gas or formation water with the mud which may be observed when circulating bottoms up, yielding a gas peak or mud contaminated with formation water.

Generally speaking, each formation is made up as a composite structure of several distinct layers. The overall stratigraphy of formations will contain multiple layers of sediment, each with their own unique composition of formation fluids formed within its own heat and pressure regime.

To understand how microfractures can occur, we first need to study the relationship of pressures within the wellbore (Fig. 1.15). As we know,

- HP = 0.052 × MW ppg × TVD ft (or HP = 0 0.134 × MW (pcf) × TVD (ft))
- Equivalent Circulating Density = BHP (psi) + Annular Pressure Loss (psi)
- At Steady State Conditions = BHP = ECD,
- Ballooning occurs when BHP + APL exceeds microfracture pressure
- APL is much, much less than Circulating Pump Pressure

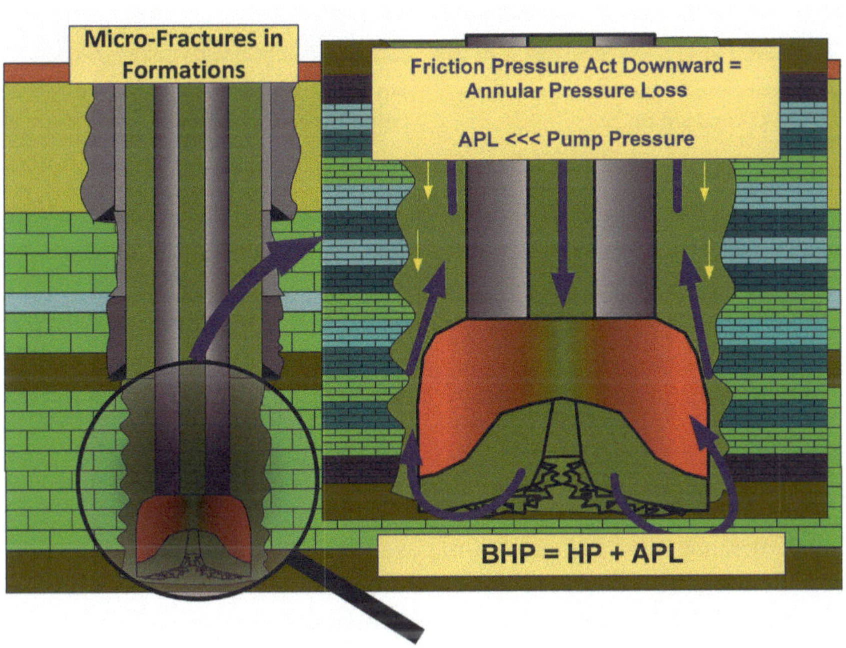

Fig. 1.15 Close-up of stratigraphic formations.

When hole sections are being drilled, with the pumps running, the friction pressure of fluid flowing up the annulus is additive to the BHP. If the BHP pressure increases to the point where formations plastically deform (crack), whole mud will be forced into the microfractures and stored at a pressure of HP + APL.

Once the pumps are shut-off, BHP = HP (as APL approaches zero). When the differential between stored fluid pressure and BHP becomes sufficient, whole mud is displaced from the microfractures into the wellbore.

This means the well will flow with the pumps off. Unlike a well kick, given sufficient time, the flowing formation fluids should decrease over time. Therefore, well ballooning will give a false indication that the well is flowing (Figs. 1.16 and 1.17).

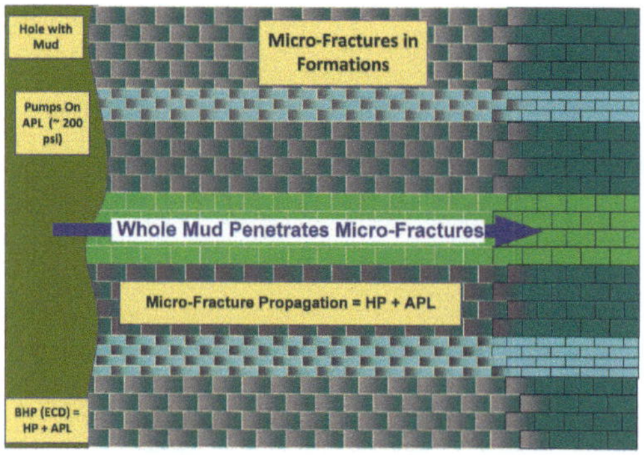

Fig. 1.16 Whole mud penetrating micro-fractures when pumps are on.

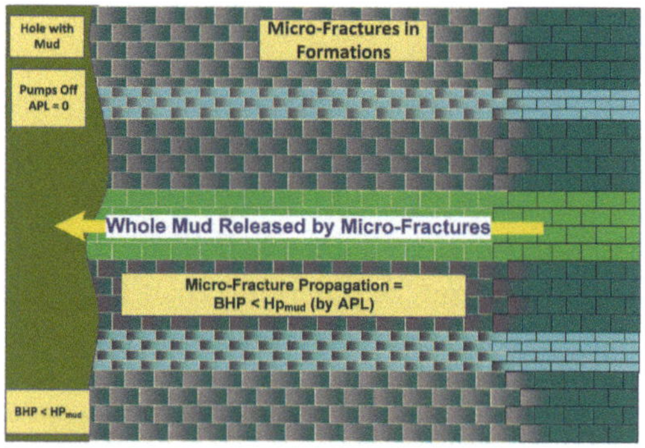

Fig. 1.17 Whole mud released into wellbore when pumps are shut-off.

The following represents an overall list of positive Indicators of well ballooning.

- Drilling fluid must have been lost during the current drilling sequence.
- Pit gain normally occurs with pump(s) off.
- If the well is shut-in the shut-in drillpipe and shut-in casing pressures approach same value.
- The shut-in drillpipe pressure must be equal to or less than the annular friction loss.

Kick tolerance

Kick Tolerance is the highest volume of an influx kick volume which can be contained after securing the well (shut-in well) without causing casing shoe formation breakdown or damage. Kick tolerance is controlled by maximum kick size and fracture pressure (as determined by LOT/FIT).

Kick Tolerance can be expressed as either the maximum intensity or, more commonly, the maximum influx volume that can be taken into the wellbore and subsequently be circulated out without breaking down the formation.

WORST CASE – GAS INFLUX DUE TO COMPRESSIBILITY
ASSUME GAS GRADIENT OF 0.104 PSI/FT (2 PPG DENSITY)

Kick Tolerance can be determined at two stages:

1. During well planning as a single point for each hole section. Estimates during each hole section include casing point, formation pressures, designed mud weights, and hole section total depth/true vertical depth.
2. While drilling determined as needed. Calculated using actual casing point, estimated formation pressures, actual mud weights, and hole section total depth/true vertical depth

Simplified well diagramming (U-tube)

In order to simplify understanding of the complexities of fluids and pressures along with gas expansion, a simple U-tube is used to describe a well with tubulars. A U-tube can be described as a simple hose in which both sides terminate at the same level. When filled with fluid, the fluid will seek a level within the vertical sections of the hose. If one side of the U-tube contains a less dense fluid, the heavier density side will fall to a level where the air plus heavier fluid are at the same hydrostatic pressure as the light density fluid. For this process to work, both tubes must be interconnected at the base (Figs. 1.18–1.20).

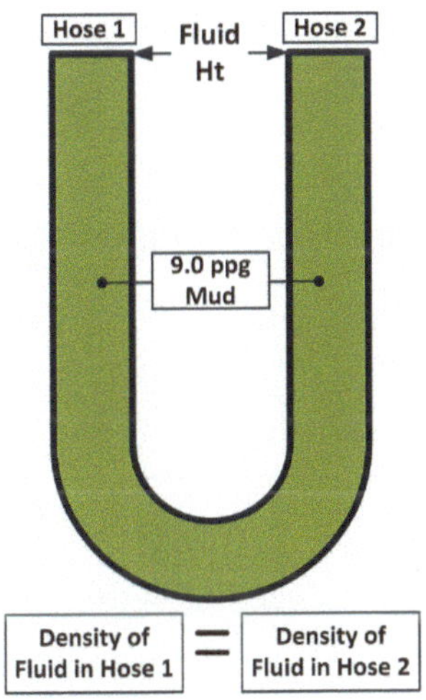

Fig. 1.18 U-tube with equal density fluids.

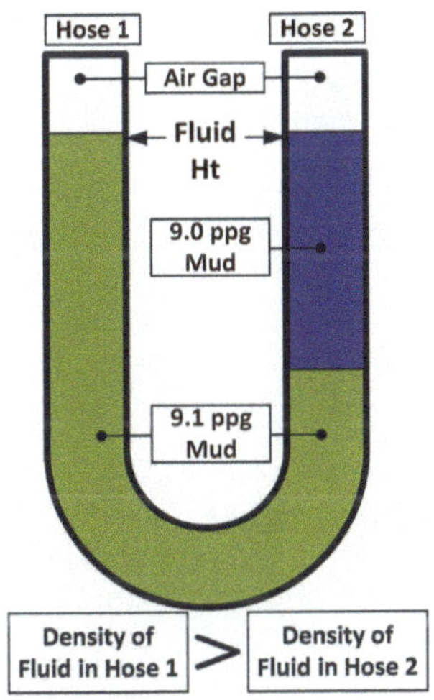

Fig. 1.19 U-tube with unequal density fluids.

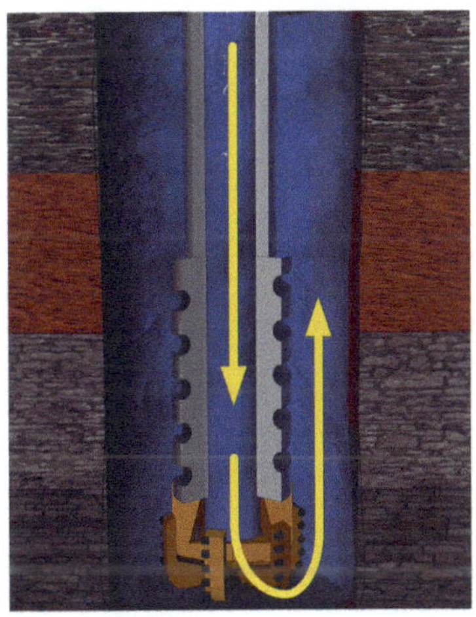

Fig. 1.20 DP/DC/bit U-tube example.

The U-tube can be used to describe tubulars within a well. Normally, the left-side represents in internal diameter of the tubular string, while the right-hand tube describes the annular volume between hole/casing and tubulars. The inter-connectivity is represented by the nozzles within the bit or an open bottom-hole assembly (Figs. 1.21–1.23).

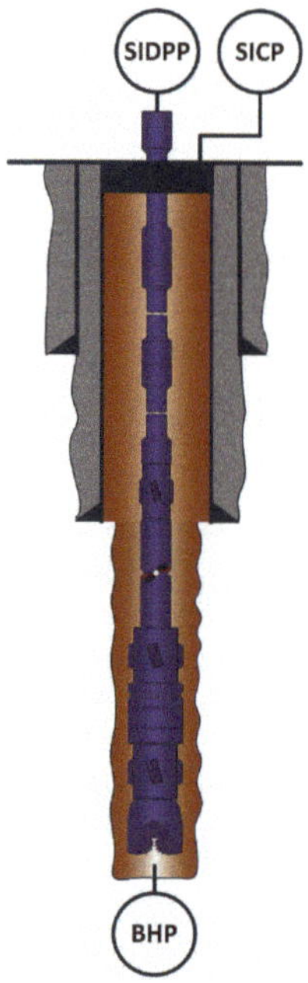

Fig. 1.21 Well schematic with pressure.

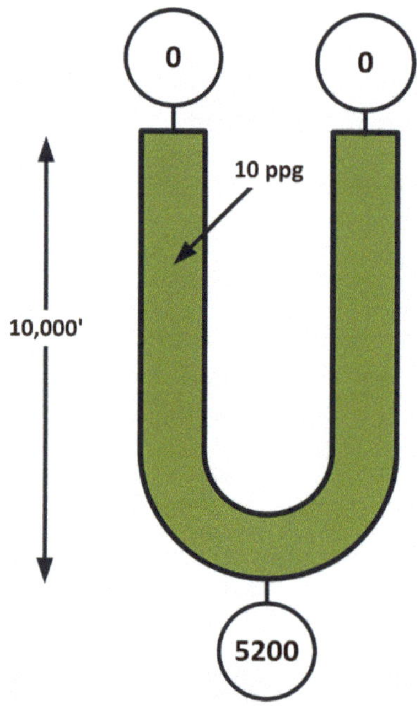

Fig. 1.22 U-tube diagram with pressures.

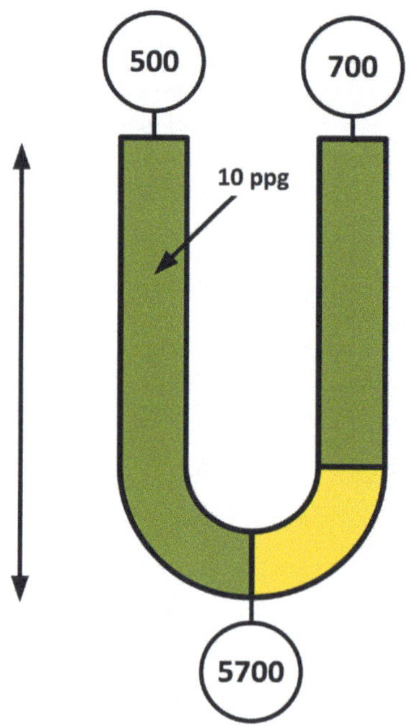

Fig. 1.23 U-tube diagram with kick.

Simplified well control equations

To understand Well Control, we begin by understanding forces acting within the wellbore which are observed from surface read-out gauges and kick volumes. During each phase of well control, the simple formula shown below can be used throughout the well control circulation to better understand what is happening.

At **static** conditions (Fig. 1.24).

$$HP_{Drillpipe} = BHP = HP_{Annulus}$$

Fig. 1.24 Static condition equation.

At static **shut-in** conditions (Fig. 1.25).

$$HP_{Drillpipe} + SIDPP = BHP = HP_{Annulus} + SICP$$

Fig. 1.25 Static shut-in condition equation.

At **dynamic** conditions (i.e., while staging pumps up to speed at constant bottom-hole pressure).

For drillpipe side, only a few strokes will be needed to meet slow circulating rate. Therefore, the HP in drillpipe and the shut-in casing pressure will not be greatly affected. Since the majority of circulating friction pressure is generated while circulating down the drillpipe, the drillpipe friction pressure will dramatically increase. Therefore, it would be difficult to maintain constant circulating drillpipe pressure by manipulating the choke (Fig. 1.26).

$$\overline{HP}_{Drillpipe} + \overline{SIDPP} + Friction\uparrow = \overline{BHP} = \overline{HP}_{Annulus} + \overline{SICP} + Friction\uparrow$$

Fig. 1.26 Dynamic conditions at beginning of staging pumps up to speed equation.

For the annulus side, the hydrostatic pressure and shut-in casing pressures will remain relatively constant. Knowing the friction pressures in the annulus will be magnitudes smaller than DP friction, and in order to keep BHP constant, the choke is opened while keeping circulating **casing pressure constant**. By keeping the casing pressure constant through manipulating the choke, the opening of the choke will automatically compensate for the small annular circulating friction pressure (Fig. 1.27).

$$\overline{HP}_{Drillpipe} + \overline{SIDPP} + Friction\uparrow = \overline{BHP} = \overline{HP}_{Annulus} + SICP\downarrow + Friction\uparrow$$

Fig. 1.27 Dynamic conditions while staging pumps up to speed equation.

ANYTIME DURING A WELL CONTROL KILL, IN ORDER TO DECREASE OR INCREASE PUMP RATES, CASING PRESSURE MUST BE HELD CONSTANT

Equivalent mud weight

The term equivalent mud weight (EMW) is used to describe a static fluid column mud weight which equals the circulating downhole pressure at a given depth.

$$EMW_{ppg} = \frac{\text{Circ Downhole Pressure}_{psi}}{(TVD_{ft}) \times (0.052)} \quad (1.23)$$

where

EMW_{ppg} = Equivalent Mud Weight (ppg)
Circ Downhole Pressure_{psi} = Circulating Downhole Pressure at given depth (psi)
TVD_{ft} = rue Vertical Depth for given depth (ft)
0.052 = Conversion Constant (psi/ft/ppg)

Slow circulating rate

The most accurate means of determining "annulus friction pressures" is to accurately measure these friction pressures at rates used to kill the well during well control operations. These friction pressures are recorded at predetermined slow circulation rates, typically in the range of 2 to 5 bpm. These slow circulating rates are normally selected at a rate to which a 1 ppg increase in mud weight can be accomplished using rig mud hoppers. The rates will vary for rigs and normally fall within the range of 20, 30, and 40 strokes per minute (spm).

These rates are to be taken at:

- Casing shoe before drilling out float equipment
- At the start of each tour
- After increasing mud weight of 0.2 ppg
- During long hole sections
- After mud pump maintenance (especially if the piston and swabs have been resized)

SCR PRESSURES ARE USED TO ACCURATELY DETERMINE INITIAL CIRCULATING PRESSURES (ICP AND FINAL CIRCULATING PRESSURES (FCP)

Gas expansion

Gas which enters a well containing water-based drilling fluids will migrate or begin to rise to the surface due to its' light density compared to the mud. Gas which is not allowed to expand in volume as it rises in the well will bring the higher bottom-hole pressure up the hole. This higher pressure can cause breaches at weak zones below casing shoes or failure of BOPE/Wellheads at surface. Although the study of gases is a complex science, the following equation can be used for normal oilfield operations.

$$\text{Boyle's Gas Law} = P_1 V_1 = P_2 V_2 \quad (1.24)$$

where

P_1 = Pressure at Condition 1 (psi)
V_1 = Volume at Condition 1 (bbl)
P_2 = Pressure at Condition 2 (psi)
V_2 = Volume at Condition 2 (bbl)

In practical terms, Boyle's law shows when the pressure between two conditions is decreased in half, the volume of the gas doubles. A **10 barrel** gas bubble at 10,000 ft. with 10 ppg EMW would expand to 20 barrels as it reaches 5000 ft. As the pressure in the bubble decreases to 14.7 psi, atmospheric pressure at surface the bubble would expand to 3500 barrels. This calculation is for a single bubble which travels up the wellbore and allowed to expand uncontrolled. Rarely do kicks behave in this manner as migrating gas strings out and does not remain as a single bubble.

Gases encountered during drilling operations can be categorized as follows:

- Drilled Gas—Enters wellbore from drilled formation with sufficient voids and interconnectivity of voids.
- Background Gas—Background gases are a combination of gases entering from a drilled formation and liberation of gases within the cuttings as they are circulated uphole. As a lagging indicator, background gases are observed after cuttings have been circulated from the bottom of the hole to the surface. Normally, background gas trends are closely tracked. In high-pressure, high-temperature applications, background gas trends may be used to shut-in the well in order to minimize size of kicks.
- Connection Gas—After drilling to a point where another drillpipe section needs to be added. At this connection point, rig pumps are turned off in order to install a new section of drillpipe. During this time, the ECD is effectively removed from the well. If gases enter the wellbore during the connection, this lagging indicator will be seen when circulating bottoms up. At the surface, connection gas reading is added to the same trend line as background gas.
- Trip Gas—Gases enter wellbore during a tripping exercise where the bit and BHA are removed from the well and rerun to bottom. Upon circulating bottoms up, another lagging indicator would be gas breaking out at surface.
- Gas Breakout from OBM—Gases may enter the wellbore full of oil-based muds. Under pressure, the gases are dissolved in the OBM and will migrate extremely slowly. With solubility properties, gases within OBM will break out of solution when the wellbore pressure reaches the bubble point pressure near surface and can easily overload surface equipment, if preparations are not made. A sufficient MGS, multiple chokes within the choke manifold should be readied as gas will break out near surface. Choke manipulation may become more critical and might require slowing pump rates.

Kick detection

Early detection of kicks and limiting the size of an influx are critical in minimizing chances of equipment or formation failure. A kick is an event where an influx enters the wellbore. In order to facilitate safety, influxes are assumed to be gaseous requiring a kill procedure where bottom-hole pressures are held constant while the gas is allowed to expand which reduces the influx pressure as it is circulated out.

Industrial and governmental bodies have defined two groups of kick indicators. **Positive indicators** represent a group of kicks in which immediate action for securing and shutting-in the well must be conducted. An additional kick indicator group is called **secondary indicators**, which may or may not be an indicator of a kick. Secondary indicators of kicks may require verification of flow before shutting-in and securing the well (Table 1.1).

Table 1.1 Kick indicators.

Positive indicators	Secondary indicators
Increase in Flow & Pit Gain While Drilling	Increase in Rate of Penetration
Increase in Flow & Pit Gain While Tripping In	Cutting size
Increase in Flow & Pit Gain While Tripping Out	pH, Chlorides
	Background and Connection Gas
	Mud weight
	Lost Circulation
	Abnormal Formation Pressure
	Shallow Gas
	Pump Pressure

Positive indicators

Increase in flow & pit gain while drilling

- Additional flow while circulating—While circulating, an increase is observed in the flow returns from well.
- Flows with Pumps off—After spacing out, the pumps are shut-off and the well flow and pit gain are observed.
- Pit gain eliminating surface additions—Additional gain in pits without adding surface volumes (switching pits)

Increase in flow & pit gain while tripping in

- Pit gain while tripping—While tripping into the well without means of circulation, flow initiates from the well.

  - Ported drillpipe float—Return flow volume exceeds the drillstring displacement volume.
  - Nonported drillpipe float—Return flow volume exceeds the drillstring capacity volume.
- Pit gain while washing down: While washing down near bottom or through tight spot, the return flow volume exceeds the drillstring displacement volume and the added flow volume from circulating. Results in a pit gain.
- Pit gain eliminating surface additions—Additional gain in pits without adding surface volumes (switching pits).

Increase in flow & pit gain while tripping out

- Short Fill-up—While tripping out with no circulation, the hole volume necessary to replace the displacement of the drillstring is less than calculated. This results in loss of HP and may lead to the well flowing.
- Pit gain while pumping out of hole—Increase in flow returns when compared to the displacement of drillpipe removed from the well. This results in a pit gain.
- Pit gain eliminating surface additions—Additional gain in pits without adding surface volumes (switching pits)

FILL-UP FREQUENCY IS AS SIMPLE AT 5, 3,1

EVERY 5 STANDS OF DRILLPIPE OR LOSS OF 75 PSI, WHICHEVER COMES FIRST
EVERY 3 STANDS OF HEAVIWATE OR LOSS OF 75 PSI, WHICHEVER COMES FIRST
EVERY 1 STAND OF DRILLCOLLARS OR LOSS OF 75 PSI, WHICHEVER COMES FIRST.

Slugging pipe

Normal operating practices include allowing slugging of the drillpipe before tripping. By pumping a heavy pill down the drillpipe, the drillpipe can be pulled in dry conditions. If a slug is pumped, sufficient time should be allowed to ensure the pill has completely fallen and the well has equalized. After equalization, the trip out may begin minimizing any potential of mistaken well control problems.

Secondary indicators

Increase in rate of penetration (drilling break)

While drilling at a constant weight on bit, pump rate, and rotational ROP, the drilling rate suddenly increases. Most companies state drill no more than 5 ft. into a drilling break, then pick-up and check for flow. The 5 ft. is a general rule and may be altered to shorter intervals due to localized requirements. The increase in rate of penetration with constant weight, pump rate, and rotational ROP may be an indication of drilling into a new horizon with more void areas (porosity) and interconnectibility (permeability). An increase in porosity and permeability may be an indication of an abnormally pressured zone and may lead to an influx entering the wellbore.

Cutting size

Cutting size and shape may indicate a change in pressure regime. A gradual increase in the number of cuttings and changes in cutting shape may indicate penetration into a higher-pressure interval. As a lagging indicator, as cuttings must be circulated from bottom to surface, higher-pressure zones may consist of sharp, splintered, and angular cuttings. These cuttings represent formation slivers which are popping off into the well.

pH, chlorides

Another lagging indicator may be a sharp change in pH and chlorides. High-pressure zones may have more void areas as the fluids have become charged and help support the rock matrix. An increase in chlorides may indicate drilling into higher-pressure/higher-temperature fluids with more saturated salts. A decrease in pH may indicate the presence of salt water or carbonates.

Background and connection gas

As stated previously, increasing trends in background and connection gases may indicate the formations being penetrated have higher pressure. Once again, a lagging indicator as samples must be circulated from the bottom of the well, and these trends should be followed during all routine drilling operations.

Mud weight

Variances in mud weights can occur to accidental dilution or improper pit transfers. If a light slug of mud is introduced into a well, the hydrostatic pressure may be lowered enough for an influx to occur. Communication is the key to identifying mud weight problems. Any variances observed should be communicated to the Person-in-Charge.

Lost circulation

Lost circulation may create complications to well control. If losses are great enough, whole densified mud may discharge into a zone and lower the hydrostatic pressure of well. If the hydrostatic pressure is lessened to an amount where the mud does not contain sufficient positive differential pressures for higher-pressure zones, they may kick and allow an influx in the well. In a well control situation, the lost circulation must be cured while attempting to kill well.

Abnormal formation pressure

Drilling parameter trends are normally monitored for changes which may result in identifying problems. Abnormal pressured formations normally are "traps" with a dense sealing formation located above the high pressure. Monitoring rate of penetration may be helpful in observing a transition zone into high-pressure zones.

Ballooning

Ballooning from an overpressured formation can look like the well is flowing with the pumps off. The well should be fingerprinted to develop a normal curve for flow back after the pumps are shut off. If there is not a decreasing trend, the well should be shut-in to determine if there is an influx.

Pump pressure changes

A decrease in pump pressure while drilling at a constant pump rate may indicate that there is an influx flowing into the well. The pumps should be shut down and the well checked for flow. If there is no flow, then most likely there is a washout in the drillstring or a pump problem.

Well construction process

Well planning

The planning process begins with a combination of information provided by several disciplines including geology, geophysics, facilities, drilling, completion, and production. Geological data identifying horizons and problem areas (i.e., shallow gas, zones of lost circulation, etc.) are developed into a detailed subsurface map. Offset well data are reviewed to identify structural problems (faults) and problem zones (shallow gas, etc.).

Once the information set has been developed, a cost proposal is normally submitted with estimated production values to ensure the asset meets the organizations target goals for profitability. If this threshold is met, the information package is submitted to the drilling, completion, and production disciplines for refinement. The drilling discipline develops detailed casing, cement, and drilling programs. Completion disciplines outline procedures and methodologies necessary to complete the well for maximum recovery. The facilities/production discipline develops tie-in plans for pipelines, surface equipment, etc.

Well design

Pore pressure predictive modeling as well as fracture gradient modeling is conducted to better understand downhole environment and problems. From these models, a detailed casing/cement program is developed as well as the directional plan. Drilling fluids are defined by composition per hole section along with weight and performance characteristics. Hydraulic modeling follows with selection of rates, pressures, BHA, and bit selections performed. Finally, a step-by-step Drilling Prognosis is completed, and the drilling package is submitted for permitting.

Well construction

With a rig on site, each individual hole section is drilled, casing run, and cemented. Surface equipment, such as wellhead spools, are installed and tested. The well is constructed with overall safety to crews and the environment as a top priority. Upon conclusion of the drilling phase, the drilling rig may be converted for the completion phase or a separate completion rig may be brought onto the well after it has been temporarily suspended. The completion phase will minimize damage to the formation while stimulating the well for maximum recovery. Once complete, the rig is moved and surface equipment is installed for production.

Multiple well pads

Multiple well pads have been developed to minimize the overall environmental footprint for well construction, as well as reduce overall well construction time and costs. The unitized pad allows use of "walking rigs" which employ means to quickly move from one well slot to another. The pad also allows for unitized operations such as fracturing, acidizing, and production operations.

Isolation of wells

Well control activities may employ a degree of complexity, as multiple wells may be impacted by the control activities of just one well. Simultaneous operations such as production operations, completion, and/or remedial operations may be impacted by a single well control activity. As in any simultaneous operation, communication between work disciplines must be established and ongoing throughout the well control activity.

Shutting down simultaneous operations, such as producing wells, may include automatic or manual closure of these wells. For more complex multiple well pads, shut down operations may include using a centralized Emergency Shutdown Device (ESD). Most ESD systems employ installed Surface-Controlled Subsurface Safety Valves (SCSSV) for each completed well. These valves require energized medium (or fluid) to keep the well open and flowing. During emergency operations, these valves are de-energized (pressure is bled off) and the valve closes, shutting in the well. As a failsafe valve, SCSSV are designed for automatic closure during emergency operations, shutting off flow and securing the well.

For packer-less completions, an alternative solution will be to employ usages of a Casing Surface-Controlled Subsurface Safety Valve. These valves are permanently installed during the construction of the well and are housed within the casing. They serve the same function as SCSSV and offer a means to shut off open hole completions.

Since most multiple well pads include simultaneous operations, manual shut-in safety devices are not suggested. Manual interfacing with these valves during emergency operations may not be possible. Without proper shut-in of flowing wells, a single well control event may lead to multiple well failures.

MANUAL SHUT-IN SAFETY DEVICES ARE DISCOURAGED FOR MULTIPLE WELL PADS

Multiple well pad design considerations

Multiple well pads have been developed to minimize the overall environmental footprint for well construction, as well as reduce overall well construction time and costs. The unitized pad allows use of "walking rigs" which employ means to quickly move from one well slot to another. The pad also allows for unitized operations such as fracturing, acidizing, and production operations.

Wellhead setting depths

To minimize radiant heat damage, considerations should be made to house key wellhead valves below ground level and housed within an individual or common cellar. The two lowermost production valves (commonly referred to as the master and swab valves) should be located below ground level within a reinforced anticollapse structure. By locating these key components below ground level, if well control is lost on one well, all other remaining multiple well cellars can be flooded with water to protect the elastomer sealing surfaces from radiant heat damage. If water sources are not sufficient or are not available, the surrounding wells can be covered with loose soil to act as an insulator to prevent radiant heat damage. The spacing of each well should be long enough to prevent fire cascading from one well to another during a catastrophic event (Figs. 1.28–1.38).

Fig. 1.28 Multiple "single cellar" wells.

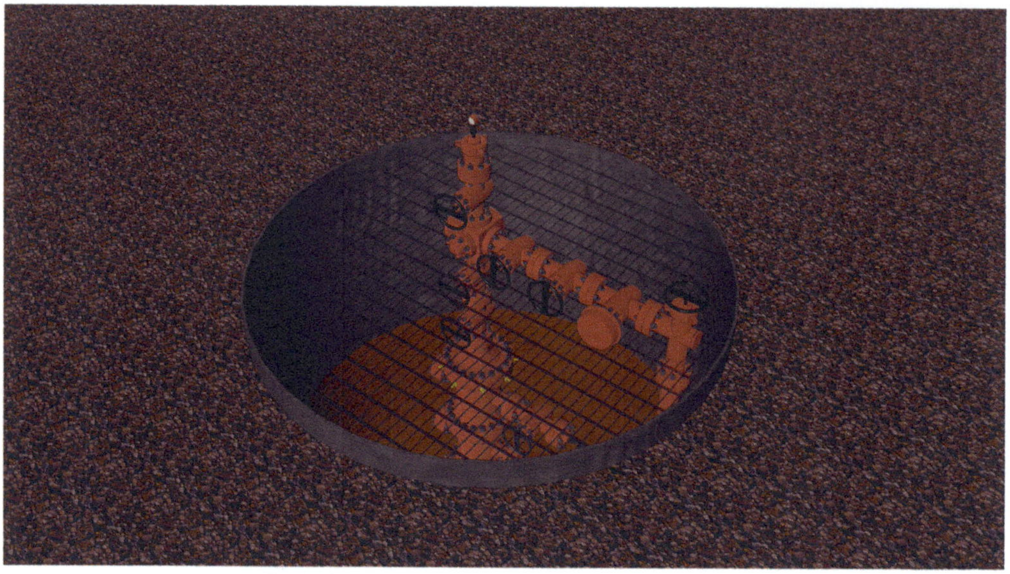

Fig. 1.29 Close up of single cellar.

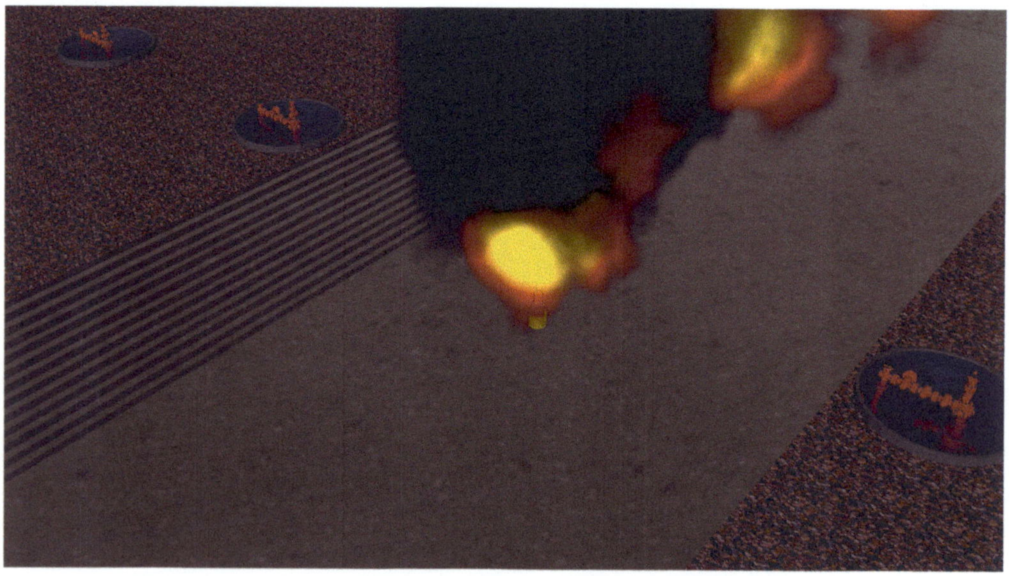

Fig. 1.30 Diagram of sufficient well spacing with cooling water added during well fire.

Fig. 1.31 Single cellar with water. Note Master and swab valves are within water, while pneumatic valve is located above ground.

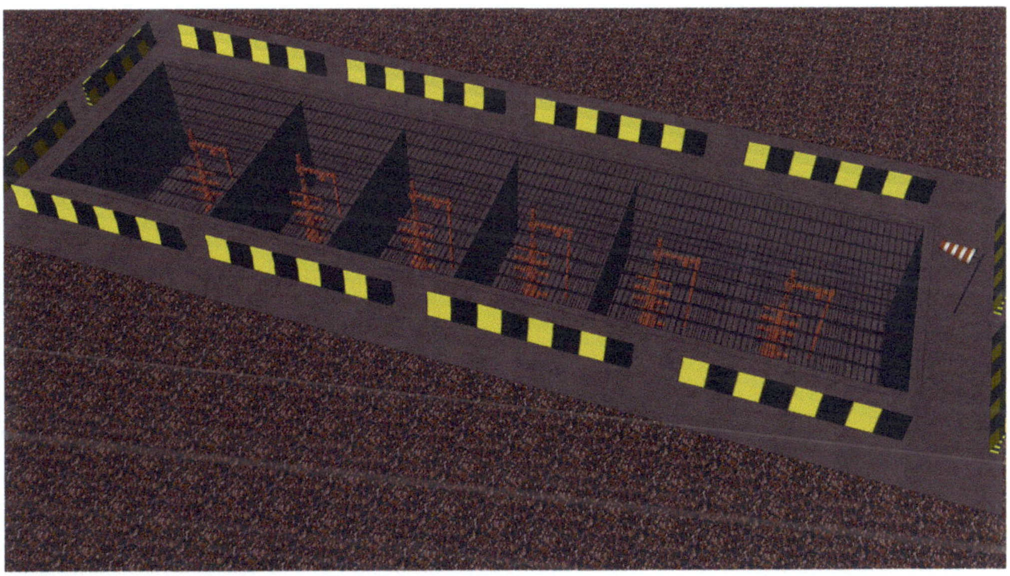

Fig. 1.32 Common cement cellar with six wells.

Fig. 1.33 Close-up of single wellhead within common cellar with blast wall weirs.

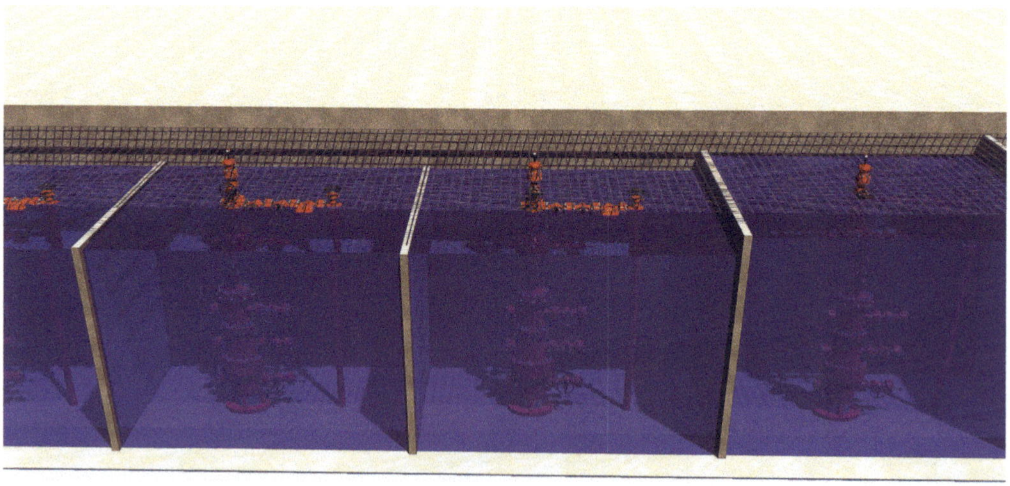

Fig. 1.34 Common cellar flooding close-up shown top weir overflow.

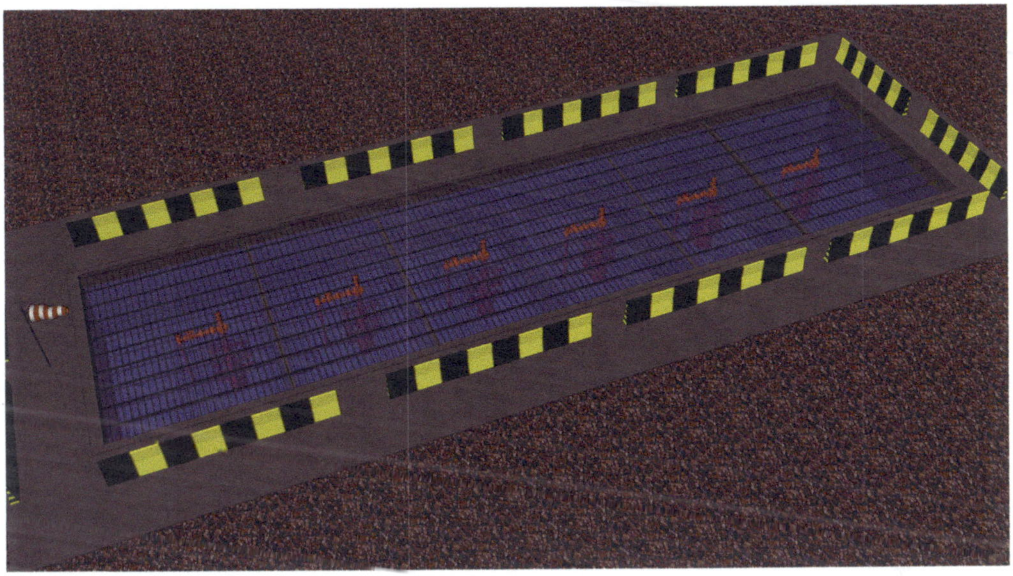

Fig. 1.35 Common cellar flooding with weir blast walls. Flow source from surface can be constructed as part of the cellar.

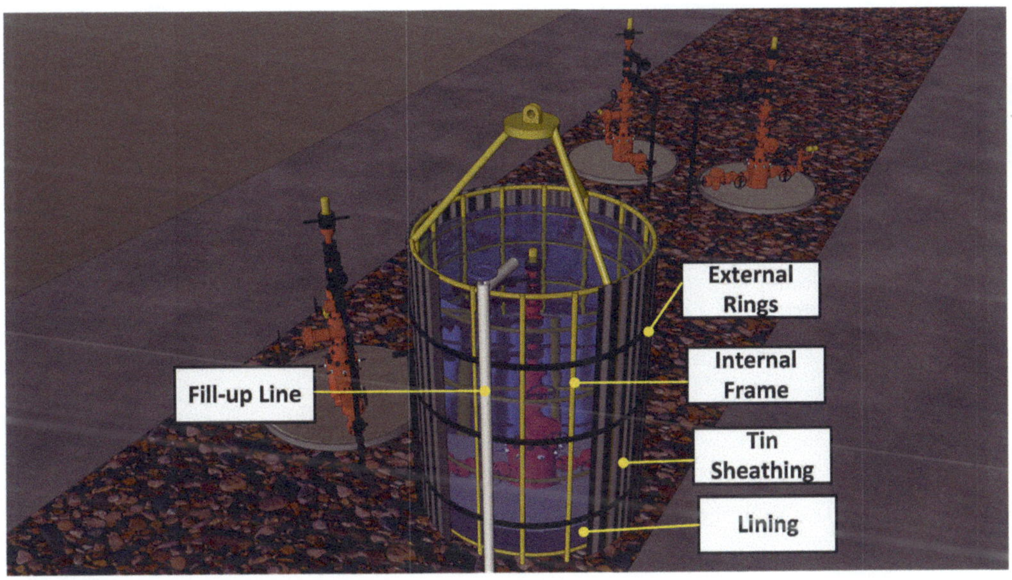

Fig. 1.36 For wellheads above ground, a simple water containment vessel can be constructed before rig operations are to be performed.

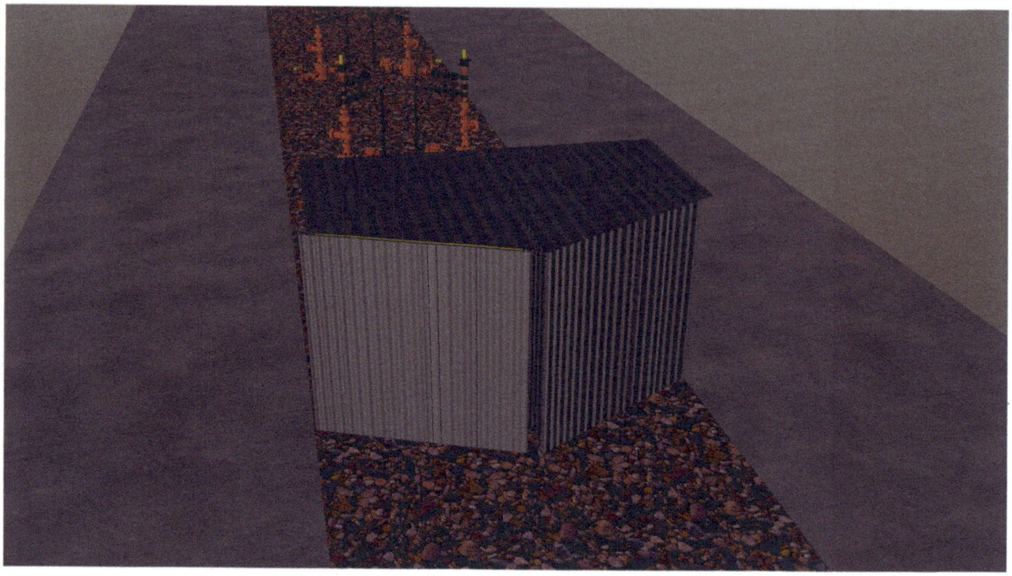

Fig. 1.37 For nondirect flow impact, a radiant heat barrier can be constructed over nearby above ground wellheads.

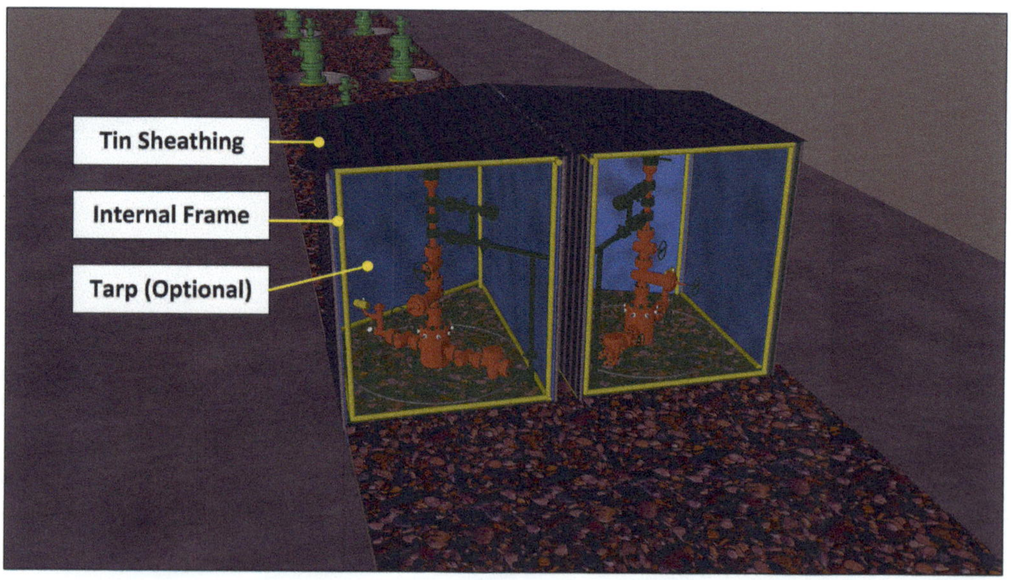

Fig. 1.38 From the reverse side, the radiant heat barrier can be constructed of tubing and tin sheeting, and provide elevated work area for personnel. Access may include using water monitors to provide cooling water from this side.

Blast walls

For common cellar or trench designs, considerations should be made to include "blast walls." Blast walls can be installed between individual wellheads to provide means of mitigating radiant heat from a catastrophic event. These walls may be constructed as thick steel paneling or steel-reinforced concrete paneling. For multiple well pads near ocean fronts where concrete paneling may be used, the concrete should be designed to resist corrosive effects of saltwater.

The thickness of the panel should provide sufficient radiant heat service over an extended period of time. For concrete panels, a thickness of 12 in with integrated lifting brackets on the surface edge is suggested. During an emergency, these brackets can be used to assist in lifting blast walls for improved access to multiple wellheads. This type of design should also include a staggered weir system allowing flooding from both ends of the cellar.

For surface applications, an integrated fall arrestor along with steel plate blast shield is suggested. The fall arrestor provides protection against dropped objects (such as a falling derrick) to impact nearby wellheads. The blast shields also provide means against radiant heat damage from nearby wells. The structure can be designed to be removed individually, for easier access to independent well (Figs. 1.39 and 1.40).

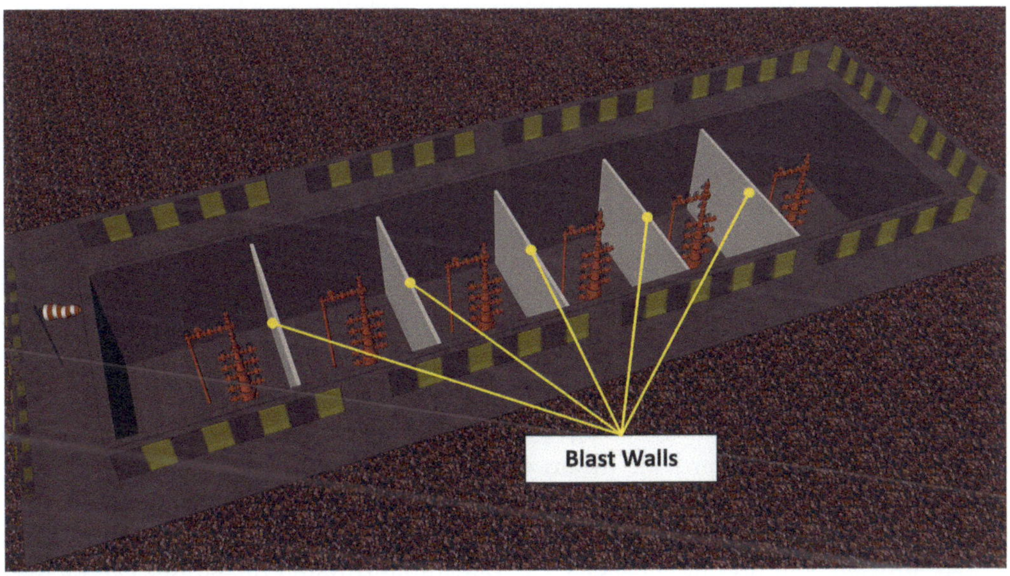

Fig. 1.39 Common cellar with concrete blast walls.

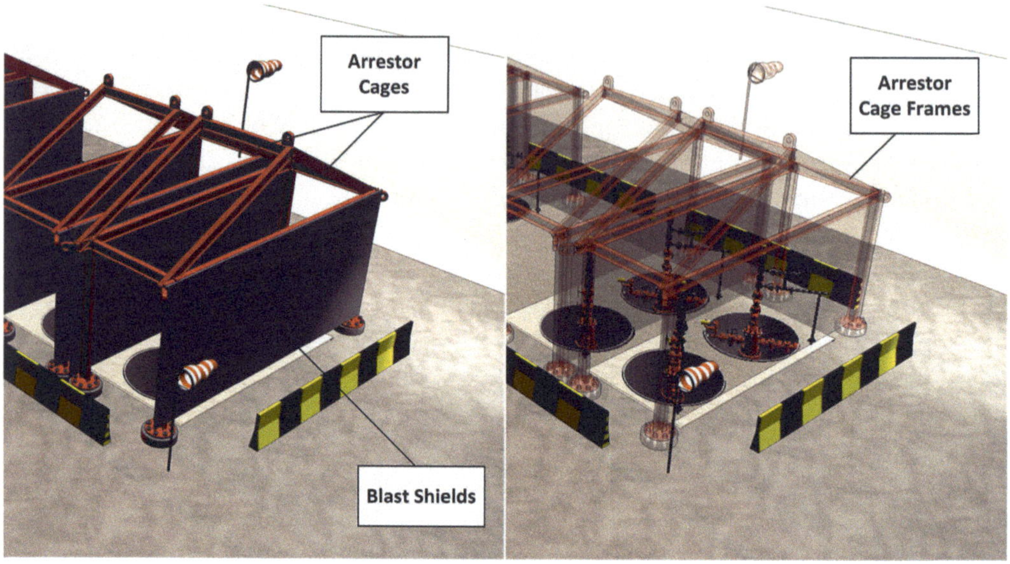

Fig. 1.40 Common cellar with fall arrestor cage (steel beams with steel blast shields).

Control room and casing SCSSV

For larger pad designs with integrated production operations, an automated control system (i.e., SCADA) is normally installed. The SCADA system provides means of simultaneously monitoring each well and production facilities while providing remote setting and operation of alarm, flowing, and shut down systems. Most control rooms are continually manned during production operations providing a means to direct products fluids as well as direct emergency systems.

The installation of SCSSVs within permanent casing strings for open hole or slim hole well designs allows means to oversight during drilling operations. The Casing SCSSV is normally installed within a selected production string at a predetermined depth near the casing collar. Control lines are installed on the exterior of the casing during installation and are cemented in place. The Casing SCSSV allows isolation of the wellbore during tripping operations and can also be used in Managed Pressure Drilling (MPD) applications. MPD designs incorporate a rotating head which is used to isolate the wellbore and provide means of stripping of the drillstring during tripping operations. The use of the Casing SCSSV will isolate the bottom of the hole, after components have been stripped above the casing shoe. Once the Casing SCSSV has been closed, isolated, and pressure checked, the remaining tripping operations can be performed without stripping.

Casing SCSSV are also used to isolate wells during fracking operations. "Frack bashing" may occur when one well's fracture path intersects a nearby well. This may increase downhole pressures, resulting in unanticipated flows which may lead to an unanticipated well control event. Once again, the Casing SCSSV can be closed to isolate the wellbore of a nearby well and prevent an unanticipated well control event from occurring.

Flaring system and H_2S considerations

For uncontrolled flows containing H_2S, consideration may be given to ignite the existing hydrocarbons especially in locations near populations or transportation routes. Considerations may be given for the installation of an automated flare system to be on site during critical rig operations. If needed, the automated flaring system will allow ignition to take place from a suitable safe distance.

With operations containing H_2S and knowing H_2S is colorless, heavier than air, and can easily kill the sense of smell at higher concentrations, common cellar and single cellar designs should be considered hazardous confined space structures (more than 3 walls). As such, operations must include safety procedures for allowing personnel access to these areas and requiring Confined Space entry permits. As with any confined space structure, personnel should have access to SCBA gear, detection monitors, and adequate PPE. Pad designs may include the use of safe rooms, constructed throughout the site, to provide an emergency positive pressure escape center to allow proper donning of the SCBA when H_2S alarms are sounded.

FOR MULTIPLE WELL PADS, SUGGEST EMPLOYING RADIANT HEAT AND DISPERSION COMPUTER MODEL STUDIES TO UNDERSTAND EFFECTS OF RADIANT HEAT AND H_2S IMPACT

Traffic control

Depending on the size and location of the multiple well pad, traffic egress routes should be studied. Pads which suffer from traffic congestion may impact overall safety of the structures, wellheads, and pedestrians. During well control events, traffic congestion problems may become even more problematic. Traffic control plans should be developed and practiced during routine safety drills (Fig. 1.41).

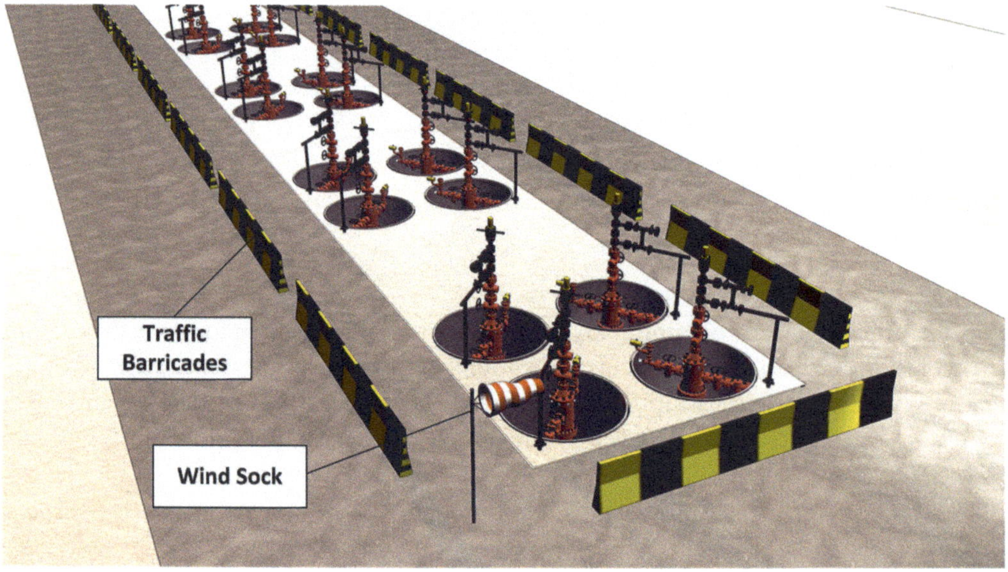

Fig. 1.41 Concrete traffic barriers around wellheads prevents unwanted vehicular traffic. Multiple well pads should contain a visual wind reference such as wind sock (right foreground).

Classifications of blowouts

Blowouts can occur on any well and at any time. To better understand blowouts, the following classifications have been developed.

Surface blowouts (land and subsea mudline)

- Shallow gas intervals (large prolific flows)
- Normally-pressure reservoirs
- Abnormally pressured (high-pressure) reservoirs
- Subnormally pressured intervals with large volume lost circulation

Underground blowouts

- Broaching to surface (cratering)
  - Flow is outside all the casing strings.

Classification standards

Class A or Class 1: Consists of a minor event in which the well is NOT on fire and may be leaking. Results in minor pollution and minimal hazards.

Class B or Class 2: A larger event with flow rates of 5 to 20+ MMSCF/D and 100 to 5000 BPD of produced hydrocarbons/water. Well is NOT on fire, and flow may be underground or at surface (mudline). Access to wellheads is possible. Pollution of fluids can be controlled.

Class C or Class 3: A larger event with flow rates of 20 to 50+ MMSCF/D and 5000 to 20,000 BPD of liquid production. The well is NOT on fire, and flow may be underground or at the surface (mudline). Access to wellhead is possible. Pollution fluids may not be easily contained. Pollution fluids may be hazardous (i.e., H_2S entrained, corrosive, etc.)

Class D or Class 4: A medium event with flow rates of 50 to 100+ MMSCF/D and 20 to 50,000 BPD of produced hydrocarbons/water. The well MAYBE on fire, and flow MAYBE underground or at the surface (mudline). Access to wellhead maybe limited or difficult. Pollution fluids may not be easily contained. Pollution fluids can be hazardous (i.e., H_2S entrained, corrosive, etc.)

Class E or Class 5: A major event with flow rates of 100+ MMSCF/D and 50,000 BPD of produced hydrocarbon/water. The well MAYBE on fire, and flow MAYBE underground or at the surface (mudline). Access to wellhead is extremely limited to impossible due to structural damage in/around wellbore. Pollution fluids cannot be contained and are considered hazardous (i.e., H_2S entrained, corrosive, etc.)

As blowouts are extremely hazardous and expensive, the Universal Well Control Manual was developed to lower the risk through improved understanding of equipment and techniques needed to mitigate kicks before a blowout can occur.

Causes of kicks

A kick may be defined as the unwanted flow (influx) of formation fluids into the wellbore as a result of formation pressure being greater than the hydrostatic pressure. If this flow can be successfully controlled, the kick will be circulated out and the wellbore hydrostatic pressures will be equalized to those of the formation. Once hydrostatic pressure of the fluid column equals the formation pressure, the well is "dead." If the flow into the well is not controlled, a blowout will result. This is an uncontrolled flow of fluids either to the surface or underground.

Not keeping the hole full during trips

Much progress has been made in the prevention and understanding what causes kicks, but most kicks are still man induced. The first step in controlling a well is to maintain a constant column of fluid. Failure to keep the hole full will result in a decrease in wellbore hydrostatic pressure. If the wellbore hydrostatic pressure becomes less than formation pressure, a kick will develop.

In order to maintain a constant fluid level when pipe is being pulled from the well, a volume of fluid equal to the volume of steel removed must be pumped into the annulus. Several methods can be used to fill the hole, but all must be able to measure accurately the amount of fluid required. The two most commonly used methods used to monitor hole fill up are:

Trip tank method

Trip Tank Method employs using any small tank and a pump to continuously fill the well with a calibration device used to monitor the precise volume of mud entering the hole. The main advantages of using the trip tank are:

(a) The hole is always full and continuously monitored for changes in volume indicating swabbing or kicks.

(b) The volume of fluid required to replace the steel being removed from the well is accurately measured.

Stroke counter method

Stroke Counter Method (either human or mechanical) is a volumetric approach to keeping the well full. For each volume of steel removed (i.e., stand of pipe), an equal volume of fluid must be added to the wellbore. This method employs accurately calculating the volume of steel removed, the volume of fluid to be added, and determining the number of strokes. If the number of strokes is greater than calculated, the wellbore might be losing returns. If the number of strokes is less than calculated, a kick may have entered the wellbore.

DO NOT CONTINUE TRIPPING OUT OF HOLE IF REQUIRED STROKES DO NOT EQUAL ACTUAL STROKES AFTER 10 STANDS.

For example:

Drillpipe	5″ IEU, 19.50 lbs./ft.
DP Displacement	0.0076 bbl/ft.
Pump	National 12P160 Triplex with 6–1/2″ liner.
Pump Displacement	0.1106 bbl/stroke

Required: Find the number of strokes per five (5) stands of drillpipe required to fill the hole.

Answer:

(1) Steel Displacement = 5 stds x 93.6 ft/std x .0076 bbl/ft = 3.56 bbls.

(2) Pump Strokes = 3.56 bbl ÷ 0.1106 bbl/ft = 32 strokes

(3) Therefore, the required fill up is 32 strokes per 5 stands pulled.

The advantages of using the pump stroke counter as a means of keeping the hole full are:

- An added ability to keep the hole full.
- A simplistic method of determining the amount of strokes needed to fill the hole requires only counting the number of strokes per specified amount of pipe removed.

The disadvantages of using the stroke counter for keeping the hole full are:

- The hole is not completely full between pumping cycles.
- The hole is not being continuously monitored for swabbing.
- Pump stroke displacement varies due to efficiency of the pump, wear of pump swabs, and liners.

Swabbing or surging

Swabbing is the temporary reduction in bottom-hole pressure, which results from the upward movement of pipe. The drillstring acts like a piston moving upward and pulling formation fluids into the wellbore by reducing the hydrostatic pressure of the wellbore. This is called the "piston-effect." Unfortunately, the piston effect can cause kicks to occur.

The most reliable method of detection of swabbing is proper hole filling. The volume of steel removed from the well must be replaced by an equal volume of fluid. For example, if 5 bbls of steel is removed from the well then 5 bbls of mud must be added to the well to maintain correct hydrostatic pressure and control. If the wellbore takes less volume than calculated, a kick may have entered the wellbore (swabbed). If swabbing is indicated, even if no flow is seen, the pipe should be immediately run back to bottom. The fluid should be circulated out and conditioned before resuming the trip.

A short trip is often made to determine the combined effects of bottom-hole pressure reductions caused by swabbing. When under or near balanced conditions, a short trip is particularly important since it would quickly indicate a need to raise fluid density or slow pulling speeds.

Controlling swabbing

Swabbing is controlled by managing pipe movement, improving fluid properties, and reconfiguring bottom-hole assemblies.

Pulling the pipe too fast

The determination of pulling speed may be made during the short trip. By pulling the pipe too fast, the piston effect may become exaggerated lowering the hydrostatic pressure on bottom and pulling formation fluids into the wellbore.

Poor fluid properties

High viscosity gels/polymers can cause fluid to "cling" to the drillpipe/tubing as it is pulled from the wellbore. This increases the piston effect, lowering the hydrostatic pressure on bottom and pulling formation fluids into the wellbore.

Restricted annular clearances

Large OD tools such as fishing tools or drill collars can enhance the piston effect especially in smaller holes. Extra care should be taken whenever pulling equipment with close tolerances out of the hole. Pulling speed should be slowed to prevent swabbing.

The key to reducing swab pressures are controlling fluid properties and tubular movement velocities. Good practices to prevent or minimize swabbing are aimed at keeping the fluid in good condition and pulling pipe at a reasonable speed.

Surging is the opposite of swab pressure. Surge pressure is the temporary increase in bottom-hole pressure, resulting in the downward movement of pipe. This induced pressure can cause the bottom-hole pressures to exceed the fracture pressure limits of the formation resulting in lost circulation. The key to reducing surge pressure is controlling fluid properties and tubular movement velocities.

Lost circulation

Lost circulation is another leading cause of well kicks throughout the world. Fractured consolidated formations with large vugular cavities are one of the leading causes of lost circulation. If heavier wellbore muds exert a pressure greater than the formation or fracture pressure, losses of whole mud into the formation can occur.

In order to maintain control of a well, the first step is keeping the hole full. Lost circulation causes a shortened fluid column in the wellbore which decreases hydrostatic pressure and can lead to a kick. The lost circulation problem must be rectified first by adding a measured amount of fluid to the annulus along with lost circulation material (if available). Pump rates can be slowed, but should be of sufficient velocity as settling of LCM does not occur.

General procedure for lost circulation recovery

- Do not look down the annulus with a flashlight to determine whether the fluid has "gone south."
- If severe losses have occurred, switch to lower density fluid and add LCM (40–100 ppb of medium and fine course). Medium and fine course material should be able to be pumped through downhole motors and telemetry BHA.
- For total losses, the hole can be filled with water or base oil to determine what hydrostatic pressure the formation can handle.
- Total losses can lower the hydrostatic pressure in the well low enough to induce a kick.
- If partial losses have occurred, continue to pump current mud weight and add LCM (20–40 ppb of medium and fine course).
  - **Note:** Beware of sudden pump pressure increases signifying plugged circulation ports.
- Vary pump rate as needed, but ensure sufficient velocity to prevent solids settling.

- Reduce pump rates if circulation cannot be established after lost circulation material has had time to work.

DO NOT STOP PUMPS WHEN ADDING LOST CIRCULATION MATERIAL. WITH PUMPS OFF, THIS MATERIAL MAY SETTLE AND PLUG CIRCULATION PORTS AND/OR TUBULARS

Insufficient fluid weight

The density of the fluid is normally monitored and adjusted to provide the hydrostatic pressure necessary to balance or slightly exceed the formation pressure. Drilling and workover programs contain weighing schedules for maintaining adequate fluid densities to balance formation pressures. These procedures must be diligently followed unless well conditions dictate otherwise. Insufficient densities can be controlled by ensuring no accidental dilution to surface pits and understanding the forces behind gas cut fluid.

Gas cut fluids

Gas cut fluid is often misconstrued to be a well control problem, resulting in unnecessary additions of weight material to the fluid. Since gas is very compressible and a very small volume of gas has a significant effect on fluid density, the volume of gas downhole will approximately double in size each time the hydrostatic pressure is reduced by 50%. Near the surface, the small volume of gas would have expanded many times resulting in a pronounced reduction of surface hydrostatic pressure.

Some important rules for handling gas cut fluids are:

- Gas cut muds DO NOT cause a large reduction in the bottom-hole pressures.
- Be cautious about raising fluid weight just because of gas cut fluid.
- Ensure the degasser is operating correctly. The best defense against gas cut fluid is a good degasser.
- If in doubt, shut off pump and check for flow. If flow continues, shut-in the well.
    - Gas cut fluids would register SIDPP $= 0$ psi and SICP ≈ 0 psi.

Abnormal formation pressure

If abnormal pressured zones are penetrated with insufficient fluid weights, a kick situation develops. This occurs when the pressure in the formation exceeds the hydrostatic pressure exerted by the fluid column. The formation fluids flow into the wellbore and must be controlled.

A normal formation pressure would be equal to the depth of burial multiplied by the density of saltwater of average salinity for an area. In other words, fluids within the pore spaces must be interconnected to the surface. Abnormal formation pressures would exceed the normal pressure of a formation at the same depth. Abnormal pressures are most likely to be formed by geological events or induced by man such as:

1. Stratigraphic Traps (Large Structures)
2. Tectonic Movement
3. Rapid Compaction
4. Salt
5. Steam and Water Injection

Stratigraphic traps (large structures)

Stratigraphic Traps may consist of large structures of similar pressure divided by different formation fluid densities. Aquifers are also stratigraphic traps which are normalized to surface pressures by a means of outcropping. Charged upper sands may also be abnormally pressured stratigraphic trap.

Some stratigraphic traps are large structures containing a similar formation pressure throughout the entire reservoir. If the traps contain gas, the hydrostatic pressure needed to control formation pressures on a structural high will be significantly higher than if the structure contains water or oil. Abnormal pressures of such structures are common, and caution should be exercised when drilling and working over (Fig. 1.42).

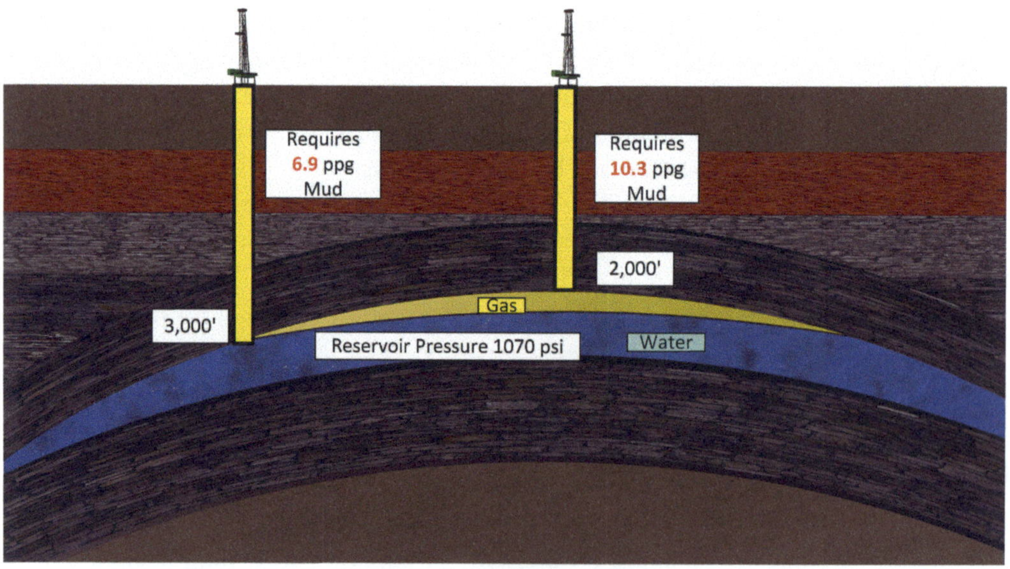

Fig. 1.42 Comparing flank to crest drilling on stratigraphic trap.

By definition, an aquifer is a formation containing mobile water much like an artesian well. An aquifer may become abnormally pressured as this shallow sand may outcrop in nearby mountains at elevations appreciably higher than the elevation of the well. The increase of true vertical elevation of the formation will yield an abnormal pressure.

The aquifer example below demonstrates a geographic hydraulic connection between a submerged outcrop at 1700′ which charges shallow formations at zero-foot datum. When penetrated at 500′, the charged formation at 500′ will need a 20 ppg fluid in order to control the well (Fig. 1.43).

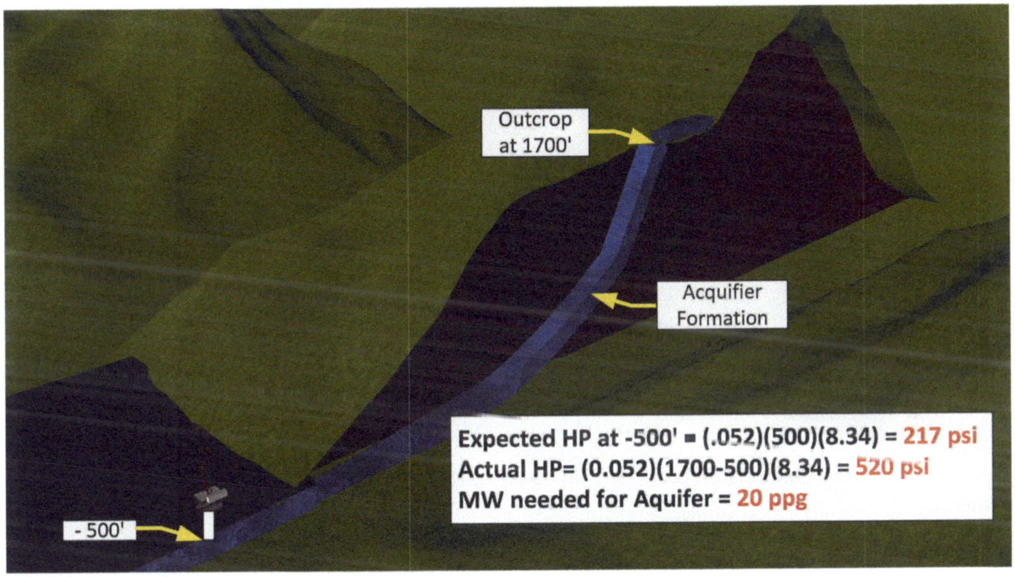

Fig. 1.43 Aquifer charging shallow formation.

High pressures can occur in shallow sands if they are charged by gas from a poor surface casing cement job, casing leaks, or a blowout in a nearby well.

Tectonic movement

Tectonic movements involve the subsurface movement of formations caused by natural forces of the Earth. These movements can be uplifting, faulting, or folding. Folding may be described by Stratigraphic Traps.

Formations normally compacted at great depth can be uplifted to a shallower depth. Should the original pressure be retained, abnormally high pressure can result. If followed, the uplift can generate abnormal pressures only when accompanied by other geological processes which reduce the relief between the buried rock and surface. The magnitude of the pressure is a function of the depth of burial and the degree of uplift.

The diagram below shows a strike/slip fault where the left strategic block has slipped or is downthrown 700′ from its original depositional environment. The right strategic block has struck or is up-thrown 700′ from its original depositional environment along a fault line. The producing zone (red) contains original trapped pressure deposited within a normal 9.0 ppg environment. With a total of 1400 ft. throw, the strike or up-thrown block represents abnormal pressure equal to a 12.2 ppg environment (Fig. 1.44).

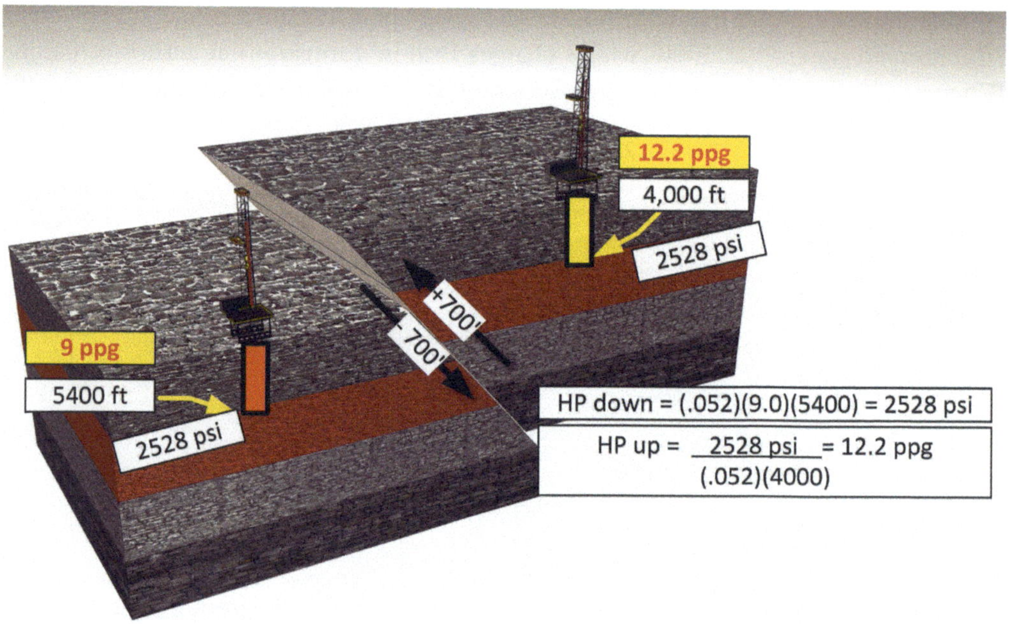

Fig. 1.44 Strike/slip faulting (left block downthrown 700 ft., right block up-thrown 700 ft. = 1400′ throw distance).

Subsurface movement can create severe faulting; deeper charged formation gas/oil/water can travel along the fault and charge shallower formations. This phenomenon will result in abnormal pressure in shallower formations. If the upper formations will not allow abnormal pressure to dissipate within the formation or escape to surface, the shallow formation will remain charged with abnormal pressure.

Rapid compaction

Compaction is the most common and best understood phenomenon causing abnormal pressure. As long as the sedimentary process allows water in the pore space to escape to the surface with the addition of overburden formation layers, the formation pressure will remain equal to the hydrostatic pressure.

When the water becomes "trapped" and unable to move, the water begins to support overburden sediment (layers of sediment and water above). As the trapped water is buried deeper and deeper, the "trapped" water becomes abnormally pressurized (Fig. 1.45).

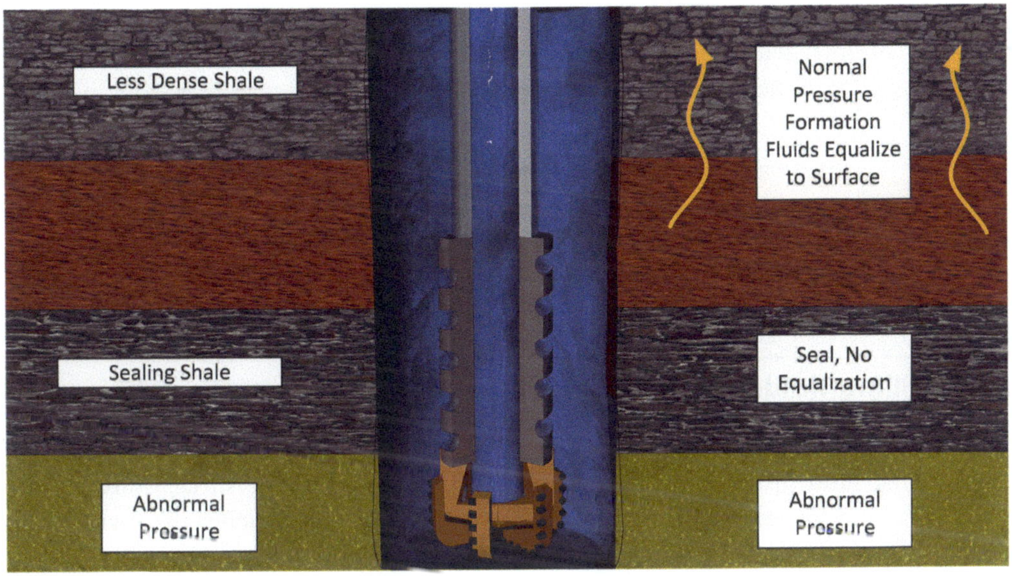

Fig. 1.45 Rapid deposition with sealing shales can lead to abnormal pressure.

Salt

Massive salt beds are known to cause high pressure. Salt is plastic and transmits the overburden pressure to the formation fluids. The pressure within and below a thick and continuous salt may generate an abnormal overburden pressure.

Steam or water injection

Just as in charged upper sands, man-made pressures for recovering hydrocarbons can develop abnormal pressures. Through injection, a reservoir formation pressure is raised by artificial means.

Geothermal gradient

Geothermal gradient is the rate of temperature increase with increasing true vertical depth. Geothermal gradients will vary for localized geological areas and will depend upon heating and cooling effects within this geological area. The most widely used average temperature gradient is 1.4°F per 100-ft TVD. High-temperature reservoirs are most often associated with high pressure with geothermal wells being the exception.

Downhole temperatures can be determined by the following equation:

$$\text{Well Temp}_{\text{TVD}} = \text{Ambient Temp} + \left(1.4 \times \text{TVD}_{\text{Depth}}\right) \tag{1.25}$$

where

$\text{Well Temp}_{\text{TVD}}$ = Average Well Temperature at True Vertical Depth of Point of Interest (°F)

Ambient Temp = Average Surface Ambient Temperature (°F)
1.4 = Normalized Temp Gradient (°F/ft)
TVD_{Depth} = True Vertical Depth for given depth (ft)

IF UNKNOWN, THE NORMALIZED TEMPERATURE GRADIENT CAN BE USED. AVERAGE TEMPERATURE GRADIENT IS 1.4°F PER 100-FT TVD.

Drilling fluids

Functions of drilling fluid

The drilling fluid (mud) is mixed, tested, and pumped from the surface through the workstring and bit, returning up the annulus. Once the fluid reaches the surface, it is routed through a series of processing equipment (i.e., flowline cleaners, shakers, desanders, desilters, and vacuum degassers into a series of processing mud pits). These "returns" are treated for proper density and mud characteristics, and then returned to the suction pit where the pumping process is started again. These drilling fluids perform the following functions:

1. Provide sufficient density to prevent formation fluid flow into the wellbore
2. Cool and lubricate bit, motor, and BHA
3. Provide gel strengths sufficient to suspend and carry cutting from hole
4. Allow efficient separation of cutting when circulated to surface
5. Provide nondamaging characteristics to formation, environment, and workers
6. Provide means to prevent wear and tear on well and tubulars
7. Minimize drilling problems and maintain wellbore stability

Maintain sufficient density and prevent flow into the wellbore

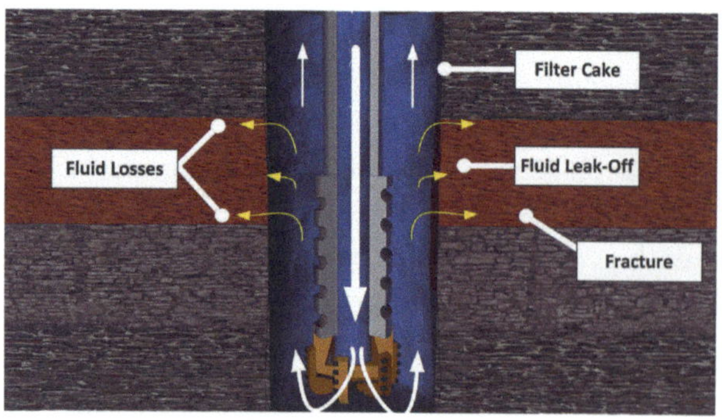

Fig. 1.46 Fluid effects while drilling.

Of the many multitudes of performance characteristics for a drilling fluid, one of the most important is to provide a constant density fluid that stabilizes the hole to prevent well collapse (i.e., swelling formations, sticking, etc.) Drilling fluids are designed to provide sufficient hydrostatic pressure to prevent charged formation gas/oil/water from entering the wellbore as an influx. Designed to be slightly greater than formation pressure, the density of the fluids must be controlled as not to fracture the well (Fig. 1.46).

Cooling and lubricate the bit, motor and BHA

The crushing and shearing of formations will generate large amounts of heat. As the bit drills deeper, for the most part, the temperature of the formations encountered will also increase in temperature. These two heat sources when combined will become detrimental to metal and rubber components.

As colder drilling fluids are pumped from the surface, they will enter the wellbore via the drillstring. By introducing a much lower temperature drilling fluid into the well, the drilling fluid acts to cool the well and transfer generated heat into the drilling fluid. The wetting characteristics of the drilling fluid will also act as a lubricant to the bit, motor, and BHA by assisting in reducing friction. Categories of drilling fluids may be selected due to high-performance characteristics such as lubricity including oil-based or synthetic oil-based drilling fluids.

Gel strengths

In order to efficiently displace cuttings from the well, drilling fluids are designed with specific particles used to assist in providing carrying capacity. These particles are also selected to minimize damage to penetrated formations. Gel strengths are measured in ability to carry and suspend particles, if pumping were to cease. Static particle suspension is extremely important and prevents particles from separating, falling, and packing off downhole.

With most of the pump pressure exerted to deliver drilling fluids to and through the bit (90%), the annular flow rates and pressures are greatly reduced and are smooth like those of a surface meandering creek or stream. Therefore, drilling fluids must provide carrying characteristics within this smooth circulating environment.

Separation of cuttings

Once at surface, the drilling fluids must allow efficient separation of liquid and solid phases. If the liquid phase adheres to the cuttings, large amounts of drilling fluids may have to be disposed. To prevent this from occurring, the drilling fluids are designed to separate from cuttings and be recirculated (after treatment) back into the well. The wetting characteristics of the liquid phases of drilling fluids are controlled to ensure efficient separation of solids aided with mechanical separators such as flowline cleaners, shakers, desanders, and desilters. Degassers are used to aid in the separation of gases from the drilling fluids.

Protecting personnel, environment, and minimize formation damage

Drilling fluids represent large portions of inert earthen soils for density and gel strengths. Drilling fluids are designed and treated to prevent formation damage. As the bit penetrates deeper formation, dissolved solids within formation liquids may be damaging to environment and personnel. pH levels within the fluid are designed to be controlled to be nondamaging to personnel and the environment. Drilling fluids are also designed to establish a thin, tough wall cake lining newly cut formations. This wall cake prevents whole mud to enter the formation by establishing a medium of sized solid particles which adheres to the newly cut formation and plugs off formation voids. By preventing whole mud from entering formations, damage to the formation is minimized.

Minimizing wear and tear on tubulars and well

The lubricating capabilities of the drilling fluids also extend to protecting wear and tear of tubulars and well protective casing. Drilling fluids aids with the rotation of the workstring by ensuring fluid is present to minimize friction. Drilling fluids also provide the same cooling effect as the bit, by lowering temperatures caused by metal to metal friction contact.

Minimizing drilling problems

If a weak formation zone is penetrated with a drilling fluid with sufficient density, the hydrostatic pressure of the well may exceed the formation fracture gradient of the zone. The loss of whole volumes of drilling fluid into the well may result in partial loss of well control and allow an influx to enter the wellbore. At this time, treatments of the drilling fluids, penetration rates, and pump velocities may be altered to assist in reducing/curing losses.

Pipe sticking chances can be reduced through the establishment of a thin, tough wall cake. If the wall cake becomes too thick, nontreatable formations can swell and stick the pipe.

Water-based drilling fluids

Water-based drilling fluids are readily available, low cost, easy-to-maintain, and nondamaging. Water-based drilling fluids are used to drill surface hole sections to prevent damage to fresh water intervals. Once fresh water formations are sealed off with surface casing and cement, formation salinities will change during penetration of deeper formations.

As salinity and formation characteristics change, water-based fluids may be altered to provide sufficient salinities to prevent leaching of formation fluids into the well, formation instability, and swelling. Saltwater drilling fluids will be treated with specific inert solids to provide gel strength and maximize carrying capacities. Water-based drilling fluids can be separated into two following categories.

Nondispersed systems

Nondispersed water-based drilling fluids are categorized by absence of inert carrying solids (solids used to provide gel strength). Cuttings and subsequent solids entering the non-dispersed system are treated by dilution or treatments, which causes the solids to become larger and easier to treat by surface separation systems, sometimes called flocculation. These solids also might be treated to stay within the water-based system, if they are of extremely small size by changing wettability characteristics, sometimes called encapsulation.

Nondispersed water-based systems rely on a group of specially designed particles of long and narrow structure developed from a form of lightweight biodegradable plastic or organic cell wall from plants (cellulose). These particles provide viscosity and fluid loss control.

Dispersed systems

Water-based drilling fluids use heavier density organic materials to break apart formation cuttings (i.e., clays) into extremely small size which can be suspended within the fluid, sometimes called deflocculation. To maintain the ability to break apart particles, the pH levels and breakdown of organic material within a high-temperature environment will require additions of additional system treatments. Dilution of the dispersed system is the most effective method for treatments.

Saltwater drilling fluids

As stated previously, as formations are drilled deeper, naturally occurring formation gas/oil/waters become increasingly more saline. Saltwater drilling fluids are used to drill high salinity formation (i.e., salt formations) and prevent swelling of shales and the prevent formulations of ice-like hydrates in subsea operations. The formulation of saltwater drilling fluids is dependent upon the type of salt used (i.e., chlorides, bromides, and formates). The chosen saltwater system will be selected on performance characteristics, formation characteristics, availability, and cost.

Brines

Formate brines represent a series of high concentration dissolved salt solutions which have performance characteristics beyond water-based drilling fluids. Used primarily in horizontal wells, these brines offer the ability to extend the life of downhole motors by limiting solids within the system and increase the rate of penetration these motors provide. As a solids free system, brines must be closely monitored to provide sufficient downhole densities. Increasing downhole temperatures may cause the densities to drop as the carrying medium (water) will expand. Brines are considered to have good lubricity, be biodegradable, and be nontoxic.

Polymer drilling fluids

When drilling reactive shales (i.e., swelling), polymer drilling fluids may be employed. As the name suggests, polymers employ narrow structure of lightweight biodegradable plastic or organic cell wall from plants (cellulose). These fluids are controlled for fluid loss, stability, and inhibitive properties, allowing a low-cost alternative for troublesome shales. Polymer drilling fluids offer improved inhibition when compared to water-based drilling fluids or brines.

Drill-in drilling fluids

A special drilling fluid which is specifically designed to drill through the production zone of a well is called a drill-in fluid. Normally used for drilling horizontal producing sections, this fluid is specifically designed to improve success of drilling the horizontal section and ensure maximum production by minimizing interconnectivity of voids (pore throats). These fluids also assist in reducing risks when installing complex completions. The drill-in fluid is similar to the completion fluid and may consist of only selected solids of appropriate particle size ranges and polymers.

Oil-based drilling fluids

Oil-based drilling fluids are used in wells where formation clays swell, slough, and react to water-base fluids and brines. Oil-based drilling fluids' liquid medium consists mainly of diesel or mineral oil. An invert emulsion is created when these mediums are mixed with water in oil/water ratios from 65%/35% to 95%/5%. Most common ratios range from 70%/30% to 90%/10%. For shale inhibition, the water phase will be treated with salts to produce a high salinity. Additional system treatments include fluid loss, thinning, pH, and gel strengths.

Oil-based fluid discharges (whole fluid or cuttings) are not permitted in offshore environments. Cuttings and whole mud must be processed and shipped to shore for disposal to a certified site. Land-based operations also follow the treatment provision to eliminate (or severely curtail) discharges with whole mud, with cuttings and mud transported and disposed to certified site.

Synthetic-based drilling fluids

Due to increase regulatory requirements, an alternative to oil-based drilling fluids was needed. Synthetic-based drilling fluid is an invert emulsion mud where synthetic fluids make up the largest portion of the emulsion when combined with water. Common ratios for synthetic oil/water emulsions are 60%/40% to 90%/10%. These fluids are designed to have the same shale inhibition, prevention of formation damage, and drill ability characterized as the fluids they've replaced. Unfortunately, the synthetic drilling fluids are initially more expensive, but possess environmental acceptance allowing for easy disposal of cuttings (i.e., into the water).

Air/aerated fluid/foam drilling fluids

In formations not suitable for supporting a column of fluid, air drilling may be used. This specialized service uses air, aerated fluid, or foam as the drilling fluid in order to circulate cuttings out of the well and prevent the wellbore from collapsing. As a specialized service, the rig is outfitted with unique sets of compressors and wellhead equipment which will allow pumping of air/aerated fluid/foam.

Compressed air is used in formations which cannot withstand the hydrostatic pressure of fluid or may be highly reactive to fluid. Compressed air is pumped down the workstring and exits the bit much like conventional fluid drilling. As the compressed air exits the jet orifices in the bit, a substantial temperature decrease occurs while the compressed air expands. This expansion greatly increases the velocity of the gas allowing cutting to be blown from the hole.

Aerated fluid and foam are used when the use of "air" as a drilling fluid cannot provide sufficient pressure to control formation pressures. These systems blend the performance of conventional fluid systems with the expansion velocities of gas in the annulus. The ratios of liquid phases must be continually monitored to ensure the well does not pool liquid at the bottom of the well.

Lost circulation materials

Lost circulation materials are added to drilling fluids to help prevent whole mud in the well from entering the formation. Lost circulation materials range from organic fibrous materials (walnut shells, seeds, cellulose), to salt flakes and to inert solids (mica). These solid materials are designs in specific size ranges, which will allow displacement through mud motors and bits and bridge off against pore throats within the lost circulation zone. Lost circulation material must be thoroughly mixed before pumping; otherwise, the material may settle out of the drilling fluid and plug off the BHA.

Lost circulation materials may also consist of pill that becomes more viscous after displacing through BHA jet orifices. These fluids are displaced to enter, thicken, and become hard when pumped into the formation. Displacement techniques are critical to proper placement of these types of lost circulation pills.

Spotting fluids

If the wall cake becomes too thick, the wall cake may be instrumental in sticking the BHA. Spotting fluids are designed to penetrate, break-up, and dissolve wall cake. Obviously, for these fluids to work, circulation up the annulus must be available. After displacement, the spotting fluids are allowed to soak. The time requirement for soaking will be dependent upon hole conditions. One or more treatments may be necessary to ensure success. When combined with jarring activities, spotting fluids may enhance chances for getting the drillstring unstuck.

Solid lubricants

In sensitive areas, tiny glass or polymer beads may be added to the drilling fluid to increase lubricity. These solid lubricants can be used in areas where chemical spotting fluids are restricted or prohibited. Solid lubricants act as mini-ball bearing and are designed to reduce friction in the metal-to-metal contact (drillstring inside casing) and formation-to-metal contact (open hole again drillstring). These systems may be used in high angle deviated wells or wells with long deviated sections to minimize contact between the drillstring and well.

Diverting

Diverting is not a well control method. Diverting is a method of directing uncontrolled flow of formation fluids away from the rig and allowing rig crews the chance to evacuate. With weak uncased formations near surface, if the well flows, the well cannot be shut-in without causing failure of the surface casing shoe or damaging the surface equipment. If the shoe fails, the charged fluid may continue exiting at the casing shoe and broach at the surface outside all casing strings. Subsequent well control operations become extremely difficult as the rig may be compromised. If exiting formation fluids contain flammable hydrocarbons or toxic gases, personnel safety may be compromised (Figs. 1.47 and 1.48).

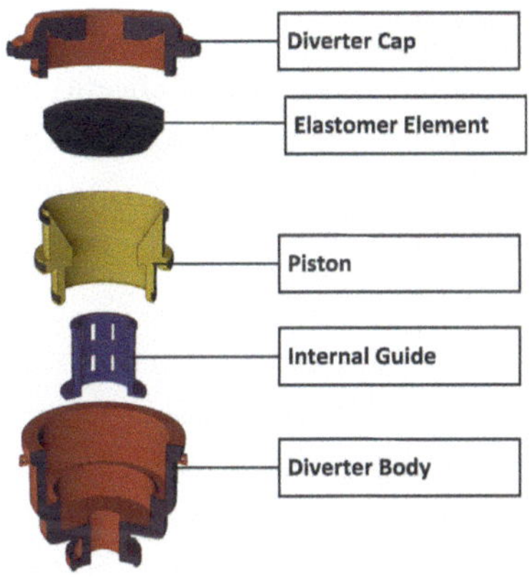

Fig. 1.47 Diverter system.

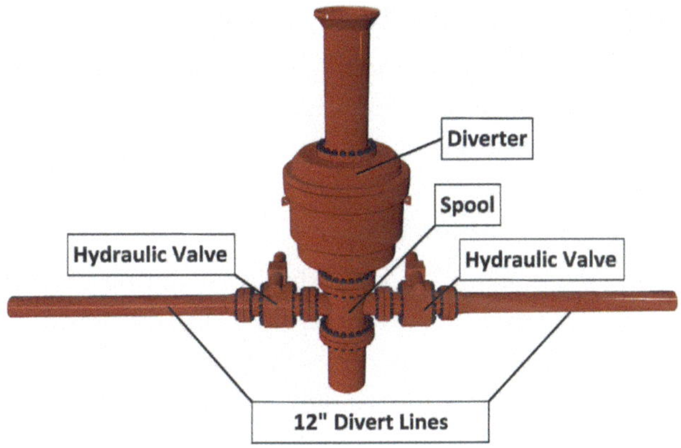

Fig. 1.48 Diverter diagram.

Shallow formations must be closely studied before drilling operations commence. The following conditions relating to shallow formations should be identified in order to mitigate problems. This mitigation may require selecting an alternative site.

Shallow overpressured formations

In older fields where pressure maintenance has occurred, shallow formations may be charged by external forces such as a past blowout or man-made such as water flooding or chemical and steam drive activities. For man-made activities, surrounding injection wells should be shut-in, and sufficient time should be ensured to allow artificial pressure to dissipate. Shallow formations may become charged due to varying stratigraphy (i.e., artesian, rapid compaction, etc.). Investigate surrounding wells to identify geologic problems.

Lost returns

As previously discussed, if the hydrostatic pressure of wellbore is significantly higher than the formation fracture gradient, the muds circulated within the wellbore may induce a fracture into the formation. If the fluid in the well drops significantly, the penetrated formation fluids may enter the well. Lost returns will complicate well control, as the well will have to be diverted while trying to rectify and mitigate losses to the formation.

Swabbing/surging

Rate of penetration must be optimized to ensure bit balling does not occur. Tripping speeds must be controlled as not to induce potential kicks while tripping in/out of the wellbore. Mud properties and type of mud must be optimized. If swabbing cannot be controlled, an alternative of drilling a smaller pilot hole may be employed.

Not keeping hole full/trip logs

Maintaining proper hole fill-up during trips must be ensured. A fill-up chart should be provided and closely followed during all trips. If the drilling fluid is allowed to drop, the charged formations will increase.

Flow after cementing

Once the cement has been displaced during casing cementing operations, the cement transitions from a liquid to solid. During this transition, the cement's hydrostatic pressure will be reduced. During this time, the loss of hydrostatic pressure may allow a charged formation to flow gas to surface. This operation may be further complicated if the diverter has been rigged down in preparation to installing blowout prevention equipment. If the diverter is not affixed, no method is available to divert escaping formation fluids. It is extremely important the diverter is not rigged down until sufficient time is allotted to observe surface sample hardening. The cement transition period must be designed to minimize the length of this transition time period and prevent loss of hydrostatic pressure from occurring.

Surface hole drilling practices

To minimize chances of a diverting situation while drilling the surface hole, the following generalized practices should be reviewed and employed:

- Review surrounding well drilling files for past surface section well control problems. Identify any blowouts which have occurred within the same structure.
- Have a surface volume of kill mud available before spudding.
- For exploratory locations without a shallow seismic survey or locations where shallow flows have occurred, consider drilling a smaller pilot hole. An 8–1/2″ hole is used extensively in the industry as the best option for a pilot hole.
- Use the maximum possible mud weight to provide the maximum possible overbalance (consistent with the goal of avoiding lost circulation).
- Ensure drilling fluids are properly conditioned prior to tipping out. Consideration should be given to pumping out of the hole in problem areas or where bit balling is suspected. Ensure trip volumes are correct throughout the operation.
- Control drilling rates to minimize chances of bit balling and alleviate annulus cutting loads.

Diverter system

A diverter system is comprised of a diverter (annular type blowout preventer) and a diverter spool with side outlet(s) diverter valve(s) and diverter line(s). The diverter valve(s) and line(s) should be comprised of the largest diameter available in order to reduce chances of abrasion while diverting and to reduce back pressure. To further reduce chances of abrasion, the diverter line(s) should be made as straight as possible. Diverter lines should be thoroughly secured to eliminate harmonic movement as gas-laden fluids exit the divert lines at high velocity.

Plumbing of the system and closure sequence is critical to successful divert operations. The system should be plumbed where both overboard divert valves open first, followed by closure of the diverter. After closure, the last sequence will be manually overridden and close the upwind valve to ensure diverting occurs downwind.

ENSURE FLOW THROUGH THE DIVERTER IS NOT DIRECTED TO THE MUD GAS SEPARATOR.

Since the diverter system is not meant to shut-in pressurized fluids, pressure testing is normally limited to function testing whereupon hydraulic control devices are closed and opened with operating pressures of 250 psi. The diverter element closure time must occur in 30 s or less for elements of 20″ in diameter or less. For elements greater than 20″, the maximum closure time must occur in 45 s or less.

Basic diverter schematic for land rigs

See Fig. 1.49.

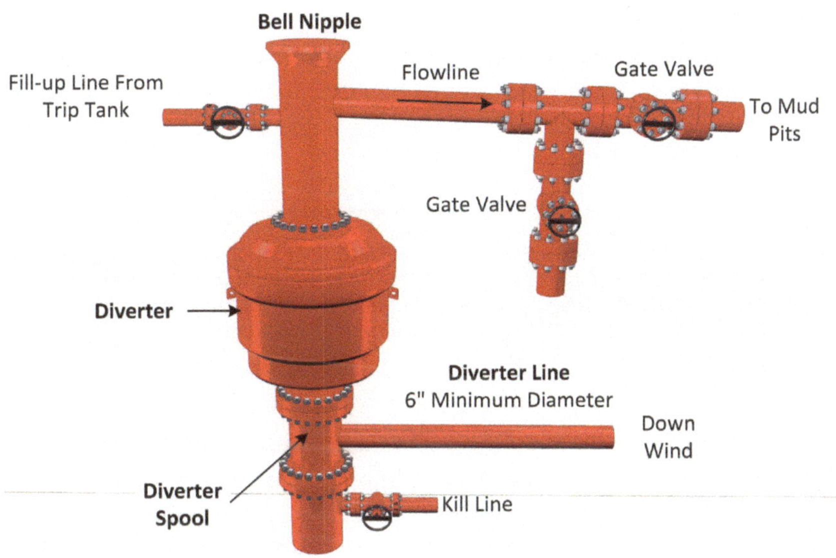

Fig. 1.49 Diverter diagram for land rigs.

Basic diverter schematic for platform, barge and jack-up rigs

See Fig. 1.50.

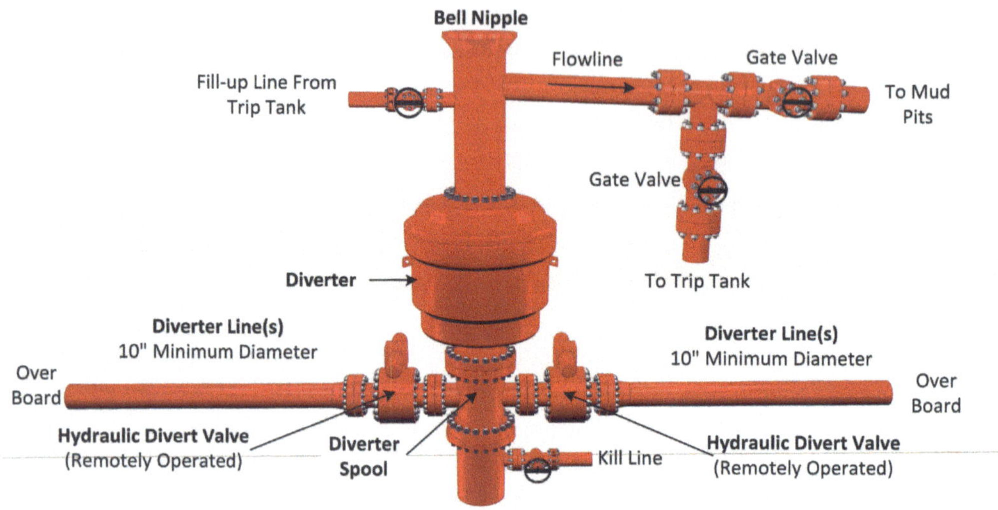

Fig. 1.50 Diverter diagram for platforms, barges, and jack-up rigs.

Diverter lines

Diverter lines are designed to direct charged wellbore fluids away from rig and allow time for personnel to effectively evacuate the rig. These large internal-diameter monobore diverter lines are used to minimize back pressure on formation. The following highlight lists the minimum internal diameter sizes of diverter lines. The largest diameter is recommended in order to reduce chances of abrasion while diverting and to reduce system back pressure.

DIVERTER VENT LINES SHOULD BE STRAIGHT AS POSSIBLE, WITH NO BENDS OR SLOPE CHANGES
FOR LAND, API RECOMMENDS 6 INCH MINIMUM INTERNAL DIAMETER VENT LINE
FOR OFFSHORE, API RECOMMENDS 10-INCH MINIMUM INTERNAL DIAMETER VENT LINES
LARGER DIAMETER LINES ARE RECOMMENDED

As stated previously, the diverter line(s) should be made as straight as possible and any turns must include targeted tees. The diverter lines must be securely anchored with large barite bags, large cement blocks, or other large surface area barriers. Posts or steel rods driven over the divert lines in "X" formations should not be used, as these types of anchors may fail with movement. Flexible couplings are not be used in lieu of hard piping as flexible coupling may fail due to erosion and movement. Diverter lines should be free of 45 degree or 90 degree ells on the end of diverter lines to direct flow downward. Such installations may cause the diverter lines to aggressively move during operation and lead to failure.

Diverter valves

Hydraulic and pneumatic diverter valves are designed to maintain a fully opened or close position. These valves are sized and designed to be opened before the diverter element is closed. Once these valves have been opened and the diverter element closed, the upwind valve is then closed to ensure escaping wellbore fluids are not blown back onto the rig.

Diverter system operation

THE DIVERTER SYSTEM SHOULD BE DESIGNED AND PLUMBED TO ENSURE DIVERTER LINE VALVE(S) ARE ALWAYS OPENED BEFORE THE DIVERTER ANNULAR ELEMENT IS CLOSED.

Opening the divert valves before the diverter element is closed prevents accidentally shutting in the well and applying excess back pressure on the formations. Remember, diverters are used to direct wellbore fluids away from the rig and allow personnel time to evacuate the area. Failure to open the valves first may lead to catastrophic failure such as failure of casing seat and broaching of formation fluids. If dual-diverter lines are used, then the upwind valve is closed to prevent prevailing winds from blowing formation fluids (i.e., gas) back toward the rig.

The following schematic shows a surface diverter panel and proper sequencing for diverting operations (Fig. 1.51).

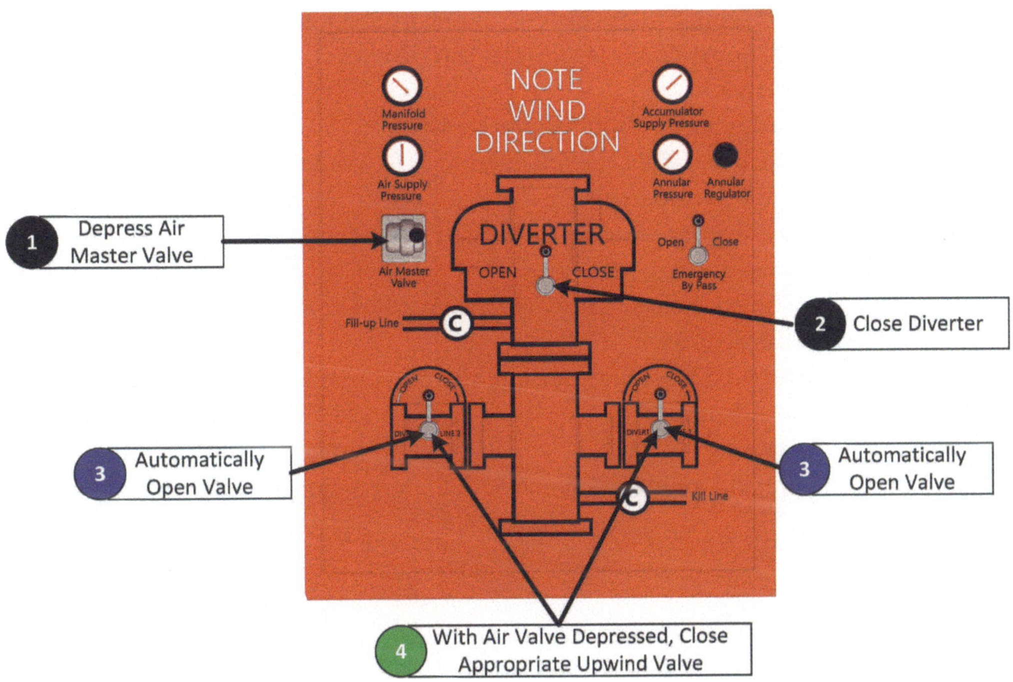

Fig. 1.51 Surface diverter panel.

Diverter function test

1. To be performed at main diverter panel and remote stations
2. Function diverter closed which should open diverter valves first and then close annular diverter
3. Close upwind diverter valve
4. Flush entire system with water to ensure no blockages to valves or lines
5. Close diverter valves and open diverter annular
6. If possible, perform low-pressure test to 80% of burst rating of surface casing. This pressure test may be conducted with the conductor casing pressure test and before drilling out conductor shoe.

Generic diverter drill

1. Upon detecting a shallow well flow, the driller will:
   - **(a)** Alert crews by sounding the alarm.
   - **(b)** Stop the ongoing operation.
     - **(i)** If drilling, stop pipe rotation. Keep pumping at full circulating rate to maintain ECD.
     - **(ii)** If tripping, stop pipe movement.
   - **(c)** Space out
     - **(iii)** While drilling: Position tooljoint to ensure it can be accessible by the rig floor crew.
     - **(iv)** While tripping (Top Drive): Position tooljoint to ensure accessibility by rig floor crew. Connect top drive or stab/close full opening safety valve.
     - **(v)** While tripping (Kelly): Position tooljoint to ensure accessibility by rig floor crew. Stab and close the full opening safety valve. Install and torque Kelly. Open full opening safety valve.
2. Function diverter system following prescribed procedure.
   - **(a)** Open both vent valves.
   - **(b)** Close flowline, shale shaker, flowline fill-up valves, and isolate mud gas separator.
   - **(c)** Close annular diverter.
3. Close the upwind vent line valve, leaving the downwind vent line valve open.
4. Notify Person-in-Charge.
5. Prepare to evacuate rig.
6. Prepare to pump kill mud for dynamic kill or prepare for rig shut down/full evacuation.

Shallow gas

Shallow gas is defined as overpressured gas encountered at a shallow depth within a formation with low fracture gradient which will not allow a kick to be shut-in and controlled with conventional techniques without risk of fracturing the casing shoe, broaching, or wellbore collapse. If the well cannot be shut-in, the influx could blow all the drilling mud out of the hole, leading to a loss of well control. A diverter is commonly used when drilling shallow formations and is used to direct the flow safely away from the rig.

These shallow zones onshore or offshore generally are encountered from surface/mudline down to the depth of conductor or surface casing. Diverter systems are employed until such time as the penetrating formations provide competent structural integrity for installation and rig-up of the heavy BOP stack.

Misconceptions about Shallow Gas Hazards include the following:

- Shallow gas flows have low to moderate flow rates.
- Shallow gas zones will quickly deplete due to small gas volumes.
- Zones are low pressure.
- Hole will bridge due to soft shallow sediments.
- Gas in the water underneath a floating rig will sink the rig.
- Diverter equipment remains intact during diversion.

Shallow gas kicks

Prolific flows from shallow zones can bring unconsolidated solids to the surface. These solids can erode parts of the surface diverter system (especially any turns) and cause catastrophic failure, endangering crews. For immovable rigs, diversion gives time for the rig to be safely evacuated while the kill mud is being pumped. If this is not successful, the rest of the crew will have to abandon the rig. If the well continues to flow or bridges off, a Well Control company should be consulted prior to personnel reboarding the rig.

Causes of shallow gas kicks include, but are not limited to, the following:

- Well has been planned and located over an unknown shallow gas zone.
- Failure to keep the hole full of proper weight mud.
- Swabbing
- Insufficient mud weight
- Lost returns
- Loss of hydrostatic head after cementing

Preplanning for shallow gas diverting

Preplanning for shallow gas should begin before rigging-up on location occurs or prior to spudding the well. All responses to shallow gas/water events are to be discussed with rig crews. During these communication meetings, preparations should be ensured to pump at high rates with both pumps on the well to attempt a dynamic kill of the shallow gas flow using the kill mud, if shallow gas is penetrated. Plans should include preparing to switch to reserve water (or seawater), if the kill mud supply is depleted. During all diversion activities, all nonessential personnel are to be evacuated. Observation spotter(s) are to be assigned to alert rig personnel of any signs of broaching. Finally, all plans should include a final evacuation plan of essential personnel in case the shallow flow does not cease.

Response to a shallow gas influx

Proper emergency response of rig personnel may be ensured through effective planning, proper equipment installation, and testing. The following general responses have been developed for fixed land rigs or bottom supported rigs (i.e., barge and jack-up rigs).

With string in hole

The following general response procedure has been developed for situations whereupon the drillstring is in the hole.

1. Operate diverter system to open downwind valve and close diverter element.
2. Evacuate all nonessential people on rig.
3. Increase pump rate and switch to kill mud until all the kill mud has been pumped.
4. Switch to reserve water/seawater and pump at a high rate.
5. If any part of the diverting system fails or if bubbling appears around the drilling unit/platform, STOP and de-energize rig and proceed to rig abandonment.

String out of the hole

The following general response procedures have been developed for situations whereupon the drillstring is out of the hole.

1. Immediately try to run a stand in hole and space out.
2. Make up top drive.
3. Activate diverter system.
4. Pump kill mud at high rate.
5. If a strong flow does not allow the previous operation, activate the diverting system, STOP and de-energize rig, and proceed to rig abandonment.

Dynamic kill

When a gas kick enters the wellbore and the diverter is activated, a dynamic kill should be attempted by pumping kill weight mud. In a dynamic kill, drilling fluid is pumped at a high flow rate into the wellbore through the drillstring with returns up the annulus. The high flow rate of the kill weight mud increases frictional pressure drop in the annulus.

The bottom-hole pressure is thus increased by the additional pressure of heavier mud plus the additional frictional back pressure. If the increase in bottom-hole pressure from pumping mud is large enough to overcome the influx pressure, the gas flow can be stopped. Responding quickly reduces the chance of unloading the wellbore and losing well control.

CHAPTER TWO

Routine well control methods

Warning signs

The most accurate sensing device is the rig crew

Rig crews may become complacent in regard to kick indicators, as they perform difficult work tasks requiring long tour hours and often times in harsh climates. Crew kick awareness can be increased and reaction times can be decreased with performance of unannounced kick drills. The kick drill may uncover deficiencies in accuracy of mechanical sensing devices as well as improve crew awareness.

It is virtually impossible for a kick situation to develop without the well exhibiting some type of warning signal. If the crew can learn to identify these warning signals and quickly react, the well can be quickly shut-in minimizing the volume of influx. A small kick volume will result in lower kill pressures and increase the margin of safety to the rig crews and environment.

Kick indicators are classified into two groups: immediate action indicators and indicators requiring confirmation. Any time an immediate action indicator is observed, the well should be shut-in immediately. When a confirmation indicator is observed, steps should be taken to verify the well is flowing.

IMMEDIATE INDICATORS

Increase in Flow Rate
Increase in Pit Volume
Flow with Pumps Off

Any time these indicators are observed, **SHUT-IN THE WELL**. With quick identification and reaction, the kick volume will be minimized.

Increase in flow rate

The first indication of a kick is an increase in the rate of mud returning from the well beyond the normal circulating rates. An increase in flow rate might indicate a possible influx of fluid into the wellbore or gas expanding in the annulus. Only calibrated measuring devices such as flowline flow rate sensors can be used to detect small increases in flow, as small increases cannot be discernable by the naked eye. These devices must be maintained and calibrated to manufacturer's specifications. If the device is not operating properly or is in need of calibration, immediate service is to be conducted.

Universal Well Control
https://doi.org/10.1016/B978-0-323-90584-8.00010-1

Increase in pit volume

Assuming no materials or water are being added to the surface pits, a gain in the pit level indicates either an influx of formation fluids into the wellbore or the expansion of gas in the annulus. Recognizing an additional gain in the pits is made easier with a calibrated measuring device (i.e., pit volume totalizers, calibrated trip tanks, etc.). These devices must be observed on a timely basis to familiarize rig crews with circulating volumes.

Flow from the well with the pumps off

Any time there is flow coming from the well and the pumps are off, the well should be shut-in. Monitor for pressure build up. If the well has experienced losses ballooning may account for flow back after the pumps are shut off. Compare flow back to fingerprint chart to determine if the well is flowing.

Shut-in procedures and crew responsibilities

Discussion

Fig. 2.1 Approved reaction.

When one or more warning signs of a kick are observed, the well is to be shut-in. If there is any doubt as to whether the well is flowing, react first by shutting in the well. Observe the shut-in pressures (Fig. 2.1).

For well control purposes, there is no difference between a "trickle" and "flow" with pumps off. This is a definite immediate indicator and the well should be shut-in.

If there are other indicators of possible flow, the pumps should be shut off and a flow check is to be performed.

UNDER NO CIRCUMSTANCES SHOULD THE PIPE BE RUN BACK TO BOTTOM OF THE WELL WITH THE ANNULAR PREVENTER OPEN AND THE WELL FLOWING.

Crew awareness, diligent observation, and rapid reaction will minimize the size of the kick. Shut-in procedures need to be developed and practiced for every type of rig activity. Crew proficiency for shut-in operations should be assured by performing unannounced drills.

(1) Drilling
(2) Tripping
(3) Running Casing
(4) Fishing
(5) Out of the hole
(6) Wireline Operations
  (a) with lubricator (without surface pressure)
  (b) with only a pack off

The development of shut-in procedures will vary for type of operation and individual complement of rig equipment. The following shut-in procedures for fishing, wireline operations, and testing are offered as general guidelines and are subject to modification for individual rig equipment complement and unique operating circumstances.

Crew responsibilities have been developed to assist crew interaction with Person-in-Charge. Once again, these are general responsibilities which are subject to modification for rig equipment complement and unique operating circumstances.

False positive kicks

False positive kicks

There are instances when an indication of a kick turns out to be a "False Positive Kick." False positive kicks may resemble an actual kick, but can be caused by drain back, transfer of fluids, stopping/starting fluid handling equipment, or ballooning.

Drain back

Drain back occurs when the mud pumps are shut down from a normal operating rate. When the pumps are shut off, fluid in the circulating system gradually slow down and will come to a stop. This stoppage may take several minutes to observe. The operating pump rate will affect the rate and volume of the drain back. This rate and volume of drain back should be known by the driller. Drain back tests should be conducted prior to drilling out of casing to establish a baseline trend. Each connection should be compared to the baseline and previous connections to help distinguish normal drain back from a kick. There should not be a sudden increase in drain back rate after a single or stand has been drilled down.

When first establishing trends, caution should be taken to ensure the drain back is not a kick. Continuations of the established trend are clearly evidence of drain back. Deviations from the trend may or may not be indications of a kick; however, drain back is to be treated as a kick until proven otherwise. When monitoring the drain back, limit the volume of drain back in the event an actual kick has occurred. Tracking where the connections are in the wellbore by the mud logger will assist the Driller in mitigating the risk associated with expansion of any entrained gas as the connections nears the surface.

Mud transfers

Mud volumes can be mixed in pits separate of the active mud system. If this mud is pumped into the active system without the driller being notified, the mud gain will replicate a pit gain of a possible kick. No fluids should be added to the active system without proper communications to the driller.

Certain mud handling equipment such as desanders, mud cleaners, degassers, and centrifuges can utilize mud from the active system. On equipment start-up, the pit level may show a drop and conversely when shut down pit levels will increase. The increase in pit level will replicate an actual kick; therefore, the driller must be notified prior to shut down of treating equipment.

Ballooning and flow back

Ballooning is a term used to identify a type of false indication of a kick. This is sometimes referred to as "flow back." Ballooning is the acceptance of drilling fluid into the formation while circulating, only to have the fluid return into the well when pumps are stopped. The return of this fluid will appear as a kick because the well is flowing and the pits are showing a gain. If shut-in, the casing and drillpipe may show imposed pressure.

Ballooning is NOT a kick. Ballooning may be difficult to positively identify prior to circulating bottoms up. During flow back, the flow continuously decreases until the formations are in static equilibrium with the mud weight in the hole and the flow stops. Ballooning formations require knowledge of how much fluid volumes are lost to the formation (losses), as flow back into the well should not exceed the volume of fluid lost to the formation while drilling.

When monitoring the well for ballooning, caution should be used as successive gains may represent actual kicks.

Therefore, each flow back should be carefully monitored and compared to previous flow backs. If the flow back rate is seen to increase, shut the well in immediately. An increase in flow rate means formation fluids are entering the wellbore, lowering the hydrostatic pressure, and allowing a greater rate of flow.

Ballooning check procedure

The following is a generic procedure to review if ballooning is occurring. In order to understand the complexities of ballooning, an accurate volume of losses must be recorded and analyzed. If ballooning is occurring, when fluids are allowed to flow back into the wellbore, the flow back MUST decrease over time. At no time should more fluid volume be allowed to flow back into the well than the volume of recorded losses.

(1) Upon noticing flow, shut-in well, record stabilized SIDPP and SICP, influx volume.
  - **(a)** Accurately determine if Drilling Mud has been lost during the drilling of the hole section.
  - **(b)** Ensure the pit gains occur with pumps off.
    - **(i)** Note if SIDPP $\approx$ SICP
    - **(ii)** Note if SIDPP < Annular Pressure Losses

(2) If the possibility of well ballooning has been indicated, circulate bottoms up using the Driller's method to control BHP.
  - **(a)** Analyze bottoms up formation for hydrocarbons, salinity, and density. Establish presence of formation fluids (if applicable).

(3) While Step 2 is being performed, gather as much information as possible from team members (Drilling/Production/Geology members)
  - **(a)** Research if recently drilled wells had similar problems within the same formation with similar TVD.
  - **(b)** If applicable, determine maximum mud weight used within same formation for field.
  - **(c)** Determine porosity, permeability of exposed formations.
  - **(d)** Research if exposed formations have been fractured, gravel packed, or stimulated within close proximity of current wellbore.
  - **(e)** Team to review all available information and recommend to Drilling Superintendent on whether they believe addition increases in MW are warranted or formation should be treated as a "Ballooning" formation.

(4) Establishing well ballooning may require several circulations of bottoms up with the well open and no indication of formation fluids.
  - **(a)** **Record volume of mud lost**. Accurate volumes of losses are needed to verify ballooning.
  - **(b)** Allow flow back until flow back volume approaches the mud lost volume.
  - **(c)** During flow back, the flow should continuously decrease until the micro-fractured formation is at a static equilibrium with mud weight in the hole and the flow stops.

(5) Once circulations have been complete, shut down holding casing pressure constant.
  - **(a)** Bleed off trapped pressure in small increments. Bleed off pressure in small increments until two successive bleeds yield zero psi
  - **(b)** Direct returns to trip tank for accurate volumetric recordings.
  - **(c)** Monitor well for pressure build up. If there is no flow, resume operations.

(6) If the MW had been raised during drilling of the hole section or use of well control procedures to minimize ballooning effects, the mud weight should be reduced in small gradual steps until all losses are stopped.
  (a) If no losses occur, no flow back can occur.
  (b) If the hole conditions will not allow reducing the mud weight, setting of casing may be used as an alternate to isolate the ballooning shale.

Shut-in procedure while drilling

The following is the procedure for shutting-in the well during drilling operations with a surface BOP (procedure assumes the remote choke is closed and a hard shut-in will be performed):

(1) Stop string rotation
(2) Sound the alarm
(3) Continue pumping and pick up Top Drive or Kelly and position the pipe to ensure no tooljoints extend across the appropriate ram preventers and shear/blind ram. Set slips
(4) Shut down the pumps
(5) Close the annular preventer (or the previously selected ram preventer)
(6) Open HCR (if applicable, open upstream manual valve, then open HCR)
(7) Confirm the well is shut-in and verify flow has stopped
(8) Record and monitor the following:
  (a) SICP
  (b) SIDPP (will be zero with DP float installed in drillstring)
  (c) Pit Gain and Time of Kick
  (d) Record Bit Depth
(9) Notify PIC of the well control situation
  (a) Obtain SIDPP by pumping down drillstring until SICP increases or follow bumping the float procedure)
(10) Complete Kill Sheet

Crew responsibilities are included so the PIC may instruct Rig personnel. These responsibilities are subject to modification for a particular Rig unit and crew.

Each member of the crew should perform the following duties then report to the PIC.

I. Driller
- Shut-in the Well
- Record shut-in drillpipe or drillpipe/tubing pressure and shut-in casing pressure.
- Record depth of BHA.
- Measure and record Pit Gain

- Check choke manifold for valve positioning and leaks
- Ensure proper torque on all connections above rotary table

II. Derrickman
- Weigh mud available on location
- Check volumes of water/base oil and chemical stocks on location
- Report to mud pits/shakers

III. Floorhand #1
- Check accumulator pressures and pumps
- Check BOP stack for leaks and proper valve positioning

IV. Floorhand #2
- Assist the Derrickman

V. Floorhand #3
- Assist the Derrickman

VI. Mud Specialist
- Verify pit mud volumes, mud density, and review chemical stock

Shut-in procedure while tripping

The following procedure is used for shutting-in the well during tripping operations with a surface BOP (procedure assumes the remote choke is closed and a hard shut-in will be performed):

(1) Upon noticing the first indication the well is flowing, stop operations
(2) Sound the alarm
(3) Set the slips on the top tooljoint of the stand
(4) Install, torque, and close the FOSV
(5) Position the pipe to ensure no tooljoints extend across the appropriate ram preventers and shear/blind ram.
(6) Close the annular preventer (or the previously selected ram preventer)
(7) Open HCR (if applicable, open upstream manual valve, then open HCR)
(8) Confirm the well is shut-in and verify flow has stopped
(9) Install Top Drive (Kelly) and open FOSV
(10) Record and monitor the following:
  - **(a)** SICP
  - **(b)** SIDPP (will be zero with DP float installed in drillstring)
  - **(c)** Pit Gain and Time of Kick
  - **(d)** Estimate Assembly Depth

(11) Notify PIC of the well control situation
  - **(a)** Obtain SIDPP by pumping down drillstring until SICP increases or follow bumping the float procedure)

(12) Complete appropriate kill sheet

Crew responsibilities are included so the PIC may instruct Rig personnel. These responsibilities are subject to modification for a particular Rig unit and crew.

Each member of the crew should perform the following duties then report to the PIC.

I. Driller
- Shut-in the Well
- Record shut-in drillpipe or drillpipe/tubing pressure and shut-in casing pressure
- Record depth of BHA
- Measure and record Pit Gain
- Check choke manifold for valve positioning and leaks
- Ensure proper torque on all connections above rotary table

II. Derrickman
- Weigh mud available on location
- Check volumes of water/base oil and chemical stocks on location
- Report to mud pits/shakers

III. Floorhand #1
- Stab FOSV
- Check accumulator pressures and pumps
- Check BOP stack for leaks and proper valve positioning

IV. Floorhand #2
- Stab FOSV
- Close safety valve
- Assist the Derrickman

V. Floorhand #3
- Assist the Derrickman

VI. Mud Specialist
- Verify pit mud volumes, mud density, and review chemical stock

Shut-in procedure while out of the hole

The following procedure is used for shutting-in the well while out of the hole with a surface BOP (procedure assumes the remote choke is closed and a hard shut-in will be performed):

(1) Upon noticing the first indication, the well is flowing close the blind rams or blind-shear rams.

(2) Open HCR valve

(3) Confirm the well is shut-in and verify flow has stopped

(4) Record the following:

  (a) SICP

(b) SIDPP
(c) Pit Gain and Time of Kick

(5) Notify PIC of the well control situation

(6) Monitor trip tank fluid level to ensure the well is shut-in and the rams are not leaking

Each member of the crew should perform the following duties then report to the Driller.

I. Driller
- Shut-in the Well.
- Record shut-in casing pressure.
- Measure and record Pit Gain.
- Check choke manifold, if in use, for valve positioning and leaks.

II. Derrickman
- Weigh mud available on location
- Check volumes of water/base oil and chemical stocks on location
- Report to mud pits/shakers

III. Floorhand #1
- Check BOP Closing Unit's pressures and pump operation
- Check BOP stack for leaks and proper valve positioning

IV. Floorhand #2
- Assist the Derrickman

V. Floorhand #3
- Assist Derrickman

VI. Mud Specialist
- Verify pit mud volumes, mud density, and review chemical stock

Shut-in procedure while fishing

The following is the procedure for shutting-in the well during fishing operations with a surface BOP (procedure assumes the remote choke is closed and a hard shut-in will be performed):

If the fish minimizes circulation ability, disengage from fish so the ability to circulate may be restored. **If BHA has circulation ports, then disregard this Step.**

(1) Upon noticing the first indication the well is flowing, stop operations
(2) Sound the alarm
(3) Set the slips on the top tooljoint of the stand
(4) Install and close the FOSV
(5) Position the pipe so there are no tooljoints across the ram preventers
(6) Close the annular preventer (or the previously agreed preventer)
(7) Open HCR
(8) Confirm the well is shut-in and verify flow has stopped
(9) Install Top Drive (Kelly) and open FOSV

(10) Record and monitor the following:
  (a) SICP
  (b) SIDPP
  (c) Pit Gain and Time of Kick
  (d) Estimate Assembly Depth
(11) Notify PIC of the well control situation
(12) Fill out kill sheet

Each member of the crew should perform the following duties, then report to the Driller.

I. Driller
- Shut-in the Well.
- Record shut-in drillpipe or drillpipe/tubing pressure and shut-in casing pressure.
- Record depth of BHA.
- Measure and record Pit Gain.
- Check choke manifold, if in use, for valve positioning and leaks
- Ensure proper torque on all connections above rotary table

II. Derrickman
- Weigh mud available on location
- Check volumes of water/base oil and chemical stocks on location
- Report to mud pits/shakers

III. Floorhand #1
- Stab FOSV
- Check accumulator pressures and pumps
- Check BOP stack for leaks and proper valve positioning

IV. Floorhand #2
- Stab FOSV
- Close safety valve
- Assist the Derrickman

V. Floorhand #3
- Assist Derrickman

VI. Mud Specialist
- Verify pit mud volumes, mud density, and review chemical stock

Wireline lubricator guidelines

Lubricators are to be used for wireline work where reservoir pressure is open to the wellbore (i.e., open hole or perforated casing). Lubricators can be outfitted with simple pack-offs to more sophisticated wireline BOPs, depending on application. Lubricators are manufactured for quick connection/high-pressure hammer unions with a crossover to a flanged connection. A flanged connection is needed to ensure the lubricator can withstand the extreme forces within the lubricator riser with maximum wellbore pressure. The lubricator

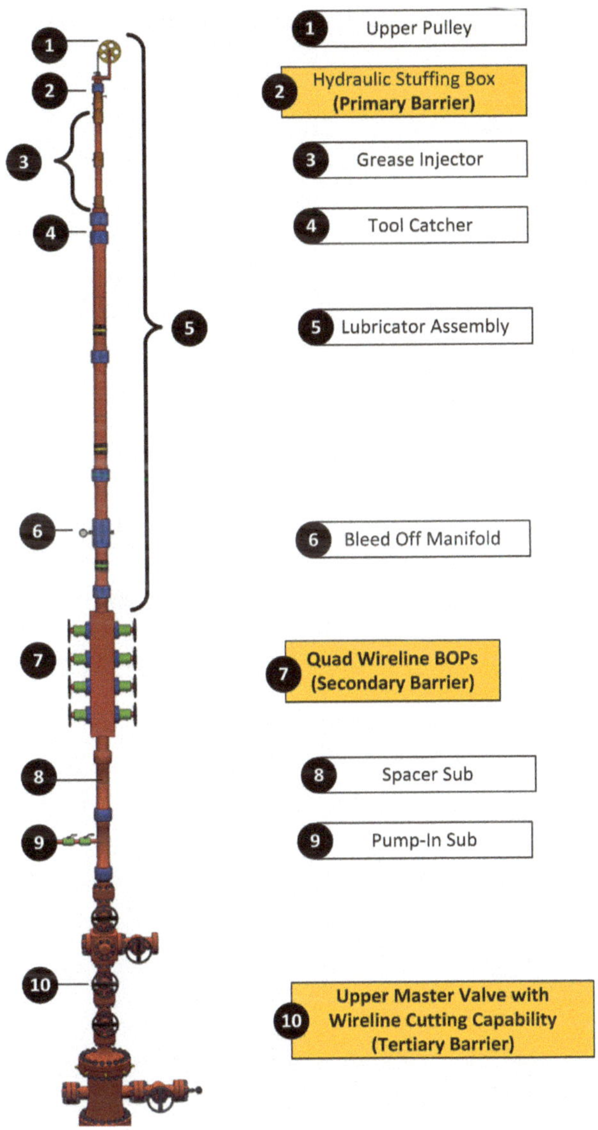

Fig. 2.2 Wireline lubricator

should be pressure rated to match the BOP stack. It should also be pressure tested to its design pressure after installation on the wellhead (Fig. 2.2).

- All attempts should be made to prevent pressure testing the lubricator against the top of the blind ram. This procedure can potentially damage the blind ram.
- However, when a tree is not nippled up and there are perforations open, this may be the only option. If required, slowly pressure test against the top of the ram to ±1000 psi or as directed by third party service contractor. Inspect the lubricator for leaks. Then slowly release the pressure and reopen the blind ram. The blind ram should be pressure tested as soon as possible after this procedure.

At any time during a wireline operation, if pressure within the wellbore increases, consideration should be given to immediately pulling tools back into the lubricator. The lubricator should be shut-in, isolating any wellbore pressure. The next step would be to develop and execute a top kill. This may not be necessary for cased hole operations.

Wireline shut-in procedure with lubricator (well without pressure)

The following procedure is used for shutting-in the well during wireline operations with a surface BOP (procedure assumes the remote choke is closed and a hard shut-in will be performed):

(1) Rigs are to have a manual wireline cutter rigged up on the rig floor prior to commencement of wireline operations.

(2) Upon noticing the first immediate indication the well is flowing, stop wireline running operations.
(3) Close pack-off with recommended hydraulic pressure
(4) Open HCR (if applicable, open upstream manual valve, then open HCR)
(5) Confirm the well is shut-in and verify flow has stopped
(6) Record the following:
  (a) SICP
  (b) SIDPP
  (c) Pit Gain and Time of Kick
  (d) Estimate Tool Depth
(7) Notify PIC of the well control situation
(8) If packing cannot hold wellbore pressure, shut the wireline BOP.
(9) If wireline BOP does not hold, cut wireline and close lubricator valve, blind ram, or wellhead master
(10) If wireline BOP has a grease injector port, attempt to pump into it to seal the well up before cutting the wireline. An intentional fishing job should be avoided if at all possible.
(11) Install circulation (kill and/or choke) lines to wellhead and pressure test lines to working pressure of preventers

Each member of the crew should perform the following duties, then report to the Driller.

I. Driller
- Shut-in the Well.
- Record shut-in casing pressure.
- Record depth of tool.
- Measure and record Pit Gain.
- Check choke manifold, if in use, for valve positioning and leaks.

II. Derrickman
- Weigh mud available on location
- Check volumes of water/base oil and chemical stocks on location
- Report to mud pits/shakers

III. Floorhand #1
- Check lubricator for leaks
- If applicable, check BOP stack for leaks and proper valve positioning

IV. Floorhand #2
- Assist the Derrickman

V. Floorhand #3
- Assist Derrickman

VI. Mud Specialist
- Verify pit mud volumes, mud density, and review chemical stock

Wireline shut-in procedure with only pack-off

The following procedure is used for shutting-in the well during wireline operations with a surface BOP (procedure assumes the remote choke is closed and a hard shut-in will be performed):

1. Rigs are to have a manual wireline cutter rigged up on the rig floor prior to commencement of wireline operations
2. Upon noticing the first indication the well is flowing, stop wireline running operations
3. If emergency situation, cut the wireline and secure the well
4. If possible, pull the tool to the wellhead and pullout of the rope socket
5. Secure the well
6. Close uppermost wellhead gate valve
7. Open HCR valve (if applicable, open upstream manual valve, then open HCR)
8. Confirm the well is shut-in and verify flow has stopped
9. Record and monitor the following:
   a. SICP
   b. SIDPP
   c. Pit Gain and Time of Kick
   d. Estimate Assembly Depth
10. Install circulation (kill and choke) lines to wellhead and test lines to rated working pressure of preventers
11. Notify PIC of the well control situation

Each member of the crew should perform the following duties, then report to the Driller.

I. Driller
- Shut-in the Well.
- Record shut-in casing pressure.
- Record depth of tool.
- Measure and record Pit Gain.
- Check choke manifold, if in use, for valve positioning and leaks.

II. Derrickman
- Weigh mud available on location
- Check volumes of water/base oil and chemical stocks on location
- Report to mud pits/shakers

III. Floorhand #1
- Check wellhead stack and lubricator for leaks.
- If applicable, check BOP stack for leaks and proper valve positioning

IV. Floorhand #2
- Assist the Derrickman

V. Floorhand #3
- Assist Derrickman

VI. Mud Specialist
- Verify pit mud volumes, mud density, and review chemical stock

Prerecorded information sheet

The Prerecorded Information Sheet is designed to assist company personnel in calculating capacities and volumes for a particular well. If this sheet is continuously filled out and updated, company personnel will be prepared prior to a well control situation. This will maximize time needed to coordinate and organize well control operations (Figs. 2.3 and 2.4).

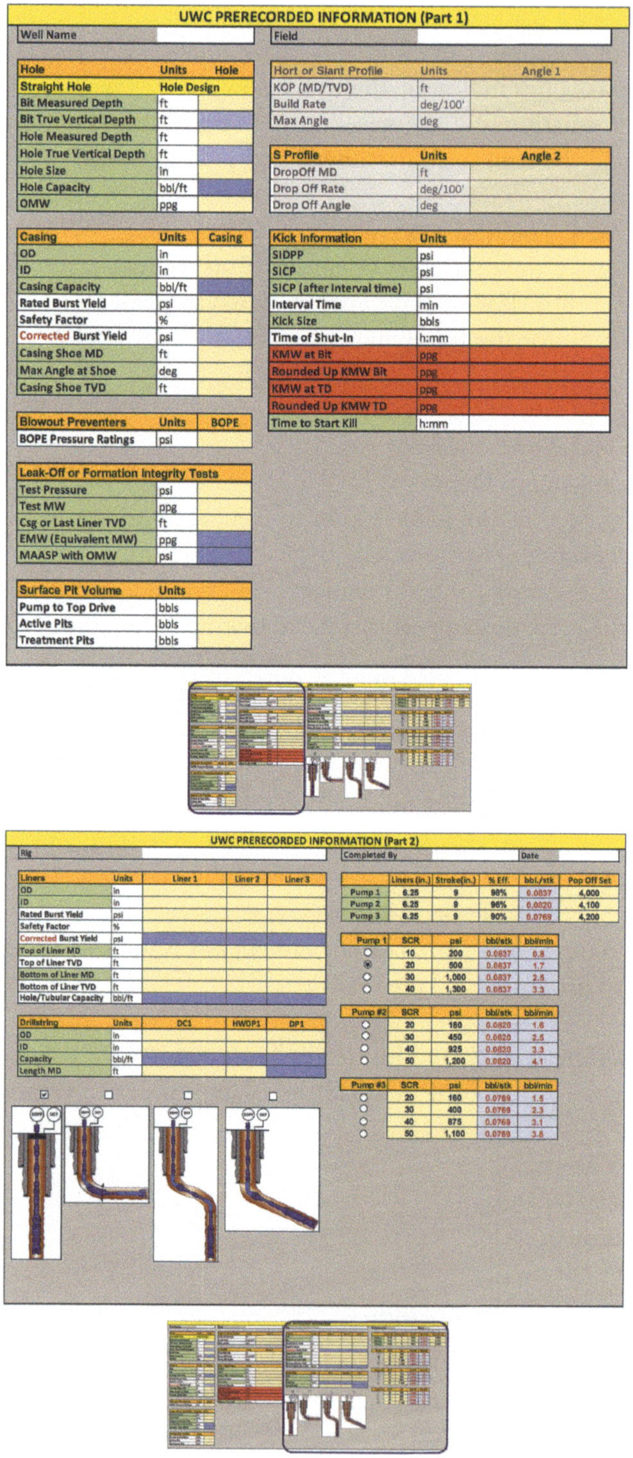

Fig. 2.3 UWC prerecorded information (Parts 1 and 2).

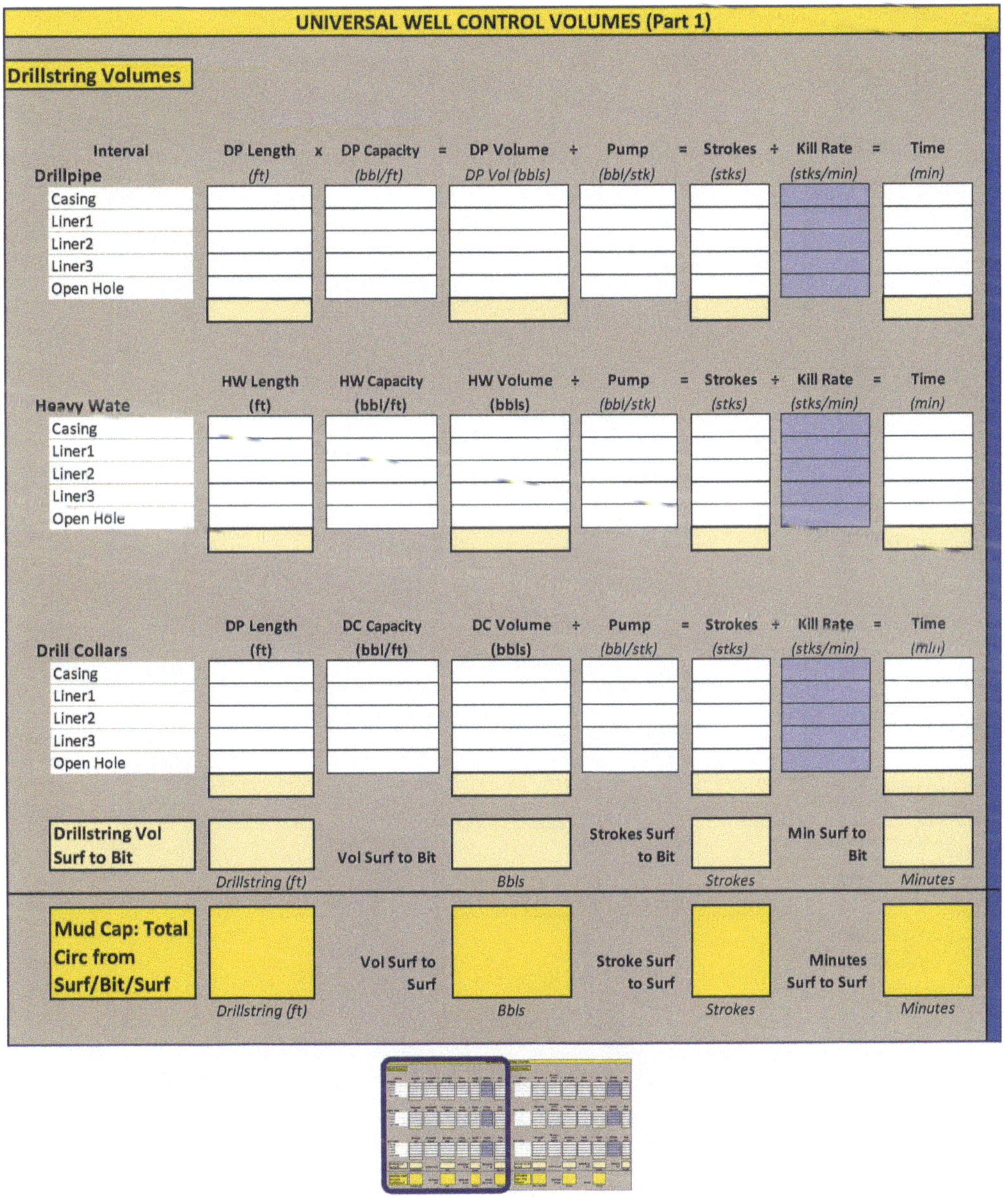

Fig. 2.4 UWC volumes (Parts 1 and 2).

Continued

UNIVERSAL WELL CONTROL VOLUMES (Part 2)

Annular Volumes

Interval Drillpipe	DP Length *(ft)*	x	DP Annular Capacity *(bbl/ft)*	=	DP Volume *DP Vol (bbls)*	÷	Pump *(bbl/stk)*	=	Strokes *(stks)*	÷	Kill Rate *(stks/min)*	=	Time *(min)*
Casing													
Liner1													
Liner2													
Liner3													
Open Hole													

Heavy Wate	HW Length (ft)	HW Annular Capacity (bbl/ft)	HW Volume ÷ (bbls)	Pump = *(bbl/stk)*	Strokes ÷ *(stks)*	Kill Rate = *(stks/min)*	Time *(min)*
Casing							
Liner1							
Liner2							
Liner3							
Open Hole							

Drill Collars	DP Length (ft)	DC Annular Capacity (bbl/ft)	DC Volume ÷ (bbls)	Pump = *(bbl/stk)*	Strokes ÷ *(stks)*	Kill Rate = *(stks/min)*	Time *(min)*
Casing							
Liner1							
Liner2							
Liner3							
Open Hole							

Annular Vol Bit to Surf	*Drillstring (ft)*	Vol Bit to Surf	*Bbls*	Stokes Bit to Surf	*Strokes*	Min Bit to Surf	*Minutes*
UnTreated Open Hole Volume	*Open Hole (bbls)*	Open Hole Strokes	*Strokes*	Minutes	*Minutes*		

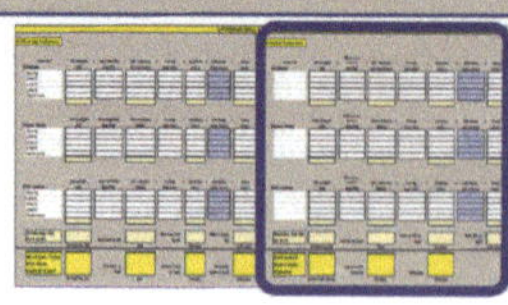

Fig. 2.4, cont'd

THE PRERECORDED INFORMATION SHEET IS TO BE UPDATED WHEN 500' OF NEW FORMATION IS DRILLED OR EVERY TIME THE FLUID WEIGHT IS CHANGED BY 0.2 PPG, OR EACH TOUR CHANGE WHICHEVER IS MOST APPLICABLE.

Well data

- Well name: Record the well name and number.
- Field: Record the field in which the well operations are performed.
- Rig unit: Record the Rig name and number.
- Completed by: Individual who completes the kill sheets.
- Date: Date kill sheet was completed.

Hole data

- Bit Measured Depth: Record the measured depth of bit (ft).
- Bit True Vertical Depth: Determine the True Vertical Depth of bit (ft).
- Hole Measured Depth: Record the Measured Depth of Hole at total depth (ft).
- Hole True Vertical Depth: Determine the True Vertical Depth of Hole at total depth (ft).
- Hole Size: Record the hole size as the diameter of the bit in the hole (in).
- Hole Capacity: Automatically determines the hole capacity when hole size is completed (bbl/ft).
- Original Mud Weight: Record the present or original mud weight (ppg).

Casing information

- Casing OD: Record the outer diameter of casing (in).
- Csg ID: Record casing internal diameter, if unknown record bit size (in).
- Rated Burst Yield: Record burst rating yield pressure for new, uncemented pipe (psi).
- Safety Factor: Record safety factor to be used (normally 80%) (percentage).
- Determine "Corrected" Burst Yield: Record multiplication of Safety Factor × Rated Burst Yield (psi).
- Casing Shoe MD: Record measured depth of casing shoe or Liner Bottom (whichever is applicable) (ft).
- Max angle at Shoe: For deviated wells, record maximum angle of casing shoe (degrees).
- Casing Shoe TVD: Determine casing shoe TVD (ft).

MASP (maximum casing pressure)

- Maximum Allowable Surface Pressure (MASP) is the maximum casing pressure which cannot be exceeded. Since BOPs and wellheads have nominal pressure ratings which greatly exceed those of the casing, the casing is considered to be the weakest burst structure at the surface. If MASP is exceeded, casing failure may occur allowing uncontrolled flow of formations into the wellbore. This may lead to a catastrophic event such as an underground blowout or blowout at surface.

Blowout preventer

- Blowout Prevention Equipment Ratings: Record maximum rating of BOPE (psi).

Leak-off test/formation integrity test (LOT/FIT)

- Test Pressure: Record actual maximum pump pressure recorded during leak-off or formation integrity test (psi)
- Test MW: Record actual mud weight used to perform leak-off or formation integrity test (ppg)
- Csg or Last Liner TVD: Record true vertical depth of the lowermost casing shoe or liner shoe (ft)
- EMW (Equivalent Mud Weight): Determine the EMW (ppg) and round down to lowest tenth of ppg
- MASP with OMW: Maximum Anticipated Allowable Annular Surface Pressure with OMW (psi). (*See more information below*)

MASP (fracture pressure at casing shoe during initial shut-in)

Maximum Allowable Surface Pressure is defined as the maximum allowable pressure exerted in the annulus at which casing shoe formations will fracture during initial closure or shut-in conditions. MASP can be determined by testing the casing shoe formation (LOT/FIT), normally performed after setting and drilling out casing.

To accurately determine this value, the well must contain a (1) homogenous (same) mud weight through entire well and (2) precise leak-off test pressure or accurate formation integrity test information. The pressure required to fracture the formation decreases, as the fluid density increases. This part of the action plan must be done for each change of 0.2 ppg of drilling fluid density.

To calculate MASP, four conditions must be met.

(1) True Vertical Depth to the Casing Shoe (ft) must be known.
(2) Current Mud Weight (ppg)—Must be same density through well. Therefore, circulate and condition mud before performing test.
(3) Maximum Mud Weight (ppg)—The equivalent mud weight from Leak-Off Test or Formation Integrity Test.
(4) Kill Mud Weight Density

MASP with original Mud Weight:

$$MASP = \left(\text{Max } MW_{\text{LOT or FIT}} - \text{Current } MW_{ppg}\right) \times 0.052 \times \text{Shoe } TVD_{ft} \tag{2.1}$$

where

MASP = Maximum Allowable Surface Pressure (psi)

Max $MW_{\text{LOT or FIT}}$ = Maximum Mud Weight from Leak-off or Formation Integrity Test (ppg)

Current MW_{ppg} = Current Mud Weight (ppg)

0.052 = Conversion Constant (psi/ft./ppg)

Shoe TVD_{ft} = True Vertical Depth to Casing Shoe (ft)

MASP with kill mud weight

$$MASP = \left(\text{Max } MW_{\text{LOT or FIT}} - KMW_{ppg}\right) \times 0.052 \times \text{Shoe } TVD_{ft} \tag{2.2}$$

where

MASP = Maximum Allowable Surface Pressure (psi)

Max $MW_{\text{LOT or FIT}}$ = Maximum Mud Weight from Leak-off or Formation Integrity Test (ppg)

KMWppg = Kill Mud Weight (ppg)

0.052 = Conversion Constant (psi/ft./ppg)

Shoe TVD_{ft} = True Vertical Depth to Casing Shoe (ft)

Surface pit volume

- Pump to Top Drive: Record volume of fluid in surface lines from pump, through standpipe to top drive (bbls).
- Active Pit: Record volume of Active pits (bbls).
- Treatment Pit: Record volume of treatment pits (bbls).

Hole Angle 1 (optional for deviated and horizontal wells)

- KOP (MD/TVD): Record kick-off point MD/TVD (ft).
- Build rate: Record build rate (deg/100′).
- EOB Max Angle: Record end-of-build maximum angle (degrees).
- EOB MD: Determine end-of-build measured depth (ft).
- EOB TVD: Determine end-of-build true vertical depth (ft).

Hole Angle 2 (optional for double angle or S shaped wells)

- KOP/DOP2 (MD): Record kick-off point #2 or drop-off point MD (ft).
- KOP/DOP2 (TVD): Determine kick-off point #2 or drop-off point TVD (ft).
- Build/Drop rate: Record build or drop-off-rate (deg/100′).

- EOB2 Max Angle: Record end-of-build point #2 maximum angle (degrees).
- EOB2 MD: Determine end-of-build point #2 measured depth (ft).
- EOB2 TVD: Determine end-of-build point #2 true vertical depth (ft).

Kick information

- SIDPP (Shut-in Drillpipe Pressure): Record stabilized shut-in drillpipe pressure (psi).
- SICP (Shut-in Casing Pressure): Record stabilized shut-in casing pressure (psi).
- SICP (Condition 2): Record stabilized shut-in casing pressure before commencing kill operations (psi).
- Interval Time: Time between recording SICP and SICP (Condition 2) (min).
- Kick Size: Record kick size volume (bbls).
- Time of Shut-In: Record time of shut-in (hours and minutes).
- KMW at bit: Determine kill mud weight at bit to three decimal places (ppg).
- Rounded up KMW at Bit: Round-up KMW at bit to one decimal place (ppg).
- KMW at TD: Determine kill mud weight at TD to three decimal places (ppg).
- Rounded up KMW at TD: Round-up KMW at TD to one decimal place (ppg).
- Time to Start Kill: Record time the kill starts (hours and minutes).

Liner data

- Liner OD: Record the outer diameter of liner, if applicable (in).
- Liner ID: Record the inner diameter of liner, if unknown record bit size (in).
- Rated Burst Pressure: Record the rated burst pressure of liners (psi).
- Safety Factor: Utilize a safety factor for burst based upon current conditions, normally 80% (percentages).
- Corrected Burst Yield: Rated burst pressure rating × safety factor yield corrected burst pressure (psi).
- Top of Liner MD: Record the measured depth of liner top (ft)
- Top of Liner TVD: Record the true vertical depth of liner top, if applicable (ft).
- Bottom of Liner MD: Record the measured depth to the bottom of the liner (ft).
- Bottom of Liner TVD: Record the true vertical depth to bottom of the liner, if applicable (ft).
- Hole/Tubular capacity: Based upon information above, automatically calculate capacity (bbls/ft).

Liner data

- DC1 OD: Outer diameter of the drill collars (in).
- HWDP OD: Outer diameter of the Heavy-weight drillpipe (in).
- DP1 OD: Outer diameter of DP (in).
- DC1 ID: Inside diameter of the drill collars (in).
- HWDP ID: Inside diameter of the Heavy-weight drillpipe (in).

- DP1 ID: Inside diameter of DP (in).
- Capacity: Automatically determined capacity based on OD/ID information (bbls/ft)
- DC1 Length MD: Measured depth of drill collars (ft).
- HWDP Length MD: Measured depth of Heavy-weight drillpipe (ft).
- DP1 Length MD: Measured depth of drillpipe (ft).

Pumps

- Liner: Liner used in pump (in).
- Stroke: Stroke length of pumps (in).
- %Eff: As measured, the efficiency of pumps (percentages). If unknown, use 98%.
- Bbl/stk: As determined, the pump output based on sizing and efficiency (bbls/stk).
- Pop Off Set: The pump "pop off" setting. Do not exceed during circulation.

Slow circulating rates (SCR)

Wells should be killed using Slow Circulating Rates for the following reasons:

1. To allow for ample time to degas mud.
2. To maintain pressures at the bottom of the hole to a minimum.
3. To allow the choke operator time to make choke adjustments.
4. To allow controlled disposal of the formation fluids at surface.
5. The handling capacity of the MGS may be limited.
6. To reduce annular velocity in the wellbore in the event that slight losses are encountered.

SCR selected normally for output between 2 and 5 bpm, or maximum flow rate at which barite can be added to the system).

- SCR 10 to 40: Range selection for slow circulating rate from 10 spm to 40 spm (spm).
- Psi: Pressure recorded at separate rates with OMW (psi).
- Bbl/stk: Volumetric Pump output at selected SCR (bbl/stk).
- Bbl/min: Pump rate at selected SCR and volumetric pump output (bbl/min).

Well volumes (from prerecorded information sheet)

For most well control worksheets, well and wellbore volumes are calculated from information provided by the Prerecorded Information Sheet. Volumes can be easily separated into Drillstring Volumes (Internal Capacity) and Annular Volumes (Annular Capacity). Once these volumes are determined along with surface equipment volume, the total circulation can be determined in barrels, strokes, and minutes.

Drillstring volumes (internal closed-in displacement)

As majority of wells are drilled with floats in the drillstring, drillstring volumes are determined using internal closed-in displacement (bbl/ft). Record the length of each drillstring and drillpipe/tubing string component by its associated internal capacity factor (bbl/ft). Treat bottom-hole assembly components (stabilizers, crossover subs, etc.) as drill collars for capacity calculations.

Calculate the total volume (bbls) for each component section by multiplying the component length by its capacity factor. Since the length of drillpipe or drillpipe/tubing will not be known until after the well kicks, the drillpipe/tubing capacity and total internal capacity will have to be calculated after the well has been secured. Verify the Measured Depth indicated is equal to the sum of the individual component lengths.

Strokes to Bit (STB): Divide the Total Internal Capacity in barrels by pump output (bbl/stk) to determine capacities in strokes.

Time from Surface to Bit: Divide the Total Internal Capacity in strokes by slow pump rate to determine capacities in minutes.

The volume action plan comprises drillstring and annular volumes. For the drillstring, the following volumes will have to be determined:

- Surface volumes: The surface volume from mud pumps through standpipe to top drive (bbls). *If unknown, estimate 10 bbls.*
- Drillpipe: Internal volumes of drillpipe within casing, liner(s), and open hole are determined (bbls, strokes, time).
- Heavy Wait: Internal volumes of Heavy wait within casing, liner(s), and open hole are determined (bbls, strokes, time).
- Drill Collars: Internal volumes of Drill Collars within casing, liner(s), and open hole are determined (bbls, strokes, time).
- Total Internal Volumes: Totalize all drillstring volumes from surface to bit inclusive of drillpipe, heavy wait, and drill collars (bbls, strokes, time).

Annular capacity

Record the length of each drillstring or drillpipe/tubing string component and its associated annular capacity factor in the given hole size. Treat bottom-hole assembly components (stabilizers, crossover subs, etc.) as drill collars for capacity calculations.

Calculate the annular capacity (bbls) opposite each component section by multiplying the component length by the annular capacity factor. Since the length of drillpipe or drillpipe/tubing will not be known until after the well kicks, the annular capacity opposite the drillpipe or drillpipe/tubing and the total annular capacity will have to be calculated after the kick has been taken, the well shut-in and secured.

Verify the Measured Depth indicated is equal to the sum of the individual component lengths.

- Add the Total Internal Capacity to the Total Annular Capacity to determine Total System Capacity (not including active pit volume).
- Total System Strokes: Divide the Total System Capacity by pump output (bbls/stk) to determine these capacities in strokes.
- Total Circulation Time: Divide the Total System Capacity in strokes by slow pump rate (SPM) to determine these capacities in minutes.

The volume action plan comprises drillstring and annular volumes. For the drillstring, the following volumes will have to be determined:

- Drillpipe: Annular volumes of drillpipe within casing, liner(s), and open hole are determined (bbls, strokes, time).
- Heavy Wait: Annular volumes of Heavy wait within casing, liner(s), and open hole are determined (bbls, strokes, time).
- Drill Collars: Annular volumes of Drill Collars within casing, liner(s), and open hole are determined (bbls, strokes, time).
- Total Annular Volumes: Totalize all annular volumes from surface to bit inclusive of drillpipe, heavy wait, and drill collars (bbls, strokes, time).

Total volumes

After the drillstring and annular volumes have been determined, the total circulating volumes can be determined. For total circulation (surface to bit to surface) can be determined by totalizing bbls, strokes, and time for circulation. Volume calculations can be done using the Well Control Action Plan shown in Figs. 2.3 and 2.4.

Since wellbore fluids (OMW) are considered to be "noncompressible" and the bit is off bottom, the total circulation time is reduced as the KMW only travels to bit depth and not total depth of the well. The KMW will exit the bit and travel up the annulus. The attached volume action plan contains a specific section entitled "Untreated Open Hole Volume" in footage, strokes, and minutes.

If the bit is off bottom and upon successful killing of the well, the bit to TD volume will have to be displaced with KMW. With the well dead (no flow), the bit is simply tripped and washed to bottom. An additional displacement of the entire annulus will be necessary to ensure no light spots within the mud system exist).

Fluids management during well control

With current rig sites presenting smaller and smaller footprints, the ability to have an active pit volume at 1.5 times the total circulating volume may prove difficult. Therefore, fluids management must be considered during most well control operations.

Since water-based drilling fluids are susceptible to gas migration rates of 50 ft./min, a kill attempt using the Weight and Wait method will include isolating an active pit with KWM and then initiating kill operations. During the time required to bring the mud weight in the pit to the KMW, the Volumetric method can be used to control casing pressure as the influx migrates. Additional KWM volume usually has to be mixed while circulating.

Slow circulating rates (SCR) have been predetermined at 20 to 50 spm (1–5 bbls/min). The SCR may be limited by the increase in mud weight needed, the mud properties needed and the effectiveness of surface mixing equipment. Throughout these operations, communication is the key in order to mix, pump, and kill with a homogeneous KMW. If crews find they cannot effectively mix the desired densities while pumping, this should be communicated to the kill team who may elect to slow pump rates. Any alterations in SCR must be accomplished holding casing pressure constant.

Using the Driller's method does not require an increase in mud weight during the first circulation. Once the kick is out of the hole the pits can be weighted to the KWM.

Depending on location volumes, active pit volumes may be decreased by pumping to portable storage tankers. Delivery of heavier density KMW would be scheduled after adequate pit storage is available. Additionally, light mud returns may be routed from settling pits to portable storage tankers. Communication during all phases of the well control operation is critical as fluid is moved between pits. Each fluid movement must be conveyed to the kill team.

For synthetic oil-based fluids and oil-based fluids, migration of a gas influx will be minimized as the gas is compressed under pressure due to hole depth. It goes into solution in the oil phase of the mud thus acts as a liquid. Minimal migration allows for improved planning and execution for developing KMW. The desires of the kill team are to pump a "homogeneous" KMW which does not have to be mixed while pumping. If time permits, it is preferable to mix 1.5 times the hole volume worth of KMW before initiating kill operations. Of course, this preference will be predicated on surface hole volumes, etc. Once again, with limited surface pit volumes available, coordination with portable tankers may be necessary.

Off-bottom well control

Commentary

Studies have shown the vast majority of kick occur during tripping operations. Yet, kicks occurring off bottom are seldom mentioned within well control literature. In today's world of horizontal drilling and frequent use of OBM/SOBM, the off-bottom kill using mud caps are widely used.

If we examine the root causes of kicks off bottom, the largest probability lies within widespread swabbing. The bottom-hole assembly can act as a piston when the drillstring is tripped out of the hole, pulling hydrocarbon kicks into the well. This piston action is largely dependent upon the speed of pulling the drillstring, viscosity or thickness of the drilling fluid, balling of the surfaces of the BHA and close tolerances between sides of the hole and bottom-hole assemblies.

Often the drillstring is circulated while pulling each stand until the bit is inside the deepest casing string or liner to help prevent swabbing. This is also referred to as pumping out of the hole.

For OBM/SOBM

Of course, the further the bit is off bottom before a kick is observed the more complicated the kill activity will likely be. As the bit is pulled further and further from the bottom of the well, the resulting kill will require a higher density mud cap kill fluid when compared to killing the well on bottom. If the bit is located near TD, well control may be as simple as circulating the well with OWM and not require increasing the fluid density.

Once the well is shut-in and secured, the top drive (Kelly) is installed and the SIDPP can be determined by bumping the float. The SIDPP should be the same as the SICP as long as the kick is below the bit. With stabilized SIDPP, SICP, and Kick Volume, several calculations may be used to determine the best way forward. These calculations include determining:

(1) If the well can be successfully killed off bottom?
(2) Does the bit need to be "stripped" back to bottom to perform the well kill?

Using any of the provided Universal Well Control Action Plans, determine the density of KMW needed for an off-bottom kill and determine if the hydrostatic pressure of the mud column will exceed the casing/liner shoe leak-off pressure or formation integrity test pressure. The next step will include estimating the reservoir pressure by adding the hydrostatic pressure of the original mud in the hole to the SIDPP. In order to kill the well, either by mud cap or stripping, the fluid column must equal or exceed the reservoir pressure to prevent more influxes from entering the hole.

Hydrostatic pressure of the fluid column within annulus and open hole can be determined in stages. Stage 1, determine the hydrostatic pressure of the length of the kick plus the hydrostatic pressure of mud from the top of the kick to the bit. Stage 2, determine the hydrostatic pressure of KMW from the bit to the surface. The combination of Stage 1 and Stage 2 must be equal to or greater than the estimated reservoir pressure. To perform the kill off bottom, only the mud from the bit to surface can be weighted up (mud cap kill).

For determining the appropriate kill weight, simply divide current SIDPP by 0.052 and the true vertical depth of the bit not the depth of the hole. This calculation estimates the additional mud weight needed to be added to current mud weight for KMW off bottom.

$$\Delta MW_{\text{Bit Off Bottom TVD}} = \frac{\text{SIDPP}}{(0.052) \times \text{Bit TVD}} \tag{2.3}$$

$$KMW_{\text{Bit off Bottom}} = OMW_{\text{ppg}} + \Delta MW_{\text{Bit Off Bottom TVD}} \tag{2.4}$$

1. If the hydrostatic pressure of calculated $KMW_{\text{bit off bottom}}$ is ***less than*** the pressure needed to fracture the casing/liner shoe based on the LOT/FIT, the well can be killed at present depth with mud cap kill. Simply displace well using appropriate constant bottom-hole pressure method along with KMW

UNDER NO CIRCUMSTANCES, SHOULD TRIPPING TO BOTTOM WITH OPEN BOP BE INITIATED WITH KICK IN THE HOLE
WITH BOP'S OPEN AND INFLUX IN WELL, THE WELL IS NOT UNDER CONTROL AS ONLY ONE BARRIER EXISTS

2. If the hydrostatic pressure of calculated $KMW_{\text{bit off bottom}}$ is ***greater than*** the pressure needed to fracture the casing/liner shoe based on the LOT/FIT, stripping may be required. Since stripping to bottom requires use of secondary barrier (i.e., annular preventer), please refer to appropriate company policy or Person-in-Charge for operational approval. If stripping is to be conducted, suggested equipment includes use of:
   a. Calibrated strip tank with ability to transfer fluid into a trip tank or mixing pit of active pit system. Since stripping volume per stand will be less than 5 bbls, a small 10 bbl graduated striping tank will provide more accurate readings during flow back and bleed off activities.
   b. Fluid exiting downstream of the choke should be routed directly to the strip tank (by-passing the MGS). Suggested tie-in points should be downstream of the buffer tank of choke manifold to strip tank.

For water-based muds

Stripping should be avoided for wells with water-based drilling fluids unless the well is horizontal or the gas has reached the surface. Gas influxes will rise through water-based fluids at about 4000 ft./h which is significantly faster when compared to the ability to strip to bottom. Manipulating casing pressure while stripping and dealing with gas migration can be difficult. To minimize the chance of fracturing the casing shoe or taking another kick, gas is allowed to migrate the surface by using the Volumetric method and then removed by using Lube and Bleed. The pipe can be stripped back to bottom or a mud cap kill may be performed at current depth. The most appropriate well control kill method can be employed to kill the well. Do not strip with dry gas at the surface. There is no way to bleed off mud volume equal to the steel volume stripped into the hole. Gas can leak while stripping and endanger personnel.

Mud cap conclusion

If the well is killed off bottom with a denser weight mud cap, the dual density mud system should be normalized before continuing to trip out of the hole. This may require tripping back to bottom, circulating bottoms up to equalize mud density. Once complete, a flow check may be performed before tripping out of hole.

Tripping well control method

Tripping well control can be classified as "off-bottom kicks." These types of kicks represent the vast majority of well control incidents during drilling and workover activities. Kicks off bottom include variances in mud density, mud properties, and hole conditions within the lower hole sections. These variances may cause the bottom-hole assembly to act as a piston when the drillstring is pulled from the well.

Swabbing

Swabbing occurs when formation cuttings adhere to the bottom-hole assembly and form a "ball." The balled BHA acts as plunger so when the drillstring is pulled, the velocity and close tolerances cause a reduction in bottom-hole pressure. If this pressure is reduced formation pressure, formation fluids will enter the wellbore. Swabbing be directly dependent upon poor mud properties, insufficient hole cleaning, thick mud cake, and poor drilling/tripping practices.

Not filling the hole properly

After the slug has been pumped and properly displaced, operations to pull the drillstring from the well are initiated. From this point forward, volumetric recording of the amount of fluid pumped into the well must be equal to or greater than the closed-in capacity of the drillstring if it is pulled wet or greater than or equal to the displacement of the pipe if it is pulled dry.

GENERAL RULE OF THUMB
FILL-UP FREQUENCY IS AS SIMPLE AT 5, 3,1

EVERY 5 STANDS OF DRILLPIPE OR LOSS OF 75 PSI, WHICHEVER COMES FIRST
EVERY 3 STANDS OF HEAVIWATE OR LOSS OF 75 PSI, WHICHEVER COMES FIRST
EVERY 1 STAND OF DRILLCOLLARS OR LOSS OF 75 PSI, WHICHEVER COMES FIRST.

IN LIEU OF FOLLOWING THE ABOVE GENERAL RULE OF THUMB, A CIRCULATING TRIP TANK MAY BE USED WHEREUPON THE HOLE IS CONTINUALLY FILLED WHILE PULLING THE DRILLSTRING FROM THE WELL.

Consequences of off-bottom kick

Generally speaking, for the well to be killed with the bit and BHA off bottom, a heavier KMW will be required as compared to an on-bottom kill. The off-bottom kill consists of pumping a KMW from bit to surface. This "mud cap" kill method is comprised of two-hole section densities (1) from bit to surface, (2) from bottom-hole depth to bit. The total hydrostatic pressure of the well with a kick on bottom will consist of (1) hydrostatic pressure of KMW from surface to the bit, (2) hydrostatic pressure of original mud weight from the bit to the top of influx, and (3) hydrostatic pressure of influx.

The major consequence of an off-bottom kick will be the decision of:

(1) Mud Cap Kill the well at current bit depth or

(2) Stripping Strip to bottom in order to kill the well

This decision to kill the well off bottom will be dependent upon:

- Fracture gradient integrity of the lowest casing or liner shoe
- Mud weight required to kill the well off bottom
- Type of mud used
- Type of influx
- Physical limitations of mixing equipment
- Availability of proper stripping equipment accessories (i.e., stripping tank, accumulator, hard-piping routes from choke manifold to stripping/tripping tank, etc.)

Of primary concern will be the type of mud used and type of influx. If the influx is gas, precautions will need to be reviewed in order to allow the gas bubble to expand while killing the well. The gas influx will begin to rise from the bottom of the hole. The rate of rise will be largely dependent upon hole design and fluid properties.

RULE OF THUMB
FOR WATER BASED FLUIDS, GAS MIGRATION RATES MAY REACH 4,000 FEET PER HOUR

FOR WATER BASED FLUIDS AND GAS INFLUX, THE VOLUMETRIC METHOD CAN BE USED IMMEDIATELY TO MINIMIZE PRESSURES EXERTED ON THE CASING SHOE WHILE PREPARATIONS ARE MADE TO KILL THE WELL.

RULE OF THUMB
FOR OIL BASED FLUIDS, GAS MIGRATIONS WILL BE NEGLIBLE AS GAS GOES INTO SOLUTION IN THE MUD WITH GAS IN SOLUTION, GAS WILL NOT MIGRATE AT EXCESSIVE RATES.

Mud cap kill

If the casing shoe has sufficient strength, mud cap kill may be used and consists of two steps.

Step 1—Pump mud cap

Circulate sufficient KMW whereupon the hydrostatic pressure of the drilling fluid from surface to bit exceeds the formation pressure.

Step 2—Kill well

Once the well flow is static with KMW completely circulated to bit and back to surface, the contaminated original mud and influx must be displaced from bottom of the hole to surface. If the well can be killed with KMW to bit and the well is static, the drillstring can be staged to various depths where the well bottoms up circulation will occur to remove any influx. The staging technique may or may not employ closing the annular preventer.

Stripping kill

If the well cannot be killed off bottom, the drillstring must be stripped to bottom with the annular closed and influx circulated out on the choke. Stripping operations are more complex and require planning and installation of specialized equipment.

The specialized equipment used for stripping consists of the following:

- Annular preventer
- Stripping bottle attached to the closing line of the annular
- Stripping tank
- Hard line from choke manifold to stripping tank

Variances to the specialized equipment include:

- Properly calibrated trip tank in lieu of stripping tank
- Use of annular without stripping bottle
- This requires minimum closing pressure on the annular to allow pipe and tooljoint passage.
- Hard piping from choke manifold to trip tank.
- Motor oil is typically used to lubricate the drillstring while stripping in the hole. Motor oil is poured on top of annular to provide lubrication and each tooljoint should be doped.
- For more information, please refer to the Stripping Section.

Calculations

Method 1: Determine KMW$_{Bit\ off\ Bottom}$

(a) Determine Incremental MW Density Increase

$$\Delta MW_{\text{Bit Off Bottom TVD}} = \frac{\text{SIDPP}}{(0.052) \times \text{Bit TVD}} \quad (2.5)$$

(b) Determine KMW $_{\text{Bit off Bottom}}$ (ppg) and compare the FIT/LOT.

$$\text{KMW}_{\text{Bit Off Bottom}} = \text{OMW}_{\text{ppg}} + \Delta\text{MW}_{\text{Bit Off Bottom TVD}} \leq \text{FIT/LOT} \tag{2.6}$$

(c) Compare KMW $_{\text{Bit off Bottom}}$ to the FIT/LOT.

- **(i)** If FIT/LOT > KWM $_{\text{Bit off Bottom}}$, proceed with kill.
- **(ii)** If FIT/LOT ≤ KMW $_{\text{Bit off Bottom}}$, stripping may be necessary. Consult PIC or Well Control Company.

Method 2: Alternate for determining KMW$_{Bit\ off\ Bottom}$

(a) Determine Vertical Influx Height (ft)

$$\text{Influx Ht (ft)} = \text{Cosine (Max Angle)} \times \left[\left(\frac{1029.4}{\text{Hole Dia}_{\text{in}}{}^2}\right) \times \text{Kick Size}_{\text{bbl}}\right] \tag{2.7}$$

Assume: 2.0 ppg Influx density

(b) Determine Influx Hydrostatic Pressure (psi)

$$\text{HP Influx}_{\text{psi}} = \text{Influx Ht}_{\text{ft}} \times \left(\text{Gas Density}_{\text{ppg}} \times 0.052\right) \tag{2.8}$$

(c) Determine Mud Hydrostatic Pressure (psi) from TD to bit

$$\text{Mud HP}_{\text{from TD to Bit}} = (\text{TVD Btm}_{\text{ft}} - \text{TVD Bit}_{\text{ft}} - \text{Influx Ht}_{\text{ft}}) \times 0.052 \times \text{MW}_{\text{ppg}} \tag{2.9}$$

(d) Determine Mud Hydrostatic Pressure (psi) from bit to surface.

$$\text{Mud HP}_{\text{from Bit to Surf}} = (\text{TVD Bit}_{\text{ft}}) \times 0.052 \times \text{MW}_{\text{ppg}} \tag{2.10}$$

(e) Determine Formation Pressure (psi)

$$\begin{aligned}\text{Form Press}_{\text{psi}} = {} & \text{HP Influx}_{\text{psi}} + \text{Mud HP}_{\text{from TD to Bit (psi)}} \\ & + \text{Mud HP}_{\text{from Bit to Surf (psi)}} + \text{SIDPP}_{\text{psi}}\end{aligned} \tag{2.11}$$

(f) Determine Formation Pressure Equivalent MW.

$$\text{Form Press}_{\text{ppg}} = \frac{\text{Form Press}_{\text{psi}}}{(0.052) \times (\text{TVD}_{\text{TD}})} \tag{2.12}$$

(g) Determine Hydrostatic Pressure at Bit with Kick in Hole (Assuming entire kick is below bit)

$$\text{HP}_{\text{Bit with Kick}} = \text{Form Press}_{\text{psi}} - \text{Mud HP}_{\text{Bit to Surf}} - \text{HP Influx} \tag{2.13}$$

(h) Determine KMW at bit needed to keep formation from flowing.

$$KMW_{\text{Off Bottom}} = \frac{HP_{\text{Bit with Kick}}}{(0.052) \times (TVD_{\text{bit}})} \tag{2.14}$$

(i) Comparison of $KMW_{\text{Bit off Bottom}}$ to Leak-Off Test/Formation Integrity Test
- **(i)** If FIT/LOT > $KWM_{\text{Bit off Bottom}}$, proceed with kill.
- **(ii)** If FIT/LOT ≤ $KMW_{\text{Bit off Bottom}}$, stripping may be necessary. Consult PIC or Well Control Company.

IF LOT/FIT IS LESS THAN KMW, PREPARE TO STRIP TO BOTTOM OR NOTIFY APPROPRIATE WELL CONTROL COMPANY

General mud cap kill procedure

1. Record all relevant well control data.
2. Perform calculations described previously.
   a. Is formation strength (at the open hole weak point) capable of withstanding the KMW?
3. Circulate the kill mud maintaining constant pressure (+ safety margin) at the bit using a drillpipe pressure schedule.
4. Once the well is killed, strip into the hole in stages until on bottom and circulated KMW. When KMW in = KMW out reading shows 3 consecutive readings after total circulation, well can be flow checked.

Stripping kill procedure

Please see Stripping subsection for more detailed procedure. The following is a general kill procedure used for stripping operations.

1. Record all relevant well control data.
2. Perform calculations described within stripping section.
   a. Is formation strength (at the open hole weak point) capable of withstanding the KMW
3. Strip to bottom adhering to stripping table information (may include multiple stages).
4. Circulate the kill mud maintaining constant pressure (+ safety margin) at the bit using a drillpipe pressure schedule.
5. Once the well is killed. When KMW in = KMW out reading shows 3 consecutive readings after total circulation, well flow can be checked.

Well control for horizontal and highly deviated wells

Commentary

Horizontal well designs represent over 50% of wells currently drilled each year in the United States. If highly deviated wells (wells greater than 60 degrees) are included, both horizontal and deviated wells comprise about 65% of well designs. Advances in these types of designs have been made possible by installation of top drives, improved hoisting, rotating and pumping systems, captain's chairs, and improved downhole telemetry tools resulting in repeatable success of these higher volume wells. Deviated and horizontal wells offer separate challenges in terms of well control, as well control kick behavior differs significantly from vertical well control.

DEFINITIONS FOR HIGHLY DEVIATED AND HORIZONTAL WELLS WILL VARY. FOR COMMONALITY, THE FOLLOWING ARBRITARY DEFINTIONS HAVE BEEN EMPLOYED WITHIN THIS MANUAL

1. **"HIGHLY DEVIATED" WELLS AS THOSE DIRECTIONAL WELLS WITH ANGLES GREATER THAN 60 DEGREES TO 80 DEGREES.**
2. **"HORIZONTAL" WELLS AS THOSE DIRECTIONAL WELLS WITH ANGLES GREATER THAN 80 DEGREES**

Based upon long horizontal hole sections, influxes within these horizontal sections are more difficult to identify when compared to vertical wells. As in all well control applications, volumetric changes to the hole volumes are still the primary method of determining when an influx has entered the well (increase in flow or pit gains). If well volumes indicate an influx has entered the wellbore and the well has been shut-in, responding SICP and SIDP may indicate minimum pressure increase with minimum gain. The smaller pressure increases may be due to the influx displacing drilling fluids in the horizontal section and not displacing the drilling fluid vertically. With small pressure increases, any calculation of KMW may be significantly less than actual density needed to kill the well.

Horizontal well kick sections may prove more difficult to efficiently remove an influx, as the influx of lighter density will gravitate and channel to the top of the wellbore. Since these influxes channel, the influx becomes elongated and will travel on the high side of the well as the kill circulation ensues. Unfortunately, the laminar displacement of fluids within the horizontal section will not efficiently remove all the influx from the well.

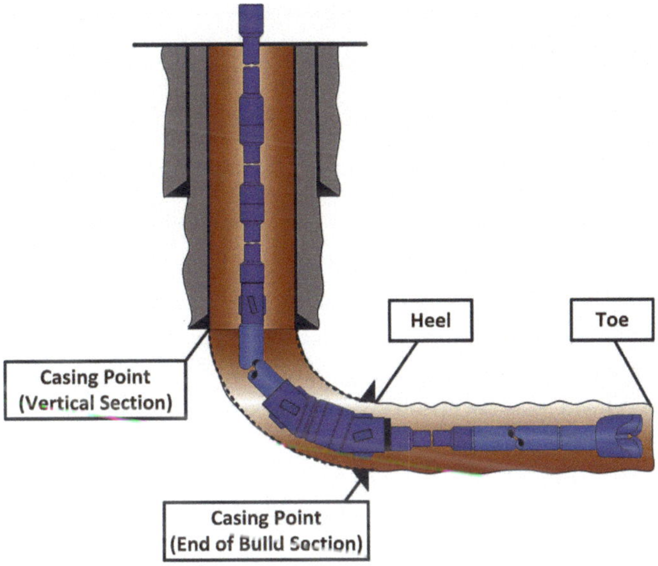

Fig. 2.5 Diagram of horizontal well.

The frequencies of kicks within horizontal wells are considered to be less when compared to vertical wells. Horizontal wells are normally constructed to land protective casing at the heel of the horizontal section. Since the horizontal section is contained within the production zone, the need to seal off weaker zones occurring within the build section minimizes chances of sticking and lost circulation. This common construction design translates to kicks within horizontal section occurring from (Fig. 2.5):

- Long horizontal lengths with varying fluid densities due to settling, cutting beds, and poor flow characteristics.
- Crossing unknown faults causing losses and/or kicks.
- Spotting of lighter density fluid pills if the BHA becomes stuck.

Secondary designs include milling windows from existing vertical casing strings and drilling a smaller directional well to penetrate production zone horizontally. These types of designs may not allow installation of protective casing to the heel of the horizontal well. Within these designs, the kick-off and build sections may be penetrating weaker zones (subnormally pressurized). If the BHA becomes stuck while drilling horizontal sections, a lighter density spotting fluid may be used. When circulated out, these fluids may reduce the overall annular hydrostatic pressure resulting in kicks. These types of kicks may become more troublesome, as increases in well density may fracture weaker zones uphole from the horizontal section. As in all well kill operations, any losses must first be cured before the well can be successfully killed. These types of wells may employ use of a simple well cap before initiating lost circulation cures and a final well control method.

THE FIRST LINE OF DEFENSE FOR KICKS IN HIGH ANGLE OR HORIZONTAL WELLS WILL BE ACCURATELY TRACKING VOLUMES DURING EACH CONNECTION AND WHILE TRIPPING.

PERFORM ROUTINE BACK-REAMING TO MINIMIZE CUTTING BEDS

MINIMIZE CHANCES OF SWABBING BY USING SHORT TRIPS, SLOW PULLING SPEEDS AND PUMPING BIT CLEANING AGENTS BEFORE TRIPPING OUT OF HOLE.

Complexities of horizontal well control

(1) Most well control methods can be applied to horizontal wells and can be used with success.
(2) If kicks occur off bottom or if the drillstring becomes plugged, well control operations may become somewhat limited as volumetric method or bullheading in a horizontal section may not be useful.
(3) For horizontal wells with an influx, the influx can be assumed to be strung out throughout the entire horizontal drilled section as the influx has been elongated and travels on the high side of the hole.
(4) Due to the influx migrating to the high side of the horizontal hole section, the removal of the influx at the slow pump rates used in well control will be less efficient and therefore increasing overall circulating times dramatically.

DUE TO THE COMPLEXITIES OF HORIZONTAL INFLUXES, THE DRILLER'S METHOD HAS BECOME THE PREFERRED WELL CONTROL METHOD FOR HORIZONTAL WELLS.

Common causes of kicks horizontal wells

Swabbing

Once the formation in which the horizontal section will be drilled is penetrated by the bit, the mud weight needed to control the zone has been defined. While drilling horizontally in the target zone, the most likely cause of a kick will be due to swabbing. Since the majority of kicks occur during tripping operations, this is typically the most likely scenario for a kick. Swabbing is largely due to cuttings beds which can develop in horizontal wells and pulling the pipe at a fast rate.

Lost circulation

Drilling long reach horizontal sections tends to produce high ECDs for the toe end of the hole. These high ECDs may induce lost return problems by overpressuring a weak or fractured formation. These losses may reduce hydrostatic pressure below formation pressure allowing an influx into the well.

Reduction in mud weight for stuck pipe

If a light-weight fluid pill is pumped in an attempt to work a stuck workstring free, the well may become underbalanced when the pill is circulated to the vertical section of the hole. Subsequently, the lighter density of the pill may reduce the bottom-hole pressure resulting in an underbalanced kick.

Secondary kicks

Horizontal wells are vulnerable to secondary kicks. With variances in mud density due to cuttings beds and long circulation times, maintaining constant BHP through choke manipulations becomes more difficult. If BHP decreases without proper choke response, formation pressure can be decreased to the point where a kick ensures.

Drilling past a pressure barrier/fault

Although horizontal sections are normally considered to be homogenous, a possibility remains the production zone may contain a sealed fault. If this condition exists during drilling operations, the fault may cause penetration from a lower-pressure zone into a higher-pressure zone. If such an environment exists, changes from a faulted structure may range from total or partial losses (subnormally pressured environment) up to large kicks. If the kicks are large enough, conditions for undergrown blowout may occur.

Complex kick detection

A pit gain or increases in flow rate out of the well are the best indicators of an influx. The only indication of kick size is the pit gain. Increase in rate or kick volume does not define whether the influx is oil or gas. In OBM and dependent of well TVD, lighter hydrocarbon influxes (gas, oil, gas/oil) will remain in solution within the mud thus potentially masking a kick. Therefore, swabbed kicks may be difficult to detect and confirm.

Light hydrocarbons may accumulate within formation pockets at the top of the horizontal section of the hole. This influx may not be recognized until the influx is pumped into the vertical section of the hole. As the hydrostatic pressure of the fluid is reduced and the influx is circulated out, when the bubble point is reached, gas will break out of solution and exit from the fluid. Therefore, gas influxes may be larger than anticipated due to the amount of gas held in solution and being released within the vertical hole sections of the well.

For horizontal sections containing fractures, influxes may enter the fractures due to high ECD and reenter the wellbore after the pumps are shut off and the back pressure reduced.

Even swabbing of the well may be more complicated. If a kick has been swabbed into the well and the pipe stops moving, the influx movement may stop and balanced conditions return. The influx may remain in the horizontal portion of the hole and flow checks will show drillpipe and casing pressures reading zero psi. Under normal circumstances, the casing pressure should be higher when compared to the drillpipe pressure. But since the influx is in the horizontal leg, the "slight increase" may not register on the gauges. If shut-in pressures exist, these reading will not yield the volume or magnitude of the kick.

If the kick is not easily recognized, results of misdiagnosed may lead to greater well control problems. If the kick remains not recognizable and additional pipe is pulled from the hole, a swabbed kick can become much larger. When this gas reaches the vertical section of the hole, the gas will begin to migrate. If the kick volume and migration rate is not properly mitigated, the entire hole can easily become overpressured resulting in lost returns. To minimize kick identification problems, ensure flow back volumes are consistent with fingerprinted volumes.

TO MINIMIZE KICK IDENTIFICATION PROBLEMS, ENSURE FLOWBACK VOLUMES ARE CONSISTENT WITH FINGERPRINTED VOLUMES.

Influx volumes

Pit gain volumes may be larger in horizontal wells due to long lengths of exposed permeability. The long lengths of permeable surface area result in greater flow area allowing influx to enter the wellbore. SOBM/OBM mud masks or "hides" the additional kick volumes. Due to solubility of the influxes, pit gains will remain small until the influx has entered the vertical hole section.

Exposure time

Low levels of under balance may create large influx volumes as large permeable area may allow flow with a "small pressure" imbalance. The longer exposure time of the pressure imbalance, the greater the chance of a well kick forming.

Effects of mud type

Most horizontal wells are drilled with SOBM/OBM for lubricity and compatibility with the formation. Kick fluids comprised of oil or gas stay in solution within the oil-phase of the mud and mask any pit gain or increase on the flow indicator.

A hydrocarbon influx within the oil-phase of the mud may not show any signs of migration until it reaches the bubble point and comes out of solution in the upper part of the hole. This can happen rapidly and seemingly without warning.

A kick in WBM should be recognizable on the PVT equipment unless the feed influx rate is very slow. Shutting in the well still may not show any kick pressures especially if the kick is residing in only the horizontal section.

Gas expansion

Gas expansion may not occur (minimal) until the influx reaches the vertical portion of the well, thus the entire horizontal section may be filled with influx before being detected by flow indicator. Pit gains via the rig's PVT system are the most effective means of recognizing horizontal kicks.

Effects of hole angle

As a kick moves from the horizontal section through the curve and into the vertical section the length of the kick which contributes to casing pressure increases as the angle decreases. Even without kick expansion, the transition from horizontal to vertical will cause a rise in the casing pressure.

At 89 degrees a 1000 ft. kick only has a vertical height of 17 ft. At 30 degrees, it has a vertical height of 866 ft. **If the kick is water** at normal pressure of 0.465 psi/ft., the hydrostatic pressure of the kick at 89 degree is 40 psi and 30 degree is 403 psi. As the influx is circulated from the horizontal section into the vertical section, a casing pressure rise could be interpreted as the kick expanding. Opening the choke to account for this pressure increase will decrease bottom-hole pressure and may allow a secondary kick to occur. Maintaining proper drillstring pressure maintains constant bottom-hole pressure (Table 2.1).

Table 2.1 Influx comparison.

	Casing pressure increase	Pit gain
Gas Expansion	YES	YES
Secondary Kick	YES	YES
Kick Moved from Horizontal to Vertical	YES	NO

Corrective actions for swabbed kicks

Swabbing is often not easily detected and is complicated by poor reaction response. If the well does not take the proper fill-up, the Driller normally performs a flow check. If no flow is detected, the Driller may shut-in the well to check for SIDPP and SICP. With horizontal wells, it is highly probable that the stabilized SIDP and SICP readings are zero. If a swabbed kick is in the horizontal portion of the hole it will have no appreciable vertical length.

The key indicator for swabbing is identified as the hole not taking the correct amount of mud. After shut-in, the drillpipe and casing pressure will more than likely read zero psi.

If the hole is not taking correct amount of mud, perform horizontal well flow check.

(1) Shut-in well, record stabilized pressures and pit volumes.
   - **(a)** If SIDPP and SICP readings are greater than zero and SICP > SIDPP, the kick may be above horizontal leg.
   - **(b)** If SIDPP and SICP readings are greater than zero and SICP = SIDPP, the kick may be in horizontal leg.

(2) With well shut-in on annular, strip pipe back to bottom.
   - **(a)** Do not attempt to run back to bottom with well open, as this may lead to a blowout.

(3) Circulate out through the choke.

WITH ANY POSITIVE INDICATOR OF WELL FLOW,
DO NOT ATTEMPT TO RUN BACK TO BOTTOM WITH WELL OPEN AS THIS MAY LEAD TO A BLOWOUT
KEEP WELL SHUT-IN AND STRIP BACK TO BOTTOM, THEN CIRCULATE OUT ON CHOKE

Reduce swabbing risk

To help avoid swabbing, keep the pumps running when moving the pipe off bottom (pump out of hole).

- Keep clearances between the hole and drilling tools as large as possible.
- Pump and back ream **as necessary** to clean cuttings beds on trips out of the hole. Pumping out to the casing shoe also helps reduce swabbing.
- Minimize trip speed until the bit is above the casing shoe.
- Keep mud properties in line with mud program.
- Believe the trip tank readings. If the well is not taking proper fill-ups, strip back to bottom and circulate on choke.

Differences in horizontal kicks

Kicks caused by an underbalanced mud column will result in a positive pressure because the formation has pushed fluid into wellbore. This influx into wellbore will result in a positive reading of pressure when the well is shut-in, whereas a swabbed kick may not show shut-in pressure.

The difference between the drillpipe and casing pressure may not be detectable because of the small vertical height of the influx within the horizontal leg.

Kill mud will not become effective until the annular volume within the horizontal section has been displaced. Conversely, for a deviated or vertical well, the kill mud begins to have an effect as soon as it reaches the bit.

For horizontal wells, the lighter density hydrocarbon influxes will migrate to the high side of the hole and may remain in the horizontal section without showing any sign at the surface. Displacing all the influx with KMW may be difficult in the horizontal section of the hole. A second circulation at a higher rate is usually needed to remove all the gas. Influx volume expansion, casing pressures and pit gains will be minimal while the kick remains within the horizontal section.

Well kill methods for horizontal wells

Driller's method

As the majority of horizontal kicks are created while tripping pipe (i.e., swabbed), no additional mud weight will be needed to kill the well. For swabbed kicks, the Driller's method is suggested. This method uses a constant DP pressure until the influx is circulated to the surface. Since accurate horizontal kick volumes are difficult to identify, this method allows for the influx to enter the vertical section of the well and expand without impacting BHPs. This method can be used for swabbed or pressure kicks.

DUE TO THE COMPLEXITIES OF HORIZONTAL INFLUXES, THE DRILLER'S METHOD HAS BECOME THE PREFERRED WELL CONTROL METHOD FOR HORIZONTAL WELLS.

Wait and weight method

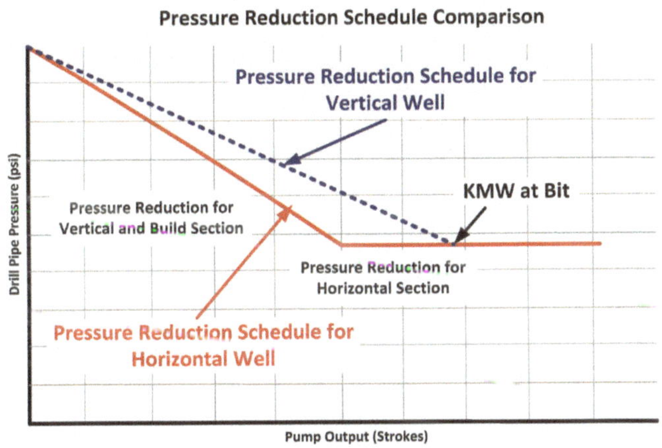

Fig. 2.6 Prerecorded information sheet for horizontal wells.

Alternately, the Wait and Weight method can be employed for horizontal kicks but may be more complicated. For accuracy, this method requires a two-step pressure reduction schedule. The first portion of the schedule honors pressure reductions within the vertical section of the well. Whereupon the second portion of the schedule honors pressure reductions within the build to horizontal sections (Fig. 2.6).

This chart illustrates how using a vertical drillpipe pressure reduction schedule based on a straight pressure reduction from the surface to the bit will overpressure the wellbore. Only the vertical component of depth will contribute to the hydrostatic pressure. Once kill mud weight reaches the beginning of the horizontal section or the heel, the drillpipe pressure should be at final circulating pressure (FCP).

FOR CREATION OF HORIZONTAL STEP-DOWN CHART, USE PUMP STROKES FOR KMW TO REACH HEEL OF HORIZONTAL LEG

Bullheading

Bullheading may be used to pump influx fluids back into the formation. For horizontal wells without protective casing set at the heel, bullheading may be problematic as there is no way to determine which formation will break down and take fluids. If all of the influx is pushed back into the formation near the heel, the rest of the open hole may still be filled with the influx. Bullheading may also induce lost returns. Heavy mud used for bullheading may bypass lighter mud at top of hole.

If the bit has been pulled off bottom or out of the hole, an annular bullhead with KMW to the casing seat may be needed to reduce pressure prior to stripping.

Kill off bottom

If the bit is off bottom and above the casing shoe when a kick occurs, it may be possible to perform a mud cap kill. This involves circulating mud which is heavy enough to kill the well from the current depth of the bit. The hydrostatic pressure of the kill weight mud at the TVD of the bit plus the hydrostatic pressure of the mud from the bit to the casing shoe must be more than the formation pressure to prevent additional kicks. But, the KMW hydrostatic pressure must be below that of the casing shoe LOT/FIT to prevent an underground blowout.

Well kill methods for highly deviated wells

One of the main differences between horizontal wells and highly deviated wells in respect to well control is the highly deviated wells experience faster influx migration rates when compared to either vertical or horizontal wells. The phenomenon appears to be due to segregation developing in the bore hole with the lighter fluid tending to rise on the high side of the hole, whereas heavier mud segregates to the lower hole sections. This stratification of fluid densities translates to significant changes in the CP as the kick is circulated out the well.

Cuttings beds may also develop in highly deviated wells and may cause problems during tripping such as swabbing or stuck pipe. Particular care must be taken to reduce any cuttings beds by use of frequent back-reaming and short trips. Lower the chances of swabbing during trips by maintain good mud properties, performing frequent back-reaming and short trip while implementing use of trip sheets for accurate tacking of hole fill-ups.

In highly deviated wells, daily operations may include frequent back-reaming (reciprocation of the drillstring) and short trips (trip to casing shoe while circulating and conditioning mud) in order to avoid sticking the pipe.

PRIMARY CONCERN WILL ALWAYS BE WELL CONTROL, WHILE STUCK PIPE REMAINS A SECONDARY CONCERN.

In highly deviated wells, small influx detections (i.e., swabbed kicks) may be difficult to recognize within OBM/SOBM. After shut-in isolation is performed, kicks should show a difference in SIDPP and SICP but the difference may be slight at high angles until such time the kick reaches the lower angle section of the well.

Wait and weight method

The Wait and Weight method is the most commonly used form of well control for highly deviated well. This method aids with the prevention of overpressuring the well by using an accurate drillpipe pressure step-down chart in order to maintain constant bottom-hole pressure as kill mud weight is pumped to the bit. With KMW in annulus, the maximum pressure when influx reaches the surface is reduced. With both KMW to bit and in the annulus, this method provides the largest margin for safety when compared to other well control methods.

DUE TO THE COMPLEXITIES OF HIGHLY DEVIATED INFLUXES, THE WAIT AND WEIGHT METHOD HAS BECOME THE PREFERRED WELL CONTROL METHOD FOR HIGH ANGLE WELLS.

Driller's method

For those rigs where the volume of the kick is large and requires a substantial increase in KMW, the Driller's Method can be safely used as an alternative. If the rig cannot effectively weight up for KMW, crews cannot keep up even with lower pump rates, or if WBM is being used, the Driller's method may be used as a means to circulate influx to surface, then properly prepare KMW in sufficient volumes suitable to kill the well.

Summary of equations for horizontal and deviated wells

1. Determine vertical influx height (ft)

$$\text{Influx Ht (ft)} = \text{Cosine (Max Angle)} \times \left[\left(\frac{1029.4}{\text{Hole Dia}_{\text{in}}{}^2}\right) \times \text{Kick Size}_{\text{bbl}}\right] \quad (2.15)$$

Assume 2.0 ppg Influx Density

2. Determine influx hydrostatic pressure (psi)

$$\text{HP Influx}_{\text{psi}} = \text{Influx Ht}_{\text{ft}} \times \left(\text{Influx Density}_{\text{ppg}} \times 0.052\right) \quad (2.16)$$

3. Determine mud hydrostatic pressure (psi) from TD to bit

$$\text{Mud HP}_{\text{from TD to Bit}} = (\text{TVD Btm}_{\text{ft}} - \text{TVD Bit}_{\text{ft}} - \text{Influx Ht}_{\text{ft}}) \times 0.052 \times \text{MW}_{\text{ppg}} \quad (2.17)$$

4. Determine mud hydrostatic pressure (psi) from bit to surface

$$\text{Mud HP}_{\text{from Bit to Surf}} = (\text{TVD Bit}_{\text{ft}}) \times 0.052 \times \text{MW}_{\text{ppg}} \quad (2.18)$$

5. Determine formation pressure (psi)

$$\begin{aligned}\text{Form Press}_{\text{psi}} = {} & \text{HP Influx}_{\text{psi}} + \text{Mud HP}_{\text{from TD to Bit (psi)}} \\ & + \text{Mud HP}_{\text{from Bit to Surf (psi)}} + \text{SIDPP}_{\text{psi}}\end{aligned} \quad (2.19)$$

6. Determine formation pressure equivalent MW (ppg)

$$\text{Form Press}_{\text{ppg}} = \frac{\text{Form Press}_{\text{psi}}}{(0.052) \times (\text{TVD}_{\text{TD}})} \quad (2.20)$$

7. Determine hydrostatic pressure at bit with kick in the hole (***assume entire kick below bit***)

$$\text{HP}_{\text{Bit with Kick}} = \text{Form Press}_{\text{psi}} - \text{Mud HP}_{\text{Bit to Surf}} - \text{HP Influx} \quad (2.21)$$

8. Determine KMW at bit needed to keep formation from flowing

$$\text{KMW}_{\text{Off Bottom}} = \frac{\text{HP}_{\text{Bit with Kick}}}{(0.052) \times (\text{TVD}_{\text{bit}})} \quad (2.22)$$

Safety Check

9. Maximum CP when influx reaches the surface

$$\text{Max CP} = \frac{\text{SIDPP}}{2} + \sqrt{\begin{aligned}&\left(\frac{\text{SIDPP}}{2}\right)^2 + \left(\text{SIDPP} + (0.052)(\text{MW})\left(\text{TVD}_{\text{TD}}\right)\right) \times \left(\frac{\text{Kick Size}}{\text{Capacity Factor}}\right) \\ &\times \left(\left((\text{MW})(0.052) - \left(\text{Influx}_{\text{ppg}}\right)(0.052)\right) \times \left(4.03 - \left[0.38 - \ln\left(\text{SIDPP} + (0.052)(\text{MW})\left(\text{TVD}_{\text{TD}}\right)\right)\right]\right)\right)\end{aligned}} \quad (2.23)$$

10. Does Max CP exceed the burst rating of the casing? Yes or no, if yes…, bullhead

$$Q = 0.008085 \times \text{SICP}_{\text{psi}} \times \text{SCR}_{\text{spm}} \times \text{Pump Output}_{\text{bps}} \times \text{Pump Efficiency}_{\%} \quad (2.24)$$

The following Prerecorded Information Sheet has been developed for **Horizontal wells**. Note: Hole Angle 1 is highlighted for completion (Fig. 2.7).

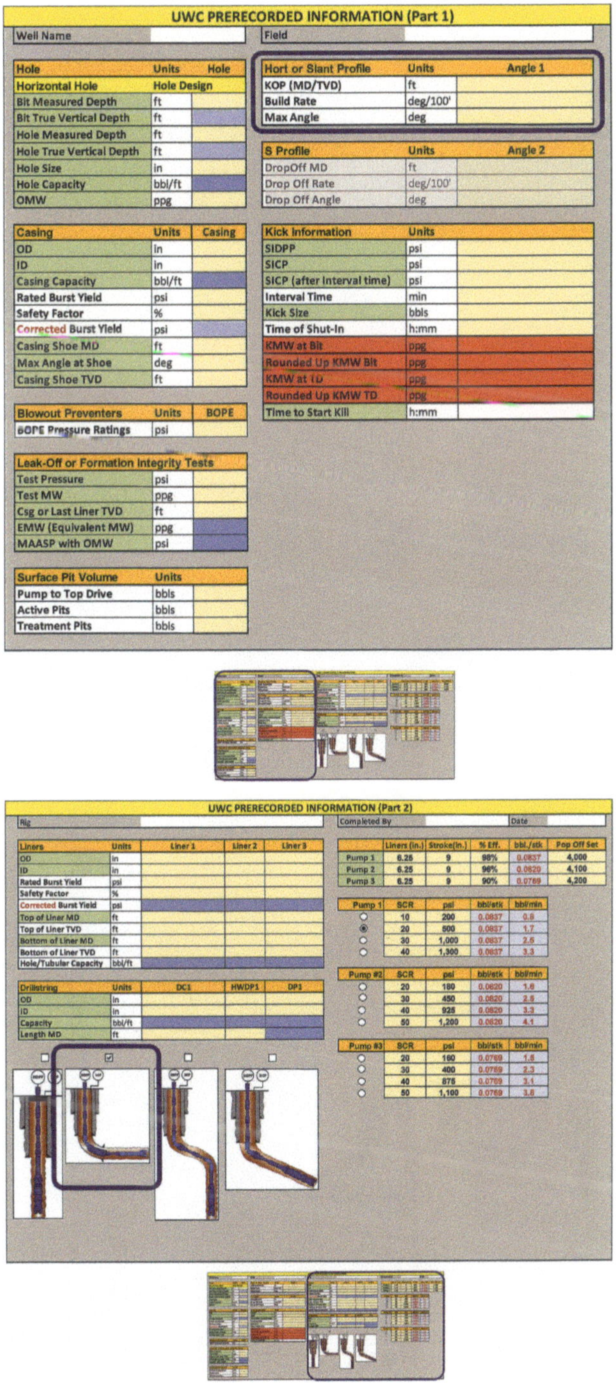

UWC PRERECORDED INFORMATION (Part 1)

Well Name		Field	

Hole	Units	Hole
Horizontal Hole	Hole Design	
Bit Measured Depth	ft	
Bit True Vertical Depth	ft	
Hole Measured Depth	ft	
Hole True Vertical Depth	ft	
Hole Size	in	
Hole Capacity	bbl/ft	
OMW	ppg	

Hort or Slant Profile	Units	Angle 1
KOP (MD/TVD)	ft	
Build Rate	deg/100'	
Max Angle	deg	

S Profile	Units	Angle 2
DropOff MD	ft	
Drop Off Rate	deg/100'	
Drop Off Angle	deg	

Casing	Units	Casing
OD	in	
ID	in	
Casing Capacity	bbl/ft	
Rated Burst Yield	psi	
Safety Factor	%	
Corrected Burst Yield	psi	
Casing Shoe MD	ft	
Max Angle at Shoe	deg	
Casing Shoe TVD	ft	

Kick Information	Units	
SIDPP	psi	
SICP	psi	
SICP (after interval time)	psi	
Interval Time	min	
Kick Size	bbls	
Time of Shut-In	h:mm	
KMW at Bit	ppg	
Rounded Up KMW Bit	ppg	
KMW at TD	ppg	
Rounded Up KMW TD	ppg	
Time to Start Kill	h:mm	

Blowout Preventers	Units	BOPE
BOPE Pressure Ratings	psi	

Leak-Off or Formation Integrity Tests		
Test Pressure	psi	
Test MW	ppg	
Csg or Last Liner TVD	ft	
EMW (Equivalent MW)	ppg	
MAASP with OMW	psi	

Surface Pit Volume	Units	
Pump to Top Drive	bbls	
Active Pits	bbls	
Treatment Pits	bbls	

UWC PRERECORDED INFORMATION (Part 2)

Rig		Completed By		Date	

Liners	Units	Liner 1	Liner 2	Liner 3
OD	in			
ID	in			
Rated Burst Yield	psi			
Safety Factor	%			
Corrected Burst Yield	psi			
Top of Liner MD	ft			
Top of Liner TVD	ft			
Bottom of Liner MD	ft			
Bottom of Liner TVD	ft			
Hole/Tubular Capacity	bbl/ft			

Drillstring	Units	DC1	HWDP1	DP1
OD	in			
ID	in			
Capacity	bbl/ft			
Length MD	ft			

	Liners (in.)	Stroke(in.)	% Eff.	bbl./stk	Pop Off Set
Pump 1	6.25	9	98%	0.0837	4,000
Pump 2	6.25	9	96%	0.0820	4,100
Pump 3	6.25	9	90%	0.0769	4,200

Pump 1	SCR	psi	bbl/stk	bbl/min
○	10	200	0.0837	0.8
◉	20	500	0.0837	1.7
○	30	1,000	0.0837	2.5
○	40	1,300	0.0837	3.3

Pump #2	SCR	psi	bbl/stk	bbl/min
○	20	180	0.0820	1.6
○	30	450	0.0820	2.5
○	40	925	0.0820	3.3
○	50	1,200	0.0820	4.1

Pump #3	SCR	psi	bbl/stk	bbl/min
○	20	160	0.0769	1.5
○	30	400	0.0769	2.3
○	40	875	0.0769	3.1
○	50	1,100	0.0769	3.8

Fig. 2.7 UWC prerecorded information (Parts 1 and 2).

The following Prerecorded Information Sheet has been developed for **Deviated (Build and Hold)** wells. Note: Hole Angle 1 is highlighted for completion (Fig. 2.8).

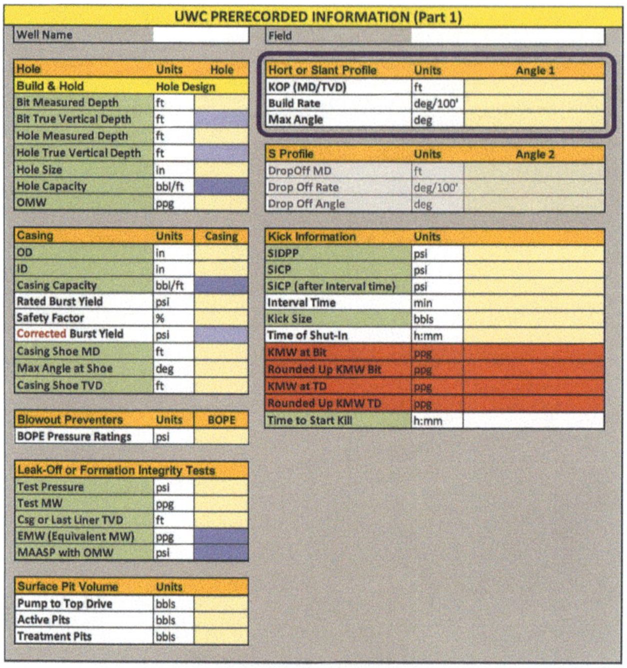

UWC PRERECORDED INFORMATION (Part 1)

Well Name		Field	

Hole	Units	Hole
Build & Hold	Hole Design	
Bit Measured Depth	ft	
Bit True Vertical Depth	ft	
Hole Measured Depth	ft	
Hole True Vertical Depth	ft	
Hole Size	in	
Hole Capacity	bbl/ft	
OMW	ppg	

Hort or Slant Profile	Units	Angle 1
KOP (MD/TVD)	ft	
Build Rate	deg/100'	
Max Angle	deg	

S Profile	Units	Angle 2
DropOff MD	ft	
Drop Off Rate	deg/100'	
Drop Off Angle	deg	

Casing	Units	Casing
OD	in	
ID	in	
Casing Capacity	bbl/ft	
Rated Burst Yield	psi	
Safety Factor	%	
Corrected Burst Yield	psi	
Casing Shoe MD	ft	
Max Angle at Shoe	deg	
Casing Shoe TVD	ft	

Kick Information	Units	
SIDPP	psi	
SICP	psi	
SICP (after interval time)	psi	
Interval Time	min	
Kick Size	bbls	
Time of Shut-In	h:mm	
KMW at Bit	ppg	
Rounded Up KMW Bit	ppg	
KMW at TD	ppg	
Rounded Up KMW TD	ppg	
Time to Start Kill	h:mm	

Blowout Preventers	Units	BOPE
BOPE Pressure Ratings	psi	

Leak-Off or Formation Integrity Tests		
Test Pressure	psi	
Test MW	ppg	
Csg or Last Liner TVD	ft	
EMW (Equivalent MW)	ppg	
MAASP with OMW	psi	

Surface Pit Volume	Units	
Pump to Top Drive	bbls	
Active Pits	bbls	
Treatment Pits	bbls	

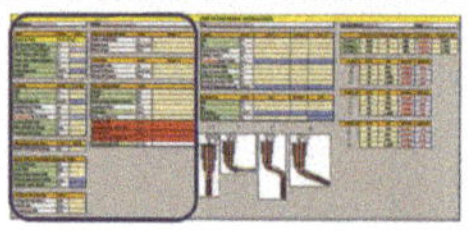

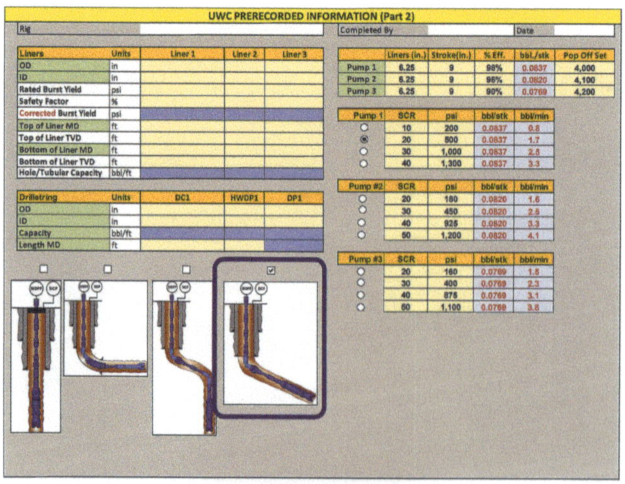

UWC PRERECORDED INFORMATION (Part 2)

Rig		Completed By		Date	

Liners	Units	Liner 1	Liner 2	Liner 3
OD	in			
ID	in			
Rated Burst Yield	psi			
Safety Factor	%			
Corrected Burst Yield	psi			
Top of Liner MD	ft			
Top of Liner TVD	ft			
Bottom of Liner MD	ft			
Bottom of Liner TVD	ft			
Hole/Tubular Capacity	bbl/ft			

Drillstring	Units	DC1	HWDP1	DP1
OD	in			
ID	in			
Capacity	bbl/ft			
Length MD	ft			

	Liners (in.)	Stroke(in.)	% Eff.	bbl./stk	Pop Off Set
Pump 1	6.25	9	98%	0.0837	4,000
Pump 2	6.25	9	96%	0.0820	4,100
Pump 3	6.25	9	90%	0.0769	4,200

Pump 1	SCR	psi	bbl/stk	bbl/min
○	10	200	0.0837	0.8
◉	20	500	0.0837	1.7
○	30	1,000	0.0837	2.5
○	40	1,300	0.0837	3.3

Pump #2	SCR	psi	bbl/stk	bbl/min
○	20	180	0.0820	1.6
○	30	450	0.0820	2.5
○	40	925	0.0820	3.3
○	50	1,200	0.0820	4.1

Pump #3	SCR	psi	bbl/stk	bbl/min
○	20	160	0.0769	1.5
○	30	400	0.0769	2.3
○	40	875	0.0769	3.1
○	50	1,100	0.0769	3.8

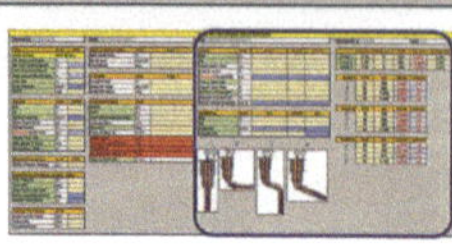

Fig. 2.8 UWC precoded information (Parts 1 and 2).

The following Prerecorded Information Sheet has been developed for **"S" Shaped** wells. Note: Hole Angle 1 and Hole Angle 2 are highlighted for completion (Fig. 2.9).

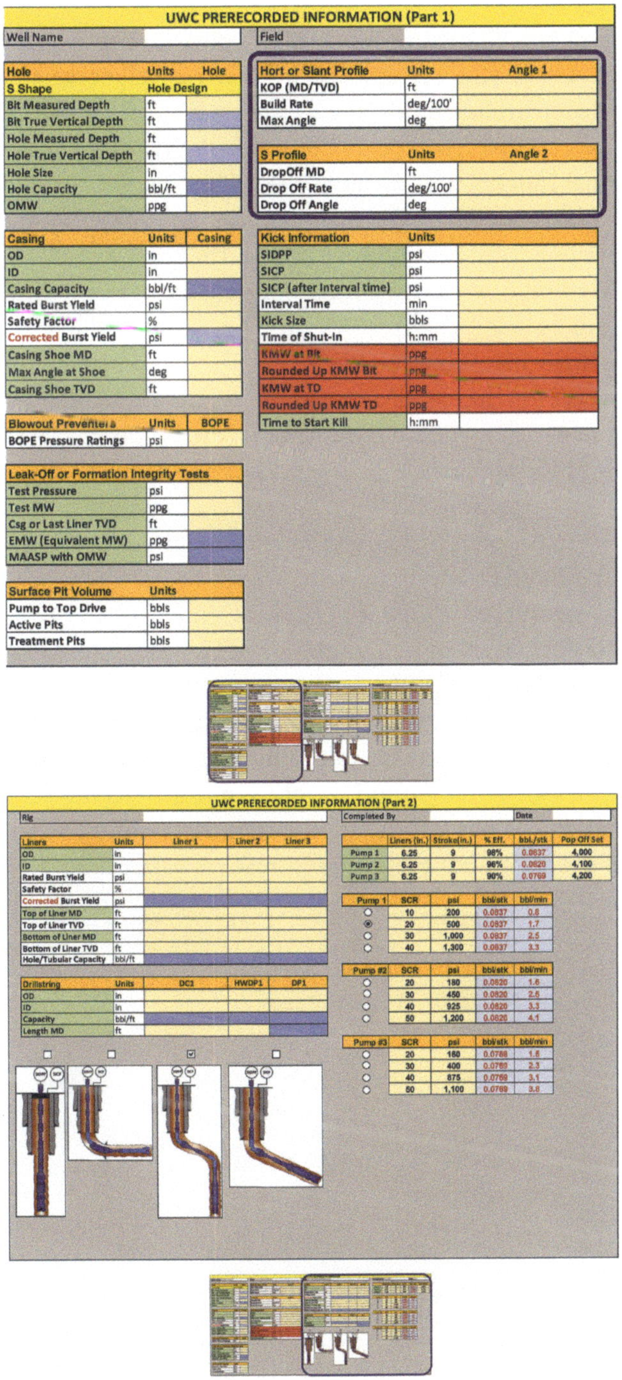

Fig. 2.9 UWC precoded information (Parts 1 and 2).

Driller's method

Driller's method commentary

The Driller's method of well control is one of the oldest, simplest, and widely used well control techniques. The Driller's Method is a preferred well control technique for wells drilled with water-based drilling fluids, as gas influxes may rapidly rise in the wellbore and cause significant well control problems. The Driller's Method allows the individual to execute well control quickly and remove the influx, before introducing kill weight mud. Requiring little or no delays for computations and kill mud mixing, well control operations can be immediately established (Fig. 2.10).

The Driller's Method of well control has several advantages over other forms of circulating well control.

- Simple with few calculations.
- Quick initialization of circulation.
- No weighting material needed on location.
- Preferred method for high-angle holes and horizontal wells.
- Less chance of gas migration.
- No complicated step-down pressure schedules required for tapered drillstrings.

Disadvantages of the Driller's Method

- Higher annulus pressures for longer periods of time.
- Higher shoe pressure in many situations.
- Greatest pressure on surface equipment.

This is a two-circulation method (minimum). During the first circulation, the drillpipe pressure is maintained at a constant value until the influx is circulated from the wellbore. During the second circulation, kill mud weight is pumped to the bit while maintaining casing pressure constant. When the kill mud enters the annulus, Final Circulating Pressure is maintained constant until the kill mud reaches surface.

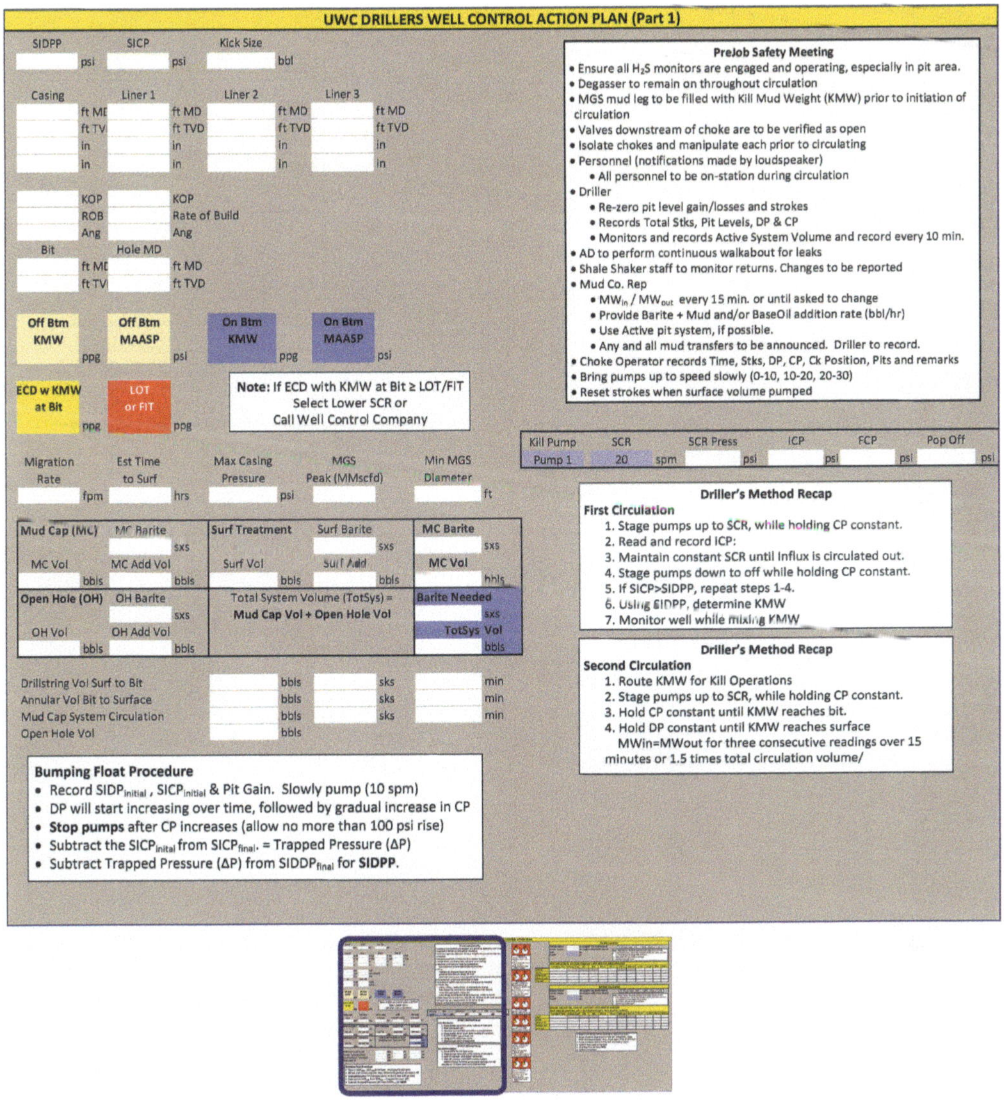

Fig. 2.10 UWC drillers well control action plan (Parts 1 and 2).

Continued

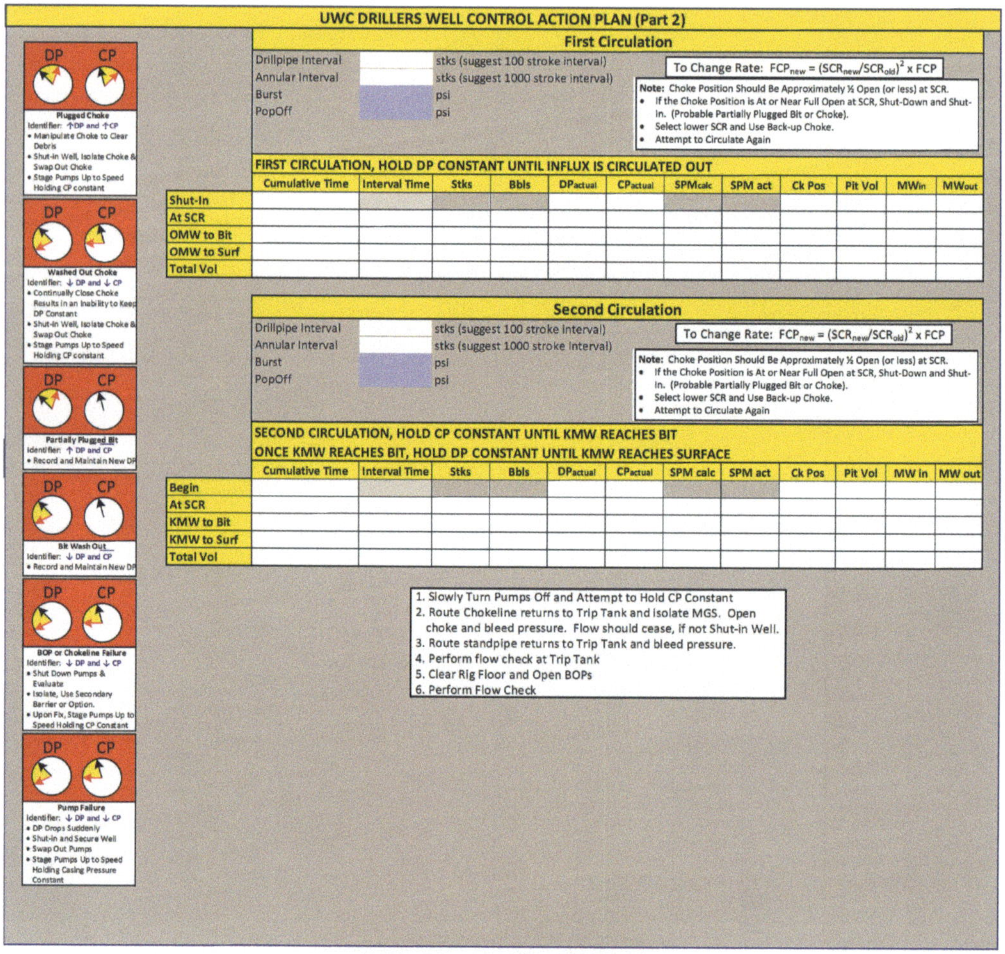

UWC DRILLERS WELL CONTROL ACTION PLAN (Part 2)

First Circulation

Drillpipe Interval		stks (suggest 100 stroke interval)
Annular Interval		stks (suggest 1000 stroke interval)
Burst		psi
PopOff		psi

To Change Rate: $FCP_{new} = (SCR_{new}/SCR_{old})^2 \times FCP$

Note: Choke Position Should Be Approximately ½ Open (or less) at SCR.

- If the Choke Position is At or Near Full Open at SCR, Shut-Down and Shut-In. (Probable Partially Plugged Bit or Choke).
- Select lower SCR and Use Back-up Choke.
- Attempt to Circulate Again

FIRST CIRCULATION, HOLD DP CONSTANT UNTIL INFLUX IS CIRCULATED OUT

	Cumulative Time	Interval Time	Stks	Bbls	DP_{actual}	CP_{actual}	SPM_{calc}	SPM act	Ck Pos	Pit Vol	MW_{in}	MW_{out}
Shut-In												
At SCR												
OMW to Bit												
OMW to Surf												
Total Vol												

Second Circulation

Drillpipe Interval		stks (suggest 100 stroke interval)
Annular Interval		stks (suggest 1000 stroke interval)
Burst		psi
PopOff		psi

To Change Rate: $FCP_{new} = (SCR_{new}/SCR_{old})^2 \times FCP$

Note: Choke Position Should Be Approximately ½ Open (or less) at SCR.

- If the Choke Position is At or Near Full Open at SCR, Shut-Down and Shut-In. (Probable Partially Plugged Bit or Choke).
- Select lower SCR and Use Back-up Choke.
- Attempt to Circulate Again

SECOND CIRCULATION, HOLD CP CONSTANT UNTIL KMW REACHES BIT
ONCE KMW REACHES BIT, HOLD DP CONSTANT UNTIL KMW REACHES SURFACE

	Cumulative Time	Interval Time	Stks	Bbls	DP_{actual}	CP_{actual}	SPM calc	SPM act	Ck Pos	Pit Vol	MW in	MW out
Begin												
At SCR												
KMW to Bit												
KMW to Surf												
Total Vol												

1. Slowly Turn Pumps Off and Attempt to Hold CP Constant
2. Route Chokeline returns to Trip Tank and isolate MGS. Open choke and bleed pressure. Flow should cease, if not Shut-in Well.
3. Route standpipe returns to Trip Tank and bleed pressure.
4. Perform flow check at Trip Tank
5. Clear Rig Floor and Open BOPs
6. Perform Flow Check

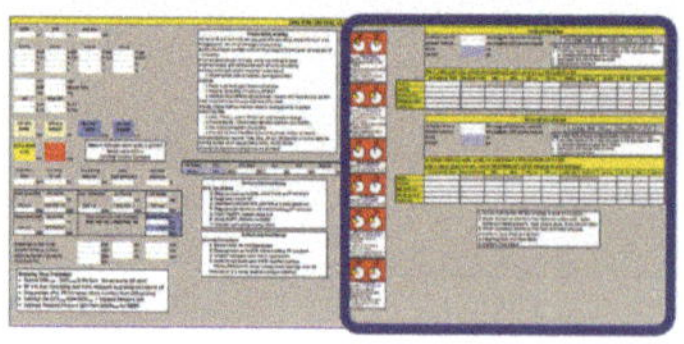

Fig. 2.10, cont'd

First circulation (getting the influx out)

Original Fluid Weight (FW) is pumped to displace the influx from the wellbore while maintaining a constant bottom-hole pressure equal to or slightly greater than the formation pressure. The constant bottom-hole pressure will prevent an additional influx from entering the well while the kick is being circulated out. By displacing the influx from the wellbore, the safety of the crew and Rig unit will be maximized.

It must be understood the well is not dead at the end of the first circulation. Rather, the influx has been safely circulated out.

Second circulation (getting the kill fluid in)

Weighted kill fluid is circulated throughout the well while maintaining a constant bottom-hole pressure equal to or slightly greater than the formation pressure. At the end of the second circulation, the well should be dead.

Step 1: Detection of the kick and proper shut-in

Upon noticing an immediate indicator of a kick, the well should be shut-in and flow stopped. Use the correct shut-in procedure for operations being performed. The Driller is responsible for shutting-in the well using the correct procedure.

Step 2: Record stabilized pressures (bumping the float)

Once the kick has been detected and the well is shut-in, allow drillpipe or drillpipe/tubing pressure and casing pressure to stabilize before recording.

The choke manifold valves should be configured properly and open the choke line valve at the preventer stack. Open the appropriate choke manifold valve to read the shut-in casing pressure.

Examine the pit level measuring device and estimate the volume gained during the kick. Verify this increase in volume with the Driller.

If the drillpipe or drillpipe/tubing does not have a drillpipe float, record the Shut-In Pressure (SIDPP).

If the drillpipe or drillpipe/tubing does contain a float, use appropriate bumping float procedure. Record pressure "lull" to estimate Kill Weight Fluid (KWM) for second circulation.

Keep in mind if the influx is gas, migration will take place and seen as a gradual increase in casing pressure.

Step 3: Prejob safety meeting (PJSM)

Hold a short prejob safety meeting (PJSM) and ensure all questions have been answered. Effective communication is the **KEY** to successful well control operations.

Fig. 2.11 When in doubt, ask person-in-charge.

Eliminate all sources of ignition to prevent fires (Fig. 2.11).

Designate the Rig unit and surrounding areas as **NO SMOKING AREA**.

Designate safe meeting area up wind of rig site (Muster Area). This area should contain auxiliary safety equipment such as fire extinguisher and all available first aid kits.

Review each individual's responsibility during the well control operation.

Person-in-charge

Will be located on the rig floor during the kill operations and will advise the choke operator when necessary during the circulation. Review the hand signals so there is no miscommunication of given directions.

Stress "well control" is a TEAM effort and everyone's performance of their assigned responsibilities is the only guarantee of safe well control operations.

Choke operator

Normally assigned by the senior drilling contractor representative. He/she should be experienced at operating the choke in a well kill situation. His/her responsibility will be to open and close the choke according to the directions given by Person-in-Charge. He/she is to keep the casing pressure constant as the pump is brought up to speed.

Driller

Will remain at his console to operate the rig pump and reciprocate the drillstring. Will coordinate with choke operator during pump start up and shut down as well as keeping pump rate constant. Will monitor pumps, pit levels, and his crew.

Driller's responsibility is to SLOWLY bring the pumps up to speed and to maintain a constant pump rate. He is to communicate when the pumps are up to speed and alert the Person-in-Charge of any changes.

Stress if the pump rate varies, well control cannot be maintained.

Mud engineer

Works with pit crew to maintain mud at correct weight and viscosity. A second mud engineer may be required for 24-h operations.

Derrickman

Maintains pumps and constant fluid density throughout circulation process. Be on alert for pump stroke changes indicating problems.

Stress well control can only be maintained if fluid density is constant.

Floorhand #1

Maintains accumulator and BOPE equipment. Visually inspects all valves, lines, and BOP equipment for leakage periodically throughout circulation to ensure pressure integrity is maximized (≈ every 10 min).

Stress if leaks are detected, immediately notify Driller.

Alternate personnel

Monitors emergency communications (i.e., field radio) to relay important messages. In the event of an evacuation, his responsibility is immediately notifying proper personnel.

Stress job importance and need to be alert for changing conditions.

Finally, the meeting should conclude with a review of the first circulation of the Driller's Method.

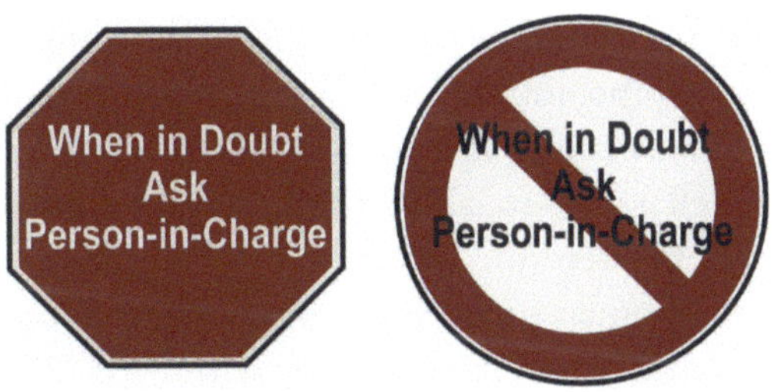

Fig. 2.12 When in doubt, ask person-in-charge

Ensure and stress the following important points (Fig. 2.12):

(a) Casing pressure will increase (if influx contains gas) as the influx reaches the surface.

(b) The amount of circulating returns should increase as the gas expands.

(c) Noise levels of escaping gas will become so high, verbal communication may become impossible. Review hand signals.

(d) Pit levels should start decreasing once the influx reaches the surface.

(e) When in doubt, ask questions.

Step 4: First circulation of Driller's method

Slowly bring pumps up to Kill pump rate (if possible) **holding casing pressure constant**. Constant casing pressure is maintained through choke manipulation.

At a stabilized Kill pump rate, record initial circulating drillpipe or drillpipe/tubing pressure. This is the actual ICP. Compare to the calculated ICP to be sure they are close to each other.

IF THE ACTUAL CIRCULATING PRESSURE IS SIGNIFICANTLY DIFFERENT FROM THE CALCULATED ICP, STOP THE PUMP, SHUT-IN THE WELL AND INVESTIGATE THE REASON

VERIFY CHOKE MOVEMENT. BEFORE BEGINNING FIRST CIRCULATION, ENSURE THE CHOKE MANIFOLD IS CORRECTLY LINED UP AND THE CHOKE CAN BE EASILY OPERATED (EITHER MANUALLY OR HYDRAULICALLY). IF TWO CHOKES ARE PRESENT, VERIFY BOTH CHOKES CAN BE MANIPULATED

MARK CASING PRESSURE GAUGE. BEFORE BEGINNING FIRST CIRCULATION, MARK THE CASING PRESSURE GAUGE.
THIS WILL PRESENT A VISUAL REMINDER OF THE CASING PRESSURE THROUGHOUT THE CIRCULATION. REMARK EVERY 10 MINUTES.

Maintain constant drillpipe or drillpipe/tubing pressure throughout circulation at a constant kill pump rate.

Step 5: Shut-in

Slowly decrease pump rate while holding casing pressure constant by closing the choke. As the pump rate decreases, the drillpipe or drillpipe/tubing circulating pressure will decrease. In order to keep bottom-hole pressure constant, the casing pressure must be held constant. Once the pumps are off and the well is shut-in, allow pressures to stabilize.

If SIDPP = SICP, everything is o.k. Proceed to Step (6).

If SIDPP < SICP, this indicates there is still some contaminant in the annulus or another kick has been taken. Repeat first circulation procedure starting with Step (2).

Calculate Driller's Kill Sheet and thoroughly mix as much of the total system capacities of kill fluid as possible. For weighted kill brines, be sure to adjust for temperature corrections.

Step 6: Second circulation of Driller's method

Hold a short prejob safety meeting (PJSM) and answer all questions. Effective communication is the **KEY** to successful well control operations. When in doubt, ask your supervisor.

Reaffirm the designated **NO SMOKING AREAS**.

Reaffirm the designated safe meeting area up wind of rig site (Muster Area). This area should contain auxiliary safety equipment such as fire extinguisher and all available first aid kits.

Quickly review any changes of individual responsibility during the second circulation:

Person-in-charge

Will be located on the rig floor during the kill operations and will advise the choke operator when necessary during the circulation. Review the hand signals so there is no miscommunication of given directions. For high-noise locations or locations without ability to electronically, review the hand signals with driller and crew to ensure there is no miscommunication of given directions.

Stress "well control" is a TEAM effort and everyone's performance of their assigned responsibilities is the only guarantee of safe well control operations.

Driller

Will assist by monitoring the equipment and personnel during the well control operations.

Driller's responsibility is to SLOWLY bring the pumps up to speed and to maintain a constant pump rate. He is to communicate when the pumps are up to speed and alert the Person-in-Charge of any changes.

Stress if the pump rate varies, well control cannot be maintained.

Derrickman

Mixes kill fluid at constant density throughout second circulation process. Always be on alert for pump stroke changes indicating problems.

Stress importance the Derrickman/mud engineer must be able to mix kill fluid at the Kill pump rate. If they are unable to maintain a constant density of Kill pump rate, the pump rate should be slowed. They must accurately report fluid weights being pumped.

Floorhand #1

Will assist the Derrickman in the mixing of the kill fluid. He will report any changes in density to the Driller. Stress importance of assisting the Derrickman at all times.

Finally, the meeting should conclude with a review of the second circulation of the Driller's Method. Ensure and stress the following important points (Fig. 2.13):

(a) If everyone performs his duties precisely, no additional circulations will be necessary. At the end of this circulation, the well should be dead.

(b) The pumps will be brought up to speed holding casing pressure constant.

Fig. 2.13 When in doubt, ask person-in-charge

(c) Casing pressure will be held constant until kill fluid reaches the bit. Reaffirm the drillstring capacity in strokes, barrels, and minutes.

(d) Once the kill fluid has reached the bit, record the circulating drillpipe or drillpipe/tubing pressure. This will be held constant until well is completely displace with kill fluid.

(e) When in doubt, ask questions.

VERIFY CHOKE MOVEMENT. BEFORE BEGINNING FIRST CIRCULATION, ENSURE THE CHOKE MANIFOLD IS CORRECTLY LINED UP AND THE CHOKE CAN BE EASILY OPERATED (EITHER MANUALLY OR HYDRAULICALLY). IF TWO CHOKES ARE PRESENT, VERIFY BOTH CHOKES CAN BE MANIPULATED

BEFORE BEGINNING SECOND CIRCULATION, MARK THE CASING PRESSURE GAUGE. THIS WILL PRESENT A VISUAL REMINDER OF THE CASING PRESSURE UNTIL KMW REACHES BIT.

AT THAT TIME, MARK DRILLPIPE PRESSURE GAUGE HOLDING THIS CONSTANT UNTIL KMW REACHES SURFACE (CASING PRESSURE GAUGE SHOULD DECREASE TO ZERO DURING THIS TIME).

Slowly bring pumps up to kill pump rate (if possible) holding casing pressure constant. Constant casing pressure is maintained through choke manipulation.

Maintain constant casing pressure until kill fluid reaches the bit. If there is gas remaining in the annulus after the first circulation and shut-in pressures are not equal, maintaining constant casing pressure could allow another kick to enter the hole. To avoid another possible kick in this situation, use a precalculated drillpipe pressure reduction schedule. The calculated circulating pressures should closely agree to the actual pressures. If they do not agree, use the calculated schedule which ensures the bottom-hole pressure is kept constant.

Then read and record circulating drillpipe or drillpipe/tubing pressure. Maintain constant circulating drillpipe or drillpipe/tubing pressure until system is displaced.

Step 7: Shut-in and check for flow

Slowly decrease pump rate while holding casing pressure constant by closing the choke. As the pump rate decreases, the drillpipe or drillpipe/tubing circulating pressure will decrease. In order to keep bottom-hole pressure constant, the casing pressure must be held constant. Once the pumps are off and the well is shut-in, allow pressures to stabilize.

- The SIDPP = SICP should **= 0 psi**, WELL IS DEAD. Proceed to Step (8): Circulate and Condition Fluid.
- If SIDPP < SICP, this indicates there is still some contaminant in the annulus or another kick has been taken.
- Line up choke manifold to trip tank. Open choke slowly to monitor flow.
- If it bleeds off, the well is dead.
- If it flows continuously, it is not dead. Recalculate new kill fluid density using the new SIDPP and repeat second circulation procedure starting with Step (7).
- Confirm well is dead by cracking open the choke; well should not flow.
- If mud was lost to the formation during the kill procedure, it is possible that residual flow from the well could be due to ballooning.

CLEAR RIG FLOOR OF PERSONNEL BEFORE OPENING BOPE AS A SMALL VOLUME OF GAS MAY BE TRAPPED BETWEEN THE PREVENTER AND CHOKE LINE.
THE SUDDEN VENTING OF THE TRAPPED GAS MAY BLOW BUSHINGS OUT OF ROTARY TABLE.

Step 8: Circulate and condition fluid

After the BOPs are opened, circulate the kill fluid and condition it to the tripping properties. A suitable trip margin may be necessary in order to trip out of the hole.

Wellbore Deviation: Hydrostatic pressure is calculated on True Vertical Depth (TVD) not on Measured Depth (MD). Friction pressure is measured by MD. If a linear pressure decline is used in a highly deviated well, the pressures could be too high and may result in formation fracture and mud loss downhole. A kill sheet should be used which takes deviation into account.

Tapered Drillstring: The BHA is not considered a tapered string due to the short length. However, if there is a long length of a second size drillpipe, any drillpipe pressure reduction schedules should take into account that there is not a linear increase in friction pressure from top to bottom.

Swabbed Kick: If the kick was swabbed in, the first circulation with the existing mud weight should kill the well since there is no need to increase the mud weight.

Driller's method action plan

The following Driller's Method Action plan has been developed by expanding Driller's Kill sheets with additional information necessary to safely kill a well. Beyond normal kill information, our Driller's Method Action Plan contains information to review process for bumping the float, holding a prejob safety meeting, as well as recap of the Driller's Method along with information need while circulating.

A review of the Driller's Method Action Plan begins with reviewing information within the Prerecorded Information Sheet and calculated volumes. This information is transferred to the Driller's Method Action Plan to ensure all appropriate information is available within this action plan.

ECD considerations

Note: If ECD with KMW at Bit ≥ LOT/FIT
Select Lower SCR or
Call Well Control Company

Fig. 2.14 Impact of ECD of KMW on LOT/FIT.

A comparison of KMW at TD and KMW at bit depth (if bit is off bottom), is conducted. Next, the ECD selected KMW at SCR is compared to the LOT/FIT. If the ECD of KMW at selected SCR is deemed to be within limits, an analysis of migration rate and MGS is conducted (Fig. 2.14).

A review of influx processing rate is analyzed. The migration rate is determined from information entered within the Prerecorded Information Sheet. If the migration rate shows a fast-rising migration rate (4000 ft./h), the Driller's Method would most likely be the preferred method to kill the well. From this information, an estimation of the Maximum Casing Pressure when gas reaches the surface is conducted. If the Maximum CP ≥ 80% of Annular Preventer rating, close appropriate set of pipe rams to minimize chances of equipment failure.

The final influx processing rate analysis is the determination of the minimum diameter of MGS to ensure adequate processing and separation of OMW/gas when gas reaches surface. The main control for processing rate is the diameter of the MGS, so a minimum diameter is calculated based upon KMW and SCR. If processing rate requires a larger diameter than MGS, thought should be given to slowing down the SCR and replacing MGS with rental, if possible.

The next analysis will be a review of surface systems needed to process KMW and receive OMW from well during circulation. This section is separated into Mud Cap barite needs (for bits off bottom). These volumes are followed by the Open Hole volumes with total barite needed and volume gained while mixing barite.

Finally, volumes as predetermined within the Prerecorded Information Sheet and Volume Sheet are transferred to the Action Plan for future review.

Bumping the float

Bumping Float Procedure

- Record $SIDP_{initial}$, $SICP_{initial}$ & Pit Gain. Slowly pump (10 spm)
- DP will start increasing over time, followed by gradual increase in CP
- **Stop pumps** after CP increases (allow no more than 100 psi rise)
- Subtract the $SICP_{inital}$ from $SICP_{final}$. = Trapped Pressure (ΔP)
- Subtract Trapped Pressure (ΔP) from $SIDDP_{final}$ for **SIDPP**.

Fig. 2.15 Review of bumping the float.

Since bumping the float is required to determine an accurate SIDPP and used to develop accurate KMW, a quick review of the bumping the float process is included for review (Fig. 2.15).

Prejob safety meeting

PreJob Safety Meeting

- Ensure all H_2S monitors are engaged and operating, especially in pit area.
- Degasser to remain on throughout circulation
- MGS mud leg to be filled with Kill Mud Weight (KMW) prior to initiation of circulation
- Valves downstream of choke are to be verified as open
- Isolate chokes and manipulate each prior to circulating
- Personnel (notifications made by loudspeaker)
  - All personnel to be on-station during circulation
- Driller
  - Re-zero pit level gain/losses and strokes
  - Records Total Stks, Pit Levels, DP & CP
  - Monitors and records Active System Volume and record every 10 min.
- AD to perform continuous walkabout for leaks
- Shale Shaker staff to monitor returns. Changes to be reported
- Mud Co. Rep
  - MW_{in} / MW_{out} every 15 min. or until asked to change
  - Provide Barite + Mud and/or BaseOil addition rate (bbl/hr)
  - Use Active pit system, if possible.
  - Any and all mud transfers to be announced. Driller to record.
- Choke Operator records Time, Stks, DP, CP, Ck Position, Pits and remarks
- Bring pumps up to speed slowly (0-10, 10-20, 20-30)
- Reset strokes when surface volume pumped

Fig. 2.16 Prejob safety meeting.

With volumes, KMW, influx processing complete, a quick look recap of the information required to review within a Prejob Safety Meeting are displayed. This comprehensive recap was developed as an outline and must be amended for individual rig and personnel requirements. The Prejob Safety Meeting recap should be used as a reminder to perform specific duties prior to initializing kill operations (Fig. 2.16).

First circulation of Driller's method

The next section contains specific information which can be used during the well kill operations, such as the SCR, circulating CP, and KMW. Information contained for performing the first circulation is provided by the following recap (Fig. 2.17).

Driller's Method Recap

First Circulation

1. Stage pumps up to SCR, while holding CP constant.
2. Read and record ICP:
3. Maintain constant SCR until Influx is circulated out.
4. Stage pumps down to off while holding CP constant.
5. If SICP>SIDPP, repeat steps 1-4.
6. Using SIDPP, determine KMW
7. Monitor well while mixing KMW

Fig. 2.17 Driller's method recap for first circulation.

Second circulation of Driller's method

The second series of information provides a reminder of what is to occur during the second circulation of the Driller's Method including SCR, circulating CP, Volume, and 1.5 times complete circulating volume. Information contained for performing the first circulation is provided by the following recap (Fig. 2.18).

Driller's Method Recap
Second Circulation
1. Route KMW for Kill Operations
2. Stage pumps up to SCR, while holding CP constant.
3. Hold CP constant until KMW reaches bit.
4. Hold DP constant until KMW reaches surface
   MWin=MWout for three consecutive readings over 15 minutes or 1.5 times total circulation volume/

Fig. 2.18 Driller's method recap for second circulation.

Choke position

The following information is included to remind choke operator of operational range of choke at SCR and complications if the choke does not operate correctly. During the first circulation of the Driller's Method, if the pumps are staged to the SCR and the choke is opened greater than ½ of the full open position on the remote choke panel, it is possible the bit has become partially plugged. It is probable that the choke has been opened too far and the BHP has dropped to a point in which another kick has been taken. The best course of action will be to shut down and shut-in the well. It is advisable to select a lower SCR and repeat staging pumps up to speed. Record ICP at lower SCR and observe if choke position is less than ½ of fully open (Fig. 2.19).

Note: Choke Position Should Be Approximately ½ Open (or less) at SCR.
- If the Choke Position is At or Near Full Open at SCR, Shut-Down and Shut-In. (Probable Partially Plugged Bit or Choke).
- Select lower SCR and Use Back-up Choke.
- Attempt to Circulate Again

Fig. 2.19 Driller's method recap for first circulation.

Change circulating rate

Attached is an equation to determine new Final Circulation Pressure, if the SCR is sped up during the circulations. Under normal kill operations and after the influx has been circulated out, discussions may center on speeding up the kill rate in order to displace the hole with KMW as quickly as possible. Crews must ensure KMW at newer SCR can be properly mixed before initiation of any changes to the SCR (Fig. 2.20).

To Change Rate: $FCP_{new} = (SCR_{new}/SCR_{old})^2 \times FCP$

Fig. 2.20 Driller's method recap for second circulation.

Common problems during well control

If problems occur during well kill operations, be prepared to cease pumping operations. If influx is in annulus and SICP continues to rise, suggest switching to Volumetric Control Method, until circulation can be reestablished.

1. **Gas blows out the mud leg on the MGS:**
   - **a.** Cease pumping and shut-in well.
   - **b.** Recalculate pressures using a slower pump rate and attempt to circulate after filling the MGS mud leg with mud and attempt to circulate out kick.
   - **c.** If the same problem occurs, use the Lube and Bleed Method to remove gas from the wellbore. Circulation can then be resumed.
2. **If the choke line freezes downstream of the choke**
   - **a.** Switch to back-up choke.
   - **b.** Cease pumping and shut-in the well. Use Volumetric Method to safely bring gas to the surface (bleed off mud using the manual choke) while thawing line.
3. **Partial or Total lost Circulation**
   - **a.** Slow pump rate to see if it has any effect on circulation.
   - **b.** Shut-in well and mix LCM pill to pump through BHA and bit. Use as heavy a concentration as possible for the BHA in use.
   - **c.** If LCM and a slower pump rate does not help, a different type of LCM may be needed such as a reactant plug.
   - **d.** Consideration should be given to contacting a well control specialist with well control company and mud specialist with mud company.

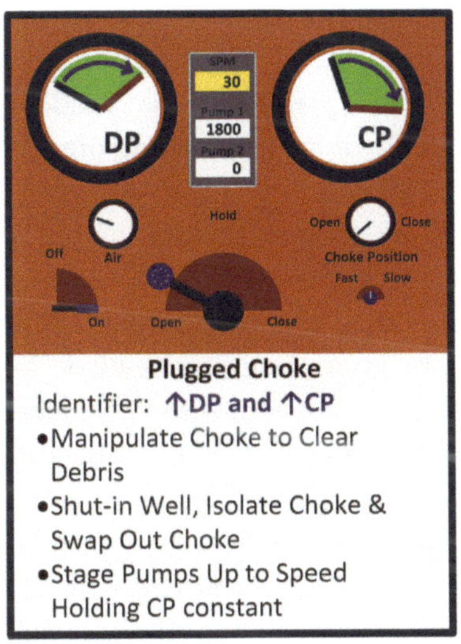

Fig. 2.21 Plugged choke.

4. **Plugged choke**—Can occur at any time; therefore, choke opening values and DP and CP must be continually recorded. If the casing pressure starts rising followed by rise in drillpipe pressure, plugging of the choke may be occurring. This may happen quickly or gradually (Fig. 2.21).

   Gradual Plugging: For gradual plugging while circulating, small opening manipulations are made to keep BHP constant. If pressures continue to rise as choke is opened, the well is to be secured by stopping pumps and shutting the well in. The current choke is to be isolated by closing isolation valves, while a secondary choke is lined up and readied for service. The pumps are staged up to speed holding previously recorded CP before plugging occurred.

Complete Plugging: For complete plugging while circulating, large opening choke manipulations will be unable to keep BHP constant. If pressures suddenly rise with the choke continually opened. The well is to be secured by stopping pumps and shutting the well in. The current choke is to be isolated by closing isolation valves, while a secondary choke is lined up and readied for service. The pumps are staged up to speed holding previously recorded CP before plugging occurred.

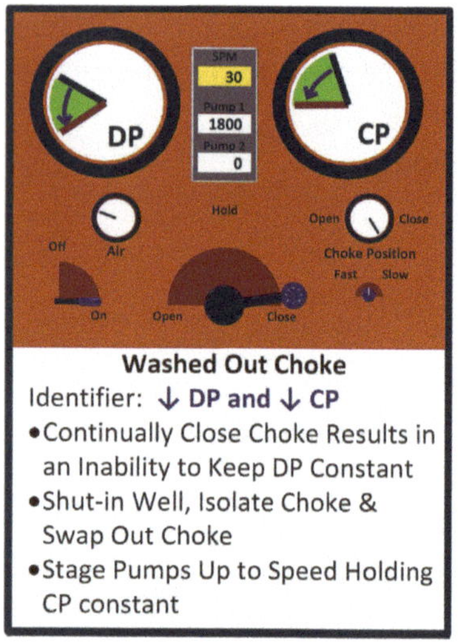

Fig. 2.22 Washed out choke.

5. **Washed out choke**—Can occur at any time. During circulation, small closing manipulations are made to keep BHP constant. If pressures continue to drop as choke is closed, the well is to be secured by stopping pumps and shutting the well in. The current choke is to be isolated by closing isolation valves, while a secondary choke is lined up and readied for service. The pumps are staged up to speed holding current casing pressure constant (Fig. 2.22).

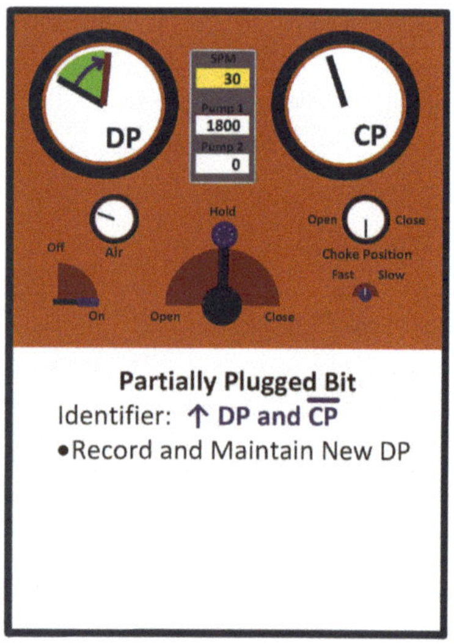

Fig. 2.23 Partially plugged bit.

6. **Partially plugged bit**—Can occur at any time and can be expected any time pump is shut off. Drillpipe pressure will suddenly increase without any increase in casing pressure (Fig. 2.23).

At ICP (KMW has not reached the bit). If pumps are staged up to speed holding casing pressure constant, record actual ICP. Recalculate step-down chart as needed. Do not open choke to achieve desired calculated ICP, as this action will cause BHP to become low enough to initiate additional kicks.

At FCP (KMW has reached the bit). If kill mud weight has reached the bit, record new drillpipe pressure and maintain this pressure throughout circulation.

Fig. 2.24 Washed out bit.

7. **Bit wash out**—Can occur at any time. Drillpipe pressure will suddenly decrease without any increase in casing pressure (Fig. 2.24).

   At ICP (KMW has not reached the bit). If pumps are staged up to speed holding casing pressure constant, record actual ICP. Recalculate step-down chart as needed. Do not close choke to achieve desired calculated ICP, as this action will cause BHP to increase and may fracture casing seat.

   At FCP (KMW has reached the bit). If kill mud weight has reached the bit, record new drillpipe pressure and maintain this pressure throughout circulation.

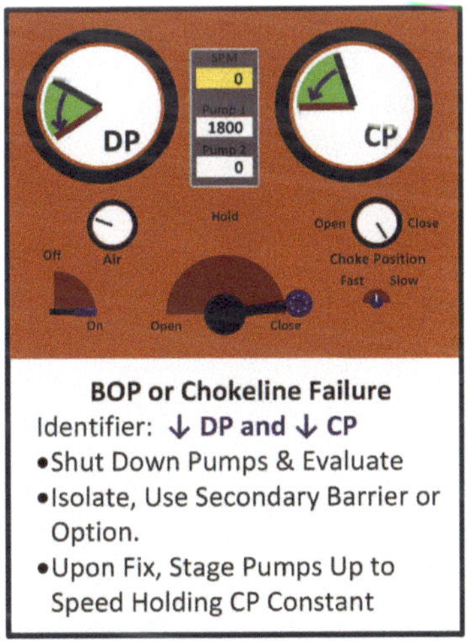

Fig. 2.25 BOP/choke line failure.

8. **BOP or choke line failure**—Can occur at any time. Drillpipe pressure and casing will decrease without changes in SPR (Fig. 2.25).

   Shut down pumps immediately and shut-in well. Investigate equipment failure, isolate and repair BOP/Choke Line. Use secondary barrier (i.e., BOP) as needed.

   Upon completing repairs, stage pups up to speed holding CP constant.

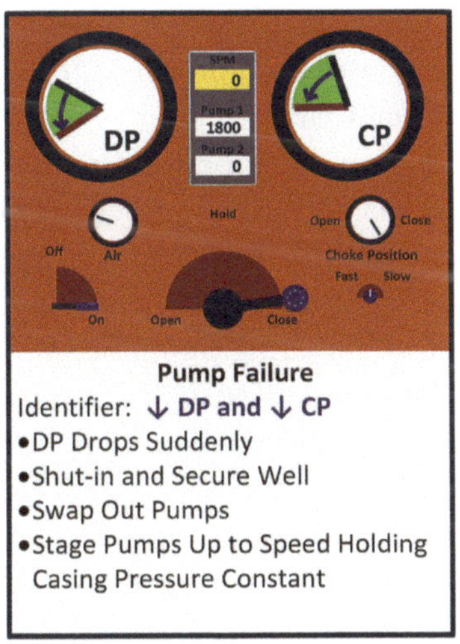

Fig. 2.26 Pump failure.

9. **Pump failure**—Can occur at any time. Pump pressure cavitation will cause Kelly/top drive hose to vibrate. Drillpipe pressure and casing will decrease without changes in SPR. If severe, Kelly hose will vibrate as pump(s) start failing (Fig. 2.26).

Shut down pumps immediately and shut-in well. Isolate defective pump and line up replacement pump.

Stage pumps up to speed holding casing pressure constant.

Conclusion of well control operations

Upon successfully circulating KMW throughout the wellbore before the pumps are shut off, the following checklist is comprised of the steps necessary to maintain control of the well while stopping the pumps. Once the well has been secured with the pumps off, perform detailed flow check of returns, before opening the BOPs. Prior to opening the BOPs, ensure the rig floor is cleared of operational personnel. Once personnel have been cleared, open the BOPs. There always remains a chance of trapped pressure below the element, which may violently expand and push particulate matter and fluid from the well. If personnel are within close proximity, this violent expansion may cause significant injuries.

After opening the BOPs, allow time for the fluids to stabilize. Perform an extended flow check to ensure the well is completely dead (Figs. 2.27).

1. Slowly Turn Pumps Off and Attempt to Hold CP Constant
2. Route Chokeline returns to Trip Tank and isolate MGS. Open choke and bleed pressure. Flow should cease, if not Shut-in Well.
3. Route standpipe returns to Trip Tank and bleed pressure.
4. Perform flow check at Trip Tank
5. Clear Rig Floor and Open BOPs
6. Perform Flow Check

Fig. 2.27 Conclusion of well control operations.

Wait and weight method

Wait and weight commentary

The Wait and Weight method for well control allows simultaneously circulating out the influx while introducing kill weight mud. These simultaneous operations require a computation derived step-down chart and requires the ability to efficiently mix KWM to desired density. The Wait and Weight Method requires closely monitoring DP and CP throughout the circulation to effectively maintain constant BHP.

The success of this method largely depends on each crew member performing their assigned tasks correctly and reporting any deviations which may occur. This is especially true for the mud mixing crews. For this method to be successful, KMW must be mixed thoroughly and at a constant rate equal to the slow circulating rate. If possible, it is preferential to weight up the hole volume via surface pits before initiating pumping. With sufficient surface volume of KMW, this negates the need for mixing KMW at a constant rate. If the mud is mixed light or heavy, variations in densities pumped downhole may result in unwanted choke manipulations and varying BHP. In turn, these variations may cause additional kicks to enter the well during circulation (Fig. 2.28).

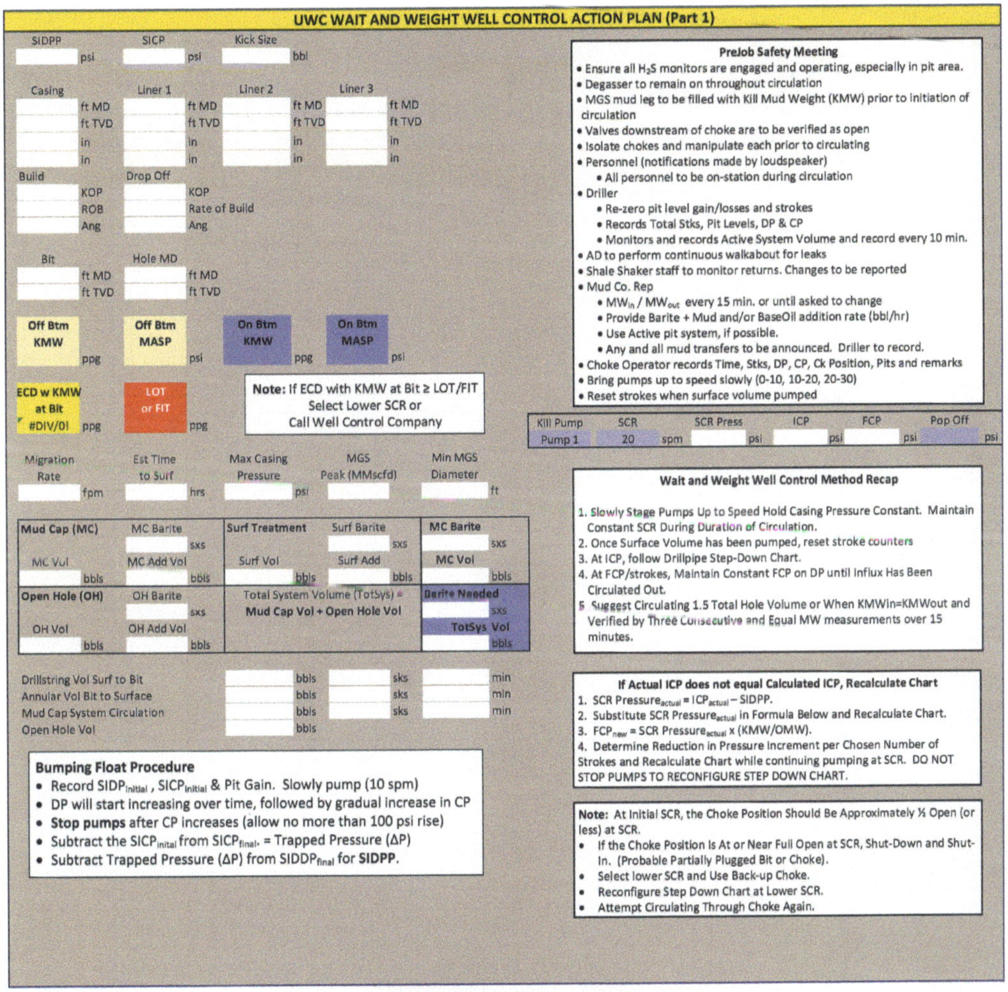

UWC WAIT AND WEIGHT WELL CONTROL ACTION PLAN (Part 1)

SIDPP psi | SICP psi | Kick Size bbl

Casing: ft MD, ft TVD, in, in | Liner 1: ft MD, ft TVD, in, in | Liner 2: ft MD, ft TVD, in, in | Liner 3: ft MD, ft TVD, in, in

Build: KOP, ROB, Ang | Drop Off: KOP, Rate of Build, Ang

Bit: ft MD, ft TVD | Hole MD: ft MD, ft TVD

Off Btm KMW ppg | Off Btm MASP psi | On Btm KMW ppg | On Btm MASP psi

ECD w KMW at Bit #DIV/0! ppg | LOT or FIT ppg

Note: If ECD with KMW at Bit ≥ LOT/FIT Select Lower SCR or Call Well Control Company

Migration Rate fpm | Est Time to Surf hrs | Max Casing Pressure psi | MGS Peak (MMscfd) | Min MGS Diameter ft

Mud Cap (MC): MC Vol bbls | MC Barite sxs | MC Add Vol bbls | Surf Treatment: Surf Vol bbls | Surf Barite sxs | Surf Add bbls | MC Barite sxs | MC Vol bbls

Open Hole (OH): OH Vol bbls | OH Barite sxs | OH Add Vol bbls | Total System Volume (TotSys) = Mud Cap Vol + Open Hole Vol | Barite Needed sxs | TotSys Vol bbls

Drillstring Vol Surf to Bit — bbls — sks — min
Annular Vol Bit to Surface — bbls — sks — min
Mud Cap System Circulation — bbls — sks — min
Open Hole Vol — bbls

Bumping Float Procedure

- Record $SIDP_{Initial}$, $SICP_{Initial}$ & Pit Gain. Slowly pump (10 spm)
- DP will start increasing over time, followed by gradual increase in CP
- **Stop pumps** after CP increases (allow no more than 100 psi rise)
- Subtract the $SICP_{Initial}$ from $SICP_{final}$ = Trapped Pressure (ΔP)
- Subtract Trapped Pressure (ΔP) from $SIDDP_{final}$ for **SIDPP**.

PreJob Safety Meeting

- Ensure all H_2S monitors are engaged and operating, especially in pit area.
- Degasser to remain on throughout circulation
- MGS mud leg to be filled with Kill Mud Weight (KMW) prior to initiation of circulation
- Valves downstream of choke are to be verified as open
- Isolate chokes and manipulate each prior to circulating
- Personnel (notifications made by loudspeaker)
  - All personnel to be on-station during circulation
- Driller
  - Re-zero pit level gain/losses and strokes
  - Records Total Stks, Pit Levels, DP & CP
  - Monitors and records Active System Volume and record every 10 min.
- AD to perform continuous walkabout for leaks
- Shale Shaker staff to monitor returns. Changes to be reported
- Mud Co. Rep
  - MW_{in} / MW_{out} every 15 min. or until asked to change
  - Provide Barite + Mud and/or BaseOil addition rate (bbl/hr)
  - Use Active pit system, if possible.
  - Any and all mud transfers to be announced. Driller to record.
- Choke Operator records Time, Stks, DP, CP, Ck Position, Pits and remarks
- Bring pumps up to speed slowly (0-10, 10-20, 20-30)
- Reset strokes when surface volume pumped

Kill Pump	SCR	SCR Press	ICP	FCP	Pop Off
Pump 1	20 spm	psi	psi	psi	psi

Wait and Weight Well Control Method Recap

1. Slowly Stage Pumps Up to Speed Hold Casing Pressure Constant. Maintain Constant SCR During Duration of Circulation.
2. Once Surface Volume has been pumped, reset stroke counters
3. At ICP, follow Drillpipe Step-Down Chart.
4. At FCP/strokes, Maintain Constant FCP on DP until Influx Has Been Circulated Out.
5. Suggest Circulating 1.5 Total Hole Volume or When KMWin=KMWout and Verified by Three Consecutive and Equal MW measurements over 15 minutes.

If Actual ICP does not equal Calculated ICP, Recalculate Chart

1. SCR $Pressure_{actual}$ = ICP_{actual} − SIDPP.
2. Substitute SCR $Pressure_{actual}$ in Formula Below and Recalculate Chart.
3. FCP_{new} = SCR $Pressure_{actual}$ x (KMW/OMW).
4. Determine Reduction in Pressure Increment per Chosen Number of Strokes and Recalculate Chart while continuing pumping at SCR. DO NOT STOP PUMPS TO RECONFIGURE STEP DOWN CHART.

Note: At Initial SCR, the Choke Position Should Be Approximately ½ Open (or less) at SCR.

- If the Choke Position Is At or Near Full Open at SCR, Shut-Down and Shut-In. (Probable Partially Plugged Bit or Choke).
- Select lower SCR and Use Back-up Choke.
- Reconfigure Step Down Chart at Lower SCR.
- Attempt Circulating Through Choke Again.

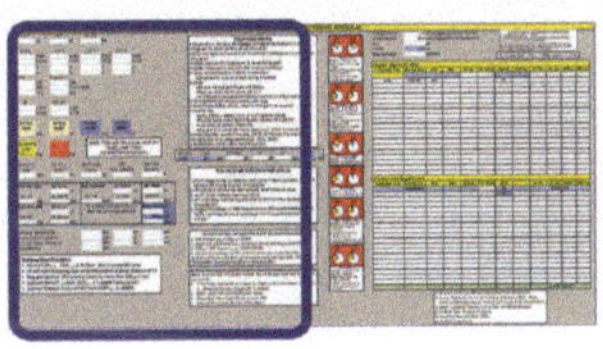

Fig. 2.28 UWC wait and weight well control action plan (Parts 1 and 2).

Continued

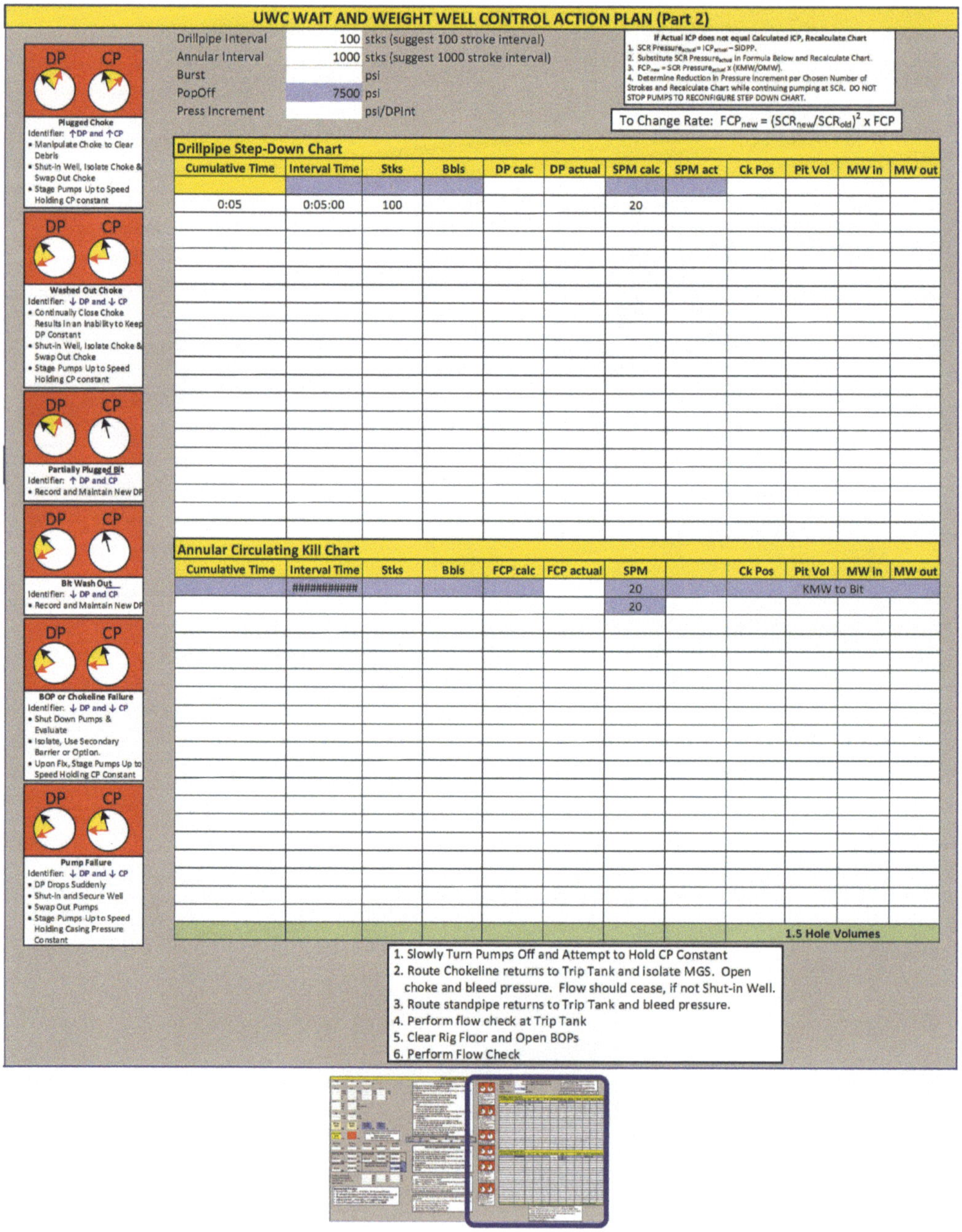

Fig. 2.28, cont'd

The Wait and Weight Method is a constant bottom-hole method comprised of one circulation, getting influx out of the well while displacing with kill weight fluid (KWM).

The Wait and Weight Method of well control has several advantages over other forms of well control.

- Although more complex, will circulate influx out quicker.
- Provides maximum amount of safety by increasing fluid density faster.
- Once kill fluid has reached the bottom-hole assembly, will require minimum choke manipulations.

Kill Weight Fluid (KWM) is pumped to displace influx from wellbore while maintaining a constant bottom-hole pressure equal to or slightly greater than the formation pressure. The constant bottom-hole pressure will prevent additional influx from entering the well while the kick is being circulated out. By displacing the influx from the wellbore, the safety of the crew and Rig unit will be maximized. At the end of the circulation, the well should be dead.

Step 1: Detection of the kick and proper shut-in

Upon noticing an immediate indicator of a kick, the well should be shut-in, and flow stopped. Use the correct shut-in procedure for operations being performed. The Driller is responsible for shutting-in the well using the correct procedure.

Step 2: Record stabilized pressures (bumping the float)

Once the kick has been detected and the well is shut-in, allow drillpipe or drillpipe/tubing pressure and casing pressure to stabilize before recording.

The choke manifold valves should be configured properly. Open the choke line valve at the preventer stack. Open the appropriate choke manifold valve to read the shut-in casing pressure.

Examine the pit level measuring device and estimate the volume gained during the kick. Verify this increase in volume with the Driller.

If the drillpipe or drillpipe/tubing does not have a drillpipe float, record the Shut-In Pressure (SIDPP).

If the drillpipe or drillpipe/tubing does contain a float, use appropriate bumping float procedure. Record pressure "lull" to estimate Kill Weight Fluid (KWM) for circulation.

Keep in mind if the influx is gas, migration will take place and seen as a gradual increase in casing pressure.

Step 3: Prejob safety meeting

Hold a short prejob safety meeting (PJSM) and ensure all questions are answered. Effective communication is the **KEY** to successful well control operations.

Fig. 2.29 When in doubt, ask person-in-charge.

Eliminate all sources of ignition to prevent fires (Fig. 2.29).

Designate the Rig unit and surrounding areas as **NO SMOKING AREA**.

Designate safe meeting area up wind of Rig site (Muster Area). This area should contain auxiliary safety equipment such as fire extinguisher and all available first aid kits.

Review each individual's responsibility during the well control operation:

Person-in-charge

Will be located on the rig floor during the kill operations and will advise the choke operator when necessary during the circulation. For high-noise locations or locations without ability to electronically, review the hand signals with driller and crew to ensure there is no miscommunication of given directions.

Stress "well control" is a TEAM effort and everyone's performance of their assigned responsibilities is the only guarantee of safe well control operations.

Choke operator

Normally assigned by the senior drilling contractor representative. He should be experienced at operating the choke in a well kill situation. His responsibility will be to open and close the choke according to the directions given by Person-in-Charge. He is to keep the casing pressure constant as the pump is brought up to speed.

Driller

Will remain at his console to operate the rig pump and reciprocate the drillstring. Will coordinate with choke operator during pump start up and shut down as well as keeping pump rate constant. Will monitor pumps, pit levels, and his crew.

Driller's responsibility is to SLOWLY bring the pumps up to speed and to maintain a constant pump rate. He is to communicate when the pumps are up to speed and alert the Person-in-Charge of any changes.

Stress if the pump rate varies, well control cannot be maintained.

Mud engineer

Works with pit crew to maintain mud at correct weight and viscosity. A second mud engineer may be required for 24-h operations.

Derrickman

Maintains pumps and constant fluid density throughout circulation process. Be on alert for pump stroke changes indicating problems.

Stress well control can only be maintained if fluid density is constant.

Floorhand #1

Maintains accumulator and BOPE equipment. Visually inspects all valves, lines, and BOP equipment for leakage periodically throughout circulation to ensure pressure integrity is maximized (≈every 10 min).

Stress if leaks are detected, immediately notify Driller.

Alternate personnel

Monitors emergency communications (i.e., field radio) to relay important messages. In the event of an evacuation, his responsibility is immediately notifying proper personnel.

Stress job importance and need to be alert for changing conditions.

Finally, the meeting should conclude with a review of the first circulation of the Driller's Method.

Ensure and stress the following important points (Fig. 2.30):

Fig. 2.30 When in doubt, ask person-in-charge.

(a) Casing pressure will increase (if influx contains gas) as the influx reaches the surface.

(b) The amount of circulating returns should increase as the gas expands.

(c) Noise levels of escaping gas will become so high verbal communication will become impossible. Review hand signals.

(d) Pit levels should start decreasing once the influx reaches the surface.

(e) When in doubt, ask questions.

Step 4: Circulation

Slowly bring pumps up to Kill pump rate (if possible) holding casing pressure constant. Constant casing pressure is maintained through choke manipulation. Read and record circulating drillpipe/tubing pressure and ensure predetermined schedule is followed until kill weight fluid reaches the bit.

At a stabilized Kill pump rate, record initial circulating drillpipe or drillpipe/tubing pressure.

VERIFY CHOKE MOVEMENT. BEFORE BEGINNING FIRST CIRCULATION, ENSURE THE CHOKE MANIFOLD IS CORRECTLY LINED UP AND THE CHOKE CAN BE EASILY OPERATED (EITHER MANUALLY OR HYDRAULICALLY). IF TWO CHOKES ARE PRESENT, VERIFY BOTH CHOKES CAN BE MANIPULATED

MARK CASING PRESSURE GAUGE. BEFORE BEGINNING FIRST CIRCULATION, MARK THE CASING PRESSURE GAUGE.
THIS WILL PRESENT A VISUAL REMINDER OF THE CASING PRESSURE THROUGHOUT THE CIRCULATION. REMARK EVERY 10 MINUTES.

Maintain drillpipe schedule as per calculated kill sheet. Once kill weight has reached bottom-hole assembly, read and record circulating drillpipe or drillpipe/tubing pressure.

Maintain constant final circulating drillpipe or drillpipe/tubing pressure until influx has been circulated out at a constant kill pump rate.

Step 5: Shut-in and check for flow

Slowly decrease pump rate while holding casing pressure constant by closing the choke. As the pump rate decreases, the drillpipe or drillpipe/tubing circulating pressure will decrease. In order to keep bottom-hole pressure constant, the casing pressure must be held constant. Once the pumps are off and the well is shut-in, allow pressures to stabilize.

- If SIDPP = SICP **= 0 psi**, WELL IS DEAD.
- If SIDPP < SICP, this indicates there is still some contaminant in the annulus or another kick has been taken. Recalculate new kill fluid density using the new SIDPP and repeat circulation procedure starting with Step (4) and new kill weight mud.
- Confirm well is dead by cracking open the choke; well should not flow.

CLEAR RIG FLOOR OF PERSONNEL BEFORE OPENING BOPE AS A SMALL VOLUME OF GAS MAY BE TRAPPED BETWEEN THE PREVENTER AND CHOKE LINE.
THE SUDDEN VENTING OF THE TRAPPED GAS MAY BLOW BUSHINGS OUT OF ROTARY TABLE.

Wait and weight method action plan

The following Wait and Weight Method Action plan has been developed by expanding Wait and Weight Kill sheets with additional information necessary to safely kill a well. Beyond normal kill information, our Wait and Weight Method Action Plan contains information to review process for bumping the float, holding a prejob safety meeting, as well as recap of the Wait and Weight Method along with information needed while circulating.

A review of the Wait and Weight Method Action Plan begins with reviewing information within the Prerecorded Information Sheet and calculated volumes. This information is transferred to the Wait and Weight Method Action Plan to ensure all appropriate information is available within this action plan.

ECD considerations

Note: If ECD with KMW at Bit ≥ LOT/FIT
Select Lower SCR or
Call Well Control Company

Fig. 2.31 Impact of ECD of KMW on LOT/FIT.

A comparison of KMW at TD and KMW at bit depth (if bit is off bottom), is conducted. Next, the ECD selected KMW at SCR is compared to the LOT/FIT. If the ECD of KMW at selected SCR is deemed to be within limits, an analysis of migration rate and MGS is conducted (Fig. 2.31).

A review of influx processing rate is analyzed. The migration rate is determined from information entered within the Prerecorded Information sheet. If the migration rate shows a slow rising migration rate (4000 ft/h), the Wait and Weight Method would most likely be the preferred method to kill the well. From this information, an estimation of the Maximum Casing Pressure when gas reaches the surface is conducted. If the Maximum CP ≥ 80% of Annular Preventer rating, close appropriate set of pipe rams to minimize chances of equipment failure.

The final influx processing rate analysis is the determination of the minimum diameter of MGS to ensure adequate processing and separation of OMW/gas when gas reaches surface. The main control for processing rate is the diameter of the MGS, so a minimum diameter is calculated based upon KMW and SCR. If processing rate requires a larger diameter than MGS, thought should be given to slowing down the SCR and replacing MGS with rental, if possible.

The next analysis will be a review of surface systems needed to process KMW and receive OMW from well during circulation. This section is separated into Mud Cap barite needs (for when the bit is off bottom). These volumes are followed by the Open Hole volumes with total barite needed and volume gained while mixing barite.

Finally, volumes as predetermined within the Prerecorded Information Sheet and Volume Sheet are transferred to the Action Plan for future review.

Bumping the float

Bumping Float Procedure

- Record $SIDP_{initial}$, $SICP_{initial}$ & Pit Gain. Slowly pump (10 spm)
- DP will start increasing over time, followed by gradual increase in CP
- **Stop pumps** after CP increases (allow no more than 100 psi rise)
- Subtract the $SICP_{inital}$ from $SICP_{final}$. = Trapped Pressure (ΔP)
- Subtract Trapped Pressure (ΔP) from $SIDDP_{final}$ for **SIDPP**.

Fig. 2.32 Review of bumping the float.

Since bumping the float is required to determine an accurate SIDPP and used to develop accurate KMW, a quick review of the bumping the float process is included for review (Fig. 2.32).

Prejob safety meeting

PreJob Safety Meeting

- Ensure all H_2S monitors are engaged and operating, especially in pit area.
- Degasser to remain on throughout circulation
- MGS mud leg to be filled with Kill Mud Weight (KMW) prior to initiation of circulation
- Valves downstream of choke are to be verified as open
- Isolate chokes and manipulate each prior to circulating
- Personnel (notifications made by loudspeaker)
    - All personnel to be on-station during circulation
- Driller
    - Re-zero pit level gain/losses and strokes
    - Records Total Stks, Pit Levels, DP & CP
    - Monitors and records Active System Volume and record every 10 min.
- AD to perform continuous walkabout for leaks
- Shale Shaker staff to monitor returns. Changes to be reported
- Mud Co. Rep
    - MW_{in} / MW_{out} every 15 min. or until asked to change
    - Provide Barite + Mud and/or BaseOil addition rate (bbl/hr)
    - Use Active pit system, if possible.
    - Any and all mud transfers to be announced. Driller to record.
- Choke Operator records Time, Stks, DP, CP, Ck Position, Pits and remarks
- Bring pumps up to speed slowly (0-10, 10-20, 20-30)
- Reset strokes when surface volume pumped

Fig. 2.33 Prejob safety meeting.

With volumes, KMW, influx processing complete, a quick look recap of the information required to review within a Prejob Safety Meeting are displayed. This comprehensive recap was developed as an outline and must be amended for individual rig and personnel requirements. The Prejob Safety Meeting recap should be used as a reminder to perform specific duties prior to initializing kill operations (Fig. 2.33).

Kill circulation using wait and weight method

The next section contains specific information which can be used during the well kill operations, such as the SCR, circulating CP, and KMW. Information contained for performing the first circulation is provided by the following recap (Fig. 2.34).

Wait and Weight Well Control Method Recap

1. Slowly Stage Pumps Up to Speed Hold Casing Pressure Constant. Maintain Constant SCR During Duration of Circulation.
2. Once Surface Volume has been pumped, reset stroke counters.
3. At ICP, follow Drillpipe Step-Down Chart.
4. At FCP/strokes, Maintain Constant FCP on DP until Influx Has Been Circulated Out.
5. Suggest Circulating 1.5 Total Hole Volume or When KMW_{in}=KMW_{out} and Verified by Three Consecutive and Equal MW measurements over 15 minutes.

Fig. 2.34 Wait and weight method recap for kill circulation.

If ICP does not equal calculated value

During the step-down chart, if the when the pumps are brought up to speed and the actual ICP does not equal the calculated value, the entire step-down chart must be recalculated. **DO NOT OPEN/CLOSE CHOKE TO ACHIEVED CALCULATED ICP**, as this will cause the well kill pressures to be inaccurate and potentially allow kick to occur during entire kill process.

Using the actual ICP, subtract the SIDPP to calculate actual SCR Pressure. With the SCR calculated from actual pressure readings, recalculate step-down chart using actual SCR Pressure x KMW/OMW. DO NOT STOP PUMPS TO RECONFIGURE STEP-DOWN CHART, as the Pressure Reduction schedule can be easily altered (Fig. 2.35).

If Actual ICP does not equal Calculated ICP, Recalculate Chart

1. SCR $Pressure_{actual}$ = ICP_{actual} – SIDPP.
2. Substitute SCR $Pressure_{actual}$ in Formula Below and Recalculate Chart.
3. FCP_{new} = SCR $Pressure_{actual}$ x (KMW/OMW).
4. Determine Reduction in Pressure Increment per Chosen Number of Strokes and Recalculate Chart while continuing pumping at SCR. DO NOT STOP PUMPS TO RECONFIGURE STEP DOWN CHART.

Fig. 2.35 What to do if actual ICP does not match calculated ICP.

Choke position

The following information is included to remind choke operator of operational range of choke at SCR and complications if the choke does not operate correctly. One of the most critical aspects of well control is establishing the correct Initial Circulating Pressure when pumps are staged to the slow circulating rate. For SCR, choke positions are normally less than ½ of the full choke position as measured at the remote choke panel. If the choke positions are greater than 1/2 full open, there is a probability the bit has been partially plugged and choke was opened in order to achieve the theoretical ICP. The best course of action would be to shut down, shut-in, and begin again. If the actual ICP is greater than the theoretical ICP and the choke position is ½ full open or less, then simply recalculate the step-down chart (Fig. 2.36).

Note: At Initial SCR, the Choke Position Should Be Approximately ½ Open (or less) at SCR.

- If the Choke Position Is At or Near Full Open at SCR, Shut-Down and Shut-In. (Probable Partially Plugged Bit or Choke).
- Select lower SCR and Use Back-up Choke.
- Reconfigure Step Down Chart at Lower SCR.
- Attempt Circulating Through Choke Again.

Fig. 2.36 Choke position during SCR.

Change circulating rate

Attached is an equation to determine new Final Circulation Pressure, if the SCR is sped up during the circulation. Under normal kill operations and after the influx has been circulated out, discussions may center on speeding up the kill rate in order to displace the hole with KMW as quickly as possible. Crews must ensure KMW at newer SCR can be properly mixed before initiation of any changes to the SCR (Fig. 2.37).

To Change Rate: $FCP_{new} = (SCR_{new}/SCR_{old})^2 \times FCP$

Fig. 2.37 Equation for determining FCP at revised SCR.

Common problems during well control

If problems occur during well kill operations, be prepared to cease pumping operations. If influx is in annulus and SICP continues to rise, suggest switching to Volumetric Control Method, until circulation can be reestablished.

1. **Gas blows out the mud leg on the MGS:**
   a. Cease pumping and shut-in well.
   b. Recalculate pressures using a slower pump rate and attempt to circulate after filling the MGS mud leg with mud and attempt to circulate out kick.
   c. If the same problem occurs, use the Lube and Bleed Method to remove gas from the wellbore. Circulation can then be resumed.
2. **If the choke line freezes downstream of the choke**
   a. Switch to back-up choke.
   b. Cease pumping and shut-in the well. Use Volumetric Method to safely bring gas to the surface (bleed off mud using the manual choke) while thawing line.
3. **Partial or Total lost Circulation**
   a. Slow pump rate to see if it has any effect on circulation.
   b. Shut-in well and mix LCM pill to pump through BHA and bit. Use as heavy a concentration as possible for the BHA in use.
   c. If LCM and a slower pump rate do not help, a different type of LCM may be needed such as a reactant plug.
   d. Consideration should be given to contacting a well control specialist with well control company and mud specialist with mud company.

Fig. 2.38 Plugged choke.

4. **Plugged choke**—Can occur at any time, therefore choke opening values and DP and CP must be continually recorded. If the casing pressure starts rising followed by rise in drillpipe pressure, plugging of the choke may be occurring. This may happen quickly or gradually (Fig. 2.38).

   Gradual Plugging: For gradual plugging while circulating, small opening manipulations are made to keep BHP constant. If pressures continue to rise as choke is opened, the well is to be secured by stopping pumps and shutting the well in. The current choke is to be isolated by closing isolation valves, while a secondary choke is lined up and readied for service. The pumps are staged up to speed holding previously recorded CP before plugging occurred.

   Complete Plugging: For complete plugging while circulating, large opening choke manipulations will be unable to keep BHP constant. If pressures suddenly rise with the choke continually opened. The well is to be secured by stopping pumps and shutting the well in. The current choke is to be isolated by closing isolation valves, while a secondary choke is lined up and readied for service. The pumps are staged up to speed holding previously recorded CP before plugging occurred.

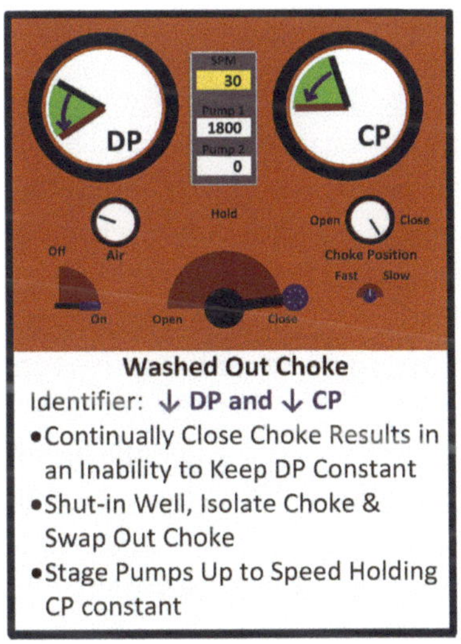

Fig. 2.39 Washed out choke.

5. **Washed out choke**—Can occur at any time. During circulation, small closing manipulations are made to keep BHP constant. If pressures continue to drop as choke is closed, the well is to be secured by stopping pumps and shutting the well in. The current choke is to be isolated by closing isolation valves, while a secondary choke is lined up and readied for service. The pumps are staged up to speed holding current casing pressure constant (Fig. 2.39).

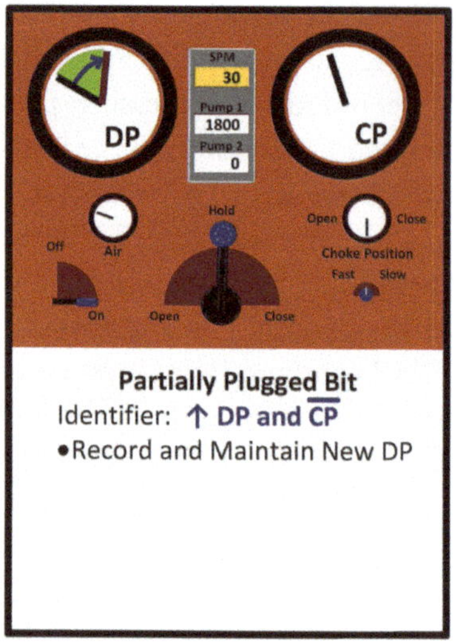

Fig. 2.40 Partially plugged bit.

6. **Partially plugged bit**—Can occur at any time and can be expected any time pump is shut off. Drillpipe pressure will suddenly increase without any increase in casing pressure (Fig. 2.40).

   At ICP (KMW has not reached the bit). If pumps are staged up to speed holding casing pressure constant, record actual ICP. Recalculate step-down chart as needed. Do not open choke to achieve desired calculated ICP, as this action will cause BHP to become low enough to initiate additional kicks.

   At FCP (KMW has reached the bit). If kill mud weight has reached the bit, record new drillpipe pressure and maintain this pressure throughout circulation.

Fig. 2.41 Washed out bit.

7. **Bit wash out**—Can occur at any time. Drillpipe pressure will suddenly decrease without any increase in casing pressure (Fig. 2.41).

   At ICP (KMW has not reached the bit). If pumps are staged up to speed holding casing pressure constant, record actual ICP. Recalculate step-down chart as needed. Do not close choke to achieve desired calculated ICP, as this action will cause BHP to increase and may fracture casing seat.

   At FCP (KMW has reached the bit). If kill mud weight has reached the bit, record new drillpipe pressure and maintain this pressure throughout circulation.

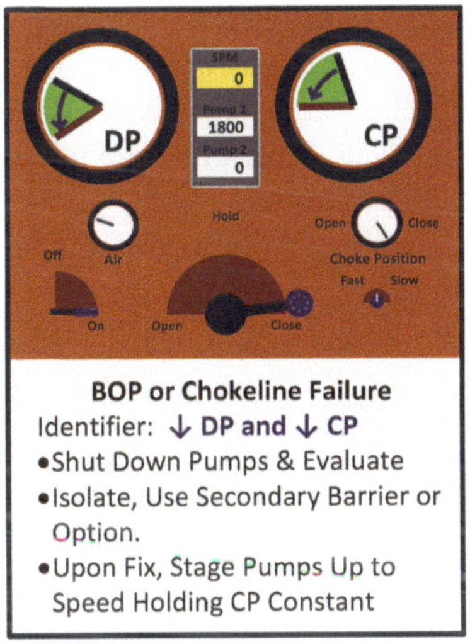

Fig. 2.42 BOP/choke line failure.

8. **BOP or choke line failure**—Can occur at any time. Drillpipe pressure and casing will decrease without changes in SPR (Fig. 2.42).

   Shut down pumps immediately and shut-in well. Investigate equipment failure, isolate and repair BOP/Choke Line. Use secondary barrier (i.e., BOP) as needed.

   Upon completing repairs, stage pups up to speed holding CP constant.

Fig. 2.43 Pump failure.

9. **Pump failure**—Can occur at any time. Pump pressure cavitation will cause Kelly/top drive hose to vibrate. Drillpipe pressure and casing will decrease without changes in SPR. If severe, Kelly hose will vibrate as pump(s) start failing (Fig. 2.43)

Shut down pumps immediately and shut-in well. Isolate defective pump and line up replacement pump.

Stage pumps up to speed holding casing pressure constant.

Conclusion of well control operations

Upon successfully circulating KMW throughout the wellbore and before the pumps are shut off, the following checklist comprises steps necessary to maintain control of the well while stopping the pumps. Once the well has been secured with the pumps off and before opening the BOPs, perform detailed flow check of returns. Prior to opening the BOPs, ensure the rig floor is cleared of operational personnel. Once personnel have been cleared, open the BOPs. There always remains a chance of trapped pressure below the element/ram, which may violently expand and push particulate matter and fluid from the well. If personnel are within close proximity, this violent expansion may cause significant injuries.

After opening the BOPs, allow time for the fluids to stabilize. Perform an extended flow check to ensure the well is completely dead (Fig. 2.44).

1. Slowly Turn Pumps Off and Attempt to Hold CP Constant
2. Route Chokeline returns to Trip Tank and isolate MGS. Open choke and bleed pressure. Flow should cease, if not Shut-in Well.
3. Route standpipe returns to Trip Tank and bleed pressure.
4. Perform flow check at Trip Tank
5. Clear Rig Floor and Open BOPs
6. Perform Flow Check

Fig. 2.44 Conclusion of well control operations.

Comparison of Driller's method and wait and weight method

	Driller's method	Wait and weight method
Calculations	Calculate ICP and step-down chart for second circulation	Calculate KMW, ICP, FCP, step-down chart
Implementation	Can begin circulation almost immediately using OMW. First circulation only removes influx. Second circulation circulates in kill weight mud.	More complex due to removing the influx and killing the well in one circulation.
Length of Initial well shut	No waiting time to increase mud weight. If gas migration may be a problem this is the better method. Helps prevent pipe sticking by resuming circulation quicker.	Must wait until KWM is mixed
Mixing kill weight mud	Does not need to be completed until after first circulation.	Must be mixed and pumped on the fly. Not all rigs are capable of this. May not get good yield on gels and chemicals when mud is mixed quickly.
Fluid Volumes	Easier to keep track of volumes. Good method for rigs with limited pit volume.	More complicated to keep track of pit volumes.
Duration of Kill	Two circulations	One circulation

Pressure at Casing Shoe	Slightly higher pressure at shoe due to original mud in annulus below kick.	Lowest pressure due to kill mud in annulus when kick reaches shoe. This is generally not a significant difference.
Pump Pressure Changes	Keep casing pressure constant while making or observing any pump pressure changes.	Recalculate pressure reduction schedule if pressure changes due to partial plugging occur while drillstring is being displaced with kill weight mud. This must be done in real time.

Casing pressures

The typical casing pressure profile for a well killed with water-based mud has a peak pressure when the gas reaches the surface. In oil-based muds, the highest casing pressure is most often seen when the well is shut-in and stabilizes. The gas may still be in solution in the mud until it goes through the choke thus the pressure when it gets to the surface is not as high.

Bullheading method

Bullheading commentary

Bullheading Kill Operations during the drilling phase are used when conventional Driller's Methods or Wait and Weight Methods cannot be utilized. Conventional well control methods may not be applicable due to close proximity of the public or in cases where the reservoirs contain high H_2S concentrations. It is not prudent to bring H_2S gas to the surface. Large kick sizes may exceed capabilities of surface equipment or jeopardize casing seats.

Bullheading Kill Operations attempt to effectively push the influx back into the same formation it came from. If the well is underbalanced kill weight mud will not only displace the influx but also the original mud. If there is a weak zone in the well, bullheading may push fluids into that zone and not the influx zone and therefore may not kill the well. If the casing seat is broken down by bullheading at too high a pressure, an underground blowout (UGBO) could occur. Bullheading is generally not the recommended initial well control method for a drilling well with open hole unless circumstances dictate this method.

Bullheading is more often used on completed wells to kill them prior to a workover or plug and abandon operation. In this case bullheading will push the fluids back into the perforated zone.

A variety of methods can be used to bullhead, but in all bullheading cases, activities must be planned to minimize compromising the integrity of the casing. All outer annuli should be monitored by separate casing gauges if possible. If outside annuli become pressurized, slower rates and pump pressures may be necessary (Fig. 2.45, 1 and 2).

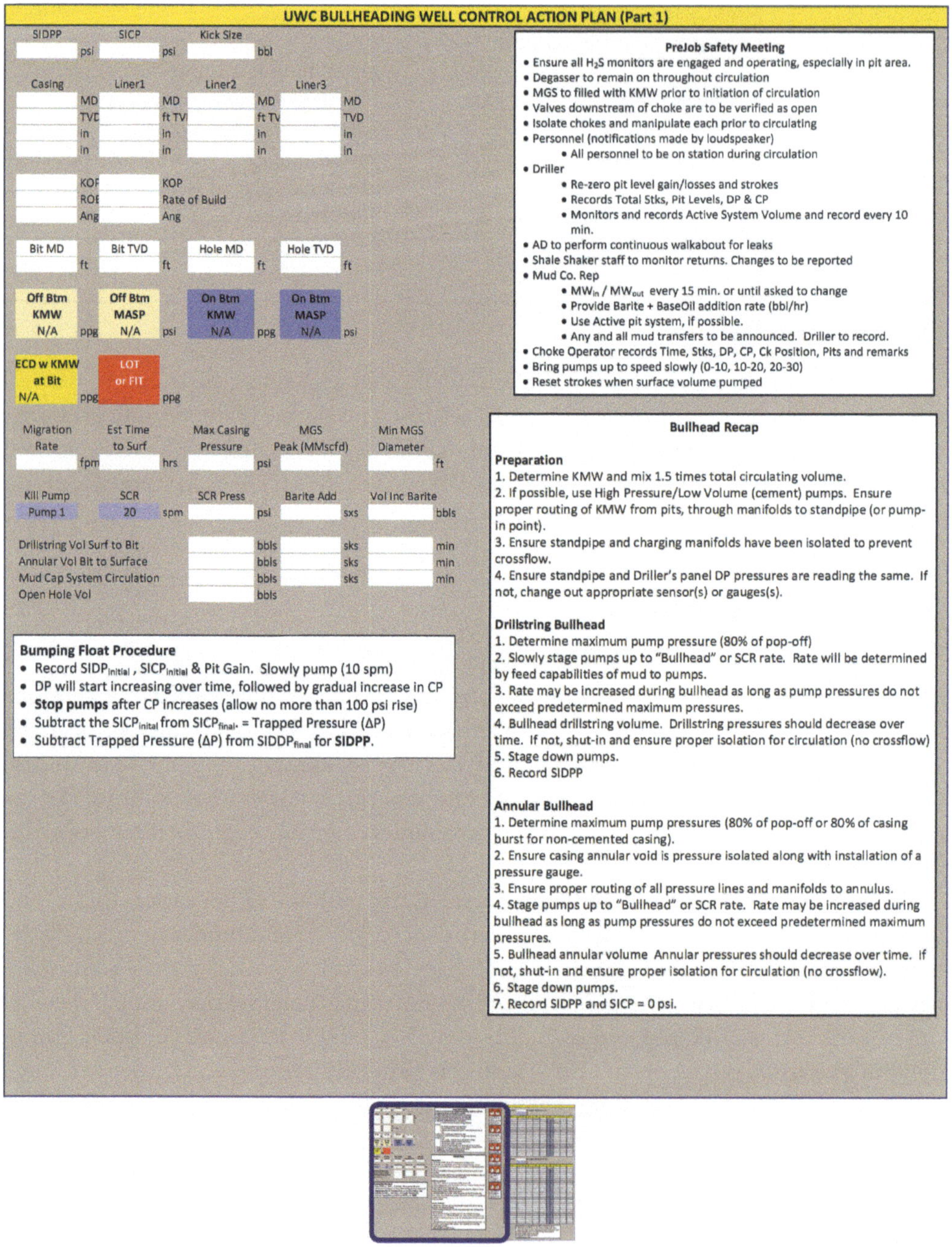

UWC BULLHEADING WELL CONTROL ACTION PLAN (Part 1)

SIDPP psi | SICP psi | Kick Size bbl

Casing: MD, TVD, in, in | Liner1: MD, ft TVD, in, in | Liner2: MD, ft TVD, in, in | Liner3: MD, TVD, in, in

KOP, ROB, Ang | KOP, Rate of Build, Ang

Bit MD ft | Bit TVD ft | Hole MD ft | Hole TVD ft

Off Btm KMW N/A ppg | Off Btm MASP N/A psi | On Btm KMW N/A ppg | On Btm MASP N/A psi

ECD w KMW at Bit N/A ppg | LOT or FIT ppg

Migration Rate fpm | Est Time to Surf hrs | Max Casing Pressure psi | MGS Peak (MMscfd) | Min MGS Diameter ft

Kill Pump: Pump 1 | SCR: 20 spm | SCR Press psi | Barite Add sxs | Vol Inc Barite bbls

Drillstring Vol Surf to Bit: bbls, sks, min
Annular Vol Bit to Surface: bbls, sks, min
Mud Cap System Circulation: bbls, sks, min
Open Hole Vol: bbls

Bumping Float Procedure

- Record $SIDP_{initial}$, $SICP_{initial}$ & Pit Gain. Slowly pump (10 spm)
- DP will start increasing over time, followed by gradual increase in CP
- **Stop pumps** after CP increases (allow no more than 100 psi rise)
- Subtract the $SICP_{inital}$ from $SICP_{final}$. = Trapped Pressure (ΔP)
- Subtract Trapped Pressure (ΔP) from $SIDDP_{final}$ for **SIDPP**.

PreJob Safety Meeting

- Ensure all H_2S monitors are engaged and operating, especially in pit area.
- Degasser to remain on throughout circulation
- MGS to filled with KMW prior to initiation of circulation
- Valves downstream of choke are to be verified as open
- Isolate chokes and manipulate each prior to circulating
- Personnel (notifications made by loudspeaker)
    - All personnel to be on station during circulation
- Driller
    - Re-zero pit level gain/losses and strokes
    - Records Total Stks, Pit Levels, DP & CP
    - Monitors and records Active System Volume and record every 10 min.
- AD to perform continuous walkabout for leaks
- Shale Shaker staff to monitor returns. Changes to be reported
- Mud Co. Rep
    - MW_{in} / MW_{out} every 15 min. or until asked to change
    - Provide Barite + BaseOil addition rate (bbl/hr)
    - Use Active pit system, if possible.
    - Any and all mud transfers to be announced. Driller to record.
- Choke Operator records Time, Stks, DP, CP, Ck Position, Pits and remarks
- Bring pumps up to speed slowly (0-10, 10-20, 20-30)
- Reset strokes when surface volume pumped

Bullhead Recap

Preparation

1. Determine KMW and mix 1.5 times total circulating volume.
2. If possible, use High Pressure/Low Volume (cement) pumps. Ensure proper routing of KMW from pits, through manifolds to standpipe (or pump-in point).
3. Ensure standpipe and charging manifolds have been isolated to prevent crossflow.
4. Ensure standpipe and Driller's panel DP pressures are reading the same. If not, change out appropriate sensor(s) or gauges(s).

Drillstring Bullhead

1. Determine maximum pump pressure (80% of pop-off)
2. Slowly stage pumps up to "Bullhead" or SCR rate. Rate will be determined by feed capabilities of mud to pumps.
3. Rate may be increased during bullhead as long as pump pressures do not exceed predetermined maximum pressures.
4. Bullhead drillstring volume. Drillstring pressures should decrease over time. If not, shut-in and ensure proper isolation for circulation (no crossflow)
5. Stage down pumps.
6. Record SIDPP

Annular Bullhead

1. Determine maximum pump pressures (80% of pop-off or 80% of casing burst for non-cemented casing).
2. Ensure casing annular void is pressure isolated along with installation of a pressure gauge.
3. Ensure proper routing of all pressure lines and manifolds to annulus.
4. Stage pumps up to "Bullhead" or SCR rate. Rate may be increased during bullhead as long as pump pressures do not exceed predetermined maximum pressures.
5. Bullhead annular volume Annular pressures should decrease over time. If not, shut-in and ensure proper isolation for circulation (no crossflow).
6. Stage down pumps.
7. Record SIDPP and SICP = 0 psi.

Fig. 2.45 UWC bullheading well control action plan (Parts 1 and 2).

continued

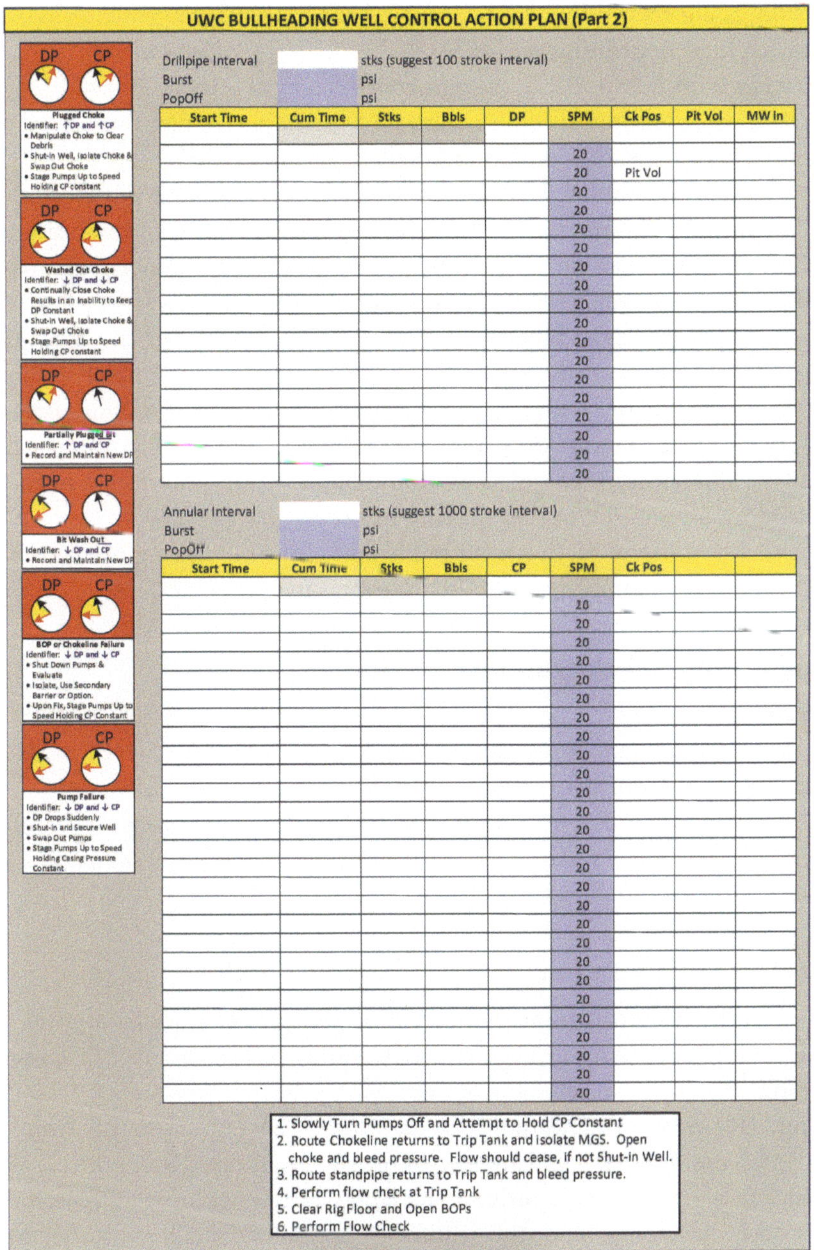

UWC BULLHEADING WELL CONTROL ACTION PLAN (Part 2)

DP CP

Plugged Choke
Identifier: ↑DP and ↑CP
• Manipulate Choke to Clear Debris
• Shut-In Well, Isolate Choke & Swap Out Choke
• Stage Pumps Up to Speed Holding CP constant

DP CP

Washed Out Choke
Identifier: ↓ DP and ↓ CP
• Continually Close Choke Results in an Inability to Keep DP Constant
• Shut-In Well, Isolate Choke & Swap Out Choke
• Stage Pumps Up to Speed Holding CP constant

DP CP

Partially Plugged Bit
Identifier: ↑ DP and CP
• Record and Maintain New DP

DP CP

Bit Wash Out
Identifier: ↓ DP and CP
• Record and Maintain New DP

DP CP

BOP or Chokeline Failure
Identifier: ↓ DP and ↓ CP
• Shut Down Pumps & Evaluate
• Isolate, Use Secondary Barrier or Option.
• Upon Fix, Stage Pumps Up to Speed Holding CP Constant

DP CP

Pump Failure
Identifier: ↓ DP and ↓ CP
• DP Drops Suddenly
• Shut-in and Secure Well
• Swap Out Pumps
• Stage Pumps Up to Speed Holding Casing Pressure Constant

Drillpipe Interval ____ stks (suggest 100 stroke interval)
Burst ____ psi
PopOff ____ psi

Start Time	Cum Time	Stks	Bbls	DP	SPM	Ck Pos	Pit Vol	MW In
					20			
					20	Pit Vol		
					20			
					20			
					20			
					20			
					20			
					20			
					20			
					20			
					20			
					20			
					20			
					20			
					20			
					20			
					20			
					20			

Annular Interval ____ stks (suggest 1000 stroke interval)
Burst ____ psi
PopOff ____ psi

Start Time	Cum Time	Stks	Bbls	CP	SPM	Ck Pos		
					20			
					20			
					20			
					20			
					20			
					20			
					20			
					20			
					20			
					20			
					20			
					20			
					20			
					20			
					20			
					20			
					20			
					20			
					20			
					20			
					20			
					20			
					20			
					20			
					20			
					20			
					20			

1. Slowly Turn Pumps Off and Attempt to Hold CP Constant
2. Route Chokeline returns to Trip Tank and isolate MGS. Open choke and bleed pressure. Flow should cease, if not Shut-in Well.
3. Route standpipe returns to Trip Tank and bleed pressure.
4. Perform flow check at Trip Tank
5. Clear Rig Floor and Open BOPs
6. Perform Flow Check

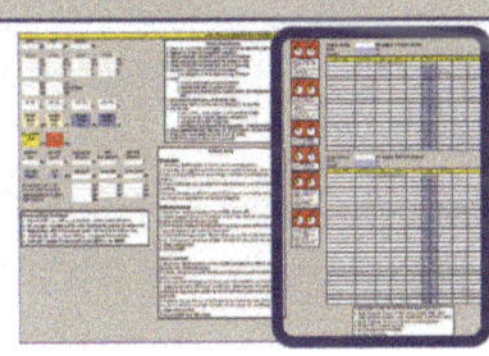

Fig. 2.45, cont'd

The bullheading method may be employed for the following situations:

- Wells with high concentrations of H_2S
- Older wells with questionable casing or wellhead integrity
- Producing wells without drillpipe/tubing (open hole completions)
- Live production/injection well initial well killing operations.

Experience has shown bullheading operations may vary due to equipment designs. As each drilling rig is a customized assembly of equipment, no one bullheading process can be developed for every rig. The following procedures are generalized, to highlight specific requirements needed for bullheading operations.

Isolation of pit volumes and fluid pumping routes may vary. For effective bullheading operations, it is preferable to isolate drillstring operations from annular volume operations. This isolation will be needed to monitor both drillstring and annulus pressures during operations. If proper isolation cannot be achieved, bullheading operations may be compromised. Therefore, after isolation of individual pump systems has been achieved, consideration for small-scale testing may be in order.

There are a number of methods for bullheading during drilling operations. Each of these bullheading operations may consist of several incremental steps.

Step bullheading kill operation

1. Bullheading ½ the drillstring, then bullheading ½ of the annular volume.
2. Stopping, evaluating, and monitoring well. If both SIDPP = SICP = 0 psi, bullheading operations may cease.
3. If SIDPP >0 psi or SICP >0 psi, additional bullheading operations may commence by bullheading remaining volume down drillstring followed by bullheading complete annular volume.

Full bullhead kill operation

1. Bullhead entire drillstring, followed by bullheading entire annular volume. This procedure requires bullheading even if the pressures reach zero before completion of bullhead.

Concurrent (Pancake) Bullheading Operations: These consist of simultaneously pumping down the drillstring and annulus at independent rates. Sometimes called "pancake bullheading," the goal is to displace the larger annular volume at the same effective rate as that of the drillstring. Since the drillstring consists of a smaller volume, the rates are determined developing a simple ratio by dividing annular volume by drillstring volume. This will develop a numerical ratio such as 5 to 1. For this example, the annulus will be bullheaded at 5 times the rate of the drillstring. Concurrent bullheading operations should only be conducted by experience personnel or with assistance by selected Well Control company personnel.

Concurrent (pancake) bullheading kill operations

1. Line up to pump simultaneously down drillstring and annulus utilizing two independent pump sources.
2. Ensure each pump system can be independently operated with isolated charging pumps and pits.
3. Ensure drillstring and casing pressure sensors can be independently isolated.
4. Ensure no possible cross-flow from one system to the other can occur by ensuring isolation valves are operating correctly. This may require testing before bullheading operations commence.
5. Stage pumps independently to desired rates using predetermined ratio.
6. Ensure maximum pump pressures do not exceed predetermined limits.
7. After fracturing pressure has been achieved, increase SCR as desired (normally limited to 100 spm or less).
8. Bullhead entire hole volume before shutting down pumps.
9. After bullhead, cease pumping by staging pumps down.
10. Perform desired bleed back into trip tanks, measure and monitor returns.
11. When SIDPP = SICP = 0 psi has been achieved, clear floor and open BOPs.
12. If fracturing has occurred, expect little or no fluids visible.
13. Initiate continuous hole filling operations. Normally, bullheading is followed by installation of isolation cement.

Concepts

When bullheading, the only controlling parameters are the pump rate and kill fluid density. Downhole pressures cannot be controlled via choke manipulation. Consequently, the friction pressures developed using bullheading method directly impact bottom-hole pressure.

When bullheading, the pump has to overcome surface and tubular friction, plus the degree of under balance which exists in the drillpipe/tubing. Furthermore, friction pressures in open hole, across perforations, gas compressibility, and pressures necessary to reinject formation fluids back into the formation will have to be taken into account. As a result, pump pressure will be significantly higher in a bullheading situation when compared to any other form of circulating well control method.

When bullheading, the hydrostatic pressure inside the drillpipe/tubing or annulus increases and the surface pressure decreases until the kill fluid displaces the formation fluids back into the perforations.

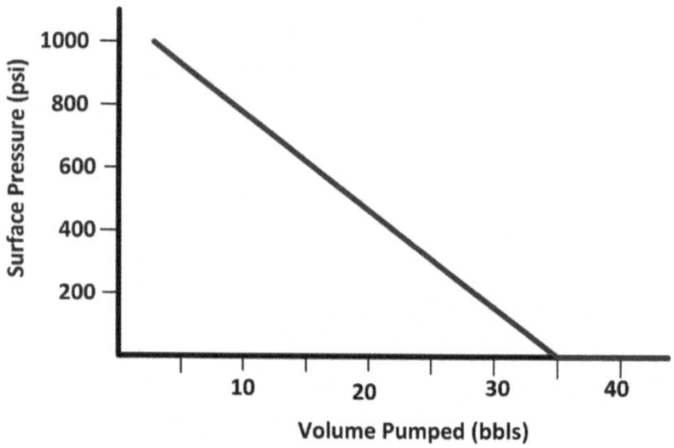

Fig. 2.46 Bullheading chart for tubing

Once all tubulars have been over-displaced, the pumps should be shut down. At this point, SITP/SIDPP and SICP should be 0 psi, if the kill weight density is sufficient to keep formation fluids from entering the wellbore. The bullheading chart (Fig. 2.46) is representative of a tubing bullhead. Any remaining surface pressure at the completion of the bullhead operation could be and indication of:

- **Trapped Pressure**—formation damage, partial plugging of perforations and/or other wellbore restrictions may lead to trapped pressure.
- **Under Balance**—the kill fluid was not dense enough to displace formation fluids from drillpipe/tubing and well.

General bullheading procedure for tubing or drillpipe

The following is a step-by-step procedure for the Bullheading Method of well control. This procedure is a general procedure and should be used as a guideline only. Specific bullheading procedures must be developed for each individual well based on unique wellbore parameters.

Step activity

1. Calculations
   - Determine the required kill fluid density (ppg) and volume (bbl) of required to bullhead.
   - Establish pump pressure limitations (psi).
   - Determine the required pump speed (SPM).
2. Pressure test all lines.
   a. Bring the pump up to the desired bullhead speed (do not exceed surface pressure limitation).
   b. Read and record the initial bullhead pressure.
   c. If pumping down the drillpipe/tubing, monitor annulus pressure throughout the procedure.
3. Monitor and record bullhead pressures and volumes during the procedure.

4. When the desired volume has been pumped, shut down the pump.
5. Read, record, and monitor shut-in surface pressure.

Step 1: Determining operating parameters

To ensure the safety of Rig personnel, various parameters should be reviewed before proceeding with utilization of bullheading techniques.

Required density and volume

If the fluid within the drillpipe/tubing is known (oil, water, etc.), the following formula can be used to determine the KF: Otherwise, utilize the most recent reservoir pressure data.

$$KMW_{ppg} = \text{Current}\, MW_{ppg} + \left(\frac{SIDPP_{psi}}{(0.052) \times (TVD_{ft})} \right) \tag{2.25}$$

where

KMWppg = Kill Mud Weight (ppg)
Current MW_{ppg} = Current MW in well (ppg)
SIDPPpsi = Stabilized Shut-in Drillpipe Pressure (psi)
0.052 = Conversion Constant (psi/ft./ppg)
TVD_{ft} = True Vertical Depth for well (ft)

If the tubing/drillpipe has a packer set in the casing, the volume needed to kill the well is the sum of the internal capacity of the tubing/drillpipe plus the casing capacity from the packer to the perforations. Volume needed is equal to the sum of the internal capacity of the drillpipe/tubing and the casing capacity between the production packer and the perforations. These capacity factors and volumes can be determined from the Prerecorded Data Sheet. The required bullhead volume is affected by the compressibility of the kick fluid, the deviation in the well, the downhole flow profile, the distance the kick has migrated, and the compressibility of the displacing (bullhead) fluid. These factors usually necessitate additional volumes (above theoretical) of bullhead fluid.

Surface bullhead pressure limitation

Prior to initiating the bullhead operation, a surface pump pressure limitation must be established. The most critical factor is the formation fracture pressure which can lead to lost circulation or formation damage. This will be site specific and determined by Reservoir Engineering community. The maximum allowable surface pressure when bullheading down drillpipe/tubing is the lesser of:

- Tubing burst rating (80%) (may be down rated due to corrosion and wear)
- Wellhead or tree rating (may be down rated due to corrosion and wear)
- BOPE rating (tubing side)
- Formation fracture pressure
- Pressure required to pump tubing out of packer or unseat the packer

These limitations are listed on the Prerecorded Data Sheet.

Another consideration in packer-type completions is the collapse rating of the casing. If there is a poor cement job, or pressure communication to the casing above the packer, bullheading could lead to collapse of the casing or a packer failure. When bullheading down the drillpipe/tubing, the casing should be filled and the pressure should be monitored to quickly detect casing collapse.

Bullheading down casing

The maximum surface pressure when bullheading down the casing will be the lesser of:

- Casing burst rating (80%) (may be down rated due to corrosion and wear)
- Wellhead or tree rating (may be down rated due to corrosion and wear)
- BOPE rating
- Formation fracture pressure
- Tubing Collapse Rating (if plugged and/or isolated)—in most cases drillpipe/tubing collapse rating is well above wellhead, BOPE, and casing burst and therefore neglected.

Fracturing the exposed formation

Experience within a production area is the best method of determining fracture limitation (Reservoir Engineering community). Formation damage, plugging, etc. will need to be considered before fracturing any formation.

DEVELOP A PRESSURE LIMITATION CHART TO ASSIST WITH BULLHEADING TECHNIQUES

A pressure limitation chart can be developed to assist in any bullheading operation.

The **maximum allowable surface pressure** (MASP) is the bottom-hole fracture pressure less the hydrostatic pressure of the kill mud weight. Plot at zero (0) strokes.

$$MASP_{psi} = Frac\ Press_{psi} - (0.052)(TVD_{ft})(KMW_{ppg}) \tag{2.26}$$

where

MASPpsi = Maximum Allowable Surface Pressure

Fracture $Press_{psi}$ = Fracture Pressure (psi)

0.052 = Conversion Constant (psi/ft./ppg)

TVD_{ft} = True Vertical Depth for well (ft)

MW_{ppg} = Kill Mud Weight (ppg)

For wells in which the production fluid density is not known, the following reference table can be used

- 0.010 ppg for gas (<10,000 ft)
- 0.015 ppg for gas (>10,000 ft)
- 0.7 ppg for oil (dependent upon API gravity)
- 8.33 ppg for freshwater
- 8.6 ppg for saltwater

- if API gravity is known use the following formula:

$$OMW_{ppg} = \left(\frac{141.5}{°API + 131.5}\right) \times 8.33_{ppg} \quad (2.27)$$

where

OMW_{ppg} = Mud or Fluid Weight in Well (ppg)
°API = Specific gravity of oil or liquid (dimensionless)
8.33_{ppg} = Weight of Freshwater (ppg)

The **maximum final surface pressure (SP_{max})** is the bottom-hole fracture pressure less the bullhead fluid hydrostatic pressure. This point is plotted at the strokes necessary to get bullhead fluid to bottom-hole or into the perforations. Mud or Fluid Weight in Well (ppg).

$$SP_{Max(psi)} = Frac\ Press_{psi} - (KMW_{ppg})(0.052)(TVD_{ft}) \quad (2.28)$$

where

$SP_{Max(psi)}$ = Maximum Final Surface Pressure (psi)
Frac $Press_{psi}$ = Fracture Pressure (psi)
KMW_{ppg} = Kill Mud Weight (ppg)
0.052 = Conversion Constant (psi/ft./ppg)
TVD_{ft} = True Vertical Depth (ft)

For a vertical well, a straight line drawn between this Initial MASP at 0 strokes and Max Final SP at displacement strokes defines the boundary of the "fracture sector." For a given number of strokes pumped, if the pressure is higher than the limit line, the formation may fracture. Establish a safety factor of 100 psi less than fracture zone for working limits (Fig. 2.47).

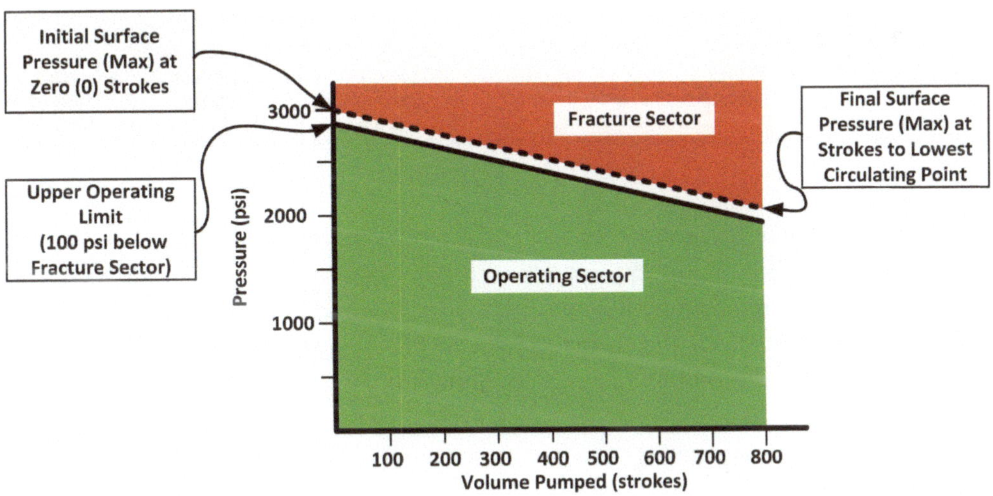

Fig. 2.47 Bullheading chart.

Required pump speed

The required pump speed for bullhead well control method is dependent upon:

- The type of kick fluid
- The type and density of kill fluid
- Gas migration rate
- The permeability of the exposed formation
- The completion equipment complexity

The type of kick fluid is also a factor which affects the pump pressure and corresponding pump rate. Viscous water-based fluids (which contain solids) create higher frictional pressures than clear brines. In addition, solids-laden fluids generate higher downhole injection pressures as they impact the perforations.

Saltwater and oil kicks create higher frictional pressures than gas. For most wells, friction pressure within the exposed formation will affect the pump pressure and rate used when bullheading. The friction pressure is largely dependent on permeability of formation (permeability is ability of formation to flow measured in cp). For completion and completion equipment, the shot density and tubular size are restrictions which can lead to higher surface pressures and lower pump rates. Any type of a restriction across the perforated interval will affect bullheading operations.

Step 2: Staging pumps up to speed

All of the lines should be pressure tested to at least the maximum surface pressure. A pressure gauge should be installed on the annulus and monitored throughout the job.

HOLD A SMALL AMOUNT OF PRESSURE ON THE BACKSIDE.
THIS INCREASES THE DRILLPIPE/TUBING BURST AND CASING COLLAPSE PRESSURE LIMITATIONS,
(FOR WORKOVERS, THIS MIGHT HELP OFFSET FORCES ACTING TO UNSET THE PACKER).
IT ALSO ENSURES THE ANNULUS IS FULL OF FLUID, WHICH WILL INSTANTLY TRANSMIT PRESSURE
SHOULD A DRILLPIPE/TUBING, CASING, OR PACKER LEAK DEVELOP

The pump should be brought up to the desired bullhead speed, and the initial bullhead pressure recorded. The pump operator should bring the pump on line **slowly**, while monitoring the pump pressure. The pump operator should be aware of the established surface pressure limitation, and in the process of bringing the pump up to the desired speed, the operator should not exceed this limitation.

Step 3: Follow pump chart

Monitor and record bullhead pressures and volumes during the procedure. The pump operator should ensure the maximum established surface pressure is not exceeded during the operation. Recording the pressures and volumes will help in recognizing trends and identifying any problems which may occur during the bullhead operation (Table 2.2).

Sample Table (Example Only)

Initial Shut-In drillpipe/tubing Pressure = 500 psi

Initial Bullheading Pressure = 2000 psi

Volume Pumped Observed Bullhead Maximum Allowable

Table 2.2 Bullheading table.

Volume pump (strokes)	Observed bullhead pressure (psi)	Maximum surface pressure (psi)
0	2000	2700
100	1938	2600
200	1875	2500
300	1813	2400
400	1750	2300
500	1688	2200
600	1625	2100
700	1563	2000
800	1500	1900
900	1800	1800

Step 4: Shut down pump and shut-in

When the desired volume has been pumped, the pumps should be shut down and the choke closed.

Step 5: Review pressure gauges

After the pump is shut down, the shut-in surface pressure should be read, recorded, and monitored. Any remaining surface pressure at this point can be an indication of:

- **Under Balance**—the kill fluid is not dense enough, or kick fluid still remains in the well.
- **Trapped Pressure**—the surface pressure can be lowered by bleeding off small increments of fluid until two consecutive pressure readings are recorded. Verification of the success of the bullhead operation, the well is indeed dead, is an important aspect of the procedure.

If the fracture gradient of the upper zone is exceeded, or the upper zone starts taking significant amounts of fluid, the success of the bullhead procedure may be limited. Often, a plugging agent (salt or gel pill) is used to bridge over the weak, uphole zone, and facilitate the kill procedure. Obviously, multiple zone completion wells are a significant concern when utilizing a bullheading procedure.

CHAPTER THREE

Nonroutine well control methods

Lubricate and bleed method well control

Lubricate and bleed commentary

The Lubricate and Bleed method is a well control technique which can be used when normal circulating methods cannot be employed and the influx is at the surface (underneath the Blowout Preventers or wellhead). For the Lubricate and Bleed method to be employed, the circulating system equipment will need to be altered from normal circulating system arrangements to kill arrangements via choke manifold. These revisions include installation of high-pressure, low-volume triplex pump system (skid or truck mounted) affixed to the kill line. The high-pressure, low-volume triplex pump system must include a graduated feed tank containing KMW.

The manual choke(s) is normally used in order to properly manipulate influx volumes from the well after KMW is pumped (lubricated) into the well. The manual choke should be isolated, operated, and greased before initiating any kill procedure. If possible, both choke and pump pressure gauges should be affixed near or at the manual choke to ensure the choke operator has an understanding of pressure thresholds to be encountered.

Situations when the Lubricate and Bleed Method can be employed are:

- On a live well prior to initiating workover operations
- On a well with conventional drillpipe/tubing string and gas at surface
- On a well without pipe in well following Volumetric Control
- "Packed Off" hole with kick below pack-off
- Plugged drillstring
- Pump failure
- Drillstring is out of the hole

The Lube and Bleed method does not maintain a constant bottom-hole pressure. Lubricate and Bleed cycles are used to maintain pressures within an acceptable range which will not fracture formations at the shoe or downhole and keep bottom-hole pressure above formation pressure.

The Lube and Bleed method does not maintain a constant bottom hole. Lubricate and Bleed cycles are used to maintain pressures within an acceptable range which will

Universal Well Control
https://doi.org/10.1016/B978-0-323-90584-8.00002-2

not fracture formations at the shoe or downhole. If the pump-in volume is undersized, over time the undersized bleed cycles may lead to the well becoming underbalanced. If underbalanced condition exists, multiple influxes may enter the well and compromise the well.

The Lubricate and Bleed Volume Method involves pumping a known volume of mud into the annulus through by means of a low-volume, high-pressure (LVHP) cement pump. Since the fluid is added into a "closed system," the hydrostatic pressure of the mud introduced increases wellbore pressure, by use of the choke, the surface pressure is reduced by an amount equal to the hydrostatic pressure of mud introduced into the well. In order not to fracture formations at the shoe (weakest point), a pressure limit is set and is not to be exceeded.

The injection/bleed cycles are repeated until the entire influx has been replaced with mud. At the conclusion of pumping operations, the well is secure but not necessarily dead. Mud caps are used to control wells, but must be displaced by homogeneous kill mud weights for long-term and operational well control.

There are basically three separate methods which can be used for lubrication and bleed. They can be separated into two categories: (A) NO losses downhole and (B) WITH losses downhole.

Lubrication and bleed method for NO losses downhole

Lubricate and Bleed for NO Losses Downhole Method is easily adaptable for a wide range of field applications, making Lubricate and Bleed for NO Losses Downhole Method the **most selected choice** for most lube and bleed solutions. This method can be accomplished in two ways:

1. The **Lube and Bleed Method for NO Losses Downhole (Variable Mud Volume)** (Most used field method). Initiate pumping of KMW into the well until Maximum Casing Pressure limit is achieved. Using this volume, the Target Reduced Pressure is determined and the well is bled. The uncontrolled volume allows pumping as much KMW as possible for each Lube and Bleed cycle resulting in less cycles needed to remove influx, stabilize, or kill the well.
2. The **Lube and Bleed Method for No Losses Downhole (Set Mud Volume)**: Predetermine the volume and hydrostatic pressure of controlled volume of KMW. Lubricate this exact volume of KMW into well, then perform bleed cycle to target reduced pressure. Using controlled volumes, for remediation of the well, a larger number of Lube and Bleed cycles may be needed. The Set Mud Volume may be used on wells with questionable cement jobs and deteriorated casing/wellheads.

If losses are occurring, the calculated hydrostatic pressure provided by a full fluid column will be compromised and the well may go underbalanced during this operation.

FOR NO FLUID LOSSES DOWNHOLE,
EITHER METHOD CAN BE SUCCESSFULLY USED TO REMOVE GAS FROM UNDERNEATH THE BOP OR WELLHEAD

THE LUBRICATE AND BLEED METHOD (VARIABLE MUD VOLUME OR SET MUD VOLUME)
CANNOT BE USED ON WELLS WITH ONGOING DOWNHOLE LOSSES

Lubrication and bleed method WITH losses downhole

3. **The lube and bleed method WITH losses downhole (pressure)** employs using surface pressure gauges to perform lube and bleed increments. Although this method can be used for situations where losses downhole exist, due to the precise nature of fluid and pumping measurement, use of the previously described methods for no losses is suggested.

All lube and bleed methods employ pumping fluid into well, allowing the fluid and gas to swap places and then bleeding influx gas from the well down to a calculated pressure. The lube and bleed processes are repeated until the gas has been removed from the well. Depending on fluid densities used, when influx has been removed from the well, the well may be shut-in with pressure (underbalanced) or successfully killed with no surface pressure.

LIMITATIONS FOR USING ANY LUBRICATE AND BLEED METHOD WILL BE DEPENDENT UPON SURFACE EQUIPMENT.
A CALIBRATED TRIP TANK MUST BE USED
TANK TO BE MANIFOLDED DOWNSTREAM OF THE CHOKE
TO ACCURATELY MEASURE THE AMOUNT OF FLUID BLED FROM WELL.

General checklist for lubricate and bleed method

Lubricate and Bleed Checklist

- ❑ Collect all relevant and pertinent well data
- ❑ Develop Detailed Wellbore Schematic with Volumes, etc.
- ❑ **Manual Choke** to be used for L&B. Service as necessary. If applicable, prepare secondary Manual Choke.
- ❑ A calibrated trip tank is to be used and manifolded downstream of the choke to accurately measure the amount of fluid bled from the well.
- ❑ Ensure CP gauge is available on Choke Manifold and easily view from Manual Choke. Optional, install Pump Pressure gauge on Choke Manifold.
- ❑ Visually inspect wellhead, flowlkines and equipment for leaks, alignment and overall condition.
- ❑ Install High Pressure / Low Volume pump for accurate Lube cycles. Ensure fluid transfer lines are installed and tested.
- ❑ Close and lock Blind Rams. Do not use Annular Preventer.
- ❑ For operations with workstring, close and lock appropriate Pipe Rams after ensuring rams will not close on tooljoints. Do not use Annular Preventer. Install Top Drive and lock in place with chains to prevent upward movement.
- ❑ Prepare the MGS for service. If applicable, fill mud leg with KMW.
- ❑ Note: As the influx is liberated through MGS, the transition from influx to mud will be difficult to observe at the MGS. Ensure MGS gauge(s) are operable. When pressure rises as mud is produced, route MGS to emergency by-pass line.

Fig. 3.1 Lubrication and bleed general checklist.

1. Collect as much well data as possible. Perform quality assurance of all data. Estimations may be needed if data are missing (Fig. 3.1).
2. Develop detailed well diagram listing as much information as possible including; KMW, Annular Volume and Gradient between Casing and Workstring (bbls and bbl/ft), Hole volume (suggest 1.5 times hole volume mud available for lube and bleed), Surface mud volumes, densities, and treating chemical stock.

Identify known or possible loss zone(s). (This will aid in determining which type of lubrication and bleed method should be used.) Determine maximum pressure to initiate formation fracturing. Record Leak-Off Test/Formation Integrity Test information.

Ensure adequate casing shoe integrity existed in order to perform Lube and Bleed. Determine maximum casing pressure limit using rule of thumb, Lessor of 80% Casing Burst or 80% of MASP. The 80% can be downrated due to condition of casing, wellhead, suspect shoe (previously squeezed), weak formations, or other criteria.

3. Manual choke to be used for Lubricate and Bleed methods as this choke provides more accurate control during lubrication and bleed cycles. With kill line closed, function test manual choke valve for smooth operation and service as necessary. If applicable, prepare secondary manual choke. Ensure the choke manifold and MGS have correct routing and have been readied for service. If applicable, mark

all valves either open or closed. Ensure Mud–Gas Separator has been displaced with KMW within the mud-leg prior to initiating operations.

4. A calibrated trip tank is to be used and manifolded downstream of choke to accurately measure the amount of fluid bled from well.
5. Ensure a well-lit and accurate casing pressure gauge is available on choke manifold. The casing pressure gauge should be located and positioned on the choke manifold to ensure it can be easily viewed from the manual choke(s). If possible, install optional pump pressure gauge from high-pressure/low-volume pump source on the choke manifold.
6. Visually inspect wellhead, flowlines and equipment or leaks, alignment, and overall condition.
7. Ensure fluid transfer lines are installed and tested. Install and properly position a high-pressure/low-volume (HPLV) pump for accurate lube cycles. HPLV units are designed to operate over a wide range of rates and pressures and offer easy control for small volume pumping. These units contain accurate calibrated mixing and holding tanks along with computer controls for heightened accuracy of pump volumes, densities, and pressures. Ensure mud transfer supply lines, pump lines, and manifolds easily assessable, installed, and tested.

IT IS NOT ADVISABLE TO USE RIG PUMPS FOR LUBE AND BLEED DUE TO INEFFICIENCIES OF THIS EQUIPMENT AT LOW SPM (0-10).

8. Close and lock Blind Rams. Do not use Annular Preventer. Install and test circulating line from LVHP pump to kill line of BOP stack drilling spool. Ensure high-pressure barriers to prevent unwanted pedestrian traffic have been installed with required signage. If BOPs are equipped with more than one circulating spool, suggest using uppermost kill line valve system for lube and bleed operations. Perform pressure test of all equipment as required to appropriate governing limits.
9. For operations with workstring, close and lock appropriate Pipe Rams after ensuring rams do not contain tooljoints. Do not use Annular Preventer. Install Top Drive and lock in place with chains to prevent upward movement. If workstring within derrick has been partially bowed, install elevators in an "inverted position" above the rotary. The elevators are to be located beneath drillpipe fish-neck (closest to rotary) and fasten (or chain down) at a 90-degree angle. The inverted elevators are used to prevent the drillstring from being pushed uphole, causing drillstring in the derrick to continue bending until it breaks or ruptures.
10. Prepare MGS for service. If applicable, fill mud leg with mud.
11. Note: As influx is liberated through MGS, the transition from influx to mud will be difficult to observe at MGS. Ensure MGS gauge(s) are operable. When pressure rises as mud is produced, route MGS to emergency by-pass line (Figs. 3.2 and 3.3).

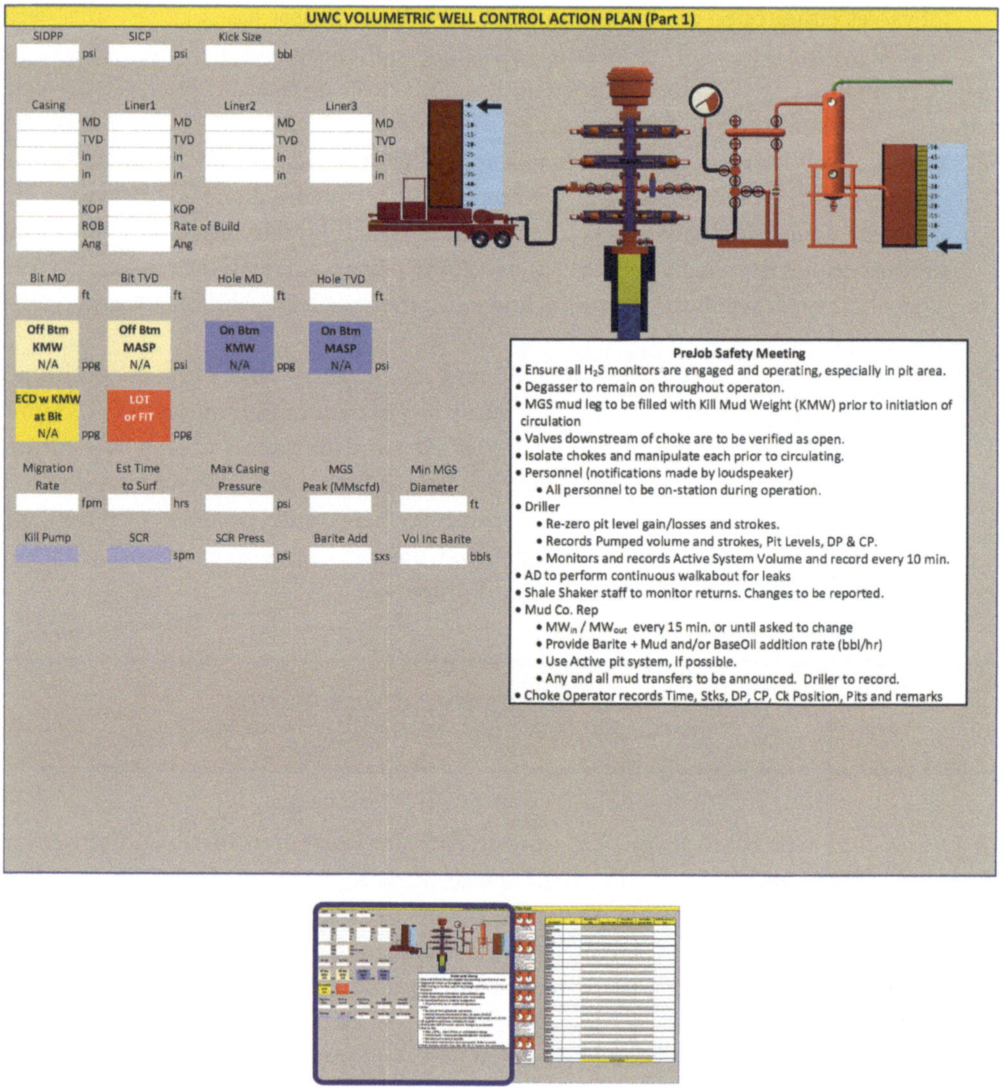

Fig. 3.2 Lube and Bleed Action Plan (Part 1).

Continued

UWC VOLUMETRIC WELL CONTROL ACTION PLAN (Part 2)

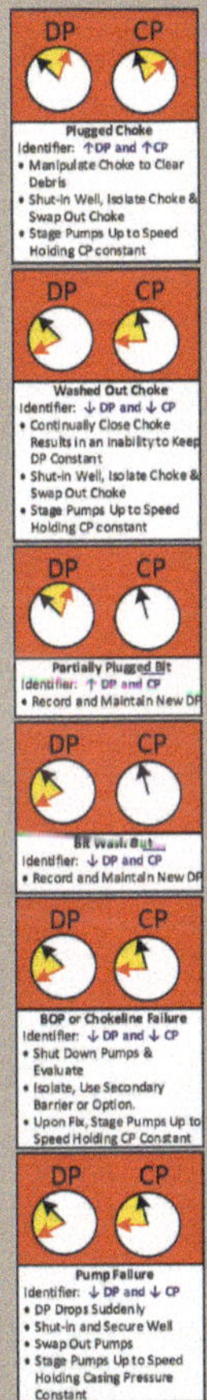

Description	SICP	Planned Vol (bbl)	Actual Vol (bbl)	Actual Bled Pressure (psi)	Cumulative Volume (bbls)	BHP Overbalance (psi)
Shut-In						
Safety Factor						
Bleed						
Migrate						
Bleed						
Migrate						
Bleed						
Migrate						
Bleed						
Migrate						
Bleed						
Migrate						
Bleed						
Migrate						
Bleed						
Migrate						
Bleed						
Migrate						
Bleed						
Migrate						
Bleed						
Migrate						
Bleed						
Migrate						
Bleed						
Migrate						
Bleed						
Migrate						
Bleed						
Migrate						
Bleed						
Migrate						
Bleed						
Migrate						
Bleed						
Migrate						
Bleed						
Migrate						
Shut-In		Gas at Surface				

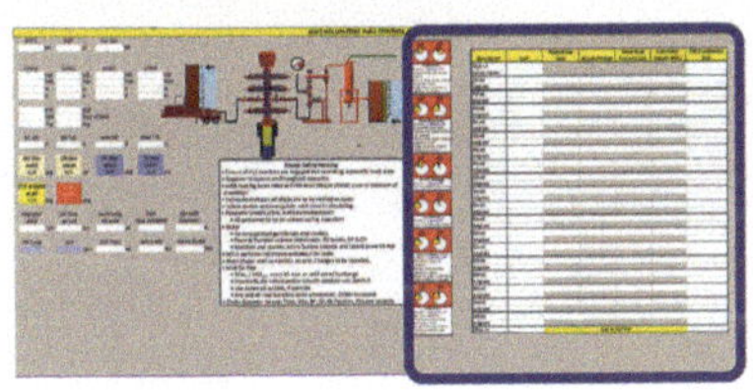

Fig. 3.2, cont'd

LUBRICATE AND BLEED WELL CONTROL ACTION PLAN (Part 2)

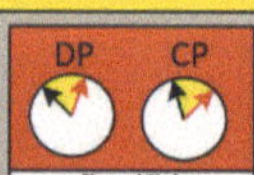

Plugged Choke
Identifier: ↑DP and ↑CP
- Manipulate Choke to Clear Debris
- Shut-in Well, Isolate Choke & Swap Out Choke
- Stage Pumps Up to Speed Holding CP constant

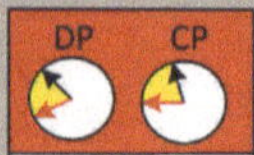

Washed Out Choke
Identifier: ↓ DP and ↓ CP
- Continually Close Choke Results in an Inability to Keep DP Constant
- Shut-in Well, Isolate Choke & Swap Out Choke
- Stage Pumps Up to Speed Holding CP constant

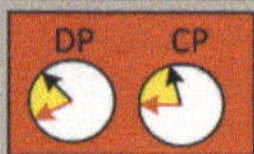

BOP or Chokeline Failure
Identifier: ↓ DP and ↓ CP
- Shut Down Pumps & Evaluate
- Isolate, Use Secondary Barrier or Option.
- Upon Fix, Stage Pumps Up to Speed Holding CP Constant

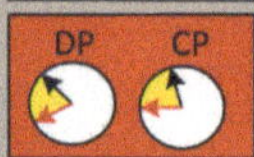

Pump Failure
Identifier: ↓ DP and ↓ CP
- DP Drops Suddenly
- Shut-in and Secure Well
- Swap Out Pumps
- Stage Pumps Up to Speed Holding Casing Pressure Constant

Volumetric Lube and Bleed Method (No Losses Downhole)

- For ***no losses*** *downhole*, the Volumetric Lube and Bleed Method with Variable Mud Volume (Most Common Method) may be used. This process employs pumping KMW into the well, waiting for the fluids to swap, then control bleeding influx from hole. Cease process when first whole mud returns are validated.
- Record stabilized SICP (if applicable, stabilized SIDPP)
- Determine Maximum Casing Pressure limit (Lessor of 80% Casing Burst or 80% of MASP). This maximum rating may be downrated due to deteriorated condition of wellhead, tubulars, type of service (i.e. sour service), quality of cement jobs, age of well, etc.
- Determine KWW (ppg). Remember to round up.

Step 1: Slowly open choke and pump KMW into well while monitoring CP and ensuring pump volumes are accurately recorded.

- Do not exceed lessor of 80% of MASP or 80% of casing burst.
- Shut-down and shut-in.
- Using the pumped volumes, determine a Target Reduction Pressure for bleed cycle.

$$\mathrm{TVD}_{\mathrm{KMW\ Lube\ Cycle\ 1}} = \frac{\mathrm{Lube\ Cycle\ 1\ Volume}_{\mathrm{bbl}}}{\mathrm{Annular\ Volume}_{\mathrm{bbl/ft}}}$$

$$\mathrm{HP}_{\mathrm{Lube\ Cycle\ 1}} = (0.052)(\mathrm{KMW}_{\mathrm{ppg}})(\mathrm{TVD}_{\mathrm{KMW\ Lube\ Cycle\ 1}})$$

$$\mathrm{Target\ Reduction\ Pressure}_{\mathrm{Lube\ Cycle\ 1}} = \mathrm{SICP}_{\mathrm{Initial}} - \mathrm{HP}_{\mathrm{Lube\ Cycle\ 1}}$$

Step 2: With pumps shut-off and well shut-in, wait and allow adequate time for the KMW to fall through the influx commonly referred to as "fluids swap".

- Adjust wait time based upon actual results

Step 3: Slowly bleed to achieve desired Target Pressure Reduction through small incremental choke adjustments. Use only manual choke.

Repeat Steps 1-3: Until all influx has been evacuated.

- See Universal Well Control manual for "Set Mud Volume", as an alternative.

Volumetric Lube and Bleed Method (With Losses Downhole)

- For wells with ***losses downhole***, the Volumetric Lube and Bleed Method with Set Mud Volume may be used. This process employs pumping a predetermined volume of mud into the well and using changes to the SICP to determine target pressure. Small bleed cycles are performed until the target pressure is achieved. Repeat process until first whole mud returns are validated and observed.
- Limitation dependent upon Surface Equipment.
- A calibrated trip tank is to be used and manifolded downstream of the choke to accurately measure the amount of fluid bled from well.
- With losses downhole, Lube and Bleed Method with Se Mud Volume can be outlined as follows:

1. Record SICP (if applicable, SIDPP)
2. Choose a safety factor of 50 – 100 psi.
3. Pump-in measured volume of fluid which will raise the casing pressure by the safety factor.
4. Record Initial SICP including the safety factor – this is P_1.
5. Pump-in a measured volume of fluid (do not exceed the lessor or 80% of MASP or 80% of Casing Burst. Maximum pressure may be downrated due to deteriorated condition of wellhead, tubulars, type of service (i.e., sour service), quality of cement jobs, age of well, etc.)
6. Allow the fluids to swap with gas (if unknown, use 1 hour to begin)
7. Record stabilized shut-in caign pressure – this is P_2
8. Calculate P_3

$$P_3 = \frac{(P_1)^2}{P_2}$$

9. Bleed off only gas (no fluid) down to the pressure of P_3.
10. P_3 becomes P_1 for the next cycle.
11. Repeat process until influx has been removed from well.

Step	Time	Position	Pressure	SCR	MW_{in}	Lube Vol	Volume	Press	Pressure
Limits				Maximum Pressure Lessor of 80% of Csg Burst or 80% MAASP					
Shut-in				Shut-in Conditions with Stabilized Pressures					
Lube									
Wait					Wait for Fluids to Swap				
Bleed			Use Small Bleed Cycles to Reach Target Reduction Pressure						
Lube									
Wait					Wait for Fluids to Swap				
Bleed			Use Small Bleed Cycles to Reach Target Reduction Pressure						
Lube									
Wait					Wait for Fluids to Swap				
Bleed			Use Small Bleed Cycles to Reach Target Reduction Pressure						
Lube									
Wait					Wait for Fluids to Swap				
Bleed			Use Small Bleed Cycles to Reach Target Reduction Pressure						
Lube									
Wait					Wait for Fluids to Swap				
Bleed			Use Small Bleed Cycles to Reach Target Reduction Pressure						
Lube									
Wait					Wait for Fluids to Swap				
Bleed			Use Small Bleed Cycles to Reach Target Reduction Pressure						
Lube									
Wait					Wait for Fluids to Swap				
Bleed			Use Small Bleed Cycles to Reach Target Reduction Pressure						
Lube									
Wait					Wait for Fluids to Swap				
Bleed			Use Small Bleed Cycles to Reach Target Reduction Pressure						
Lube									
Wait					Wait for Fluids to Swap				
Bleed			Use Small Bleed Cycles to Reach Target Reduction Pressure						
Lube									
Wait					Wait for Fluids to Swap				
Bleed			Use Small Bleed Cycles to Reach Target Reduction Pressure						
Lube									
Wait					Wait for Fluids to Swap				
Bleed			Use Small Bleed Cycles to Reach Target Reduction Pressure						

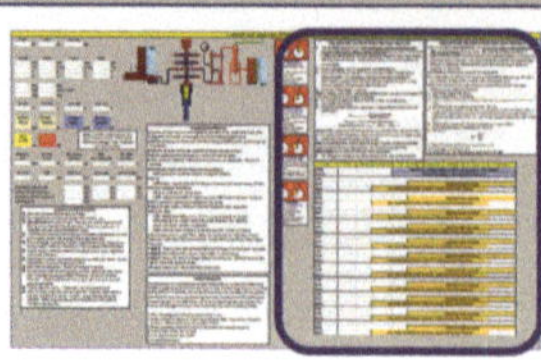

Fig. 3.3 Lube and Bleed Action Plan (Part 2).

Lube and bleed method for NO losses downhole (variable mud volume)

Preparation for variable mud volume.

THE VOLUME METHOD LUBE AND BLEED METHOD
CANNOT BE USED ON WELLS WITH ONGOING DOWNHOLE LOSSES

Begin preparations by mobilizing low-volume/high-pressure (LVHP) pump (cement unit) to location. Perform checklist as provided.

1. Record stabilized SIDPP and SICP pressure (initial conditions) (Fig. 3.4)

$$\overline{BHP} = \overline{HP}_{\text{Casing Annulus}} + \overline{SICP}$$

Fig. 3.4 Stabilized forces when shut-in.

2. Determine Maximum Casing Pressure (Lessor of 80% Casing Burst or 80% of MASP). This maximum rating may be downrated due to deteriorated condition of wellhead, tubulars, type of service (i.e., sour service), quality of cement jobs, age of well, etc.
3. Determine KMW (ppg), round up.
4. Visually inspect wellhead, flowlines and equipment or leaks, alignment, and overall condition.
5. Ensure manual choke is operable and aligned for the bleed cycle.
6. Ensure accurate casing pressure gauge is installed on choke manifold and can be easily read from manual choke position.
7. Develop Lube and Bleed Table to record process.

Step 1: Perform lube cycle and determine target reduction pressure

Target pressure reduction is determined by lubricating mud into the well until a pre-determined maximum casing pressure is achieved. The lubricated volume of mud must be accurately recorded, as this volume will determine the bleed cycle. Pump lube cycle first ensuring volumes are accurately recorded. Using the pumped volumes, determine a target pressure for bleed cycle.

Initiate variable mud volume lube cycle

1. Slowly pump KMW into the well while monitoring CP (if applicable, monitor DP)
   - Accurately record lube volume (Fig. 3.5).

$$\uparrow BHP = \uparrow HP_{\substack{Casing\\Annulus}} + \uparrow SICP + \uparrow Friction$$

Fig. 3.5 Annular Forces When Initiating Pumping.

Stop pumps once Maximum Casing Pressure has been achieved.

- During each Lube Cycle, the Lube Volume will vary (hence name Variable Mud Volume).

2. Once pumps have stopped and the well shut-in, determine Target Reduction Pressure based upon volume of KMW lubricated into the well.

$$TVD_{KMW\ Lube\ Cycle\ 1} = \frac{Lube\ Cycle\ 1\ Volume_{bbl}}{Annular\ Volume_{bbl/ft}} \tag{3.1}$$

$$HP_{Lube\ Cycle\ 1} = (0.052)\left(KMW_{ppg}\right)\left(TVD_{KMW\ Lube\ Cycle\ 1}\right) \tag{3.2}$$

$$Target\ Reduction\ Pressure_{Lube\ Cycle\ 1} = SICP_{Initial} - HP_{Lube\ Cycle\ 1} \tag{3.3}$$

Step 2: Wait

With pumps shut off and well shut-in, wait and allow adequate time for the KMW to fall through the influx commonly referred to as "fluids swap." For first lube cycle, the following rule of thumbs can be used and adjusted accordingly based upon actual results.

1. For viscous muds, for initial KMW to fall through influx is approximately 30 min.
2. For clear fluids, for initial KMW to fall through influx is approximately 5 min.
3. Adjust wait time based upon actual results.

Step 3: Perform bleed cycle based on target pressure reduction

- Perform Bleed cycle to achieve Target Pressure Reduction (Fig. 3.6).

$$\downarrow BHP = \overline{HP}_{\substack{Casing\\Annulus}} + \downarrow SICP$$

Fig. 3.6 Annular Forces When Bleeding.

<u>Initiate variable mud volume bleed cycle</u>

1. Perform bleed cycle: Slowly bleed to achieve desired target pressure reduction through small incremental choke adjustments. Use only the manual choke.

Repeat steps 1 through 3 until conclusion

Repeat Lube and Bleed cycle as many times as needed to evacuate influx from the wellbore.

a. After each lubrication cycle and with known lubrication volume, determine revised Target Reduction Pressure based upon the lubrication volume most recently pumped. Anticipate Target Reduction pressures will decrease in value over time, as with less influx, smaller volumes will be needed to achieve Maximum Casing Pressures.

b. Lube and Bleed performed with KMW will form a Mud Cap of heavy density fluid at top of well. If the Mud Cap contains sufficient hydrostatic pressure based upon its density, SIDPP = SICP = 0.
- Open choke and by-pass to trip tank to verify no flow.
- If no flow is observed, evacuate rig floor and open BOPs. Perform extended flow check.

c. Lube and Bleed performed with MW equal to well MW density, with the influx removed, the hydrostatic pressure of the lubricated mud will not offset formations pressures and SIDPP = SICP > 0 psi. Without sufficient density to kill the well, the well is controlled BUT NOT DEAD.
- Initiate operations to reestablish circulation and kill well (i.e., wireline perforating of workstring, snubbing operations, etc.) (Fig. 3.7)

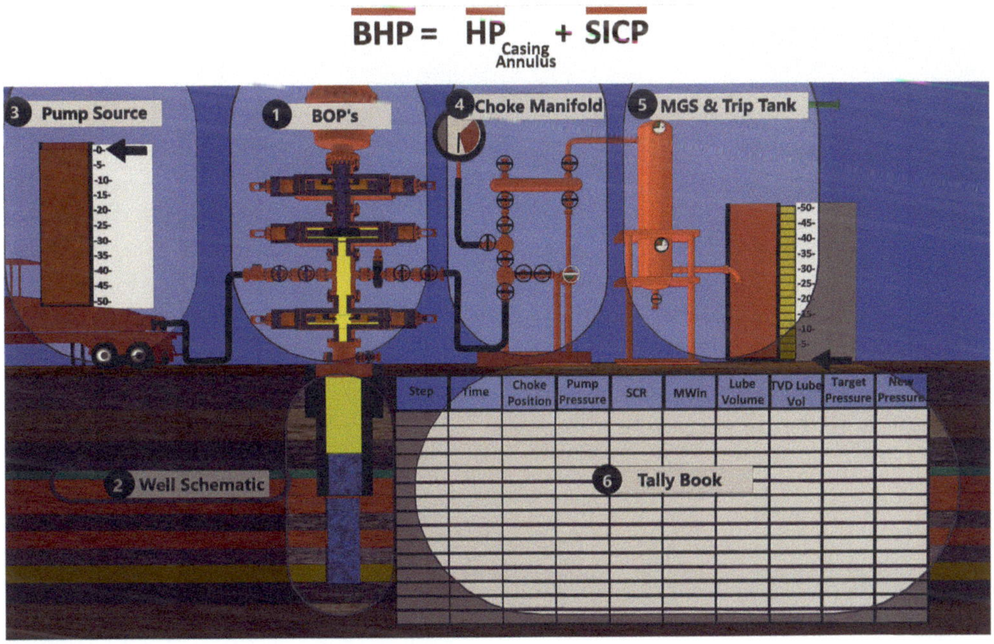

Fig. 3.7 Lube Bleed Worksheet with Diagram.

Lubricate and bleed for NO losses downhole (variable mud volume) example

The following is an example of Lubricate and Bleed for NO Losses Downhole Well Control using the Variable Mud Volume. This example represents an actual field response to a well control problem. This example of Variable Mud volume demonstrates the principle of tracking and lubricating mud until casing pressure has risen to a predetermined limit. The mud volume is used to determine the reduction in casing pressure caused by lubricating mud, then slowly venting casing pressure (in small increments) to a calculated reduction in pressure. Multiple cycles will be needed in order to stabilize and kill the well (Fig. 3.8).

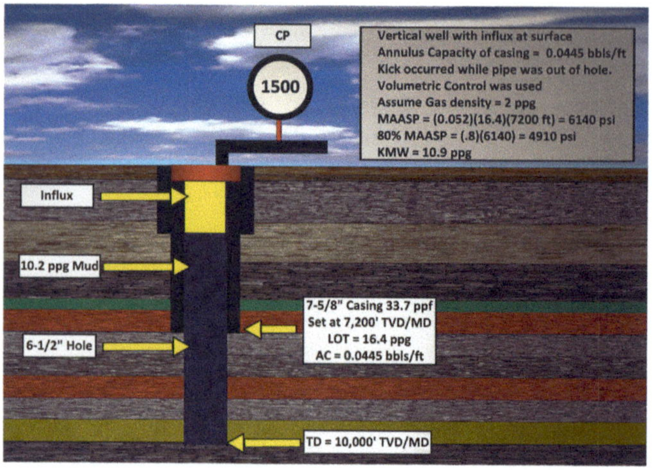

Fig. 3.8 Example Well Data Diagram

- **Vertical** well with influx at surface, with an unknown overbalance on the well
- SICP = 1500 psi (stabilized pressure)
- Casing shoe at 7200' TVD/MD
- Annulus Capacity of casing **=0.0445** bbls/ft.
- LOT = 16.4 ppg
- Kick occurred while pipe was out of hole.
- Volumetric Control was used to get kick to surface without breaking down casing shoe formation.

- Existing mud weight in the well = 10.2 ppg
- Assume Gas density = 2 ppg
- **MASP = (0.052)(16.4)(7200 ft) = 6140 psi**
- **KMW = 10.9 ppg** (Fig. 3.9)

Step	Time (min:sec)	Max Choke Position (In or %)	Max Pump Pressure (psi)	SCR (spm)	MWin (ppg)	Lube Volume (bbls)	TVD Lube Vol (ft)	Target Reduction Pr (psi)	New Target Pressure (psi)

Fig. 3.9 Lube and Bleed Worksheet.

THE VARIABLE MUD VOLUME LUBE AND BLEED METHOD CANNOT BE USED ON WELLS WITH ONGOING DOWNHOLE LOSSES

Preparation for variable mud volume

Determine Kill Mud Weight. Establish the flow route from low-volume, high-pressure cement pump and the 50 bbl graduated suction tank to kill line. From the choke line to the choke manifold and manual choke routed to MGS to trip tank. Ensure the MGS, suction tank, and trip tanks have been cleaned. Once cleaned, add 10.9 bbl mud to the MGS mud leg and suction tank. Install lights as needed to perform nighttime operations (Fig. 3.10).

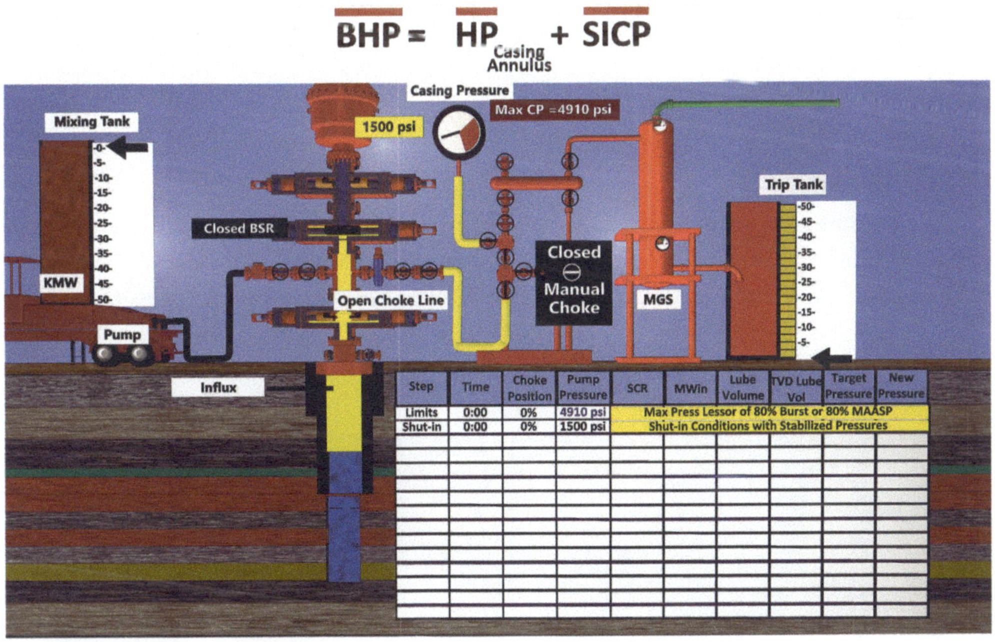

Fig. 3.10 Shut-in conditions.

Step 1: Perform lube cycle and determine target reduction pressure

Pump at slow pump lubrication rate to ensure smallest friction pressures possible. Lubricate KMW until pump pressure is near the lessor of 80% Casing Burst or 80% of MASP. Record findings in table (Fig. 3.11).

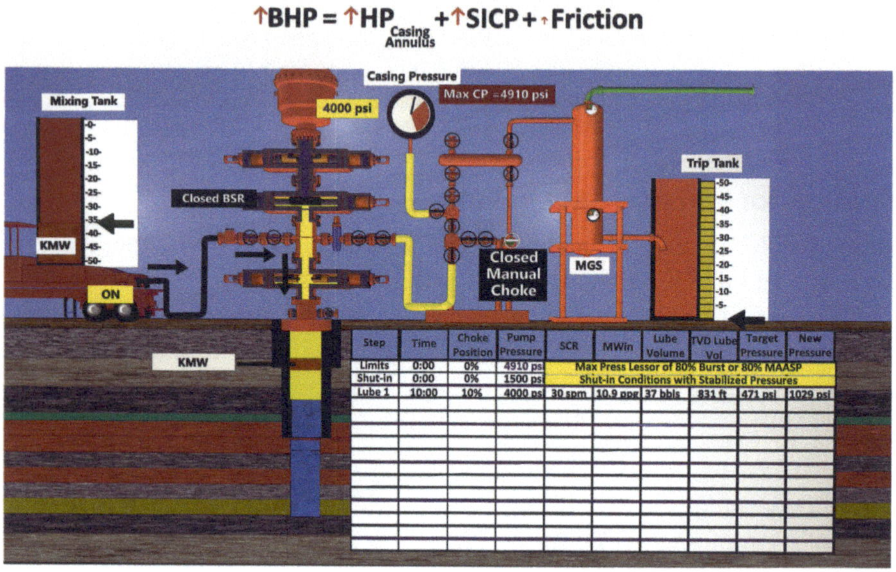

Fig. 3.11 Lubricate Until Maximum Pump Pressure is Reached. Record Lube Volume and Determine New Target Pressure.

Step 2: Wait

Allow sufficient time for the lubricated mud to fall through the influx. While monitoring pressures, refill mix tank on low-volume/high-pressure pump with KMW. May see an increase in SICP as the mud displaces influx forcing the influx to compress (Fig. 3.12).

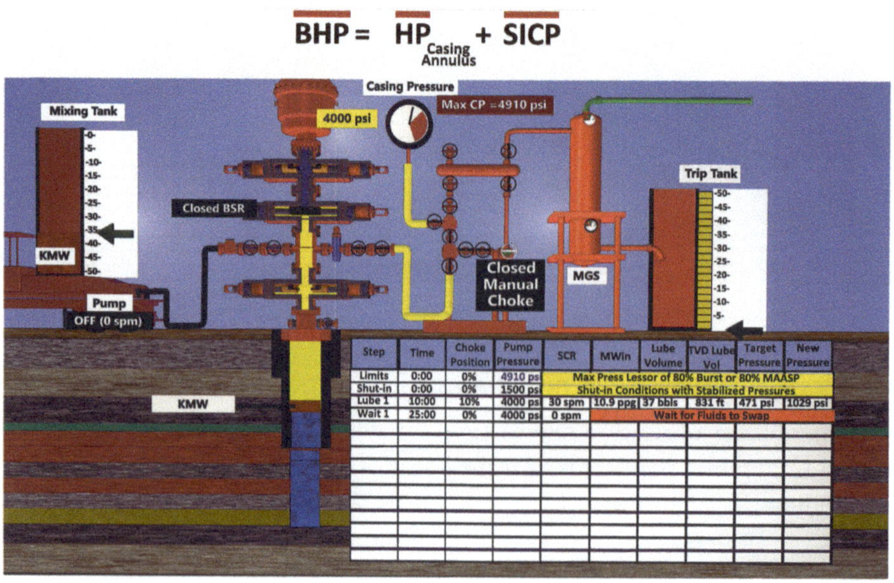

Fig. 3.12 Allow Fluids to Swap with Influx.

Step 3: Perform bleed cycle based on target pressure reduction

Given sufficient time for lubrication mud to swap with the influx, the choke is slowly (cracked) open and the casing pressure is slowly allowed to decrease. Only dry gas should be exiting from the Mud Gas Separator. At the end of the lubrication cycle, the SICP read 1040 psi. The target pressure was 1029 psi.

Based on target pressure, the current SICP of 1500 psi was reduced 471 psi by control bleed of dry influx through manual choke and exiting the MGS. The difference between 1500 and 1029 is 471 psi. The 471-psi bleed volume represents hydrostatic pressure of 37 bbls of 10.9 ppg. The difference between target pressure and actual pressure will be trapped pressure or 11 psi (1040–1029 psi) trapped pressure (Fig. 3.13).

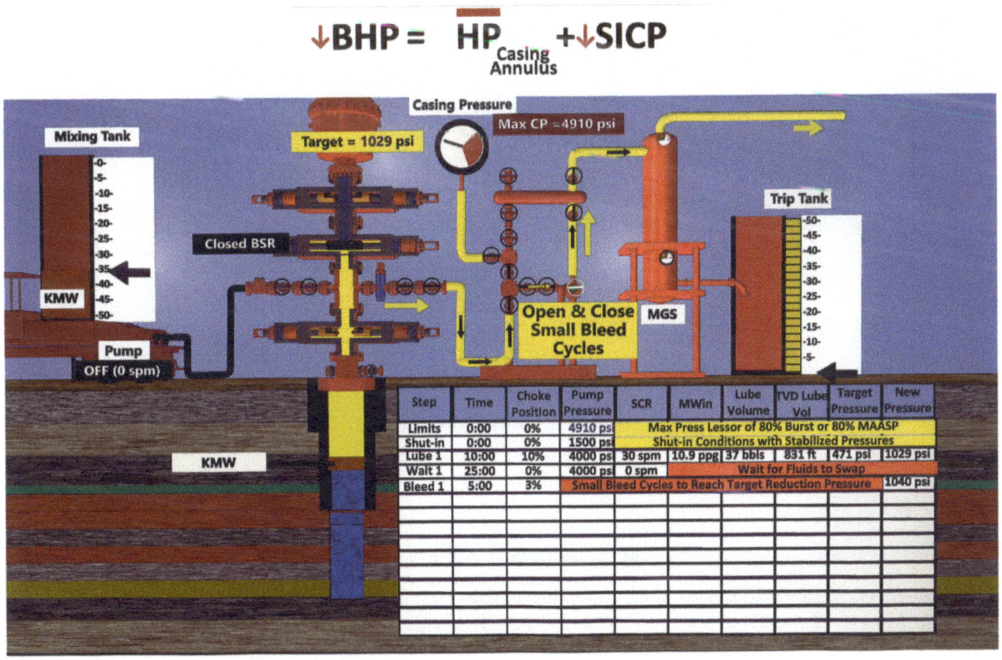

Fig. 3.13 Allow influx to migrate and BHP to increase.

Repeat steps 1 through 3 until conclusion

Continue performance of cycles of Lubricate, Wait, and Bleed until influx has been removed from well. As the gas volume in annulus becomes smaller, the effects of lubrication become larger. Therefore, during the time for each cycle should diminish as the influx is reduced within the annulus.

Lubricate and Bleed Method is used to reduce potential accidents from uncontrolled release of pressurized fluids from the wellbore. Circulation will need to be reestablished in order to finalize well control using constant bottom-hole pressure method (Fig. 3.14).

$$\overline{BHP} = \overline{HP}_{\substack{Casing\\Annulus}} + \overline{SICP}$$

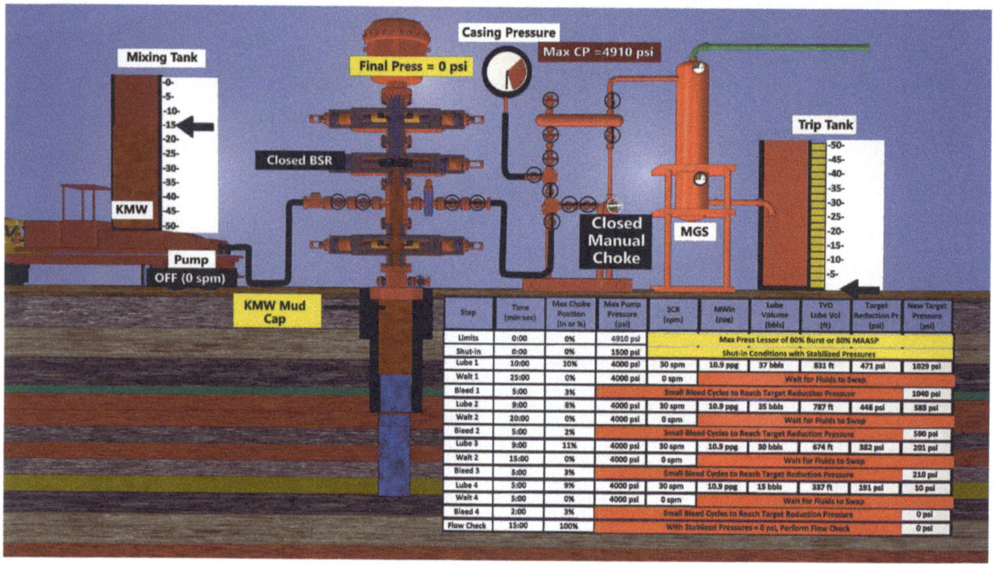

Step	Time (min:sec)	Max Choke Position (In or %)	Max Pump Pressure (psi)	SCR (spm)	MWin (ppg)	Lube Volume (bbls)	TVD Lube Vol (ft)	Target Reduction Pr (psi)	New Target Pressure (psi)
Limits	0:00	0%	4910 psi	Max Press Lessor of 80% Burst or 80% MAASP					
Shut-in	0:00	0%	1500 psi	Shut-in Conditions with Stabilized Pressures					
Lube 1	10:00	10%	4000 psi	30 spm	10.9 ppg	37 bbls	831 ft	471 psi	1029 psi
Wait 1	25:00	0%	4000 psi	0 spm	Wait for Fluids to Swap				
Bleed 1	5:00	3%	Small Bleed Cycles to Reach Target Reduction Pressure						1040 psi
Lube 2	9:00	8%	4000 psi	30 spm	10.9 ppg	35 bbls	787 ft	446 psi	583 psi
Wait 2	20:00	0%	4000 psi	0 spm	Wait for Fluids to Swap				
Bleed 2	5:00	2%	Small Bleed Cycles to Reach Target Reduction Pressure						590 psi
Lube 3	9:00	11%	4000 psi	30 spm	10.9 ppg	30 bbls	674 ft	382 psi	201 psi
Wait 2	15:00	0%	4000 psi	0 spm	Wait for Fluids to Swap				
Bleed 3	5:00	3%	Small Bleed Cycles to Reach Target Reduction Pressure						210 psi
Lube 4	5:00	9%	4000 psi	30 spm	10.9 ppg	15 bbls	337 ft	191 psi	10 psi
Wait 4	5:00	0%	4000 psi	0 spm	Wait for Fluids to Swap				
Bleed 4	2:00	3%	Small Bleed Cycles to Reach Target Reduction Pressure						0 psi
Flow Check	15:00	100%	With Stabilized Pressures = 0 psi, Perform Flow Check						0 psi

Fig. 3.14 With influx lubricated out, mud cap should provide sufficient density to kill well.

ALWAYS ASSUME STORED WELL PRESSURE REMAINS AFTER LUBRICATION AND BLEED ARE FINISHED. DO NOT OPEN BOPS, AS TRAPPED PRESSURE MAY VIOLENTLY BE RELEASED RESULTING IN LOWERING THE BHP AND INTRODUCING KICK

Lubricate and bleed for NO losses downhole (set mud volume) example

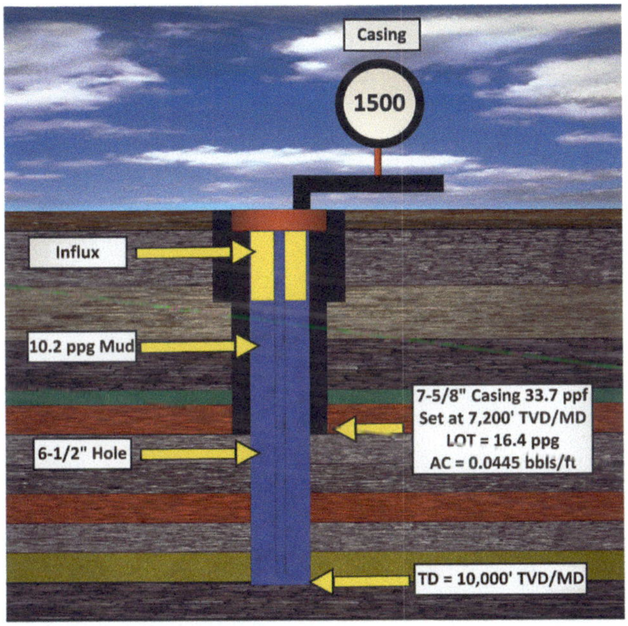

Fig. 3.15 Example well data diagram.

- **Vertical** well with influx at surface, with an unknown overbalance on the well (Fig. 3.15)
- SICP = 1500 psi (stabilized pressure)
- Casing shoe at 7200′ TVD/MD
- Annulus Capacity of casing **=0.0445** bbls/ft
- LOT = 16.4 ppg
- Kick occurred while pipe was out of hole
- Volumetric Control was used to get kick to surface without breaking down casing shoe formation
- Existing mud weight in the well = 10.2 ppg

- Assume Gas density = 2 ppg
- **MASP = (0.052)(16.4)(7200 ft) = 6140 psi**
- **KMW = 10.9 ppg** (Fig. 3.16)

Step	Time (min:sec)	Max Choke Position (In or %)	Max Pump Pressure (psi)	SCR (spm)	MWin (ppg)	Lube Volume (bbls)	TVD Lube Vol (ft)	Target Reduction Pr (psi)	New Target Pressure (psi)

Fig. 3.16 Lube and bleed worksheet.

Preparation for set mud volume

Determine and prepare Kill Mud Weight. Establish the flow route from low-volume, high-pressure cement pump and its 50 bbl graduated suction tank to kill line. From the choke line to the choke manifold and manual choke routed to MGS to trip tank. Ensure the MGS, suction tank, and trip tanks have been cleaned. Once cleaned, for our example, add 10.9 bbl mud to the MGS mud leg and suction tank. Install lights as needed to perform nighttime operations.

Determine Lube Cycle Volume from KMW and prepare Lube and Bleed Table (Fig. 3.17).

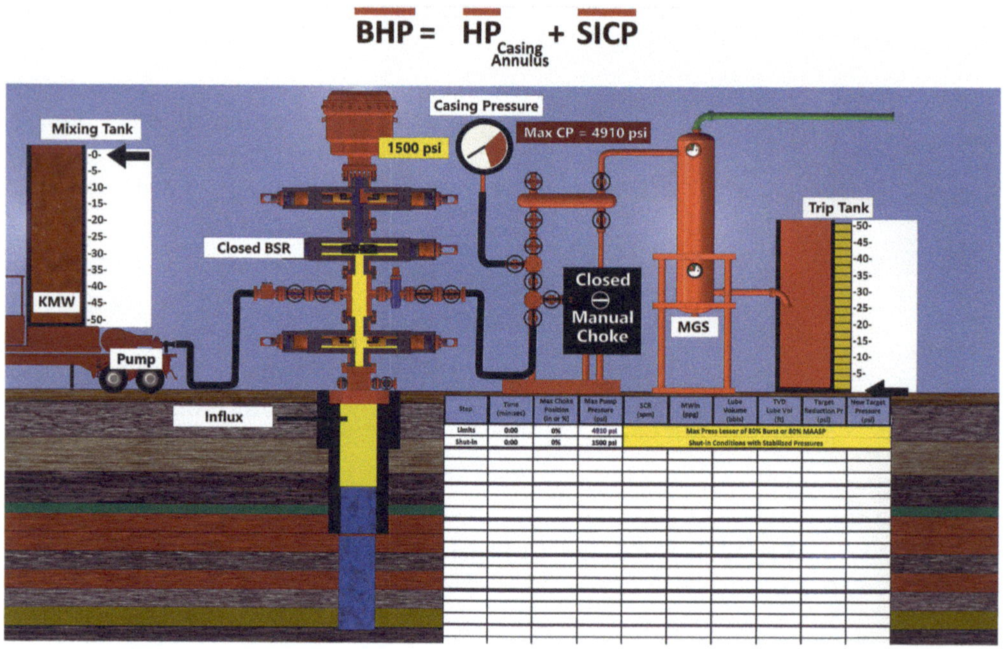

Fig. 3.17 Shut-in conditions.

$$\text{TVD Lube Cycle} = \frac{300}{(10.9)(0.052)} = 529\,\text{ft.} \tag{3.4}$$

$$\text{Lube Cycle Volume}_{\text{bbl}} = (529\text{ft.}) \times \left(0.0445\frac{\text{bbls}}{\text{ft.}}\right) = 23.5\ \text{bbls} \tag{3.5}$$

Step 1: Determine HP envelope and perform lube cycle

The selected pressure increment of 300 psi determines the amount of fluid to be lubricated into the well (23.5 bbls of 10.9 ppg). The fluid is pumped at a reduced rate to ensure casing pressure observations will allow sufficient time to stop the pumps and avoid fracturing the shoe. Lube and bleed cycle should be performed slowly to closely achieve desired lube volumes and reductions in pressure during bleed cycles. The pumps were stopped after 23.5 bbls had been pumped (Fig. 3.18).

$$\uparrow BHP = \uparrow HP_{\substack{Casing\\Annulus}} + \uparrow SICP + \uparrow Friction$$

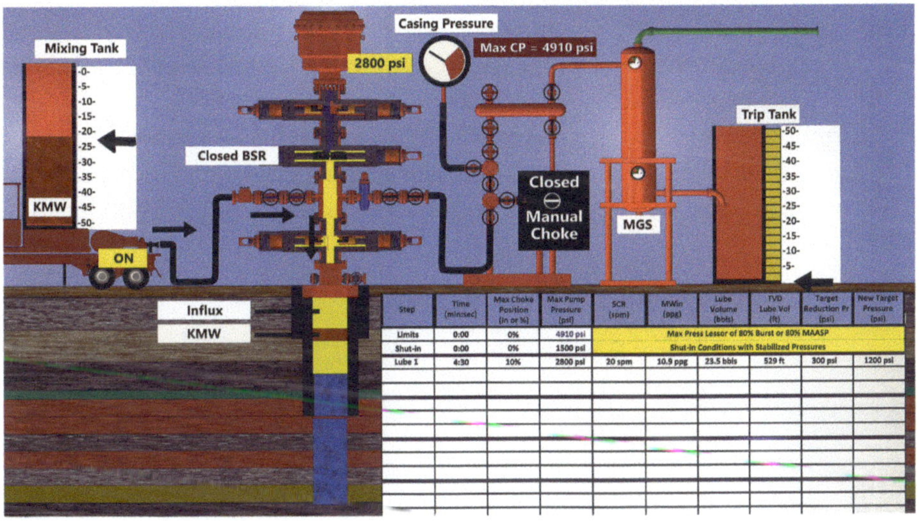

Fig. 3.18 Lubricate 4 bbls at slow pump rate.

Step 2: Wait

Allow sufficient time for the lubricated mud to fall through the influx. While monitoring pressures, refill mix tank on low-volume/high-pressure pump with KMW. May see an increase in SICP as the mud displaces influx forcing the influx to compress. The mixing tank has been "topped off" with KMW. This action is to be performed during each wait cycle (Fig. 3.19).

$$\overline{BHP} = \overline{HP}_{\substack{Casing\\Annulus}} + \overline{SICP}$$

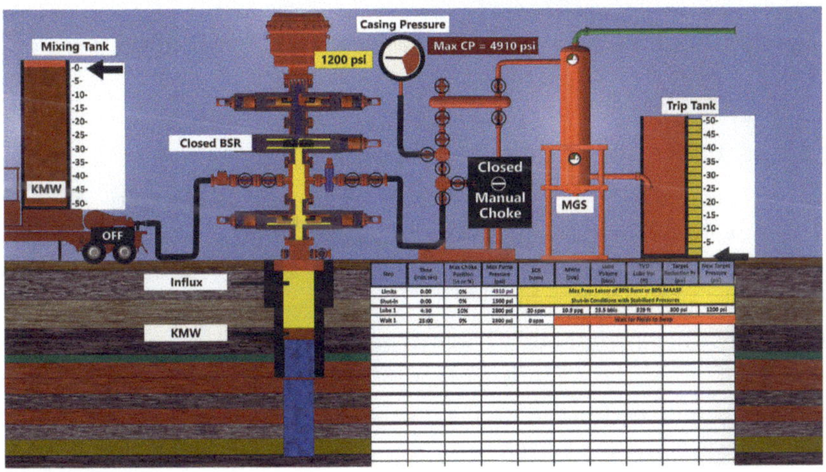

Fig. 3.19 Allow fluids to swap with influx.

Step 3: Perform bleed cycle

Given sufficient time for lubrication mud to swap with the influx, the choke is slowly (cracked) opened and the 2800 psi shut-in casing pressure is slowly allowed to decrease. Only dry gas should be exiting from the Mud Gas Separator. At the end of the lubrication cycle, the SICP read 1200 psi.

The current SICP of 1200 psi was reduced 300 psi from initial 1500 psi shut-in casing pressure by control bleed of dry influx through manual choke and exiting the MGS. The 300-psi bleed pressure represents hydrostatic pressure of 23.5 bbls of 10.9 ppg (Fig. 3.20).

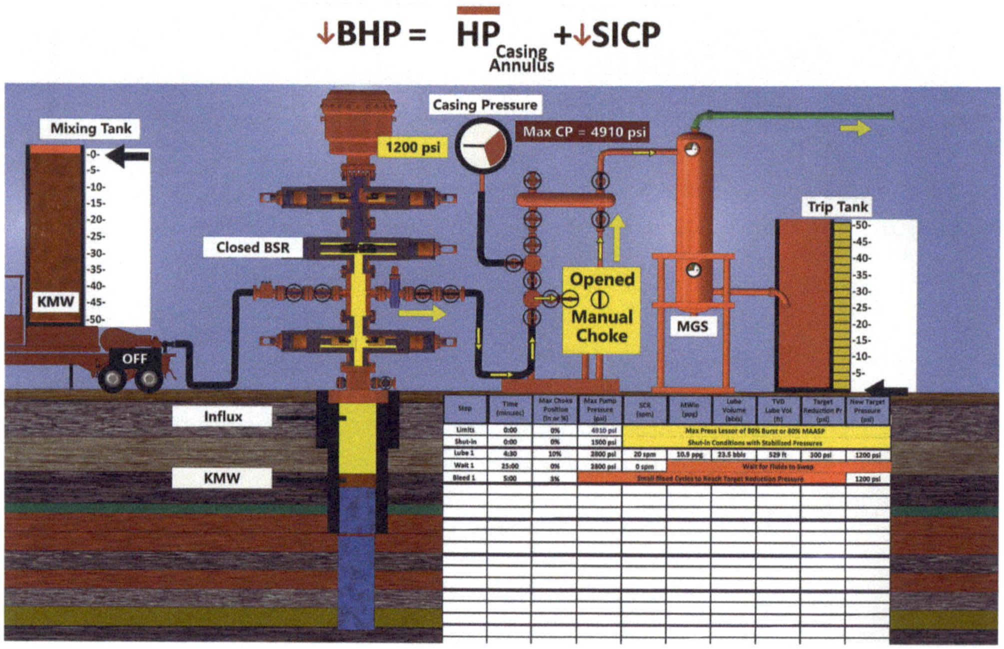

Fig. 3.20 Bleed measured volume from well in small increments to desired reduced pressure.

Repeat steps 1 through 3 until conclusion

Continue performance of cycles of Lubricate, Wait, and Bleed until influx has been removed from well. As the gas volume in annulus becomes smaller, the effects of lubrication become larger. Therefore, during the time for each cycle should diminish as the influx is reduced within the annulus.

Lubricate and Bleed Method is used to reduce potential accidents from uncontrolled release of pressurized fluids from the wellbore. Circulation will need to be reestablished in order to finalize well control using constant bottom-hole pressure method (Fig. 3.21).

$$\overline{BHP} = \overline{HP}_{\substack{Casing\\Annulus}} + \overline{SICP}$$

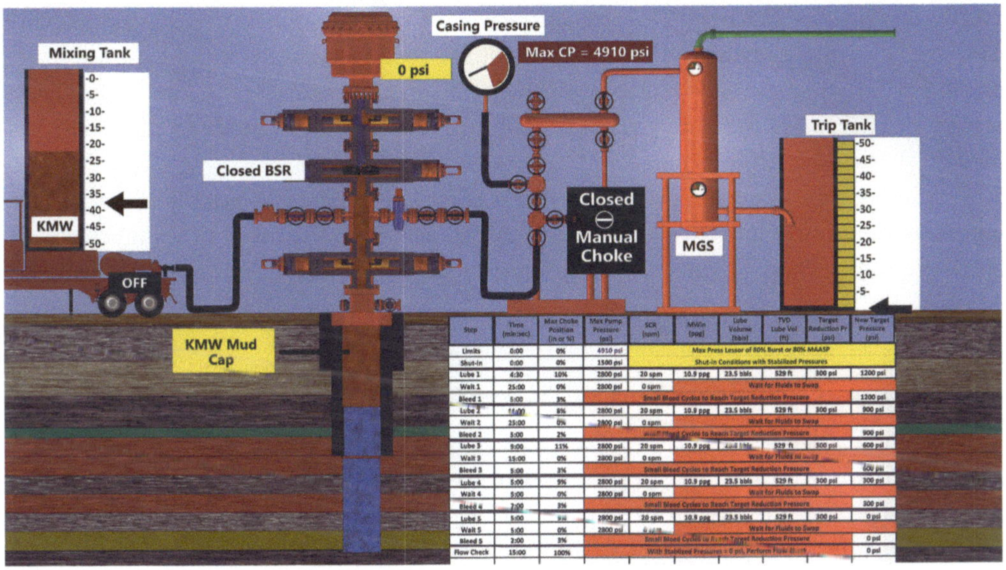

Fig. 3.21 With influx lubricated out, mud cap should provide sufficient density to control well.

ALWAYS ASSUME STORED WELL PRESSURE REMAINS AFTER LUBRICATION AND BLEED ARE FINISHED.
DO NOT OPEN BOPS, AS TRAPPED PRESSURE MAY VIOLENTLY BE RELEASED
RESULTING IN LOWERING THE BHP AND INTRODUCING KICK

Lubricate and bleed method WITH losses downhole (pressure)

Commentary

The Lubricate and Bleed Method with Losses Downhole (Pressure) Method is limited only to partial losses in which KMW can be introduced with slight pump pressures without causing total lost circulation. Therefore, this method is limited to partial losses, as total lost circulation would require a separate approach where lightweight fluids would be introduced in the annulus in an attempt to establish a fluid leg to surface. Fighting total losses may require use of additive bridging agents.

For partial losses, the use of bridging agents is not suggested as they may plug surface lines from the HPLV pump to wellhead. Original mud weight fluids (OMW) may be used in lieu of kill mud weights (KMW) in order to establish a hydrostatic fluid column in the annulus. Individual selection of OMW, KMW, and bridging agents would have to be determined on actual case examples and experience.

In most situations where losses are occurring, lubricate and bleed methods employed would utilize the smallest of pump rates in order to keep pump pressures as low as possible. Rather than concentrating on the volumes pumped, this method uses actual pump pressures and static pressures to mitigate the influx. As such, the pump gauges used to track pump pressures and casing pressures must be highly accurate at low pressures. This may require changing out gauges in order to increase accuracy of the displaced volumes.

Since low pump pressures will be used, smaller volumes will be pumped which ultimately require more lube, wait, and bleed cycles when compared to other lubricate and bleed methods. Safety factors may be limited to easy-to-read numbering on gauges such as 50 psi or 100 psi.

Lubricate and bleed method WITH losses downhole (pressure)

Lubricate and Bleed Method ***WITH*** Losses Downhole (Pressure) relies on accurately determined surface pressures. High-pressure gauges used for drilling may be replaced with more sensitive lower pressure gauges to maximize accuracy.

A determination of sufficient hydrostatic interval is determined using annular volume and KMW. Once interval volume is determined, the KMW is prepared and equipment is lined up to lubricate downhole. With the choke barely opened, the HPLV pump begins pumping the KMW down the annulus while monitoring pressures. The slow pump rate is kept to a minimum as not to further fracture the well leading to total losses.

After pumping the required volume, the pumps are stopped and the well is shut-in. A waiting period follows allowing the influx to percolate up the annulus and through the measured volume of KMW. The KMW is allowed to settle to the bottom of the influx. This occurs by holding Pump Pressure on well.

Once the waiting cycle has been complete, the returns from the choke are routed to the MGS. With all appropriate valves lined-up and opened, the choke operator begins to slowly open the manual choke and allows the SICP to decrease until the desired reduction in pressure is achieved (equal to HP of KMW pumped).

ONLY THE LUBRICATE AND BLEED METHOD WITH LOSSES DOWNHOLE (PRESSURE) SHOULD BE USED IF THE WELL IS EXPERIENCING DOWNHOLE FLUID LOSSES TO THE FORMATION.

Lubricate and bleed method with losses downhole (pressure) method calculations

For this method, a simple equation is used to calculate the bleed down pressure for each cycle.

$$P_3 = \frac{(P_1)^2}{P_2} \tag{3.6}$$

These three pressures are defined below.

P_1 = Initial Pressure for each cycle including safety factor.

P_2 = Actual Casing Pressure after fluid has been added or pumped into the well for each cycle

- Maintain Pump-In Casing Pressure less than MASP or Formation Fracture Pressure, whichever is applicable.
- Volume pumped will be dependent upon well conditions.
- Allow fluid and gas to swap (stabilize).

P_3 = Calculated Pressure which the Casing is bled down to in each cycle.

The calculated pressures provide a conservative estimate for water-based mud. Inaccuracies will be introduced if this equation is applied to oil-based mud. Due to gas solubility and compressibility of the base oil, the predictive pressures can only be managed by a multiphase simulator.

Lubricate and bleed method with losses downhole (pressure) method procedure

1. Record SICP (and if applicable, SIDPP).
2. Choose a safety factor of 50–100 psi.
3. Pump in a volume of fluid which will raise the casing pressure by the safety factor.
4. Record Initial SICP including the safety factor—this is P_1.
5. Pump-in a measured volume of fluid (do not exceed 80% of MASP or 80% of Casing Burst).
6. Allow the fluid to swap with the gas (if unknown, use 1 h to begin).
7. Record the shut-in pressure on the casing once it stabilizes—this is P_2.
8. Calculate P_3.
9. Bleed off only gas (no fluid) down to a pressure of P_3.
10. P_3 becomes P_1 for the next cycle.
11. Repeat process until influx has been removed from the well.

Lubricate and bleed method with losses downhole (pressure) method example

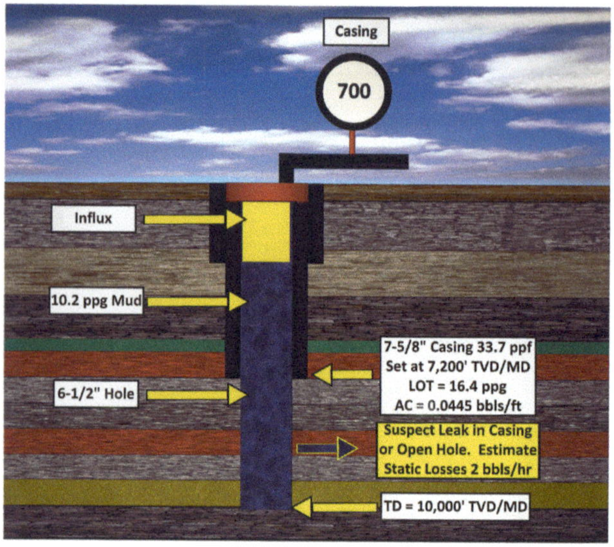

Fig. 3.22 Example well data diagram.

- **Vertical** well with influx at surface, with an unknown overbalance on the well (Fig. 3.22)
- **SICP = 700 psi** (stabilized pressure)
- Casing shoe at 7200′ TVD/MD
- Annulus Capacity of casing **=0.0445** bbls/ft.
- LOT = 16.4 ppg
- Kick occurred while pipe was out of hole
- Volumetric Control was used to get kick to surface without breaking down casing shoe formation
- Losses observed at 1–2 bph
- Existing mud weight in the well = 10.2 ppg
- Assume Gas density = 2 ppg
- **MASP = (0.052)(16.4)(7200 ft) = 6140 psi**
- **KMW = 10.9 ppg** (Fig. 3.23)

Step	Time (min:sec)	Max Choke Position (In or %)	Max Pump Pressure (psi)	SCR (spm)	MWin (ppg)	Lube Volume (bbls)	TVD Lube Vol (ft)	Target Reduction Pr (psi)	New Target Pressure (psi)

Fig. 3.23 Lube and bleed worksheet for pressure method.

Preparation for variable mud volume

Determine Kill Mud Weight. Establish the flow route from low-volume, high-pressure low-volume pump and the 50 bbl graduated suction tank to kill line. From the choke line to the choke manifold and manual choke routed to MGS to trip tank. Ensure the MGS, suction tank, and trip tanks have been cleaned. Once cleaned, add 10.9 bbl mud to the MGS mud leg and suction tank. Install lights as needed to perform nighttime operations. Establish Safety Factor of 100 psi by slowly pumping KMW into annulus with closed choke (Fig. 3.24).

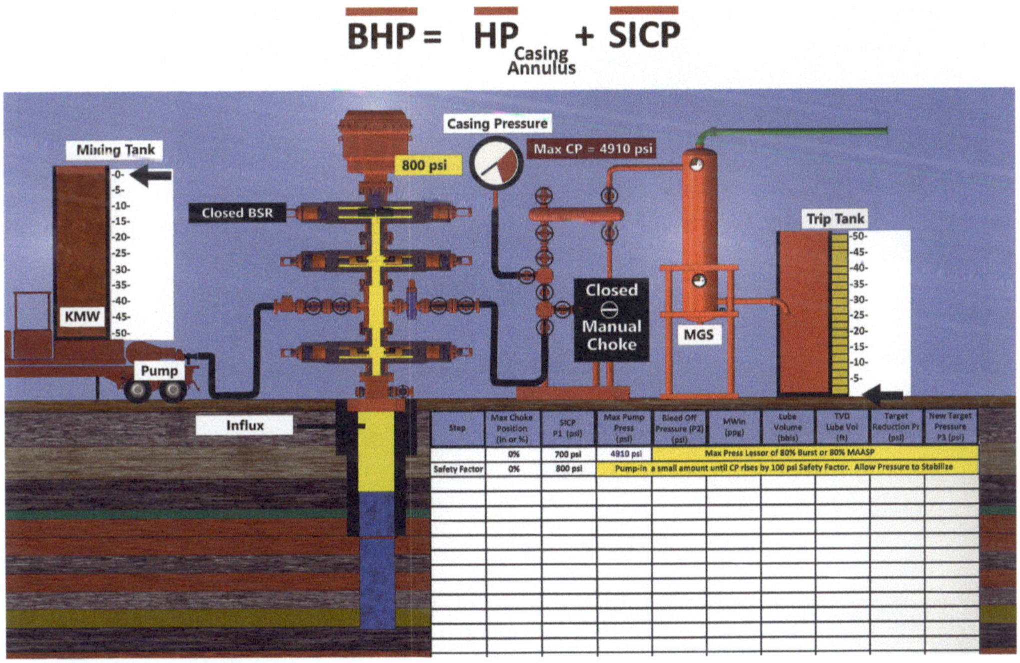

Fig. 3.24 Shut-in conditions.

Step 1: Perform lube cycle and determine target reduction pressure

Pump at **slow pump lubrication rate** to ensure smallest friction pressures possible. Lubricate KMW until pump pressure is near the lessor of 80% Casing Burst or 80% of MASP. Record findings in table (Fig. 3.25).

$$\uparrow BHP = \uparrow HP_{\substack{Casing \\ Annulus}} + \uparrow SICP + \uparrow Friction$$

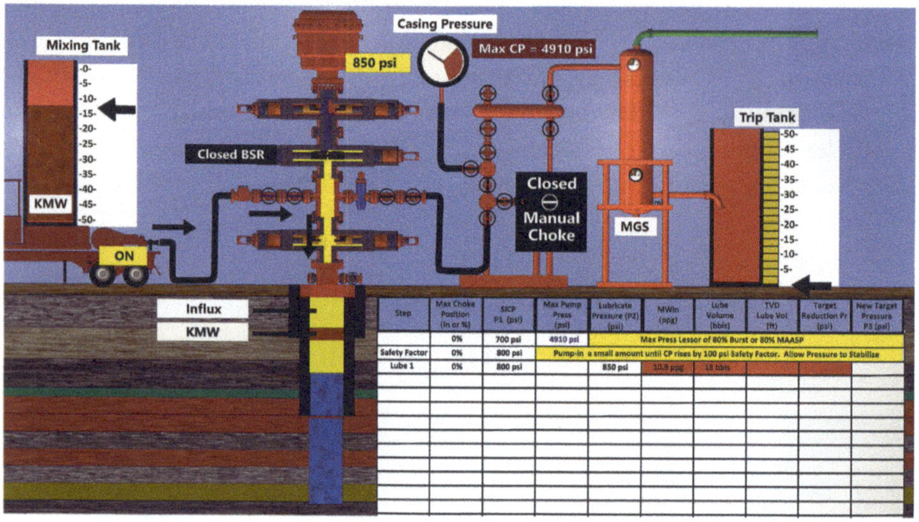

Fig. 3.25 Slowly lubricate KMW using minimal pump pressure to minimize downhole leakage. Record Lube Volume.

Step 2: Wait

Allow sufficient time for the lubricated mud to fall through the influx (first attempt around 30 min). While monitoring pressures, refill mix tank on high-pressure low-volume pump with KMW. Fluctuations in SICP may occur as the mud displaces influx forcing the influx to compress (Fig. 3.26).

$$\overline{BHP} = \overline{HP}_{\substack{Casing \\ Annulus}} + \overline{SICP}$$

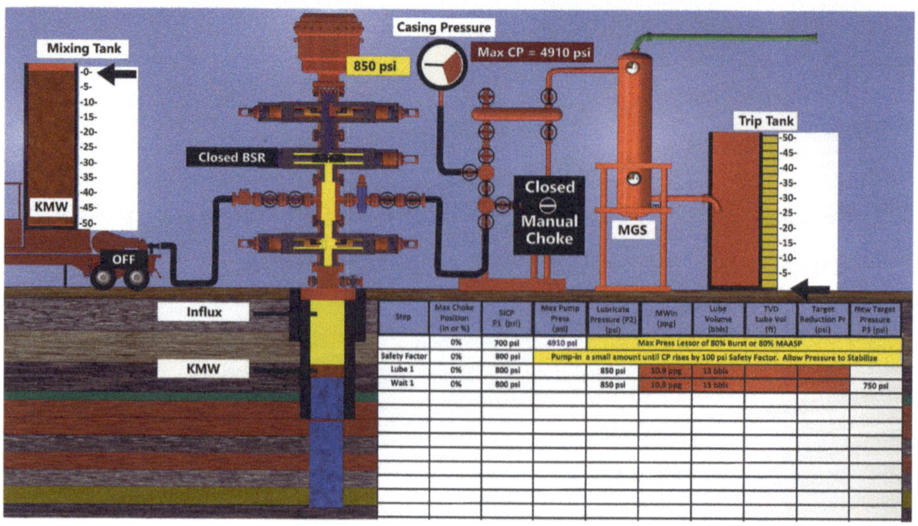

Fig. 3.26 Allow fluids to swap with influx and determine pressure reduction (P_3).

While waiting on fluid to swap, perform pressure calculations for upcoming bleed cycle.

$$P_3 = \frac{(800)^2}{850} = 750 \text{ psi} \tag{3.7}$$

Step 3: Perform bleed cycle based on target pressure reduction

Given sufficient time for lubrication mud to swap with the influx, the choke is slowly (cracked) open and the casing pressure is slowly allowed to decrease. Only dry gas should be exiting from the Mud Gas Separator. At the end of the lubrication cycle, the SICP read 750 psi.

The current SICP of 850 psi was reduced 100 psi by control bleed of dry influx through manual choke and exiting the MGS. The difference between 850 and 750 is 100 psi. The 100-psi bleed volume represents hydrostatic pressure of 13 bbls of 10.9 ppg (Fig. 3.27).

$$\downarrow BHP = \overline{HP}_{\text{Casing Annulus}} + \downarrow SICP$$

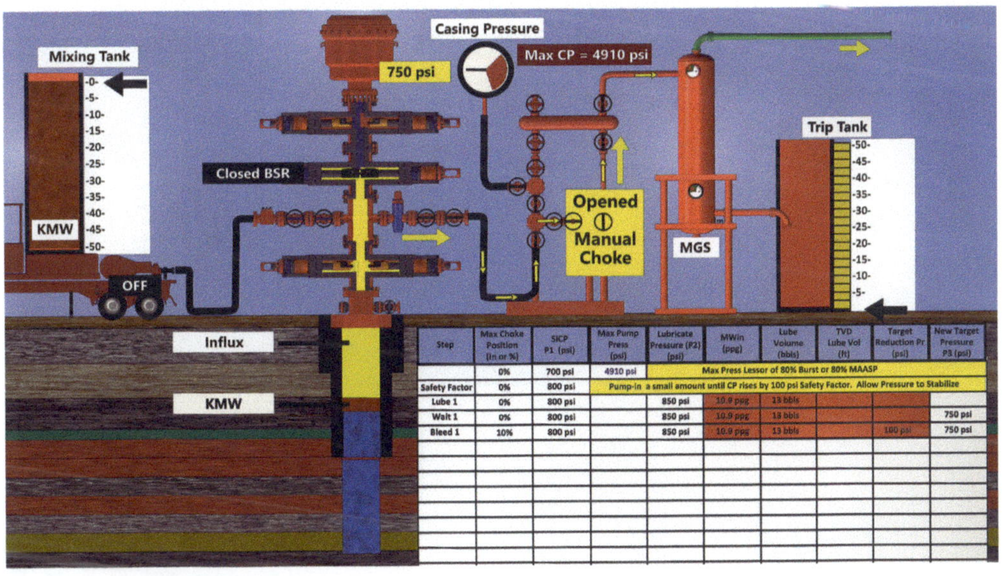

Step	Max Choke Position (In or %)	SICP P1 (psi)	Max Pump Press (psi)	Lubricate Pressure (P2) (psi)	MWin (ppg)	Lube Volume (bbls)	TVD Lube Vol (ft)	Target Reduction Pr (psi)	New Target Pressure P3 (psi)
	0%	700 psi	4910 psi	Max Press Lessor of 80% Burst or 80% MAASP					
Safety Factor	0%	800 psi	Pump-in a small amount until CP rises by 100 psi Safety Factor. Allow Pressure to Stabilize						
Lube 1	0%	800 psi		850 psi	10.9 ppg	13 bbls			
Wait 1	0%	800 psi		850 psi	10.9 ppg	13 bbls			750 psi
Bleed 1	10%	800 psi		850 psi	10.9 ppg	13 bbls		100 psi	750 psi

Fig. 3.27 Allow influx to migrate and BHP to increase.

Repeat steps 1 through 3 until conclusion

Continue performance of cycles of Lubricate, Wait, and Bleed until influx has been removed from well. As the influx volume in annulus becomes smaller, the effects of lubrication loom larger. Variances in both volumes and pressures are to be expected as lubrication continues, but overall lubrication volumes should diminish in size and pressure as the influx is evacuated from the well. Once whole KMW is circulated out the bypass line, mud may be circulated across the well before the BOPs are opened.

ALWAYS ASSUME STORED WELL PRESSURE REMAINS AFTER LUBRICATION AND BLEED ARE FINISHED.
DO NOT OPEN BOPS, AS TRAPPED PRESSURE MAY VIOLENTLY BE RELEASED
RESULTING IN LOWERING THE BHP AND INTRODUCING KICK

Any lubrication and bleed method is used to introduce a mud cap and decrease chances of uncontrolled release of pressurized fluids from the wellbore. After placement of the mud cap, circulation of the entire well will need to be reestablished in order to finalize well control using constant bottom-hole pressure method. Tubulars and reestablishing circulation may require tripping or stripping with current equipment complement or mobilizing snubbing unit (Fig. 3.28).

$$\overline{BHP} = \overline{HP}_{\substack{Casing\\Annulus}} + \overline{SICP}$$

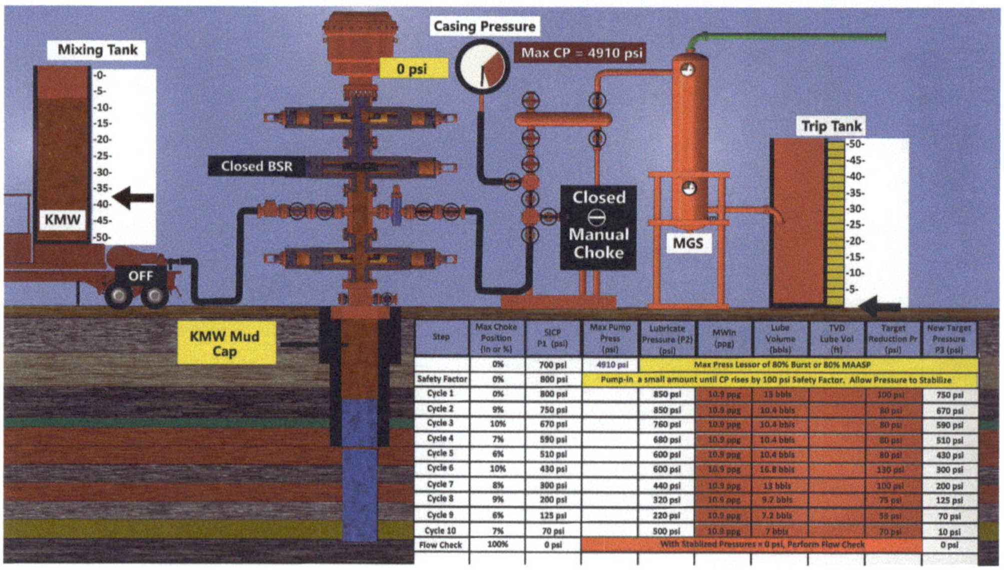

Step	Max Choke Position (In or %)	SICP P1 (psi)	Max Pump Press (psi)	Lubricate Pressure (P2) (psi)	MWin (ppg)	Lube Volume (bbls)	TVD Lube Vol (ft)	Target Reduction Pr (psi)	New Target Pressure P3 (psi)
	0%	700 psi	4910 psi	Max Press Lessor of 80% Burst or 80% MAASP					
Safety Factor	0%	800 psi	Pump-in a small amount until CP rises by 100 psi Safety Factor. Allow Pressure to Stabilize						
Cycle 1	0%	800 psi		850 psi	10.9 ppg	13 bbls		100 psi	750 psi
Cycle 2	9%	750 psi		850 psi	10.9 ppg	10.4 bbls		80 psi	670 psi
Cycle 3	10%	670 psi		760 psi	10.9 ppg	10.4 bbls		80 psi	590 psi
Cycle 4	7%	590 psi		680 psi	10.9 ppg	10.4 bbls		80 psi	510 psi
Cycle 5	6%	510 psi		600 psi	10.9 ppg	10.4 bbls		80 psi	430 psi
Cycle 6	10%	430 psi		600 psi	10.9 ppg	16.8 bbls		130 psi	300 psi
Cycle 7	8%	300 psi		440 psi	10.9 ppg	13 bbls		100 psi	200 psi
Cycle 8	9%	200 psi		320 psi	10.9 ppg	9.7 bbls		75 psi	125 psi
Cycle 9	6%	125 psi		220 psi	10.9 ppg	7.2 bbls		55 psi	70 psi
Cycle 10	7%	70 psi		500 psi	10.9 ppg	7 bbls		70 psi	10 psi
Flow Check	100%	0 psi	With Stablized Pressures = 0 psi, Perform Flow Check						0 psi

Fig. 3.28 With influx lubricated out, mud cap should provide sufficient density to kill well.

Derivation the lube and bleed pressure method

Using the ideal gas law and neglecting temperature changes, determinations of effects on gas (pressure, volume, and temperature) can be made at the beginning and end of the lubrication cycle. Based on Boyle's Law, a pressure relationship can be developed for lube and bleed with losses where pressures are the only factors which can accurately measured as whole mud volumes are lost to the formation.

$$P_3 = \frac{(P_1)^2}{P_2} \tag{3.8}$$

Lubricate and bleed action plan

The Lubricate and Bleed Method Action plan employs a mud cap to kill and control well pressures. The Lubricate and Bleed method is employed when the influx has reached the surface (after volumetric control) or on production wells currently offline with influx at surface. Using a Lubricate and Bleed Kill sheets, a checklist has been added along with notes for prejob safety meeting, as well as recap of the Lubricate and Bleed Method along with information need while lubricating and bleeding.

A review of the Lubricate and Bleed Method Action Plan begins with reviewing information within the Prerecorded Information Sheet and calculated volumes. This information is transferred to the Lubricate and Bleed Method Action Plan to ensure all appropriate information are available within this worksheet.

Considerations

> **Note:** Lubricate and Bleed operations normally occur in the annulus. Ensure Burst Pressure of Casing or LOT/FIT Shoe Pressures are not Exceeded

Fig. 3.29 Burst pressure note.

If correct gas expansion methods have been employed, the influx at the surface will exhibit a pressure less than casing burst or LOT/FIT pressure (Fig. 3.29).

Checklist for lubricate and bleed

Since Lubricate and Bleed Method is not readily used in drilling operations, the following checklist has been developed to help remind field personnel of important steps to be used before initiating operations. Collect all relevant and pertinent well data and develop detailed wellbore schematic with volumes, etc.

For precise control of lubrication and bleed cycles, a manual choke is strongly suggested. As manual chokes are not routinely used for well control, these chokes should be isolated to ensure proper operation. The chokes should be serviced to ensure smooth operations. A casing pressure should be located within easy sight of the manual choke. As an option, a drillpipe pressure gauge should also be located within easy sight of the manual choke.

A high-pressure, low-volume pump (cement pump or fracture pump) is suggested for use during lubricate cycles. This pump, lines, and manifolds should be properly installed, tested, and isolated against personnel traffic. If applicable, a supply line from the active pit system or mixing of KMW must take place before operations begin.

Lubricate and Bleed Checklist

- ☐ Collect all relevant and pertinent well data
- ☐ Develop Detailed Wellbore Schematic with Volumes, etc.
- ☐ **Manual Choke** to be used for L&B. Service as necessary. If applicable, prepare secondary Manual Choke.
- ☐ A calibrated trip tank is to be used and manifolded downstream of the choke to accurately measure the amount of fluid bled from the well.
- ☐ Ensure CP gauge is available on Choke Manifold and easily view from Manual Choke. Optional, install Pump Pressure gauge on Choke Manifold.
- ☐ Visually inspect wellhead, flowlkines and equipment for leaks, alignment and overall condition.
- ☐ Install High Pressure / Low Volume pump for accurate Lube cycles. Ensure fluid transfer lines are installed and tested.
- ☐ Close and lock Blind Rams. Do not use Annular Preventer.
- ☐ For operations with workstring, close and lock appropriate Pipe Rams after ensuring rams will not close on tooljoints. Do not use Annular Preventer. Install Top Drive and lock in place with chains to prevent upward movement.
- ☐ Prepare the MGS for service. If applicable, fill mud leg with KMW.
- ☐ Note: As the influx is liberated through MGS, the transition from influx to mud will be difficulte to observe at the MGS. Ensure MGS gauge(s) are operable. When pressure rises as mud is produced, route MGS to emergency by-pass line.

Fig. 3.30 Lubricate and bleed checklist.

For lubricate and bleed operations, strongly suggest not using the Annular Preventer due to wear and pressures which might rise above 1000 psi. Isolate the wellbore by closing and locking appropriate blind rams. For operations with work string, close and lock appropriate pipe ram. Ensure pipe rams are free of any tooljoints before closure. After closure, install Top Drive and lock same down with chains to prevent upward movement.

Prepare MGS for service by ensuring mud leg is free of debris and contains KMW. If applicable, ensure MGS pressure gauge is installed and functioning. As influx is liberated through MGS, the transition from influx to mud will be difficult to observe at MGS. Ensure MGS gauge(s) are operable. When pressure rises as mud is produced, route MGS to emergency by-pass line (Fig. 3.30).

Prejob safety meeting

Prejob Safety Meeting

- Ensure all H_2S monitors are engaged and operating, especially in pit area.
- Degasser to remain on throughout lube and bleed operations.
- MGS mud leg to be filled with Kill Mud Weight (KMW) prior to initiation of circulation.
- Prepare by-pass line to be used when mud reaches the surface.
- Valves downstream of choke are to be verified as open.
- With manual choke(s) isolated, ensure chokes are operable. Service if needed.
- Personnel (notifications made by loudspeaker).
  - All personnel to be on-station during circulation.
- Driller
  - Monitors Active System Volume and records volume every 10 minutes.
- Pump Company Service Personnel
  - Install and test all pump lines and manifolds.
  - Mix volume of KMW or install/test KMW feed lines from Active Pit System.
- AD to perform continuous walkabout for leaks.
- Shale Shaker staff to monitor returns. All Changes are to be reported.
- Mud Co. Rep.
  - Mix KMW and MW_{in} every 15 min. or until asked to change.
  - Provide Barite + Mud and/or Base Oil addition rate (bbls/hr).
  - Use Active pipe system, if possible.
  - Any and all mud transfers to be announced. Driller to record.
- Choke Operator records Time, MW_{in}, Pump Pressure, SCR, Max Choke Position, Lube Vol, TVD Lube Volume, Target Reduction Pressure, New Target Pressure.
- **Step 1:** Open choke and pump KMW while monitoring CP with volume pumped.
- **Step 2:** Shut-off pumps, close-in and allow time for fluids to swap.
- **Step 3:** Bleed off Influx to Target Reduction Pressure. DO NOT BLEED OFF MUD. Use only manual choke.
- Repeat Steps 1 to 3 until influx has been evacuated.

Fig. 3.31 Prejob safety meeting.

With checklist performed, information required to review within a Prejob Safety Meeting are displayed (Fig. 3.31). This comprehensive recap was developed as an outline and must be amended for individual rig and personnel requirements. The Prejob Safety Meeting recap should be used as a reminder to perform specific duties prior to initializing lube and bleed operations. Ensure all questions are answered.

Lubricate and bleed method for no losses downhole (variable mud volume)

Volumetric Lube and Bleed Method (No Losses Downhole)

- For ***no losses*** *downhole*, the Volumetric Lube and Bleed Method with Variable Mud Volume (Most Common Method) may be used. This process employs pumping KMW into the well, waiting for the fluids to swap, then control bleeding influx from hole. Cease process when first whole mud returns are validated.
- Record stabilized SICP (if applicable, stabilized SIDPP)
- Determine Maximum Casing Pressure limit (Lessor of 80% Casing Burst or 80% of MASP). This maximum rating may be downrated due to deteriorated condition of wellhead, tubulars, type of service (i.e. sour service), quality of cement jobs, age of well, etc.
- Determine KWW (ppg). Remember to round up.

Step 1: Slowly open choke and pump KMW into well while monitoring CP and ensuring pump volumes are accurately recorded.

- Do not exceed lessor of 80% of MASP or 80% of casing burst.
- Shut-down and shut-in.
- Using the pumped volumes, determine a Target Reduction Pressure for bleed cycle.

Step 2: With pumps shut-off and well shut-in, wait and allow adequate time for the KMW to fall through the influx commonly referred to as "fluids swap".

- Adjust wait time based upon actual results

Step 3: Slowly bleed to achieve desired Target Pressure Reduction through small incremental choke adjustments. Use only manual choke.

Repeat Steps 1-3: Until all influx has been evacuated.

- See Universal Well Control manual for "Set Mud Volume", as an alternative.

Fig. 3.32 Lubricate and method no losses downhole (variable mud volume) recap.

The Lubricate and Bleed for NO Losses Downhole Method with Variable Mud Volume (Most Common Method) may be used (Fig. 3.32). This process employs pumping KMW into well, waiting for fluids to swap, then control bleeding influx from hole. Cease process when first whole mud returns are validated.

The process begins by recording stabilized SICP (if applicable, stabilized SIDPP). Limits for Maximum Pressure Rating are determined by the minimum rating of 80% of casing burst or 80% of MASP. The Maximum Pressure rating may also be reduced or downrated due to deteriorated condition of wellhead, tubulars, type of service (i.e., sour service), quality of cement jobs, age of well, etc.

From the SICP (or SIDPP), the KMW (ppg) is determined and rounded up. Once known, mix and ready a sufficient volume of KMW.

After lubrication and bleed cycle preparations are complete, begin the three-step lubrication and bleed process.

Step 1: Slowly open choke and pump KMW into the well while monitoring CP ensuring volumes are accurately recorded.

- Do not exceed lessor of 80% of MASP or 80% of Casing Burst. Maximum pressure may be downrated due to deteriorated condition of wellhead, tubulars, type of service (i.e., sour service), quality of cement jobs, age of well, etc.
- Shut down and shut-in.
- Using the pumped volumes, determine a Target Reduction Pressure for bleed cycle.

$$\mathrm{TVD}_{\mathrm{KMW\ Lube\ Cycle\ 1}} = \frac{\text{Lube Cycle 1 Volume}_{\mathrm{bbl}}}{\text{Annular Volume}_{\mathrm{bbl/ft}}} \tag{3.9}$$

$$\mathrm{HP}_{\mathrm{Lube\ Cycle\ 1}} = (0.052)\left(\mathrm{KMW}_{\mathrm{ppg}}\right)\left(\mathrm{TVD}_{\mathrm{KMW\ Lube\ Cycle\ 1}}\right) \tag{3.10}$$

$$\text{Target Reduction Pressure}_{\mathrm{Lube\ Cycle\ 1}} = \mathrm{SICP}_{\mathrm{Initial}} - \mathrm{HP}_{\mathrm{Lube\ Cycle\ 1}} \tag{3.11}$$

Step 2: With pumps shut off and well shut-in, wait and allow adequate time for the KMW to fall through the influx commonly referred to as "fluids swap."

- Adjust wait time based upon actual results.

Step 3: Slowly bleed to achieve desired Target Pressure Reduction through small incremental choke adjustments.

- Use only the manual choke.

Repeat Steps 1 to 3: Until all influx has been evacuated.

Lubricate and bleed with losses downhole (pressure)

Volumetric Lube and Bleed Method (With Losses Downhole)

- For wells with ***losses downhole***, the Volumetric Lube and Bleed Method with Set Mud Volume may be used. This process employs pumping a predetermined volume of mud into the well and using changes to the SICP to determine target pressure. Small bleed cycles are performed until the target pressure is achieved. Repeat process until first whole mud returns are validated and observed.
- Limitation dependent upon Surface Equipment.
- A calibrated trip tank is to be used and manifolded downstream of the choke to accurately measure the amount of fluid bled from well.
- With losses downhole, Lube and Bleed Method with Se Mud Volume can be outlined as follows:

1. Record SICP (if applicable, SIDPP)
2. Choose a safety factor of 50 – 100 psi.
3. Pump-in measured volume of fluid which will raise the casing pressure by the safety factor.
4. Record Initial SICP including the safety factor – this is P_1.
5. Pump-in a measured volume of fluid (do not exceed the lessor or 80% of MASP or 80% of Casing Burst. Maximum pressure may be downrated due to deteriorated condition of wellhead, tubulars, type of service (i.e., sour service), quality of cement jobs, age of well, etc.)
6. Allow the fluids to swap with gas (if unknown, use 1 hour to begin)
7. Record stabilized shut-in caign pressure – this is P_2
8. Calculate P_3

$$P_3 = \frac{(P_1)^2}{P_2}$$

9. Bleed off only gas (no fluid) down to the pressure of P_3.
10. P_3 becomes P_1 for the next cycle.
11. Repeat process until influx has been removed from well.

Fig. 3.33 Lubricate and bleed with losses downhole (set mud volume) recap.

The Lubricate and Bleed With Losses Downhole Method (Pressure) (Fig. 3.33). This process employs pumping a predetermined volume into well, using changes to SICP to determine target pressure, then bleed off influx until the calculated target pressure is achieved. Repeat process until first whole mud returns are validated or observed.

- Limitations dependent upon Surface Equipment.
- A calibrated trip tank is to be used and manifolded downstream of choke to accurately measure the amount of fluid bled from well.

With losses downhole, Lubrication and Bleed Method with Set Mud Volume can be outlined as follows:

1. Record SICP (if applicable, SIDPP).
2. Choose a safety factor of 50–100 psi.

3. Pump-in a measured volume of fluid which will raise the casing pressure by the safety factor.
4. Record Initial SICP including the safety factor—this is P_1.
5. Pump in a measured volume of fluid (do not exceed lessor of 80% of MASP or 80% of Casing Burst. Maximum pressure may be downrated due to deteriorated condition of wellhead, tubulars, type of service (i.e., sour service), quality of cement jobs, age of well, etc.)
6. Allow the fluid to swap with the gas (if unknown, use 30 min to begin).
7. Record stabilized shut-in casing pressure—this is P_2.
8. Calculate P_3.

$$P_3 = \frac{(P_1)^2}{P_2} \tag{3.12}$$

9. Bleed off only gas (no fluid) down to a pressure of P_3.
10. P_3 becomes P_1 for the next cycle
11. Repeat process until influx has been removed from the well

Common problems during lube and bleed operations

The problems discussed in the following section are commonly found in well control situations. Many occur while circulating and are not really applicable to lube and bleed. However, as conditions change, circulating instead of lubing and bleeding may become necessary. Some problems described may occur during the lube and bleed process.

1. **Partial or Total lost Circulation**
   a. May change from Volumetric Lubrication and Bleed Method (No Losses Downhole) to Lubrication and Bleed Method with Losses Downhole.
   b. Heavy mud pills with LCM may be considered for use for losses, once the influx has been evacuated from well.

2. **Plugged choke**—Can occur at any time, therefore choke opening values and DP and CP must be continually recorded. If no liquid or gas is released during bleeding plugging of the choke may be occurring (Fig. 3.34).

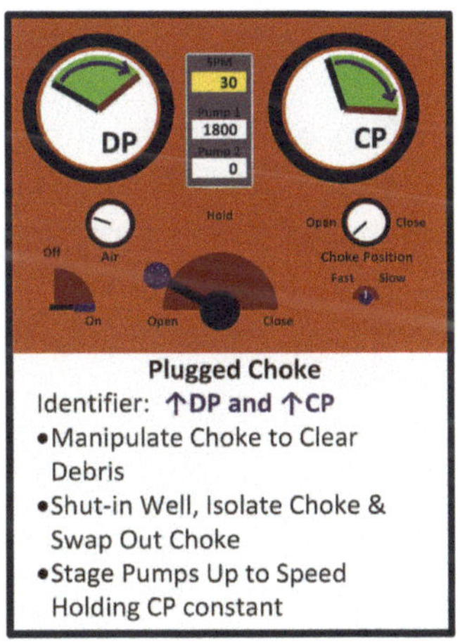

Fig. 3.34 Plugged choke.

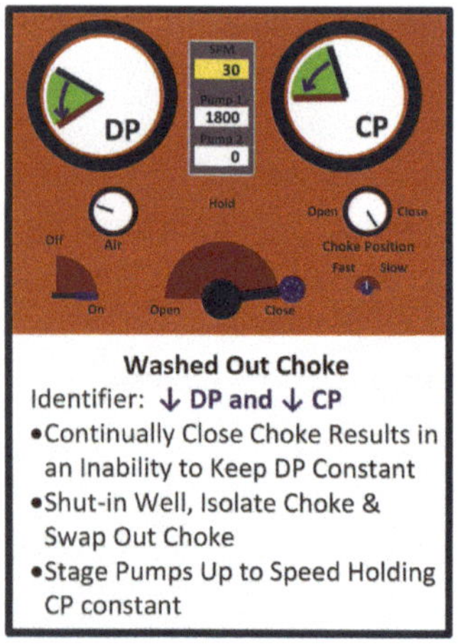

Fig. 3.35 Washed out choke.

3. **Washed out choke**—Can occur at any time. If the choke leaks in the fully closed position, close the valve in front of it and switch to a new choke (Fig. 3.35).

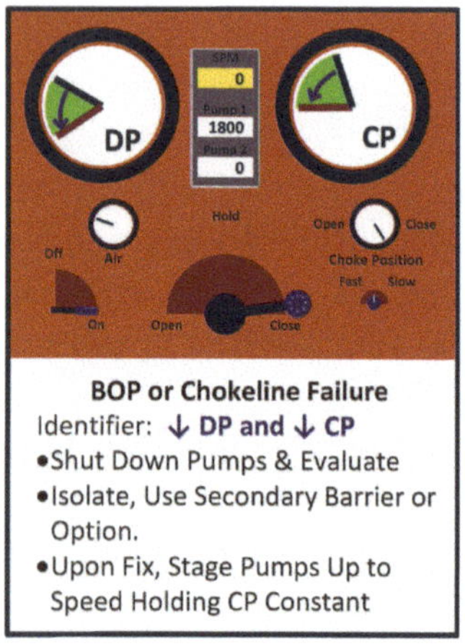

Fig. 3.36 BOP/choke line failure.

4. **BOP or choke line failure**—Can occur at any time. Drillpipe pressure and casing will decrease without warning. Shut down operations immediately and investigate equipment failure, isolate and repair BOP/choke line. Use secondary barrier (i.e., BOP) as needed (Fig. 3.36).

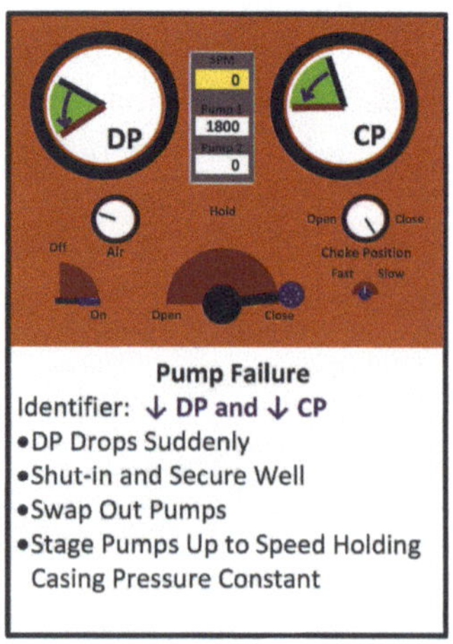

Fig. 3.37 Pump failure.

5. **Pump Failure**—Can occur at any time. Shut down pumps immediately and shut-in well. Isolate defective pump and line-up replacement pump (Fig. 3.37).

6. **Mud Won't Swap with Gas**—If mud flows from the well when the bleed cycle begins, shut the well back in and wait longer for the mud and gas to swap. In viscous muds, it may take an hour or longer for this to occur. No significant quantity of mud should be bled from the well during the bleed cycle. Any mud returned should be measured. Bleed pressures need to take into account any mud recovered. The volume of mud recovered should have the length of that volume calculated and then its hydrostatic pressure calculated. This pressure is subtracted from the hydrostatic pressure of the volume of mud pumped for each cycle in which this occurs.
7. **Annular Pressure Approaches MASP**—If the annular pressure reaches MASP with little or no fluid pumped, then either there is no gas at the surface or so little that lube and bleed is no longer possible. It may not be possible to bleed off all the pressure even if all the gas has been bled off. The well may still be underbalanced and does not have sufficient mud weight to overcome the formation pressure.

MGS observations, lube and bleed conclusion

During lubrication and bleed cycles with fluids bled to the Mud Gas Separator (MGS), the ability to determine the exact timing of when mud returns to surface will be difficult to identify. As the fluids returning into the MGS are not visible, identification may occur by observing increases in pressure on the MGS vessel pressure gauge. If the MGS does not have a pressure gauge, the by-pass may have to be engaged multiple times to identify fluid returns without influx.

Prior to opening the BOPs, ensure the rig floor is cleared of operational personnel. Once personnel have been cleared, open the BOPs. There always remains a chance of trapped pressure below the element, which may violently expand and push particulate matter and fluid from the well. If personnel are within close proximity, this violent expansion may cause significant injuries.

After opening the BOPs, allow time for the fluids to stabilize. Perform an extended flow check to ensure the well is completely dead (Fig. 3.38).

MGS/Conclusion

During lubrication and bleed cycles with fluids bled to the Mud Gas Separator (MGS), the ability to determine the exact timing of when mud returns to surface will be difficult to identify. As the fluids returning into the MGS are not visible, identification may occur by observing increases in pressure on the MGS vessel pressure gauge. If the MGS does not have a pressure gauge, the bypass may have to be engaged multiple times to identify fluid returns without influx.

1. After bleeding and verifying no influx, shut-in well.
2. Route Chokeline returns to Trip Tank and isolate MGS. Open choke and bleed pressure. Flow should cease, if not Shut-in Well.
3. Route standpipe returns to Trip Tank and bleed any trapped pressure.
4. Perform flow check at Trip Tank
5. Clear Rig Floor and Open BOPs
6. Perform Flow Check

Fig. 3.38 Conclusion of well control operations.

Volumetric well control method (passive)

Commentary

Volumetric control is not a well killing technique, rather is a way to maintain control of a well until circulation can be established. Volumetric control is constrained to emergency situations and is considered to be a passive type of well control as no pumping or circulation is required. Volumetric control can be used to overcome wellbore constraints such as:

- Plugged workstring with kick in well.
- Inoperative pumps with kick in the well.
- Gas migrating while waiting to commence or recommence well kill operations.
- Viscous water-based fluids (brines difficult to control, OBM limited influx expansion or migration).
- If at any time during the Volumetric Control, circulation can be established, use one of the more conventional circulating well control methods can be initiated to finish removing the influx from the well.

To use volumetric control effectively, three basic principles must be understood and used:

1. Boyle's Law ($P_1V_1 = P_2V_2$)
2. Hydrostatic Pressure
3. Volume and Height of Influx.

Volumetric control is the controlled expansion of a bubble as it migrates up the hole while keeping bottom-hole pressure nearly constant. If a gas kick cannot be circulated from the well, gas migration may occur resulting in high surface, casing shoe and bottom-hole pressures. To minimize this, it will be necessary to allow the influx to expand in a controlled fashion as it migrates up the well bore.

Surface pressure will be allowed to increase to give an overbalance as safety margin. Then the pressure on surface will be allowed to increase to by pressure increment (operating range), and then the corresponding volume of mud will be bled off, while maintaining the surface pressure constant. Then the surface pressure will be allowed to increase again and so on. For example, if the casing pressure is allowed to increase by 100 psi, a volume of mud equal to 100 psi of hydrostatic pressure is bled off.

When gas reaches the surface, the well should remain shut-in until circulation can be established. Depending on well complications, lube and bleed can be initiated or the first circulation of the Driller's Method can be employed to remove the influx from the well.

For volumetric well control to work effectively, the following conditions must exist.

- Surface equipment must include a calibrated trip (strip) tank tied-in downstream of the choke. As fluid exists the choke, the fluid is routed to the trip tank for collection and measurement.
- In case gas escapes from the choke to the trip tank, an emergency by-pass or MGS by-pass must be readied for emergency operation.

- Hazardous gas monitors must be operable on trip tank.
- Additional PPE protection should be made available for rig crews including portable SCBA.
- The primary (if applicable back-up) manual choke should be isolated, serviced, and readied for operation.
- Ensure trip tank pump and overflow systems are operable. For large volumes, the trip tank may have to be emptied periodically.

Volumetric control serves straight holes with viscous fluids where influx rise travels an estimated 4000 ft/h. The viscous fluids help impede influx travel allow for improved choke responses when compared to simple brines. Influx may rise faster in brines requiring faster choke manipulations along with faster bleed cycles.

Wells with high angles present additional difficulties, as influxes travel at higher rates up the high sides of these wells. For angle wells, the rise and bleed cycles may be more difficult to control.

Volumetric control procedure (blockage of DP/tubing)

The Volumetric Control procedure can be used when a restriction or blockage in the workstring prevents accurate monitoring of shut-in drillpipe/tubing pressure.

Determine safety margin (migrate)

Select appropriate Safety Margin (psi) based on SICP which is easy to read on the casing pressure gauge such as 100 psi. This safety factor is only applied once at the start of the Volumetric Method. As the influx rises with no expansion, the SICP will rise. In order to control BHP, the influx is allowed to migrate with no expansion until a Safety Margin is established. This can be represented by the following equation, where the BHP increases by the Safety Margin (Fig. 3.39).

$$\downarrow BHP = \downarrow HP_{\substack{Casing\\Annulus}} + \overline{SICP}$$

Fig. 3.39 Annular forces when establishing a safety margin.

The bleed range for holding bottom-hole pressures nearly constant will be determined by reviewing the fracture pressure range, which is the difference between Maximum Allowable Surface Pressure (as determined by Leak-Off test or Formation Integrity Limit) and the shut-in casing pressure. This can be expressed mathematically as:

$$\text{Fracture Pressure Range}_{psi} = MASP_{psi} - SICP_{psi} \tag{3.13}$$

Determine bleed volume

First, select a Bleed Increment Pressure (psi) less than the Safety Factor which is easy to read on the casing pressure gauge such as 50–100 psi. The Bleed Increment Pressure is normally one-fourth to ½ of the Safety Margin. The Bleed Increment Pressure is the range for casing pressure increase as determined by the Bleed Volume (bleed volume is determined by bleed pressure; the pressure is not determined by the volume). The Bleed Increment (psi) will also determine the amount of hydrostatic pressure reduction which will be permitted on each mud bleed/gas expansion cycle.

BLEED INCREMENT PRESSURE IS NORMALLY ¼ to ½ OF THE SAFETY MARGIN
BLEED VOLUME IS DETERMINED BY SELECTED BLEED PRESSURE

Convert Selected Bleed Increment Pressure (psi) into Bleed Volume (bbls)

$$\text{Bleed Volume}_{\text{bbl}} = \frac{\text{Bleed Pressure}_{\text{psi}} \times \left(\text{Casing Capacity}_{\text{bbls/ft}} - \text{Tubular Displacement}_{\text{bbl/ft}}\right)}{\left(\text{MW}_{\text{ppg}}\right) \times (0.052)} \tag{3.14}$$

where

$\text{Bleed Pressure}_{\text{psil}}$ = Selected Bleed Pressure (psi)
$\text{Casing Capacity}_{\text{bbl/ft}}$ = Casing capacity of casing annulus (bbl/ft)
$\text{Tubular Displacement}_{\text{bbl/ft}}$ = Displacement of tubular in well (bbls/ft)
MW_{ppg} = Mud weight in well (ppg)
0.052 = conversion constant (psi/ft./ppg).

Rig up for volumetric control

In preparation to perform volumetric control method, surface systems must be prepared. The trip tank must be emptied and well lighted for night operations. The choke manifold should be aligned to perform bleed cycles using a manual choke. If a casing pressure gauge is not currently installed on the choke manifold, one should be installed. A back-up choke should be prepared in case of plugging. Both chokes should be checked for opening and closing prior to commencing operations.

NEVER BLEED GAS LADEN OR LIBERATED GASES FROM WELL DURING VOLUMETRIC CONTROL,
SHUT-IN WELL AND ALLOW PRESSURE TO STABLIZE.

Volumetric control equipment preparations must also include preparations in case unexpected gases become liberated. Although not directly used for volumetric control, these systems must be readied for personnel and asset protection and to direct gases away from trip tank and rig.

The MGS should be circulated to ensure the mud leg contains no settled plugs. The MGS may be circulated with KMW as determined by SICP + HP annulus. Normally, MGS are routed with returns to the treatment mud tank with equalization to the trip tank. This equalization line is normally closed to isolate the trip tank.

The choke manifold emergency by-pass line (sometimes called the "blooey" line) should be readied in the event MGS becomes inoperable. The exact sequence of operation for routing gas laden fluids to the MGS or through the emergency by-pass is to be discussed as part of the Volumetric Control plan.

Step 1: Establish safety margin (migrate)

Simply monitor and wait for the SICP to rise and establish the Safety Margin (200 psi) (Fig. 3.40).

$$\downarrow BHP = \overline{HP}_{\substack{Casing\\Annulus}} + \downarrow SICP$$

Fig. 3.40 Annular forces establishing a safety margin.

Step 2: Establish bleed increment pressure PLUS safety margin

Just like establishing the Safety Margin, the well is simply observed until the SICP reaches the desired Bleed Increment Pressure plus the Safety Margin (Fig. 3.41).

$$\uparrow BHP = \overline{HP}_{\substack{Casing\\Annulus}} + \uparrow SICP + \uparrow Bleed_{Increment}$$

Fig. 3.41 Annular forces establishing a safety margin and bleed increment.

Step 3: Bleed increment

In order to maintain BHP greater than formation pressures and prevent additional influxes from entering the well, the original influx is allowed to expand under control by bleeding the measured bleed increment into the trip tank. The controlled bleeding process is accomplished by slowly opening the manual choke in small increments, while bleeding whole mud from the wellbore back into the trip tank. The bleed cycle is to be performed while attempting to hold Casing Pressure constant. To ensure bleed cycles are performed with precision, experience has shown these cycles can be best accomplished using a manual choke, if the choke manifold has been outfitted with a casing pressure gauge which can be read from the manual choke. **DO NOT BLEED MUD VOLUMES GREATER THAN THE CALCULATED BLEED INCREMENT**, otherwise an additional influx may be taken.

MANUAL CHOKES OFFER GREATER PRECISION WHEN PERFORMING VOLUMETRIC CONTROL, IF THE CHOKE MANIFOLD IS OUTFITTED WITH A CASING PRESSURE GAUGE.

After obtaining the Bleed Increment Pressure, the controlled expansion of the influx will be equal to the volume of fluid bled from the well. Subsequently, the BHP is reduced but due to the Safety Margin, the BHP is still greater than formation pressures resulting in no additional influxes entering the well (Fig. 3.42).

$$\downarrow \mathrm{BHP} = \downarrow \mathrm{HP}_{\substack{\text{Casing}\\ \text{Annulus}}} + \overline{\mathrm{SICP}}$$

Fig. 3.42 Forces when performing bleed.

Step 4: Migrate

BHP was reduced by the equivalent hydrostatic pressure of the Bleed Increment; the influx is allowed to migrate with no expansion, allowing the SICP to increase by the Bleed Increment Pressure. The rise in SICP is determined by the rate of rise of the influx, which can take minute to hours depending upon well complexities and fluid parameters. The following equation describes the migration step (Fig. 3.43).

$$\uparrow \mathrm{BHP} = \overline{\mathrm{HP}}_{\substack{\text{Casing}\\ \text{Annulus}}} + \uparrow \mathrm{SICP}$$

Fig. 3.43 Forces when influx migrates.

Step 5: Perform bleed increment/migrate cycle

The Bleed Increment and Migration cycles are repeated until gas reaches the surface or circulation can be reestablished, whichever comes first. If properly executed, the Safety Margin will ensure no additional influxes will have entered the well. With influx at surface, SICP will be at its maximum value. The well should be shut-in until such time circulation can be reestablished and the well killed. This may be accomplished by lubricate and bleed or performing constant bottom-hole pressure method for well control (Driller's Method or Wait and Weight Method).

Volumetric procedure

- Record SIDPP and SICP and kick volume.
- Determine Safety Margin (100 psi)
- Ensure high-pressure flowline from choke manifold downstream of choke is routed to the Trip Tank.
- Calculate Bleed Increment (usually about ½ of the safety factor or 50 psi).

Initiate volumetric control

1. Establish Safety Margin: Wait and allow SICP to rise by Safety Margin (100 psi).
2. Establish Pressure Increment Plus Safety Margin: Wait and allow SICP to rise by Bleed Increment Pressure plus Safety Margin (150 psi).

3. Perform Bleed: Slowly bleed the Bleed Increment Volume while attempting to keep casing pressure constant. Use manual choke (the hydrostatic pressure of the mud bled off is equal to the pressure increase on the casing) while keeping casing pressure constant.
4. Establish New Pressure Increment Plus Safety Margin. Wait and allow SICP to rise by New Bleed Increment Pressure plus Safety Margin.
5. Repeat Steps 3 and 4 Until Conclusion: Perform Bleed and Migrate cycles until influx has reached surface or circulation can be reestablished (whichever comes first).

Volumetric control method example

Given

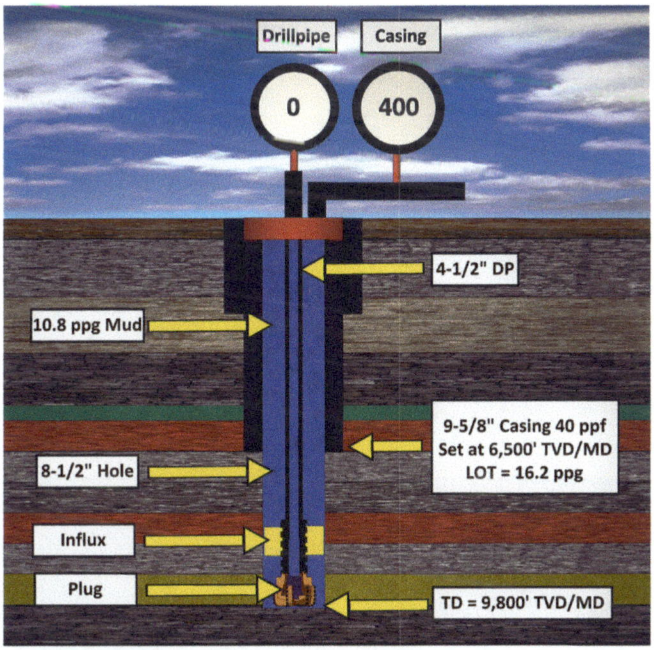

Fig. 3.44 Example volumetric control well.

While drilling an 8–1/2″ straight hole at a depth of 9800 with water-based 10.8 ppg mud, the drillpipe became plugged. Subsequent varying pump speeds and moving pipe did not clear blockage. The crews noted the well was flowing and shut-in and secured well (Fig. 3.44).

Vertical well
MW = 10.8 ppg (Water-Based Mud)
Kick size = 20 bbl
SICP = 400 psi (Stabilized)
Bit Depth = 9800 psi
9–5/8″, 40 ppf casing at 5000′ MD/TVD.
LOT = 16.2 ppg
4–1/2″ DP (Annular Casing Capacity Factor = 0.0561 bbl/ft)
Drift diameter used for ease of calculations.

$$\text{MASP} = (\text{LOT} - \text{MW})(0.052)(6500\,\text{ft}).$$
$$= (16.2 - 10.8)(0.052)(6500').$$
$$= 1825\,\text{psi}$$

$$\text{Fracture Pressure Range}_{psi} = \text{MASP}_{psi} - \text{SICP}_{psi} \tag{3.15}$$

$$\text{Fracture Pressure Range}_{psi} = 1825\ \text{psi} - 400\ \text{psi} = 1425\ \text{psi} \tag{3.16}$$

An appropriate Safety Factor is selected as 100 psi based upon a fracture pressure range of 1425 psi. During the volumetric control process, the controlled expansion will maintain the BHP nearly constant and not allow the shoe to fracture.

Calculate the bleed volume

Based on the safety margin of 100 psi, an appropriate bleed increment pressure is selected as 50 psi.

$$\text{Bleed Volume}_{bbl} = \frac{\text{Bleed Pressure}_{psi} \times \left(\text{Casing Capacity}_{bbls/ft} - \text{Tubular Displacement}_{bbl/ft}\right)}{\left(\text{MW}_{ppg}\right) \times (0.052)} \tag{3.17}$$

$$\text{Bleed Volume}_{bbl} = \frac{50\ \text{psi} \times (0.0561\ \text{bbls/ft})}{(10.8\ \text{ppg}) \times (0.052)} = 5\ \text{bbls} \tag{3.18}$$

Shut-in conditions

Shut-in Conditions include SIDPP = 0 psi, SICP = 400 psi, and 20 bbl gain (Fig. 3.45).

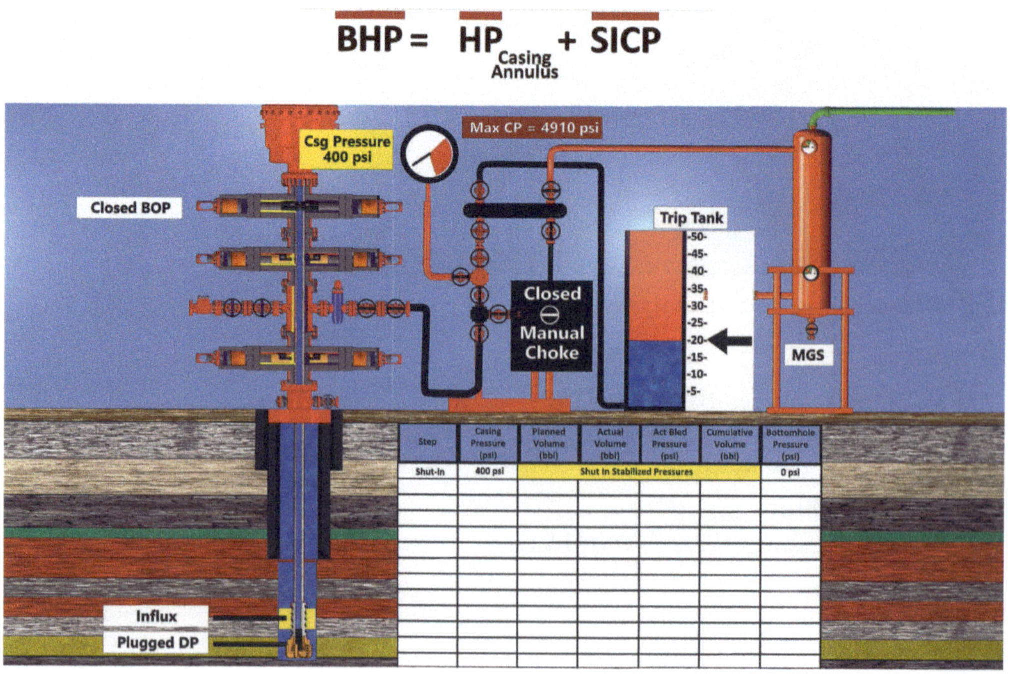

Fig. 3.45 Shut-in conditions.

Rig up to perform volumetric control method

Establish the flow route from BOP to choke manifold. Route to manual choke. Route from manual choke to trip tank. Ensure trip tank is cleaned and emptied. Install lights as needed to perform nighttime operations.

Step 1: Establish safety margin (migrate)

Simply wait and allow influx to rise without expansion by observing SICP. The SICP should rise to the predetermined Safety Margin of 100 psi (Fig. 3.46).

$$SICP = 400 \text{ psi} + 100 \text{ psi} = 500 \text{ psi} \tag{3.19}$$

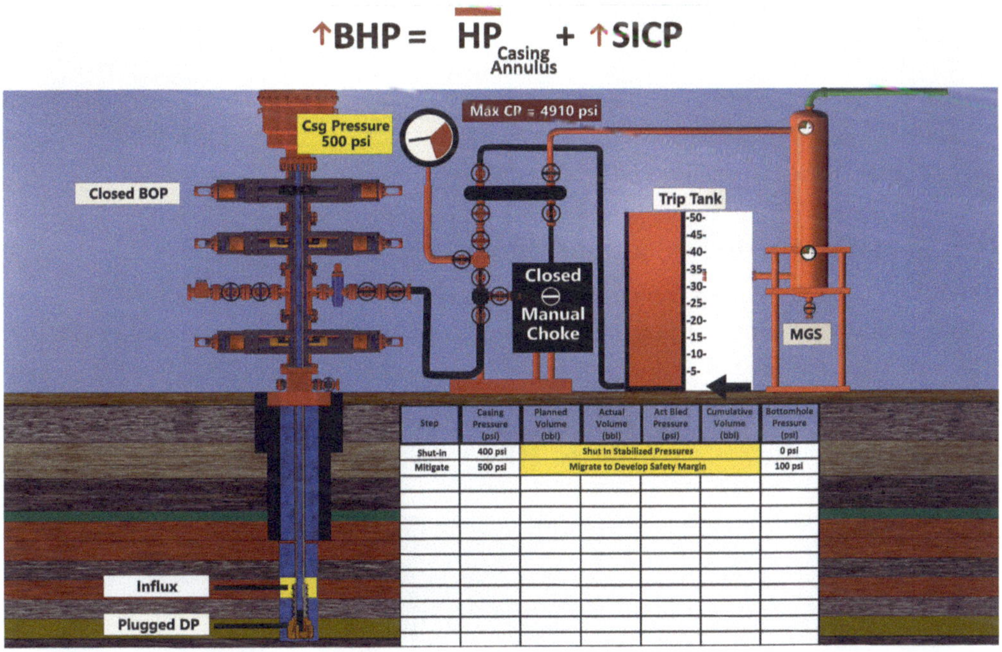

Fig. 3.46 Establish safety margin.

Step 2: Establish pressure increment plus safety margin (migrate)

Simply wait and allow influx to rise without expansion by observing SICP. The SICP should rise to the predetermined Bleed Incremental Pressure plus Safety Margin (50 psi + 100 psi) (Fig. 3.47).

$$SICP = 400 \text{ psi} + 100 \text{ psi} + 50 \text{ psi} = 550 \text{ psi} \tag{3.20}$$

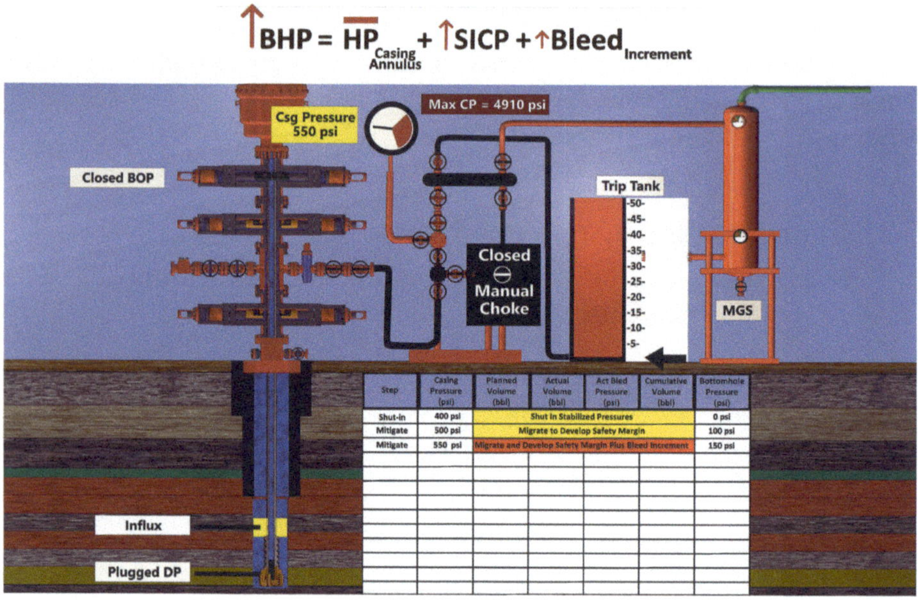

Fig. 3.47 Establish bleed increment plus safety margin.

Step 3: Perform bleed (bleed increment)

Slowly bleed whole mud to trip tank in small increments while attempting to hold casing pressure constant. Continue intermittent bleeds and closures until 5 bbls have been bled back into trip tank (Fig. 3.48).

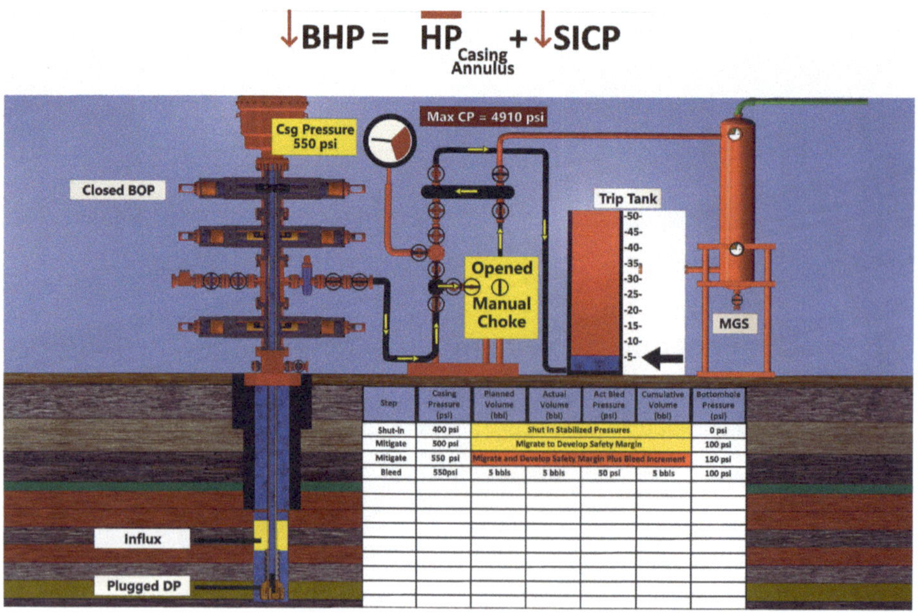

Fig. 3.48 Bleed 5 bbls to trip tank in small increments.

Step 4: Establish bleed increment pressure plus safety margin

By bleeding back 5 bbls, the annular hydrostatic pressure has been reduced by 50 psi. Allow influx to migrate without allowing expansion (choke closed) and regain the bleed increment pressure of 50 psi (Fig. 3.49).

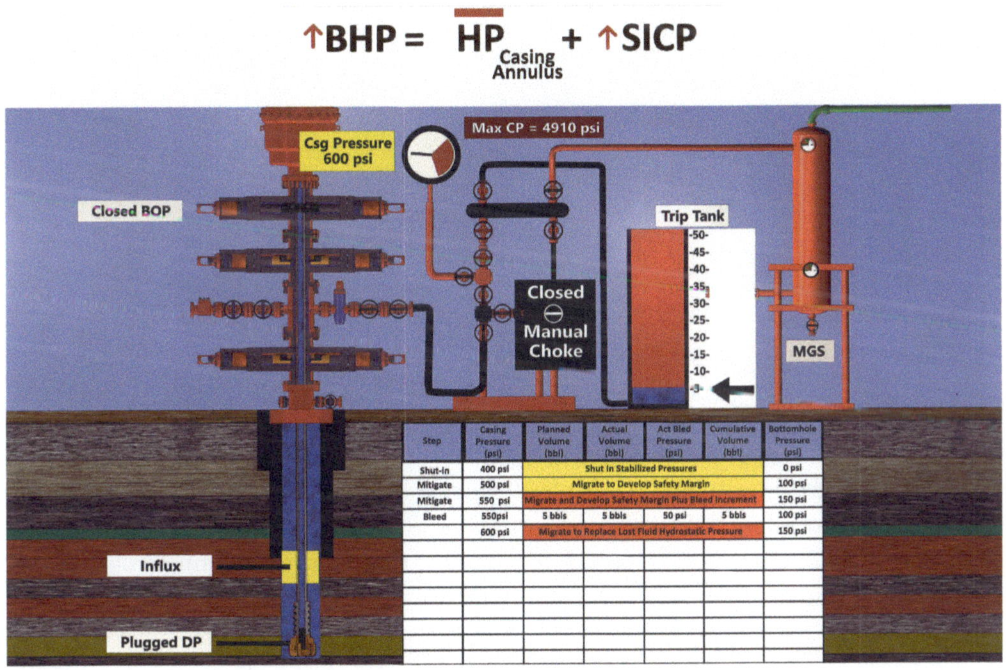

Fig. 3.49 Allow influx to migrate, reestablish SM (BHP increases).

Repeat steps 3 and 4 until conclusion

Repeat Steps 3 and 4 until conclusion has been reached. Perform Bleed and Migrate Increments until circulation can be reestablished or the influx has reached the surface (whichever comes first) (Fig. 3.50).

$$\downarrow BHP = \downarrow HP_{\text{Casing Annulus}} + \overline{SICP}$$

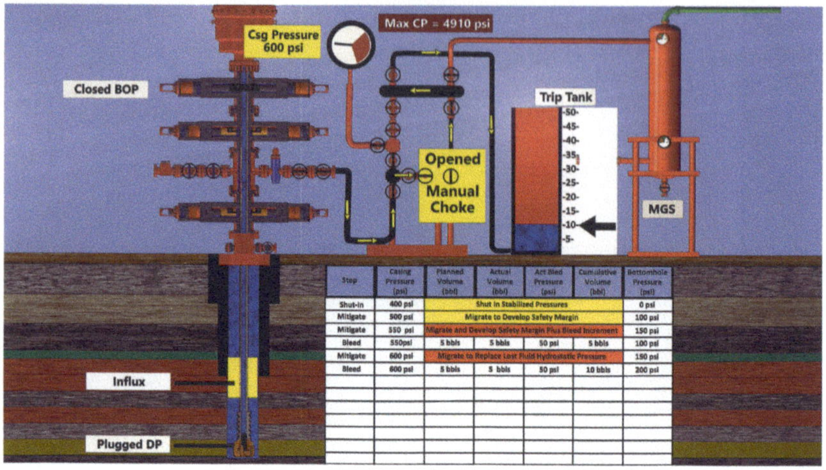

Fig. 3.50 Bleed 4 bbls to trip tank in small increments.

Conclusion

Continue to perform Bleed and Migrate increments and recording all information. As the influx begins to get close to surface, the bleed/migrate cycles will become faster. As gas starts to break out, maintaining BHP may be more difficult as gas laden fluids will need to be routed through the MGS. Once as gas is at surface, continue to monitor well until circulation can be reestablished or the influx has reached the surface (whichever comes first) (Fig. 3.51).

VOLUMETRIC METHOD IS COMPLETE WHEN GAS IS AT THE SURFACE AND CASING PRESSURE REMAINS CONSTANT

$$\overline{BHP} = \overline{HP}_{\text{Casing Annulus}} + \overline{SICP}$$

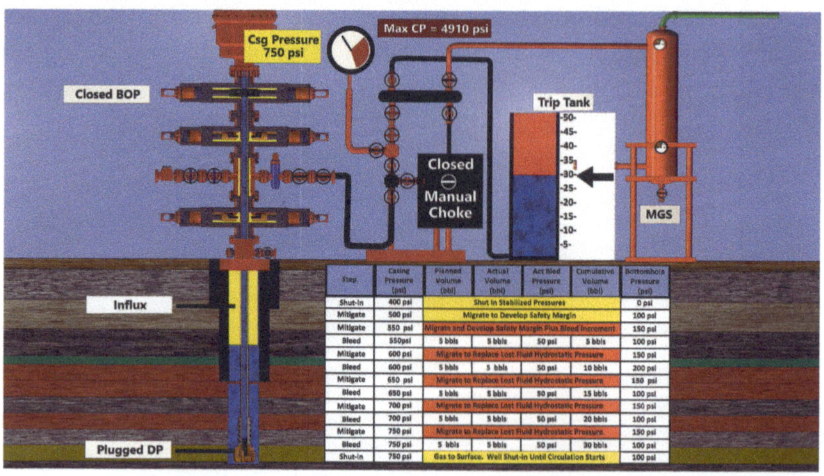

Fig. 3.51 Continue bleed and migrate cycles until influx is at surface.

The loss of hydrostatic pressure in annulus can be determined from the following formula.

$$\text{Loss of HP}_{\text{annulus}} = \left(\frac{\text{Mud Gradient}_{\text{psi/ft}} - 0.104_{\text{psi/ft}}}{\text{Capacity}_{\text{annulus}}}\right) \times \text{Total Vol Bled}_{\text{bbl}} \tag{3.21}$$

where

Loss of $\text{HP}_{\text{annulus}}$ = Loss of Hydrostatic Pressure in annulus (psi)
Mud Gradient = Mud Gradient (MW × 0.052) (psi/ft)
0.104 = Gas Gradient Constant based on 2 ppg gas (psi/ft)
$\text{Capacity}_{\text{annulus}}$ = Annular Capacity (bbl/ft)
Total Vol Bled = Total volume bled from annulus (bbl).

At the time gas reaches to surface, 30 bbls had been bled off. Substituting into the equation above yields,

$$\text{Loss of HP}_{\text{annulus}} = \left(\frac{(0.052 \times 10.8\,\text{ppg}) - 0.104_{\text{psi/ft}}}{0.0561}\right) \times 30\,\text{bbls} = 245\,\text{psi} \tag{3.22}$$

Stripping well control method

Commentary

Stripping is the process of lowering the workstring in the well with closed blowout preventers during well control. As most well control situations arise during tripping operations with bit off bottom, the need may arise where the mud cap hydrostatic pressure may jeopardize casing shoe fracture gradient. In these instances, stripping the well to total depth will allow the well to be killed with less dense kill mud weights which may not jeopardize casing shoe fracture gradients. Stripping procedures are limited to low pressure well control situations where the current pipe weight is sufficient to overcome the upward force created by well pressure. The well pressure acts against the cross-sectional area of the pipe creating forces which act to push the workstring from the well. The pressure resistance envelope can be estimated to ensure the well is a candidate for stripping operations.

RULE OF THUMB: STRIPPING OPERATIONS SHOULD BE CONFINED TO WELLS WITH LESS THAN 1000 PSI CASING PRESSURE AND CONTAIN OIL-BASED OR SYNTHETIC OIL BASED MUDS

DO NOT ATTEMPT STRIPPING ON WELLS WITH WATER-BASED MUDS UNTIL INFLUX HAS BEEN MIGRATED TO THE SURFACE AND REMOVED FROM THE WELLBORE BY THE LUBE AND BLEED PROCESS

Stripping operations should be confined to wells with less than 1000 psi casing pressure. Since stripping will be accomplished with the BOP Annular Preventer which acts as a second barrier, the probability of leakage (or failure) of the annular preventer increases with every foot stripped in the hole. For stripping operations with wells with over 1000 psi, experience has shown the need to mobilize a snubbing unit to accomplish safe operations.

Stripping operations should generally be confined to wells with oil-based or synthetic oil-based muds. In deeper horizons, if the influx is gaseous, the hydrostatic pressure of the fluid column may be sufficient to keep the gases within the liquid state of the oil-based mud and the influx will not readily rise. Conversely, water-based muds will allow gases to rise at velocities approaching 4000 ft/h, ensuring a complex operation as the gases must be allowed to expand while attempting stripping operations. Therefore, stripping operations should not be attempted on wells with water-based muds until the influx has been migrated to the surface using the Volumetric Method and then removed from the wellbore by using the Lube and Bleed Process.

After determining selection criteria for stripping operations, a specific stripping well control procedure must be developed and adjusted to suit well conditions and available equipment. There are two types of stripping operations which can be considered.

- Annular stripping (common)
- Ram-to-ram stripping (limited applications)

Annular stripping uses existing rig equipment where the annular closing pressure is manipulated to allow passage of tooljoints during stripping operations. With wide variety of pressure control equipment and arrangements on various sized workover and drilling rigs, annular stripping offers the widest coverage.

WHEN UNDER HIGH ANNULAR PRESSURE CONDITIONS, DRILL STRING WEIGHT MAY NOT BE SUFFICIENT TO PERFORM STRIPPING OPERATIONS

Ram-to-ram stripping is limited to those rigs in which an equalization loop can be installed. An equalization loop extends between two sets of pipe rams and would require two circulating (drilling) spools for installation. Since average BOP installations only include one circulating (drilling) spool, ram-to-ram stripping is normally constrained to Class V BOP stacks (four ram preventers and one annular) used in High-Pressure/High-Temperature applications. For the most part, ram-to-ram stripping should be limited to snubbing applications or managed by Well Control provider.

If the well contains significant wellbore pressure, the upward forces generated by the pressure may exceed workstring weight and if left unchecked, will force tubulars from the well.

RAM-TO-RAM STRIPPING SHOULD BE CONSIDERED A SPECIALIZED SERVICE
SHOULD ONLY BE PERFORMED BY WELL CONTROL PROVIDER

Snubbing refers to when the pipe weight is not sufficient to overcome the upward force created by well pressure on the cross-sectional area of the pipe. Snubbing requires that an external force must be applied to move the pipe through the BOPs. To ensure personnel safety and environmental protection, with higher wellbore pressures, specialized equipment and experience will be needed to successfully provide well control (see Snubbing section of manual for more information).

Two methods for stripping have been developed.

1. *Oil-Based Stripping Method*—Tracks time, footage, and return volume. This method allows for maximum footage stripped before attempting to bleed off additional volume and allow influx to expand. Tubulars are stripped with accurate return volume tracked. Based upon predetermined footage or time, tubulars are stripped until limit threshold has been achieved. At this point, an expansion bleed occurs to recover a predetermined volume. Once the hydrostatic pressure of annulus is reduced, the casing pressure is allowed to increase to offset loss in hydrostatic pressure.
2. *Volumetric Stripping Method*—Tracks footage, return volume, and additional expansion volume. This method requires accurate volume tracking to ensure tubular displacement is correct plus a small additional volume for influx expansion. The expansion bleed occurs with every stand stripped. The casing pressure is increased after the additional volume for influx expansion reaches a predetermined level.

Stripping checklist

A simple stripping checklist has been developed to assist rigsite personnel in preparations for safe stripping operations (Table 3.1).

Table 3.1 Equipment needed for stripping operations.

Equipment needed
☐ Calibrated Stripping Tank routed to Trip Tank and Choke Manifold
☐ Operation Manual Choke
☐ 0–3000 psi gauge on Choke Manifold which can be viewed from Manual Choke
☐ Surge bottles for annular
☐ Equipment to remove burrs from tooljoints
☐ DP rubber removal tool
☐ Inside BOP
☐ Oil to Lubricate Annular Element (do not use pipe dope as this lubricant contains solids which can damage the annular element)
Procedure
☐ Detailed Stripping Procedure
☐ Shut-in and monitor well
☐ Prepare and test equipment

Continued

Table 3.1 Equipment needed for stripping operations—cont'd

Equipment needed
☐ Develop Performance Criteria
☐ Review annular closing pressure Cycle
Operations
☐ Pre-Job Safety Meeting
☐ Strip to establish SM
☐ Strip while bleeding and holding casing pressure constant
☐ Make connections as needed
☐ Perform bleed cycles for casing pressure increases if necessary

OBM/SOBM and water-based mud

The type of fluid in the how will greatly impact stripping operations. The following rules of thumb can be applied:

1. For Oil-Based and Synthetic-Oil Based muds with kick in hole and off bottom, stripping operations are greatly simplified as the influx will remain in solution reflected in small changes to casing pressure. After obtaining a sufficient safety margin, the tubulars are simply stripped in the hole and observing casing pressure. If the casing pressure rises 50% more than the safety margin, shut down, shut-in with full operating pressure and perform bleed cycle-based theoretical closed-in displacement of the tubulars multiplied by number of feet stripped. Depending on parameters, it may be possible to strip to bottom without performing a bleed cycle.
2. For Water-Based muds with kick in hole and off bottom, stripping operations become more complicated as the influx rises. **Therefore, stripping operations with water-based muds are highly discouraged and should not be attempted.** In order to alleviate increasing bottom-hole pressures, the influx must be allowed to expand (4000 ft./h). By time plans are made and equipment lined-up, influx may be near surface. Casing pressures must be held constant by using Volumetric Control. Depending on hole parameters, it may be preferable to allow the influx to reach the surface under a controlled environment using Volumetric Control. Once the influx is a surface, the influx can be removed using lubricate and bleed method while maintaining nearly constant bottom hole pressure. After influx has been removed, proceed to stripping to bottom to kill well or kill well using mud cap.

Determine bleed method to be used

Traditionally, most reference materials written speak of constantly bleeding fluid while tubulars are being stripped in the hole. Field histories of using the constant bleed method have demonstrated its use for experienced crews who have previously used this method.

The premise of constant bleed method is to control bottom-hole pressures within a small envelope to ensure the casing shoe is not fractured. If such a small window exists, it would be preferable to mobilize a Well Control Company of choice. Since well control practices (specifically stripping operations) are nonroutine activities for Drilling and Workover rig crews, an experienced Well Control professional is suggested.

As an alternative, a stop and bleed method may be employed. By keeping track of the tubular footage and volume stripped into a closed-system, casing pressures are monitored. When casing pressure rises 50% above the safety margin, stripping is stopped and a bleed cycle is performed.

Determination if mud cap exceeds LOT/FIT

If a kick is taken while tripping pipe and the well is safely shut-in, a decision will need to be made whether to attempt an off-bottom kill or strip back to bottom. The distance from the bit to TD, kill weight mud needed, water- or oil-based mud, surface pressures, and hole conditions are all factors in this determination. A calculation of kill weight mud needed to affect an off-bottom kill can assist in this decision process.

Once the shut in surface pressures are known, the mud weight needed from the bit to the surface can be calculated. A denser kill mud will be necessary at a bit depth when compared to kill mud density at total depth. An elevated kill mud weight may result in producing a hydrostatic pressure (at the casing shoe) which exceeds fracture pressures as defined by Formation Integrity Test or Leak-Off Test. The result would be initiating fractures and causing losses while attempting to kill the well.

THE HIGHER OFF BOTTOM A BIT WILL BE, THE HIGHER THE KMW DENSITY WILL NEED TO BE TO KILL WELL.

Before stripping commences, a determination of whether stripping can be accomplished without fracturing the casing shoe needs to be developed. The following series of calculations will assist in this determination.

1. Divide current SIDPP by 0.052 and the true vertical depth of the bit off bottom. This calculation estimates the additional mud weight needed to be added to current mud weight for KMW off bottom

$$\Delta MW_{\text{Bit Off Bottom TVD}} = \frac{\text{SIDPP}}{(0.052) \times \text{Bit TVD}} \tag{3.23}$$

where

$\Delta MW_{\text{Bit off Bottom TVD}}$ = Change in mud weight at current bit depth (ft)
SIDPP = Shut-In Drillpipe Pressure (psi)
0.052 = Conversion Constant (psi/bbl/ft))
Bit TVD = True Vertical Depth to Bit (ft).

$$KMW_{\text{Bit Off Bottom}} = OMW_{\text{ppg}} + \Delta MW_{\text{Bit Off Bottom TVD}} \tag{3.24}$$

where

$KMW_{Bit\ Off\ Bottom}$ = KMW at current bit depth (ppg)

OMW = Original Mud Weight or Current Mud Weight (ppg)

ΔMW = Change in mud weight at current bit depth (ft).

2. Compare the hydrostatic pressure of calculated $KMW_{bit\ off\ bottom}$ to the casing/liner shoe to the LOT/FIT

$$\text{If KMW Bit Off Bottom} > \text{LOT/FIT, well is a candidate for stripping} \tag{3.25}$$

3. Determine KMW on bottom

$$KMW_{TD\ TVD} = \frac{SIDPP_{psi}}{(0.052)(TD\ TVD_{ft})} + OMW_{ppg} \tag{3.26}$$

where

$KMW_{TD\ TVD}$ = KMW at Total Depth TVD (ft)

SIDPP = Shut-In Drillpipe Pressure (psi)

0.052 = conversion constant (psi/bbl/ft)

TD TVD = True Vertical Depth to Bit (ft)

OMW = Original Mud Weight or Current Mud Weight (ppg).

Mud cap determination (bit off bottom)

The example below illustrates the calculations required to determine whether an off-bottom kill is possible. Generally speaking, a well killed off bottom will require a heavier density kill mud when compared to killing a well on bottom. If the KMW at current bit depth exceeds limitations of the casing shoe, the bit will have to be stripped in the hole where a lower KMW can be used to circulate out the influx and kill the well (Fig. 3.52).

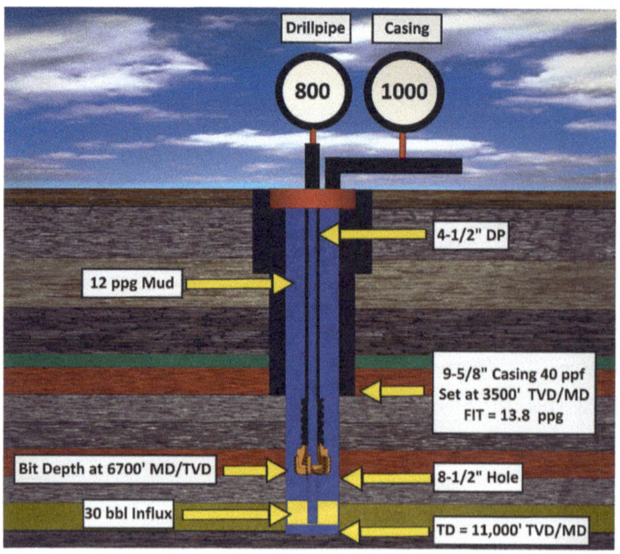

Fig. 3.52 Off-bottom kill decision example.

- **Vertical hole**
- MW = 12 ppg (Oil-Based Mud)
- SIDPP = 1000 psi
- SICP = 1200 psi
- Kick size = 30 bbl
- Bit Size = 8.5 in.
- Hole TD = 11,000′ TVD/MD
- Bit Depth = 6700 ft.
- **FIT = 13.8 ppg at 3500′ TVD/MD**
- Kicked during Trip, Off Bottom

1. Divide current SIDPP by 0.052 and the true vertical depth of the bit off bottom. This calculation estimates the additional mud weight needed to be added to current mud weight for KMW off bottom.

$$\Delta MW_{\text{Bit Off Bottom TVD}} = \frac{\text{SIDPP}}{(0.052) \times \text{Bit TVD}} \tag{3.27}$$

$$\Delta MW_{\text{Bit Off Bottom TVD}} = \frac{800 \text{ psi}}{(0.052) \times 6700 \text{ ft}} = 2.29 \text{ ppg or } 2.3 \text{ ppg} \uparrow \tag{3.28}$$

$$KMW_{\text{Bit Off Bottom}} = OMW_{\text{ppg}} + \Delta MW_{\text{Bit Off Bottom TVD}} \tag{3.29}$$

$$KMW_{\text{Bit Off Bottom}} = 12.0 \text{ ppg} + 2.3 \text{ ppg} = 14.3 \text{ ppg} \tag{3.30}$$

2. Compare the hydrostatic pressure of calculated $KMW_{\text{bit off bottom}}$ to the casing/liner shoe to the LOT/FIT
   a. If $KMW_{\text{bit off bottom}} >$ LOT/FIT, well is a candidate for stripping

$$KMW = 14.3 \text{ ppg} > 13.8 \text{ ppg FIT, candidate for stripping} \tag{3.31}$$

3. Determine KMW on bottom

$$KMW_{\text{TD TVD}} = \frac{\text{SIDPP}_{\text{psi}}}{(0.052)(\text{TD TVD}_{\text{ft}})} + OMW_{\text{ppg}} \tag{3.32}$$

$$KMW_{\text{TD TVD}} = \frac{800 \text{ psi}}{(0.052)(11,000 \text{ ft})} + 12 \text{ ppg} = 13.8 \text{ ppg} \tag{3.33}$$

Balance point

Balance point is defined as the transition depth point between stripping and snubbing. If the workstring combined weights (Bit, BHA, and Workstring) are greater than the force generated by wellbore pressure over the cross-sectional outer diameter of largest tooljoint, then the well is a candidate for stripping. If the workstring weight is less than wellbore forces, the workstring would have to be "forced" into the well which requires employing snubbing operations. For determining balance point for stripping into a pressurized fluid, a separate two-step calculation is needed.

- **IF BALANCE POINT WEIGHT > WELL FORCE, WELL IS A CANDIDATE FOR *STRIPPING* OPERATIONS**
- **IF BALANCE POINT WEIGHT < WELL FORCE, WELL IS A CANDIDATE FOR *SNUBBING* OPERATIONS**

Well force determination

Well force is the force acting upward against the largest diameter portion of the workstring to be stripped into the well.

$$\text{Well Force}_{\text{lbs}} = \text{Largest OD}_{\text{in.}^2} \times 0.7854 \times \text{Wellhead Pressure}_{\text{psi}} \tag{3.34}$$

where

Well Force = Well Force Acting Upward (lbs)
Largest OD = Diameter of Maximum Tooljoint Collar (in.^{2})
0.7854 = Conversion Constant ($\pi/4$)
Wellhead Pressure = Shut-in Well Pressure at Tree (psi).

Balance point determination

Well force is the force acting upward against the largest diameter portion of the workstring to be stripped into the well. (See Drillpipe Specifications.) If balance point weight at current bit depth is greater than well force, the string can be stripped into the well.

$$\text{Workstring}_{\text{lbs}} \downarrow = \left(\text{DP Weight}_{\text{ppf}} \times \text{DP Length}_{\text{ft}}\right) + \left(\text{DC Weight}_{\text{ppf}} \times \text{DC Length}_{\text{ft}}\right) \tag{3.35}$$

where

Workstring = Workstring Weight Acting Downward (lbs)
DP Weight = Drillpipe Weight (ppf)—See Drillpipe Specification Table
DP Length = Length of Drillpipe (ft)
DC Weight = Drill collar Weight (ppf)—See Drill collar Specification Table
DC Length = Length of Drillpipe (ft) (Tables 3.2–3.4).

Table 3.2 Drillpipe dimensions table (part 1 of 3).

Drillpipe specifications

Pipe						Tooljoint						
Size OD (in.)	Nominal weight (lb/ft)	Upset type	Grade	Wall thickness (in.)	ID (in.)	Connection type	OD (in.)	OD decimal (in)	ID (in.)	ID decimal (in)	Pin tong space[a] (in.)	Box tong space[a] (in.)
2–3/8″	6.65	EU	E-75	0.280	1.815	NC26	3–3/8″	3.375	1–3/4″	1.750	9	10
			X-95	0.280	1.815	NC26	3–3/8″	3.375	1–3/4″	1.750	9	10
			G-105	0.280	1.815	NC26	3–3/8″	3.375	1–3/4″	1.750	9	10
			S-135	0.280	1.815	NC26	3–5/8″	3.625	1–3/4″	1.750	9	10
2–7/8″	10.40	EU	E-75	0.362	2.151	NC31	4–1/8″	4.125	2–1/8″	2.125	9	11
			X-95	0.362	2.151	NC31	4–1/8″	4.125	2″	2.000	9	11
			G-105	0.362	2.151	NC31	4–1/8″	4.125	2″	2.000	9	11
			S-135	0.362	2.151	NC31	4–3/8″	4.375	1–5/8″	1.625	9	11
3–1/2″	13.30	EU	E-75	0.368	2.764	NC38	5″	5.000	2–11/16″	2.688	10	12 1/4
			X-95	0.368	2.764	NC38	5″	5.000	2–9/16″	2.563	10	12 1/4
			G-105	0.368	2.764	NC38	5″	5.000	2–7/16″	2.438	10	12 1/4
			S-135	0.368	2.764	NC38	5″	5.000	2–1/8″	2.125	10	12 1/4
	15.50	EU	E-75	0.449	2.602	NC38	5″	5.000	2–9/16″	2.563	10	12 1/4
			X-95	0.449	2.602	NC38	5″	5.000	2–7/16″	2.438	10	12 1/4
			G-105	0.449	2.602	NC38	5″	5.000	2–1/8″	2.125	10	12 1/4
			S-135	0.449	2.602	NC40	5–1/2″	5.500	2–1/4″	2.250	9	12
4	14.00	IU	E-75	0.330	3.340	NC40	5–1/4″	5.250	2–13/16″	2.813	9	12
			X-95	0.330	3.340	NC40	5–1/4″	5.250	2–11/16″	2.688	9	12
			G-105	0.330	3.340	NC40	5–1/2″	5.500	2–7/16″	2.438	9	12
			S-135	0.330	3.340	NC40	5–1/2″	5.500	2–7/16″	2.438	9	12

Continued

Table 3.2 Drillpipe dimensions table (part 1 of 3)—cont'd

Drillpipe specifications

Pipe						Tooljoint						
Size OD (in.)	**Nominal weight (lb/ft)**	**Upset type**	**Grade**	**Wall thickness (in.)**	**ID (in.)**	**Connection type**	**OD (in.)**	**OD decimal (in)**	**ID (in.)**	**ID decimal (in)**	**Pin tong space (in.)**	**Box tong space (in.)**
		EU	E-75	0.330	3.340	NC46	6″	6.000	3–1/4″	3.250	9	12
			X-95	0.330	3.340	NC46	6″	6.000	3–1/4″	3.250	9	12
			G-105	0.330	3.340	NC46	6″	6.000	3–1/4″	3.250	9	12
			S-135	0.330	3.340	NC46	6″	6.000	3″	3.000	9	12
	15.70	IU	E-75	0.380	3.240	NC40	5–1/4″	5.250	2–11/16″	2.688	9	12
			X-95	0.380	3.240	NC40	5–1/4″	5.250	2–7/16″	2.438	9	12
			G-105	0.380	3.240	NC40	5–1/2″	5.500	2–7/16″	2.438	9	12
			S-135	0.380	3.240	NC40	5–1/2″	5.500	2″	2.000	9	12
		EU	E-75	0.380	3.240	NC46	6″	6.000	3–1/4″	3.250	9	12
			X-95	0.380	3.240	NC46	6″	6.000	3–1/4″	3.250	9	12
			G-105	0.380	3.240	NC46	6″	6.000	3–1/4″	3.250	9	12
			S-135	0.380	3.240	NC46	6″	6.000	3″	3.000	9	12

[a]2″ longer than standard.

Table 3.3 Drillpipe dimensions table (part 2 of 3).

Drillpipe specifications

Pipe						Tooljoint						
Size OD (in.)	Nominal weight (lb/ft)	Upset type	Grade	Wall thickness (in.)	ID (in.)	Connection type	OD (in.)	OD decimal (in)	ID (in.)	ID decimal (in)	Pin tong space[a] (in.)	Box tong space[a] (in.)
4–1/2″	16.60	IEU	E-75	0.337	3.826	NC46	6–1/4″	6.250	3″	3.000	9	12
			X-95	0.337	3.826	NC46	6–1/4″	6.250	3–1/4″	3.250	9	12
			G-105	0.337	3.826	NC46	6–1/4″	6.250	2–3/4″	2.750	9	12
			S-135	0.337	3.826	NC46	6–1/4″	6.250	2–3/4″	2.750	9	12
		EU	E-75	0.337	3.826	NC50	6–5/8″	6.625	3–3/4″	3.750	9	12
			X-95	0.337	3.826	NC50	6–5/8″	6.625	3–3/4″	3.750	9	12
			G-105	0.337	3.826	NC50	6–5/8″	6.625	3–3/4″	3.750	9	12
			S-135	0.337	3.826	NC50	6–5/8″	6.625	3–1/2″	3.500	9	12
	20.00	IEU	E-75	0.430	3.640	NC46	6–1/4″	6.250	3″	3.000	9	12
			X-95	0.430	3.640	NC46	6–1/4″	6.250	2–3/4″	2.750	9	12
			G-105	0.430	3.640	NC46	6–1/4″	6.250	2–1/2″	2.500	9	12
			S-135	0.430	3.640	NC46	6–1/4″	6.250	2–1/4″	2.250	9	12
		EU	E-75	0.430	3.640	NC50	6–5/8″	6.625	3–5/8″	3.625	9	12
			X-95	0.430	3.640	NC50	6–5/8″	6.625	3–1/2″	3.500	9	12
			G-105	0.430	3.640	NC50	6–5/8″	6.625	3–1/2″	3.500	9	12
			S-135	0.430	3.640	NC50	6–5/8″	6.625	3″	3.000	9	12
5″	19.50	IEU	E-75	0.362	4.276	NC50	6–5/8″	6.625	3–3/4″	3.750	9	12
			X-95	0.362	4.276	NC50	6–5/8″	6.625	3–1/2″	3.500	9	12
			G-105	0.362	4.276	NC50	6–5/8″	6.625	3–1/4″	3.500	9	12
			S-135	0.362	4.276	NC50	6–5/8″	6.625	2–3/4″	2.750	9	12
			E-75	0.362	4.276	5–1/2 FH	7″	7.000	3–3/4″	3.750	10	12
			X-95	0.362	4.276	5–1/2 FH	7″	7.000	3–3/4″	3.750	10	12
			G-105	0.362	4.276	5–1/2 FH	7″	7.000	3–3/4″	3.750	10	12
			S-135	0.362	4.276	5–1/2 FH	7–1/4″	7.250	3–1/2″	3.500	10	12

Continued

Table 3.3 Drillpipe dimensions table (part 2 of 3)—cont'd

Drillpipe specifications

Pipe						Tooljoint						
Size OD (in.)	Nominal weight (lb/ft)	Upset type	Grade	Wall thickness (in.)	ID (in.)	Connection type	OD (in.)	OD decimal (in)	ID (in.)	ID decimal (in)	Pin tong space (in.)	Box tong space (in.)
	25.60	IEU	E-75	0.500	4.000	NC50	6–5/8″	6.625	3–1/2″	3.500	9	12
			X-95	0.500	4.000	NC50	6–5/8″	6.625	3″	3.000	9	12
			G-105	0.500	4.000	NC50	6–5/8″	6.625	2–3/4″	2.750	9	12
			S-135	0.500	4.000	NC50	6–5/8″	6.625	2–3/4″	2.750	9	12
			E-75	0.500	4.000	5–1/2 FH	7″	7.000	3–1/2″	3.500	10	12
			X-95	0.500	4.000	5–1/2 FH	7″	7.000	3–1/2″	3.500	10	12
			G-105	0.500	4.000	5–1/2 FH	7–1/4″	7.250	3–1/2″	3.500	10	12
			S-135	0.500	4.000	5–1/2 FH	7–1/4″	7.250	3–1/4″	3.250	10	12
5–1/2″	21.90	IEU	E-75	0.361	4.778	5–1/2 FH	7″	7.000	4″	4.000	10	12
			X-95	0.361	4.778	5–1/2 FH	7″	7.000	3–3/4″	3.750	10	12
			G-105	0.361	4.778	5–1/2 FH	7–1/4″	7.250	3–1/2″	3.500	10	12
			S-135	0.361	4.778	5–1/2 FH	7–1/2″	7.500	3″	3.000	10	12
	24.70	IEU	E-75	0.415	4.670	5–1/2 FH	7″	7.000	4″	4.000	10	12
			X-95	0.415	4.670	5–1/2 FH	7–1/4″	7.250	3–1/2″	3.500	10	12
			G-105	0.415	4.670	5–1/2 FH	7–1/4″	7.250	3–1/2″	3.500	10	12
			S-135	0.415	4.670	5–1/2 FH	7–1/2″	7.500	3″	3.000	10	12
6–5/8″	25.20	IEU	E-75	0.330	5.965	6–5/8 FH	8″	8.000	5″	5.000	10	13
			X-95	0.330	5.965	6–5/8 FH	8″	8.000	5″	5.000	10	13
			G-105	0.330	5.965	6–5/8 FH	8–1/4″	8.250	4–3/4″	4.750	10	13
			S-135	0.330	5.965	6–5/8 FH	8–1/2″	8.500	4–1/4″	4.250	10	13
	27.70	IEU	E-75	0.362	5.901	6–5/8 FH	8″	8.000	5″	5.000	10	13
			X-95	0.362	5.901	6–5/8 FH	8–1/4″	8.250	4–3/4″	4.750	10	13
			G-105	0.362	5.901	6–5/8 FH	8–1/4″	8.250	4–3/4″	4.750	10	13
			S-135	0.362	5.901	6–5/8 FH	8–1/2″	8.500	4–1/4″	4.250	10	13

[a]2″ longer than standard.

Table 3.4 Drill collar dimensions table (part 3 of 3).
Drill collar specifications

Size OD	Nominal weight (lb/ft)	Connection type	Body type (slick/spiral)	Range	Decimal size OD (in.^{2})	Decimal size ID (in.^{2})	Decimal drift (in.^{2})
1–3/4″	6.41	1–1/4″MT	Slick	2	1.750″	0.813″	0.625″
1–3/4″	6.67	1–1/4″MT	Slick	2	1.750″	0.750″	0.625″
1–13/16″	7.26	1–1/4″MT	Slick	2	1.313″	0.750″	0.625″
2–1/16″	8.7	1–1/2″MT	Slick	2	2.063″	1.000″	0.875″
2–7/8″	18	2–3/8″HT-PAC	Slick	2	2.875″	1.250″	1.125″
2–7/8″	18	2–3/8″HT-PAC	Spiral	2	2.875″	1.250″	1.125″
3–1/8″	21	2–7/8″HT-PAC	Spiral	2	3.125″	1.500″	1.375″
3–1/8″	22	2–7/8″HT-PAC	Spiral	2	3.125″	1.250″	1.125″
3–1/2″	27	NC26	Spiral	2	3.500″	1.500″	1.375″
4–1/8″	35	NC31	Spiral	2	4.125″	2.000″	1.875″
4–7/8″	50	NC38	Spiral	2	4.875″	2.250″	2.125″
4–7/8″	50	NC38	SLICK	2	4.875″	2.250″	2.125″
4–7/8″	53	NC38	Spiral	2	4.875″	2.000″	1.875″
4–7/8″	50	XT39	Spiral	2	4.875″	2.250″	2.125″
4–15/16″	52	XT39	Spiral	2	4.938″	2.250″	2.125″
5–1/4″	60	NC40	Spiral	2	5.250″	2.250″	2.125″
6–1/4″	83	NC46	Spiral	2	6.250″	2.813″	2.688″
6–1/2″	92	NC46	Spiral	2	6.500″	2.813″	2.688″
6–1/2″	99	NC46	Spiral	2	6.500″	2.250″	2.125″
6–1/2″	92	NC50	Spiral	2	6.500″	2.813″	2.688″
6–3/4″	101	NC50	Spiral	2	6.750″	2.813″	2.688″
7–1/4″	119	NC50	Spiral	2	7.250″	2.813″	2.688″
7–3/4″	139	NC56	Spiral	2	7.750″	2.813″	2.688″
8″	150	6–5/8″ Reg	Spiral	2	8.000″	2.813″	2.688″
8–1/4″	161	6–5/8″ Reg	Spiral	2	8.250″	2.813″	2.688″
9–1/2″	217	7–5/8″ Reg (Low Torq)	Spiral	2	9.500″	3.000″	2.875″
9–1/2″	217	7–5/8″ Reg	Spiral	2	9.500″	3.000″	2.875″
11″	299	8–5/8″ Reg (Low Torq)	Spiral	2	11.000″	3.000″	2.875″

Balance point example

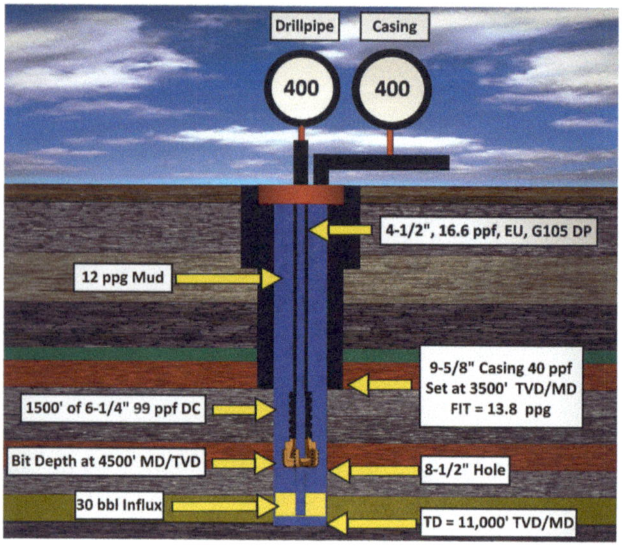

Fig. 3.53 Off-bottom kill decision example.

DP = 4–1/2″, 16.6 ppf, EU, G105 (Fig. 3.53).
Closed-in OD Capacity = 0.0197 bbl/ft.
Tooljoint Diameter = 6.625″ (from Drillpipe Specifications).
DC = 1500′ of 6–1/2″ 99 ppf (from Drill collar Specifications).
MW = 12 ppg.
SICP = 400 psi.

$$\text{Well Force}_{\text{lbs}} = \text{Largest OD}_{\text{in.}^2} \times 0.7854 \times \text{Wellhead Pressure}_{\text{psi}} \tag{3.36}$$

$$\text{Well Force}_{\text{lbs}} = (6.625)^2 \times 0.7854 \times 400 \text{ psi} = 137{,}000 \text{ lbs} \uparrow \tag{3.37}$$

$$\text{Workstring}_{\text{lbs}} \downarrow = \left(\text{DP Weight}_{\text{ppf}} \times \text{DP Length}_{\text{ft}}\right) + \left(\text{DC Weight}_{\text{ppf}} \times \text{DC Length}_{\text{ft}}\right) \tag{3.38}$$

$$\text{Workstring}_{\text{lbs}} = (16.7 \text{ ppf} \times 4500 \text{ ft}) + (99 \text{ ppf} \times 1500 \text{ ft}) = 198{,}300 \text{ lbs} \downarrow \tag{3.39}$$

$$\text{Workstring}_{\text{lbs}}\ (198{,}300 \text{ lbs}) > \text{Well Force}\ (137{,}000 \text{ lbs}), \text{Drillstring can be stripped in hole.} \tag{3.40}$$

Stripping considerations

Once the well has been determined to be a candidate for stripping operations, a review of considerations should be conducted. The types of muds used for drilling and working over wells may determine the initial stripping procedure. The types of muds may define how and when stripping operations may be considered. Since stripping operations require accurate tracking of small volumes of returning fluids, the use of common mud pits is discouraged.

Stripping tanks

Calibrated strip tanks with internal volumes from 10 to 25 bbls are suggested for use in stripping operations. Unlike 100 bbl trip tanks or 1500 bbl mud tanks, the smaller volume strip tanks are used to accurately measure the smaller volumes of mud returns while stripping workstrings into wells. Workstring closed-in displacements can range from 0.5 to 5 bbls/std. For these smaller volumes and to ensure accuracy, a smaller calibrated tank is suggested.

Manual chokes

With smaller bleed volumes, the use of manual chokes for bleed increments is suggested. Manual chokes allow for smooth operational control when compared to hydraulic chokes. Manual chokes are located on the choke manifold and are used in concert with an accurate Casing Pressure gauge which can easily be observed from the manual choke station.

Water-based muds

For water-based muds, influx migration must be considered during determination for stripping. As the workstring is stripped into the well, light density influxes (i.e., gas) will migrate. If the migration rate exceeds the time necessary to strip to total depth, stripping may be relegated to a secondary form of well control. In these cases,

1. the influx is allowed to expand using the volumetric methods,
2. when influx at surface, the lube and bleed method is employed to mitigate the influx, and
3. with the wellbore full of fluid, stripping operations can be conducted to ensure bottom-hole assembly is at total depth. At this point, a constant bottom-hole pressure method can be employed to ensure a minimum KMW is used.

Oil-based muds and synthetic oil-based muds

Oil-Based muds (OBM) and Synthetic Oil-Based muds (SOBM) utilize emulsions which allow influx gases to remain in solution. If the gases remain in solution, minimal migration will occur until such time the hydraulic pressure of the annulus falls before the bubble point pressure of the influx gases. At this time, gas will be liberated and migration will occur.

With minimum migration effects, off-bottom kicks in OBM and SOBM can effectively be controlled with stripping operations. Depending on well conditions, stripping in OBM and SOBM requires accurate tracking of volumes as well as accurate choke manipulations.

Preplanning stripping operations

For successful stripping operations, stripping should be a preplanned event complete with Stripping Drills to ensure rig crews are proficiently training for this well control activity. Stripping operational planning should be initiated during prespud and before operational rig-up and customized for the equipment of the individual rig.

By effectively developing a customized stripping plan, if needed, safe stripping execution can be assured. Stripping plans begin with following equipment options.

1. Annular Element: The type, size, and elastomeric specifications of annular elements are reviewed. These element's traceable elastomers should be verified as fit for service based upon projected wellbore temperatures and type of drilling mud to be used. (Reference IADC Safety Alert: 00-19.) If possible, a visual inspection of the annular element should occur reviewing for blemishes, cuts, abrasions, wear, and tear. If annular elements lack supporting documentation, they should be replaced with OEM elements according to service and pressure.
   a. For emergency situations, the use of nonstandard elements can be used, if needed. These nonstandard elements offer modified sizes for larger OD tooljoints and may be considered depending on the estimated snubbing length needed as well as pressure envelop. These nonstandard elements must be able to affect seals on the body of most common workstring bodies.
2. Strip Tank: Ensure a recirculating calibrated stripping tank is properly installed. The stripping tank should be manifolded to a downstream point on choke manifold with piping and valves rated for same service as choke manifold valves. Normally, these tanks are affixed downstream of the buffer tank with equalization to trip tank and mud treating tank. The strip tank should be serviced by a separate centrifugal pump suitable for pumping mud from the tank to the trip tank. If needed, fluid can be added to the strip tank by means of equalization lines or pumping service.
   a. If a strip tank is not available, a trip tank can be used. The trip tank should contain modular sections (weirs) which will allow for smaller, more accurate measurements within smaller tank volumes. If trip tank contains sections, ensure lower equalization valves are closed between modular sections.
3. Manual chokes within choke manifold should be routinely serviced to ensure proper operation during times of an emergency. If applicable, a primary manual choke as well as secondary manual choke should be operated and serviced daily.
   a. Proper operation of the manual choke should occur before initiating any stripping or well control operation.
4. Installation of Surge bottle: If surge bottles for the annular preventer are to be used, affix the surge bottle system (bottle, bottle rack, manifold) to the closing line of annular preventer. For fastest pressure reactions, installation of the surge bottle may be affixed directly on the preventer stack or within a short distance of the annular preventer (possibly within the substructure).
   a. Surge bottles automatically reduce and increase annular preventer closing pressures as larger OD tooljoints of the workstring are stripped through the annular preventer.
   b. Surge bottles are used to reduce wear and tear of the annular element through rapid pressure changes.
   c. Surge bottles alleviate problems inherent with manual pressure interfacing.

5. No Surge Bottles: If surge bottles are unavailable, annular closing pressures will have to be manually manipulated for every tooljoint which is stripped through the element. Manual interfacing can occur on some remote BOP panels, but routinely occurs with a Operator stations at the BOP Closing unit to oversee manual annular pressure regulator manipulations. For best performance, communication between Operator and Driller must be assured to match changes to regulated pressure with ongoing stripping operations of the workstring.
6. Regulated Closing Pressure: Most Original Equipment Manufacturer (OEM) suggests seepage of wellbore fluids between element and workstring body for lubrication purposes at a rate of 1 gallon per minute. During initial stripping operations, the normally regulated Closing Pressure of the annular preventer is lowered until seepage is observed. If an acceptable seepage rate cannot be established, then closing pressure is adjusted to a point of providing and effective seal.
7. Full Annular Closing Pressure: During each connection, the reduced regulated closing pressure may be increased to full annular closing pressure to ensure safety of personnel. The increase of full annular closing pressure should occur before slips are set during any connection.
8. No Rotation within Closed Annular Preventer: Annular preventer elements are not designed for rotation of workstring tubulars. Do not rotate workstring tubulars within closed annular preventers unless written concurrence is obtained by contractor and client management.
9. Inspect Equipment Fluid Routing and Alignment: All high-pressure flowlines, gate valves, strip/trip tank components, choke manifold components, choke and kill lines, choke panel, and BOP systems are inspected, calibrated, functioning, and aligned properly.
   a. Ensure accuracy of all pressure gauges (Driller's Panel, Standpipe, Choke Panel and Choke Manifold, BOP Closing Unit).
   b. Ensure accuracy of pit measurement devices.
   c. Ensure accuracy of Gas Detectors.
10. Depending on rig design, ensure manifolding of high-pressure flowlines from choke panel is routed to stripping tank and trip tank. When stripping, returns and closed-in displacement may be routed to the strip tank and measured. Then, these volumes may be pumped to the trip tank to monitor all returns throughout stripping.
11. For Subsea Operations: Within the LMRP, the upper annular is normally used for stripping operations. The lower annular preventer is used as a back-up system. The closing pressure of the annular preventer is lowered to minimum values along with compensated operating pressures.

Annular preventer closing pressure range

In the course of stripping operations, the regulated annular closing pressure for pipe body and tooljoint body will have to be determined. Before reducing the regulated annular closing pressure, the regulated "full" closing pressure will be defined for initial shut-in conditions ($Pressure_{Full}$). The fully closing pressure is used to fully close the element around the workstring body as determined by manufacturer specifications (normally 600–900 psi).

The regulated closing pressure is slowly lowered until leakage of wellbore fluids is observed. Suggested equipment standards define the maximum leak rate of 1 gal/min. The leakage rate is needed to help lubricate the annulus between annular element and workstring body. If the leak rate cannot be established, then closing pressure must be reduced to the minimum possible while maintaining a seal and stripping must be kept to a minimum. This pressure is recorded as minimum pressure across the workstring body ($Pressure_{DP\ Body}$).

As the larger outer diameter of the tooljoint passes through the annular element, the tooljoint will force the annular element from closed position toward an open position. As the element is forced open, the annular operating piston will be retracted hydraulic fluid within the closing chamber will be compressed. As the hydraulic fluid is compressed, the pressure of the closing chamber will increase.

In order for the tooljoint to pass through the annular element without damaging the element, the hydraulic pressure of the closing chamber fluid will need to be reduced in order to effect lubricating leakage. The lowering of the closing pressure can be performed by two individual methods.

- Manual Interface Without Surge Bottles—Closing pressures are reduced by partially opening the regulated annular closing pressure until leakage is observed at no more than 1 gal/min. This pressure is recorded as partial pressure ($Pressure_{Tooljoint}$).
- Automatic Interface With Surge Bottles—Closing pressures are reduced by controlled venting of set volume of hydraulic fluid into partially charged surge bottle. As the fluid expands, the pressure is reduced. No changes to the regulator will need to be made and this pressure is recorded as partial pressure ($Pressure_{Tooljoint}$).

As the tooljoint exits the annular element, the operating piston will return to original closed position and the hydraulic fluid within the closing chamber will expand. As the hydraulic fluid expands, the pressure of the closing chamber will be reduced. In order to affect a seal across the body of the workstring, the closing pressure must be increased to the original minimum pressure across the workstring body ($Pressure_{DP\ Body}$). The regulated closing pressure from Pressure Tooljoint to Pressure DP Body can be performed by two individual methods.

- Manual Interface Without Surge Bottles—Closing pressures are increased by partially closing the regulated annular closing pressure until previous minimum pressure across the workstring body ($Pressure_{DP\ Body}$) is achieved and leakage is observed at no more than 1 gal/min.
- Automatic Interface With Surge Bottles—Closing pressures are increased by allowing vented fluid under pressure to travel from precharged surge bottle to the closing chamber and increasing the hydraulic pressure of the system. As the fluid contracts, the pressure is increased. No changes to the regulator will need to be made as the minimum pressure across the workstring body pressure is achieved ($Pressure_{DP\ Body}$).

The regulated closing pressure ranges are represented as:

1. $Pressure_{Full}$: Full regulated annular closing pressure for complete seal, used during connections.
2. $Pressure_{DP\ Body}$: Partially regulated annular closing pressure for leakage seal around workstring body.
3. $Pressure_{Tooljoint}$: Partially regulated annular closing pressure for leakage seal around tooljoint body.

REGULATED ANNULAR CLOSING PRESSURE RANGES MAY HAVE TO BE ADJUSTED DUE TO WEAR AND TEAR OF THE ANNULAR ELEMENT DURING THE STRIPPING OPERATION

AFTER CONCLUSION OF WELL CONTROL, ANY ELEMENT USED FOR STRIPPING SHOULD BE REPLACED.

Closing pressure range procedure WITHOUT surge bottles

The general procedure was developed to establish operating ranges for stripping operations on rig systems which have **not been** outfitted with surge bottles.

1. Station BOP Closing Unit Operator at Annular Pressure Regulator. Ensure Driller and BOP Closing Unit Operator have means for direct communication throughout stripping operations.
2. Each tooljoint is to be liberally coated with grease before stripping through the annular element. Ensure grease used is environmentally friendly and compatible with drilling fluids. Grease is preferable to Pipe Dope, as grease contains no solids which can lead to abrasions and cuts of the annular preventer element.
3. BOP Closing Unit Operator to verify Full Regulated Annular Pressure to affect seal at shut-in conditions according to manufacturer specifications. Record as $Pressure_{Full}$.
4. Operator to slowly open annular regulator until leakage of 1 gal/min is observed by crew on rig floor. Record as $Pressure_{DP\ Body}$.

5. Driller to slowly strip grease coated tooljoint into element. Once the tooljoint has been positioned within element, stripping is stopped.
6. The BOP Closing Unit Operator will observe increase in annular closing pressure. The BOP Closing Unit Operator will open the annular regulator and reduce the annular closing pressure until leakage of 1 gal/min is observed by crew on rig floor. Record as $\text{Pressure}_{\text{Tooljoint}}$.
7. Driller commences stripping operations until tooljoint has passed through the annular preventer element. The BOP Closing Unit Operator will observe a decrease in annular closing pressure and slowly close regulator until Pressure DP Body is achieved and leakage is observed.
8. Driller commences stripping operations coordinating $\text{Pressure}_{\text{DP Body}}$ to $\text{Pressure}_{\text{Tooljoint}}$ to $\text{Pressure}_{\text{DP Body}}$ as each tooljoint passes through the annular preventer. At each connection, full regulated annular pressure is applied to ensure no leakage and to protect rig crew workers on rig floor. Operations continue until desired workstring depth is achieved.

Closing pressure range procedure WITH surge bottles

The following general procedure has been developed for BOP systems which employ annular surge bottles. The surge bottles are installed on annular closing system hydraulic lines and are located on or near the annular preventer. The surge bottles are normally fully sized to match BOP Closing Unit accumulator bottles and only differ in location and precharge. The internal accumulator bottle bladder is precharged with nitrogen to higher pressures when compared to BOP Closing Unit pressures. The precharge pressure on surge bottles will range from 300 to 500 psi according to manufacturer specifications. With precharge pressure, the bladder is fully extended, lands, and shuts the operating valve. Once closed, the surge bottle is ready to accept hydraulic fluid from the closing line.

As the hydraulic fluid enters the surge bottle, the valve opens. The fluid is forced into the bottle compressing the bladder and its nitrogen gas. As the tooljoint passes the element, the operating piston is forced open and the closing chamber fluid is forced into the surge bottle. As the fluid under pressure enters the surge bottle, the bladder is compressed and “stores” energy. When the tooljoint passes through the element, the operating piston returns to the closed position by means of charged fluid exiting the surge bottle and energizing the hydraulic fluid within the closing piston. The surge bottle system performs all regulated pressure manipulations automatically. Only manual interfacing will be needed when fully operating pressure is applied during a connection.

A surge bottle system normally consists of two (2) 11-gal accumulator bottles fasten to a manifold and housed within a dropped-objects prevention structure. The surge bottle system may be affixed to the annular or within a short distance among the substructure.

For Subsea Operations: The surge bottle precharge pressure must include water depth compensation (Fig. 3.54).

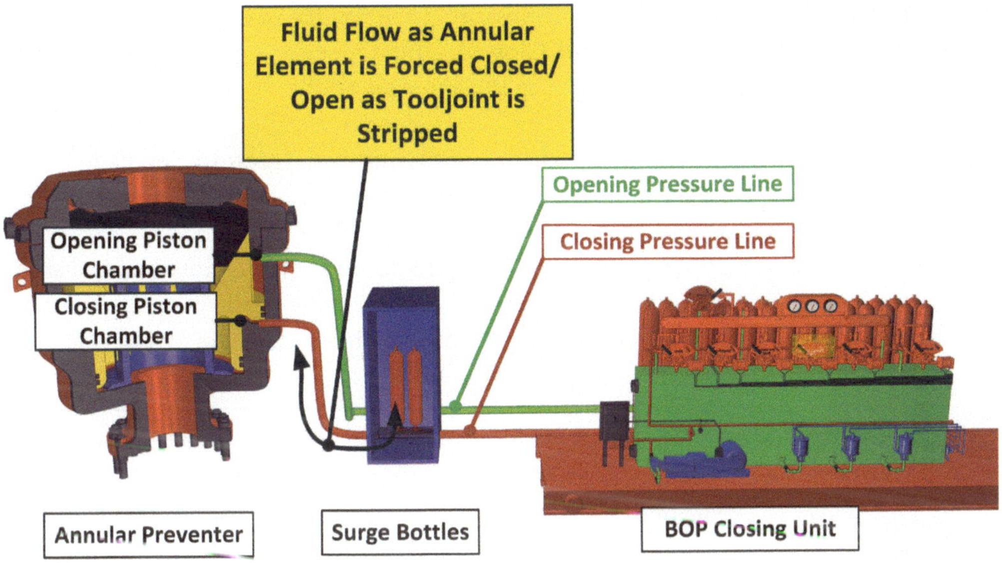

Fig. 3.54 Stripping bottle arrangement.

1. Install Annular Surge System per manufacture specifications. Suggested minimums of at least two (2) 11-gal surge bottles, manifold and protective enclosure. Install Annular Surge System within closing line of annular preventer and in close proximity of same.
2. Precharge Surge Bottle bladders to manufacturer specifications (300–500 psi).
3. Station BOP Closing Unit Operator at Annular Pressure Regulator. Ensure Driller and BOP Closing Unit Operator have means for direct communication throughout stripping operations.
4. Each tooljoint should be inspected for tong marks and sharp edges. Smooth and file off these abrasions as necessary.
5. Each tooljoint is to be liberally coated with grease before stripping through the annular element. Ensure grease used is environmentally friendly and compatible with drilling fluids. Grease is preferable to Pipe Dope, as grease contains no solids which can lead to abrasions and cuts of the annular preventer element.
6. BOP Closing Unit Operator to verify Full Regulated Annular Pressure to affect seal at shut-in conditions according to manufacturer specifications. Record as $\text{Pressure}_{\text{Full}}$.
7. Operator to slowly open annular regulator until leakage of 1 gal/min is observed by crew on rig floor. Record as $\text{Pressure}_{\text{DP Body}}$.
8. Driller to slowly strip grease coated tooljoint into element. Once the tooljoint has been positioned within element, stripping is stopped.
9. The BOP Closing Unit Operator will observe automatic reduction in annular closing pressure until leakage of 1 gal/min is observed by crew on rig floor. Regulator may need to be adjusted accordingly. Record as $\text{Pressure}_{\text{Tooljoint}}$.

10. Driller commences stripping operations until tooljoint has passed through the annular preventer element. The BOP Closing Unit Operator will observe automatic increase in annular closing pressure to $Pressure_{DP\ Body}$) has been achieved and leakage is observed. Regulator should not need adjustments.
11. Driller commences stripping operations with automatic pressure regulation from $Pressure_{DP\ Body}$ to $Pressure_{Tooljoint}$ to $Pressure_{DP\ Body}$ as each tooljoint passes through the annular preventer.
12. At each connection, manual interfacing may be needed to apply full regulated annular pressure to ensure no leakage and to protect rig crew workers on rig floor.
13. Operations continue until desired workstring depth is achieved.

Stripping recap

1. Shut-in and Monitor Well
2. Prepare Equipment
   a. SICP <60% of Annular Pressure Limit
   b. Install 0–3000 psi Casing Pressure Gauge
   c. Route from downstream of choke to strip tank
   d. Ensure pit storage for returns are greater than or equal to estimated return volume
   e. Ensure manual choke and low-volume/high-pressure pump are installed and tested
3. Develop Performance Criteria
   Determine weight of workstring. Workstring weight must **be greater than** Well Force in order to perform stripping operations
   NEVER bleed casing pressure in order to achieve desired Well Force. This will cause BHP to be lowered and may result in additional well kicks
   Define Maximum SICP
   Define Fill-up Frequency
   Define Safety Margin
   Determine Strip Volume
   Maintain constant vigilant control of all bleeds and maintain accurate records of return volumes and bleed cycles
   For operations where a drillpipe float is used, the SIDPP should remain at 0 psi
   Stripping operations may include installation of inside blowout preventer installed in the workstring to provide secondary back-up in case the drillpipe float fails
   Running of full opening safety valves within the workstring is strongly discouraged
4. Conduct Annular Closing Cycle Review
   a. $Pressure_{Full}$
   b. $Pressure_{DP\ Body}$
   c. $Pressure_{Tooljoint}$

Step 1: Strip and Establish Safety Margin. Observe SICP increases. (Suggest 200 psi.)

Step 2: Strip: Perform stripping operations while monitoring casing pressure. Monitor and record volumes.

Continue stripping until connection.

Step 3: Connection: Stop at connection and shut-in well with full annular closing pressure.

Install connection and torque to specification.

Resume stripping operations monitoring casing pressure.

Step 4: Perform Selected Bleed Cycle Method after achieving desired Casing Pressure, time or footage limit.

Bleed off Displacement Volume (in small increments) while monitoring SICP.

Step 5: Repeat Steps 2 and 4 until bit is at TD. Once at TD, initiate constant bottom-hole kill method.

Detailed stripping procedure

Shut-in and monitor well

If a kick is identified with the pipe off bottom, the well must be shut-in immediately using proper procedures. After recording the SICP, proceed with bumping the float. Monitor well and continue recording stabilized SIDPP and SICP every 5 min.

During the period after shut-in and up to the point where the pipe is actually stripped in, the well must be monitored continuously. Bottom-hole pressure must be maintained constant at all times during this period. May need to employ Volumetric Method to keep bottom-hole pressure constant.

If the casing pressure increases to the maximum allowed during stripping operations, then STOP stripping and consider circulating the well with kill weight mud or bullheading.

Equipment preparations

SURFACE PRESSURE FOR STRIPPING SHOULD NOT EXCEED 1,000 PSI.
EFFECTIVE LIMIT MAXIMUM OF 800 PSI IS SUGGESTED.
IF THE SURFACE PRESSURES ABOVE 800 PSI, ATTEMPTS TO LOWER PRESSURE SHOULD BE MADE TO ALLOW ANNULAR STRIPPING.
OPTIONS INCLUDE CIRCULATING OUT THE INFLUX, VOLUMETRIC METHOD TO GET INFLUX TO THE SURFACE AND LUBRICATE AND BLEED TO REMOVE GAS OR THE BULLHEAD METHOD.

RAM TO RAM STRIPPING IS CONSIDERED TO BE SPECIALIZED SERVICE AS SPECIALIZED PRESSURE CONTROL EQUIPMENT IS NEEDED TO ENSURE NO DAMAGE TO BOP STACK
FOR RAM-TO-RAM STRIPPING APPLICATIONS, SUGGEST USING WELL CONTROL COMPANY TO PERFORM

Surface pressure for stripping should not exceed 1000 psi. An effective upper limit is suggested at 800 psi.

1. If the surface pressure s above 800 psi, attempts to lower pressure should be made to allow annular stripping.
2. Options include circulating out the influx, volumetric method to get influx to the surface, and lubricate and bleed to remove gas or the bullhead method.

Ram-to-ram stripping should be considered a specialized service and should not be attempted without well control company oversight. Ram-to-ram stripping requires specialized equipment (i.e., flow loop) to prevent damage to existing BOP stack and endangering personnel.

Equipment Preparations include ensuring sufficient pit volume is available for return volumes which may be pushed out of the well during influx expansion. The pit volumes should also include the additional volumes generated by increasing to kill weight mud densities.

For stripping operations, use of the manual choke is suggested to allow more accurate volumes to be bled from the well. Before operations are to commence, ensure the manual choke is operable and fully tested.

Develop performance criteria

- For successful stripping operations, $Bbl_{in} = Bbl_{out}$. For each barrel of steel stripped in the well, an equal barrel of mud returns is to be recovered.
- As stripping operations enter the influx, casing pressure and drillpipe pressures may fluctuate. Adjustments should be minor in nature and will probably not require changes to the stripping operations.
- Bleed cycle stoppage may be needed to control influx expansion rates. Bleed cycles will be determined by time or amount of footage stripped (or combination).

1) Determine Stripping Application

$$\text{Well Force}_{\text{lbs}} = \text{Largest OD}_{\text{in.}^2} \times 0.7854 \times \text{Wellhead Pressure}_{\text{psi}} \quad (3.41)$$

$$\text{Workstring}_{\text{lbs}} \downarrow = \left(\text{DP Weight}_{\text{ppf}} \times \text{DP Length}_{\text{ft}}\right) + \left(\text{DC Weight}_{\text{ppf}} \times \text{DC Length}_{\text{ft}}\right) \quad (3.42)$$

$$\text{Workstring}_{\text{lbs}} > \text{Well Force?} \quad (3.43)$$

2) Determine Maximum SICP
   a) Lessor of 80% of MASP or 80% of Casing Burst. Maximum SICP can be downrated due to poor wellbore condition or cement isolation.
3) Fill workstring per regulation or industry standard such as every 5 stands or loss of 75 psi, whichever comes first.
4) Define Safety Margin
   a) Safety Margin is dependent upon shoe strength

b) To be established by stripping workstring with choke closed
c) Safety Margin suggestion = 200 psi

5) Determine Strip Volume
a) Strip Increment = Strip Volume per stand (bbls/stand).

$$\text{Strip Volume}\left(\frac{\text{bbls}}{\text{ft}}\right) = \frac{1\,\text{ft}}{\text{Closed} - \text{In DP Displacement}_{\text{bbls/ft}}} \quad (3.44)$$

$$\text{Strip Volume}\left(\frac{\text{bbls}}{\text{stand}}\right) = \text{Strip Volume}_{\frac{\text{bbl}}{\text{ft}}} \times \text{Length of Stand}_{\frac{\text{ft}}{\text{stand}}} \quad (3.45)$$

6) Determine Volume and Frequency of Bleed Cycle
a) *Perform bleed at intervals of one-third of Total Strip Volume or every 1 to 2h.*
- For 6500 ft./30 ft./min ~4 four hours to strip to bottom, therefore bleed cycles may be limited to 2–3 cycles.

b) $\text{Vol}_{\text{Bleed}}$ (bbls/stand) = ½ Safety Margin.

$$\text{Vol}_{\text{Bleed}}\left(\frac{\text{bbl}}{\text{ft}}\right) = \left(\frac{100\,\text{psi}}{(0.052) \times \left(\text{MW}_{\text{ppg}}\right)}\right) \times \text{Closed In DP Displacement}\left(\frac{\text{bbl}}{\text{ft}}\right) \quad (3.46)$$

$$\text{Vol}_{\text{Bleed}} = 160\,\text{ft} \times 0.197\frac{\text{bbl}}{\text{ft}} = 3.2\,\text{bbls where HP is decreased by } 100\,\text{psi.} \quad (3.47)$$

This means while stripping, we should perform one to three bleed intervals (3.48)

c) Bleed additional volume at predetermined intervals (every 2000 ft or 2 h).
d) Bleed off 3.2 bbls in small increments (When 100 psi of $\text{Vol}_{\text{Bleed}}$ is reached, the bubble is allowed to expand)
- Once bleed volume occurs, strip workstring into closed system until SICP rises equal to Hydrostatic Pressure of Bleed Cycle (100 psi).

e) Once new SICP has been reached, continue stripping operations keeping new SICP constant by opening manual choke.

Conduct annular closing cycle review

a) $\text{Pressure}_{\text{Full}}$
b) $\text{Pressure}_{\text{DP Body}}$
c) $\text{Pressure}_{\text{Tooljoint}}$

Step 1: Strip and establish safety margin

a. Initiate Safety Margin by stripping into closed system. (Suggest 200 psi over SICP.)
b. Place mark on workstring horizontally immediately above rotary table along with a vertical mark on workstring extending from horizontal mark up about 2 ft. These

marks are made for quick visualization of workstring torque and vertical travel distance.

c. Continually monitor and record workstring weight and wellbore pressures.

d. Ensure tooljoint vertical location is updated constantly after tooljoint passes through rig floor to prevent tooljoint from being spaced-out across the Blind/Shear rams.

Step 2: Strip and monitor casing pressure

a. Monitor Casing Pressure while stripping in well.

b. Stripping rate of 30 ft./min (3 min/std) is a desired while maintain constant CP. Displacement returns should equal the workstring stripping displacement ($Bbl_{in} = Bbl_{out}$).

c. Top-off workstring by performing fill-ups every 5 stands or loss of 75 psi if stripping out, whichever comes first.

d. All operations should be performed in a smooth manner to minimize creation of surge pressures.

e. If needed, file off and smooth any sharp edges or tong marks from the tooljoints. Perform inspection on each and every tooljoint.

f. If casing pressure approaches maximum limits, consider immediate application of a constant bottom-hole pressure well kill method at current bit depth, provided Mud Cap pressure does not exceed LOT/FIT pressure.

g. Stripping operations are to continue until connection point.

Step 3: Connection

a. At connection point, stop stripping operations and close-in well with full annular pressure.

b. Before stopping to make-up the connection, increase the regulated annular pressure ($Pressure_{Body}$) to full regulated annular pressure. By increasing the annular pressure, the well pressures are fully contained maximizing safety to personnel and protection of the environment during the connection.

c. Monitor and record all pressures, volumes, workstring weight, and bit depth.

d. Install connection and torque to specification.

e. Resume stripping operations monitoring casing pressure.

Step 4: Perform selected bleed cycle method

a. After achieving desired Casing Pressure, time or footage limit.

b. Bleed off Displacement Volume (in small increments) while monitoring SICP.

a. Displacement returns are to be routed from choke manifold to calibrated strip or trip tank Verify Displacement Volume Equals Displacement Return Volume.

b. Repeat Steps 2 and 4 until on bottom. Initiate constant bottom-hole kill method.

Step 5: Repeat steps 2 to 4 until bit is on bottom

a. Shut down and shut-in well.

b. When bit and BHA enters influx, sudden increase in CP can be expected. Increase in pressures should be minor in nature with no corrections needed.

c. At TD, bleed off Safety Margin pressures. Observe stabilized SIDPP and SICP. Initiate an appropriate constant bottom-hole pressure method to kill the well.

Example 1: Stripping with bleed example (oil-based mud)

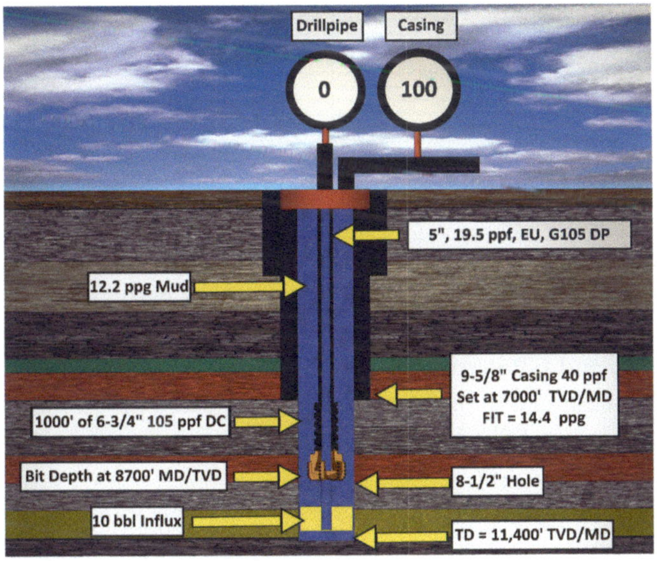

Fig. 3.55 Oil-based stripping example well.

29 stands from bottom (Fig. 3.55).

Current Bit Depth = 8700 ft.

Kick size = 10 bbls.

Float in DP, SIDPP = 0 psi.

SICP = 100 psi.

9–5/8″ 40 ppf (8.755″).

Set at 7000′ TVD, 14.5 ppg shoe test.

8–1/2″ hole with MW = **12.2 ppg OBM.**

DP = 5″, 19 ppf, EU G105 DP.

DP Tooljoint OD = 6–5/8″ from Table.

1000′ of 6–3/4″ DC 105 ppf.

Stand = 93 ft.

TD = 11,400′ MD/TVD.

DP Closed-In Displacement = 0.0243 bbl/ft.

Shut-in conditions, equipment preparations, and annular closing pressure cycle

While shut-in, prepare and complete Stripping Table. Ready equipment and prepare for stripping (Fig. 3.56).

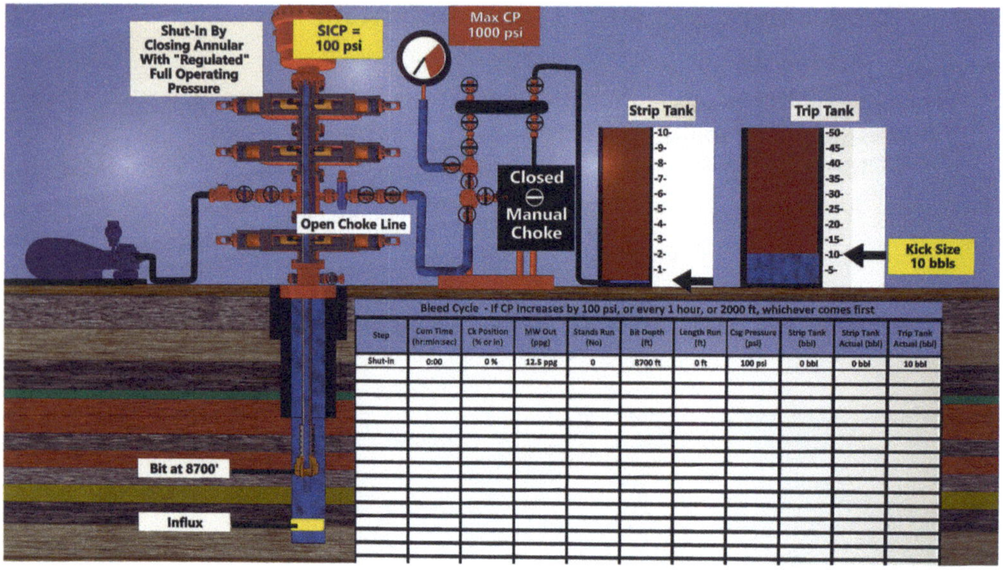

Fig. 3.56 Shut-in conditions.

The well is shut-in on Annular Preventer with closing pressure noted at 900 psi. The regulated fluid pressure is lowered to 700 psi, and leak of annular fluids is observed at less than 1 gpm. The pressure is increased to 900 psi as equipment is lined up. Bit depth (8700′) along with volumes, MW, and CP is recorded.

1. Stripping operations should be confined to oil-based or synthetic-based muds. Influx migration within water-based muds is difficult to control lends use of volumetric control, followed by lube and bleed.
2. Using rule of thumb, the maximum Annular Pressure for stripping operations is limited to 1000 psi due to unknown wear on element. If CP is over 1000 psi, stripping operations are not suggested.
3. Ram-to-ram stripping within BOP rams is not suggested for stripping operations. BOPs are designed for emergency shut off control. Ram-to-ram stripping requires specialized expertise and should only be attempted with representatives of select well control company present.
4. Stripping should not be attempted without stripping tank due to the small volumes which will be generated for DP stripped in hole.
5. Annular Preventers outfitted with stripping bottles are suggested in order to ensure smoother stripping operations with more control.

A manual choke is suggested for use for most accurate bleed volumes. Ensure manual choke is serviced, lubricated and is easy functioning. If back-up manual choke is available, ensure this choke is readied for use.

Equipment is prepared by installing a 0–3000 psi casing pressure gauge, ensuring appropriate valves are opened to direct fluids to the strip tank. The kick volume has been determined by the trip tank (10 bbls). After establishing this volume, it is pumped to the active pit and the trip tank is zeroed. The trip tank will be used for cumulative bleed volumes as determined by the strip tank.

Performance criteria

a) Determine Stripping Application

$$\text{Well Force}_{\text{lbs}} = \text{Largest OD}_{\text{in.}^2} \times 0.7854 \times \text{Wellhead Pressure}_{\text{psi}} \tag{3.49}$$

$$\text{Well Force}_{\text{lbs}} = (6.625)^2 \times 0.7854 \times 100 \text{ psi} = 3500 \text{ lbs} \uparrow \tag{3.50}$$

$$\text{Workstring}_{\text{lbs}} \downarrow = \left(\text{DP Weight}_{\text{ppf}} \times \text{DP Length}_{\text{ft}}\right) + \left(\text{DC Weight}_{\text{ppf}} \times \text{DC Length}_{\text{ft}}\right) \tag{3.51}$$

$$\text{Workstring}_{\text{lbs}} = (19.5 \text{ ppf} \times 7700 \text{ ft}) + (105 \text{ ppf} \times 1000 \text{ ft}) = 255{,}000 \text{ lbs} \downarrow \tag{3.52}$$

$$\text{Workstring}_{\text{lbs}} (255{,}000 \text{ lbs}) > \text{Well Force } (3500 \text{ lbs}), \text{Drillstring can be stripped in hole.} \tag{3.53}$$

b) Maximum Annular Pressure **during stripping operations limited to = 1000 psi**

c) Fill-up Frequency = **Every 2 stands**

d) Define Safety Margin

- **i)** Safety Margin is dependent upon shoe strength
- **ii)** To be established by pumping into annulus
- **iii)** Estimated **Safety Margin = 200 psi**
- **iv)** Bleed cycle limit = ½ **of Safety Margin or 100 psi**

e) Determine Strip Volume

- **i)** Strip Volume

$$\text{Strip Volume}\left(\frac{\text{bbls}}{\text{ft}}\right) = \text{Closed} - \text{In DP Displacement}_{\text{bbls/ft}} \times \text{Strip Length (ft)} \tag{3.54}$$

$$\text{Strip Volume}\left(\frac{\text{bbls}}{\text{ft}}\right) = 0.0243\frac{\text{bbl}}{\text{ft}} \times \text{Length Stripped (ft)} \tag{3.55}$$

$$\text{Strip Volume}\left(\frac{\text{bbls}}{\text{stand}}\right) = \text{Disp Volume}_{\frac{\text{bbl}}{\text{ft}}} \times \text{Length of Stand}_{\frac{\text{ft}}{\text{stand}}} \tag{3.56}$$

$$\text{Strip Vol}\left(\frac{\text{bbl}}{\text{stand}}\right) = 0.0243\frac{\text{bbl}}{\text{ft}} \times 93\frac{\text{ft}}{\text{stand}} = 2.2 \text{ bbl/std} \tag{3.57}$$

$$\text{Total Strip Vol (bbl)} = 0.0243\frac{\text{bbl}}{\text{ft}} \times (\text{Hole MD}_{\text{ft}} - \text{Bit MD}_{\text{ft}}) \tag{3.58}$$

$$\text{Total Strip Vol (bbl)} = 0.0243 \times (11{,}400' - 8700') = 65.6 \text{ bbls without any bleeds} \tag{3.59}$$

d) Determine Volume and Frequency of Bleed Cycle.

i) *Perform bleed at intervals of one-third of Total Strip Volume or every 1–2 h.*

- For 29 stands at 5 min/stand (30 ft./min strip time plus connection time), estimate 2–1/2 h to strip to bottom, without bleed cycles. Perform bleed check every hour, estimate 2 bleed cycles.

ii) Vol_{Bleed} (bbls/stand) = ½ Safety Margin.

$$Vol_{Bleed}\left(\frac{bbl}{ft}\right) = \left(\frac{100\,psi}{(0.052)\times\left(MW_{ppg}\right)}\right)\times \text{Closed In DP Displacement}\left(\frac{bbl}{ft}\right) \quad (3.60)$$

$$Vol_{Bleed} = 158\,ft \times 0.0243\,\frac{bbl}{ft} = 3.7\,\text{bbls where HP is decreased by } 100\,psi \quad (3.61)$$

$$\text{Two bleed cycle has been scheduled, if needed} \quad (3.62)$$

iii) Bleed off 3.7 bbls in small increments (When 100 psi of Vol_{Bleed} is reached, the influx is allowed to expand). Once bleed volume has been recovered, initiate pumping operations and allow the Back Pressure to rise above previous Aggregate Back Pressure by the Hydrostatic Pressure of Bleed Cycle. (Original SICP + Safety Margin + Back $Pressure_1$ + Back $Pressure_2$).

e) Reinitiate stripping operations by maintain constant casing pressure while maintaining constant displacement volumes.

f) When bit and BHA enters influx, sudden increase in CP can be expected. Increase in pressures should be minor in nature with no corrections needed.

g) At TD, bleed off Safety Margin pressures. Observe stabilized SIDPP and SICP. Initiate an appropriate constant bottom-hole pressure method to kill the well.

Step 1: Strip and establish safety margin

After adjusting Annular Closing Pressure to allow leakage, strip into well until Casing Pressure increases by 200 psi over SICP. In this example, 35 ft. of workstring is stripped in a closed well. SICP increases from 100 to 300 psi. The trip tank has been reset to zero (Fig. 3.57).

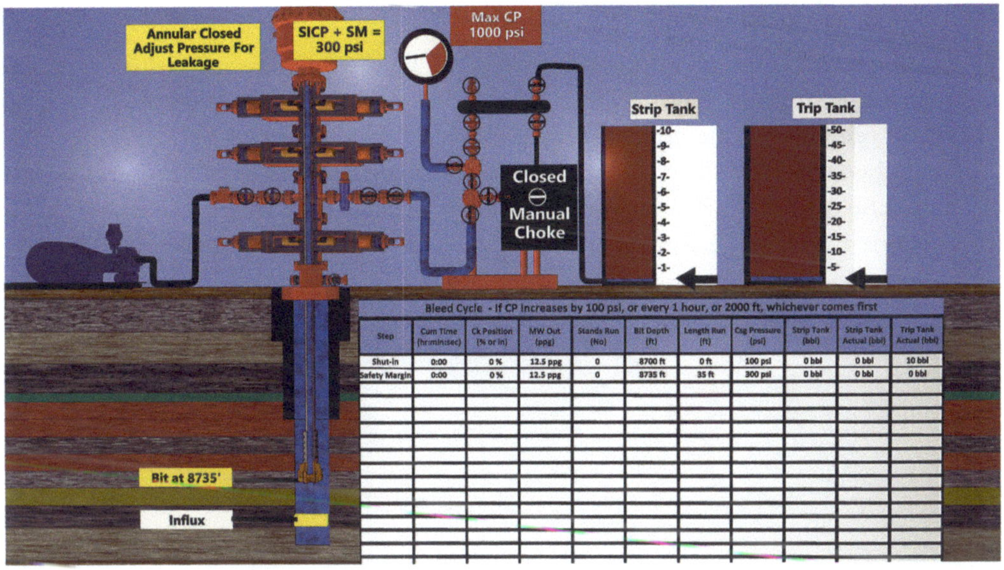

Fig. 3.57 Establish safety margin by slowly stripping into closed system.

Step 2: Strip while monitoring casing pressure

With reduced Annular Closing Pressure and leakage established, open manual choke and simultaneously strip into bleeding off 2.2 bbls/stand. Closely monitor CP, if CP increases by 100 psi, shut down and shut-in with full closing pressure and perform bleed cycle (Fig. 3.58).

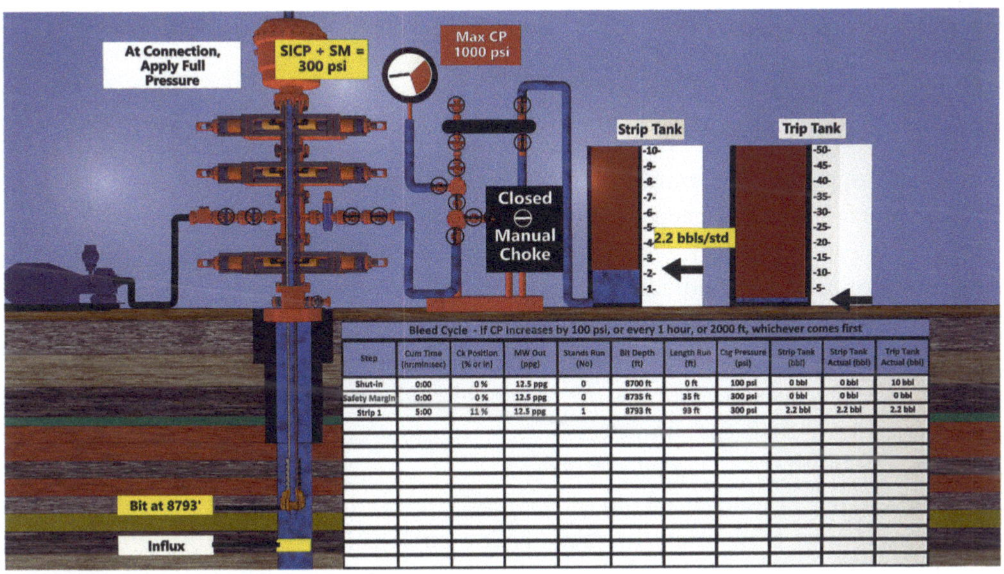

Fig. 3.58 Strip and bleed simultaneously via manual choke.

Step 3: Connection

When each stand connection point has been reached, stop and shut well-in with full annular closing pressure. Ensure DP is filled every two stands while stripping in hole. Install and properly torque each stand.

Record volumes, pressures, and bit depth. If needed, transfer mud from strip tank to trip tank. Use trip tank for cumulative return volumes and strip tank for individual stand strip volume recovery. Maintain accurate total volumes of mud returns (Fig. 3.59).

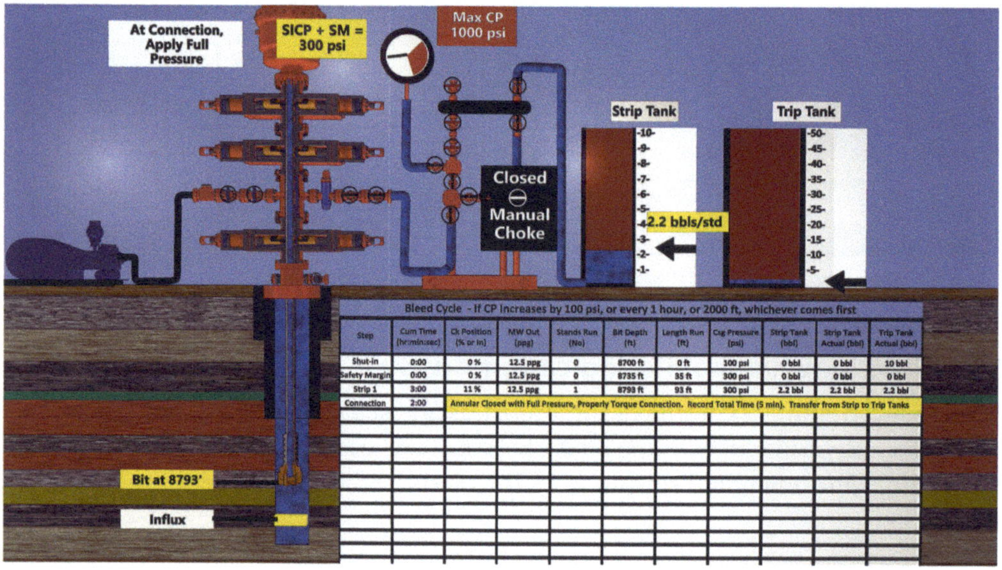

Fig. 3.59 At connection, record volumes, pressures, and bit depth. If needed, transfer mud from strip tank to trip tank.

Step 4: Perform bleed cycle

Continue stripping in well until casing pressure has increased by ½ safety margin or time limit or footage limit has been reached.

Once the well has been secured the maximum pressure on annulus, a bleed cycle determination is to be conducted.

1. During stripping operations, no casing pressure increases were observed. For this case, a bleed cycle is not needed, and the stripping operations can recommence.
2. If the CP increases but is below the threshold of ½ safety margin, a bleed cycle can be deferred until such time the overall increase in CP equals ½ safety margin. At this time, the bleed cycle can be performed.
3. If the CP increases to threshold of ½ safety margin, the well is secured and required volume (equal to ½ safety margin) if bled off KEEPING CASING PRESSURE CONSTANT. The bleed cycle is used to allow the influx to expand under controlled

conditions. If the volume increases, the influx pressure decreases, and the BHP decreases by the same amount.

The example below demonstrates a bleed cycle is needed at 10,605 ft. based upon CP rising by 100 psi. The bleed cycle is performed by bleeding 3.7 bbls (100 psi equivalent) from well while holding casing pressure constant. The bleed cycle may require multiple small bleeds to ensure CP is held constant. A secondary pump source may be used if the CP is inadvertently lowered below the step threshold.

In the example below, after stripping 12 stands and performing one bleed, total volume recovered is 30.1 bbls (3.7 bbls + 26.4 bbls) (Fig. 3.60).

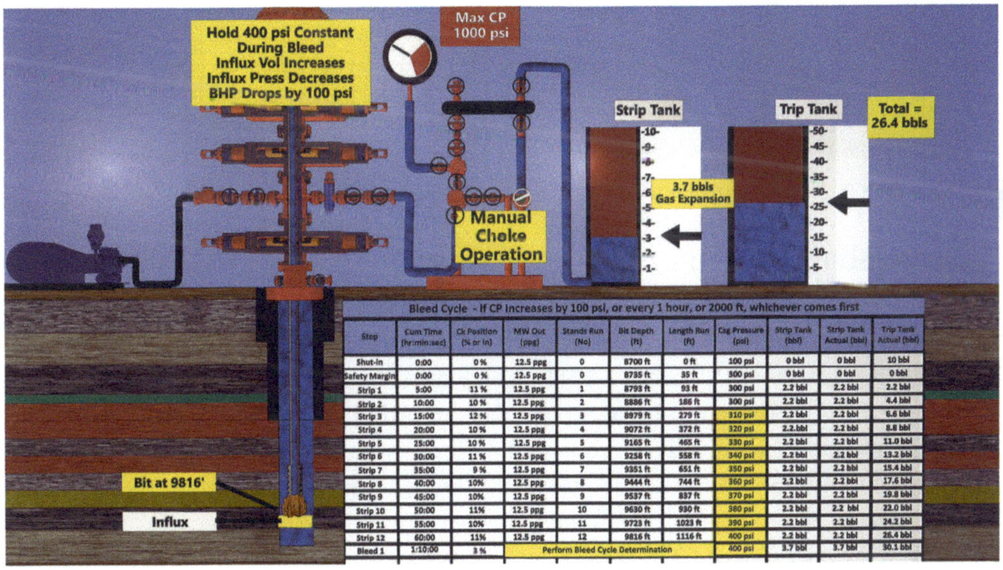

Bleed Cycle - If CP increases by 100 psi, or every 1 hour, or 2000 ft, whichever comes first										
Step	Cum Time (hr:min:sec)	Ck Position (% or in)	MW Out (ppg)	Stands Run (No)	Bit Depth (ft)	Length Run (ft)	Csg Pressure (psi)	Strip Tank (bbl)	Strip Tank Actual (bbl)	Trip Tank Actual (bbl)
Shut-in	0:00	0 %	12.5 ppg	0	8700 ft	0 ft	100 psi	0 bbl	0 bbl	10 bbl
Safety Margin	0:00	0 %	12.5 ppg	0	8735 ft	35 ft	300 psi	0 bbl	0 bbl	0 bbl
Strip 1	5:00	11 %	12.5 ppg	1	8793 ft	93 ft	300 psi	2.2 bbl	2.2 bbl	2.2 bbl
Strip 2	10:00	10 %	12.5 ppg	2	8886 ft	186 ft	300 psi	2.2 bbl	2.2 bbl	4.4 bbl
Strip 3	15:00	12 %	12.5 ppg	3	8979 ft	279 ft	310 psi	2.2 bbl	2.2 bbl	6.6 bbl
Strip 4	20:00	10 %	12.5 ppg	4	9072 ft	372 ft	320 psi	2.2.bbl	2.2 bbl	8.8 bbl
Strip 5	25:00	10 %	12.5 ppg	5	9165 ft	465 ft	330 psi	2.2 bbl	2.2 bbl	11.0 bbl
Strip 6	30:00	11 %	12.5 ppg	6	9258 ft	558 ft	340 psi	2.2 bbl	2.2 bbl	13.2 bbl
Strip 7	35:00	9 %	12.5 ppg	7	9351 ft	651 ft	350 psi	2.2 bbl	2.2 bbl	15.4 bbl
Strip 8	40:00	10%	12.5 ppg	8	9444 ft	744 ft	360 psi	2.2 bbl	2.2 bbl	17.6 bbl
Strip 9	45:00	10%	12.5 ppg	9	9537 ft	837 ft	370 psi	2.2 bbl	2.2 bbl	19.8 bbl
Strip 10	50:00	11%	12.5 ppg	10	9630 ft	930 ft	380 psi	2.2 bbl	2.2 bbl	22.0 bbl
Strip 11	55:00	10%	12.5 ppg	11	9723 ft	1023 ft	390 psi	2.2 bbl	2.2 bbl	24.2 bbl
Strip 12	60:00	11%	12.5 ppg	12	9816 ft	1116 ft	400 psi	2.2 bbl	2.2 bbl	26.4 bbl
Bleed 1	1:10:00	3 %	Perform Bleed Cycle Determination				400 psi	3.7 bbl	3.7 bbl	30.1 bbl

Fig. 3.60 With CP rising from 300 to 400 psi, perform bleed cycle.

Repeat oil-based strip/connection cycle

With reduced Annular Closing Pressure and leakage established, continue stripping in well until casing pressure has increased by 100 psi, time limit or footage limit has been reached. If CP increases by 100 psi, shut down and shut-in with full closing pressure and perform bleed cycle.

When each stand connection point has been reached, stop and shut well-in with full annular closing pressure. Record volumes, pressures, and bit depth. If applicable, transfer mud from strip tank to trip tank. Use trip tank for cumulative return volumes and strip tank for individual stand strip volume recovery.

In the example below, a total of 56.6 bbls have been recovered after stripping 24 stands and performing one bleed cycle (Fig. 3.61).

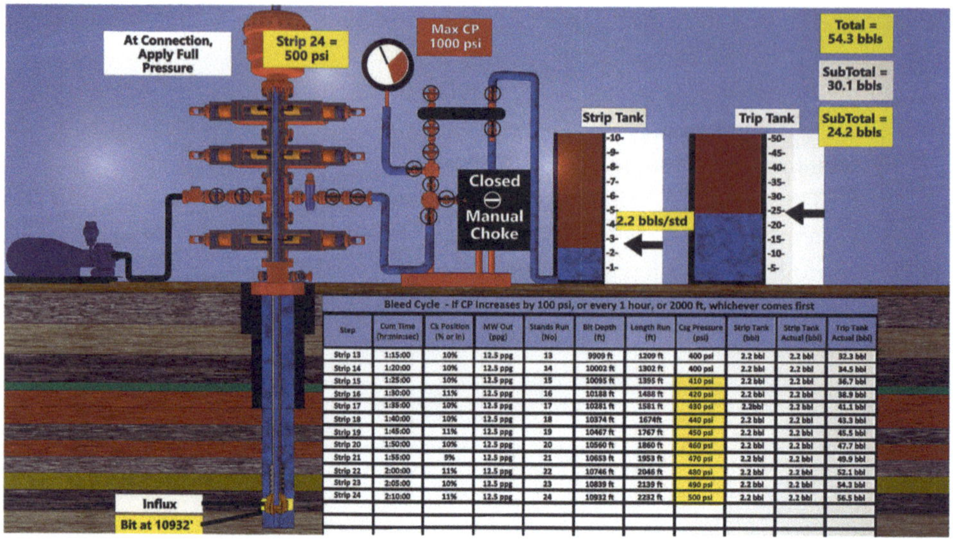

Fig. 3.61 Perform strip/connection cycles.

Repeat oil-based bleed cycle

The example below demonstrates a bleed cycle at 10,932 ft. based on increased casing pressure threshold of 500 psi. Next, the bleed cycle is performed 3.7 bbls of mud from well WHILE HOLDING CP CONSTANT. Once the bleed cycle is performed, the mud is transferred to the trip tank. In our example, the 50 bbl trip tank limits the amount of mud which can be cumulatively recorded. When trip tank becomes nearly full, the mud is then transferred to the active pit system. For our example, a total of 60.2 bbls have been recovered after stripping 24 stands and performing two bleed cycles (30.1 bbls + 26.4 bbls + 3.7 bbls) (Fig. 3.62).

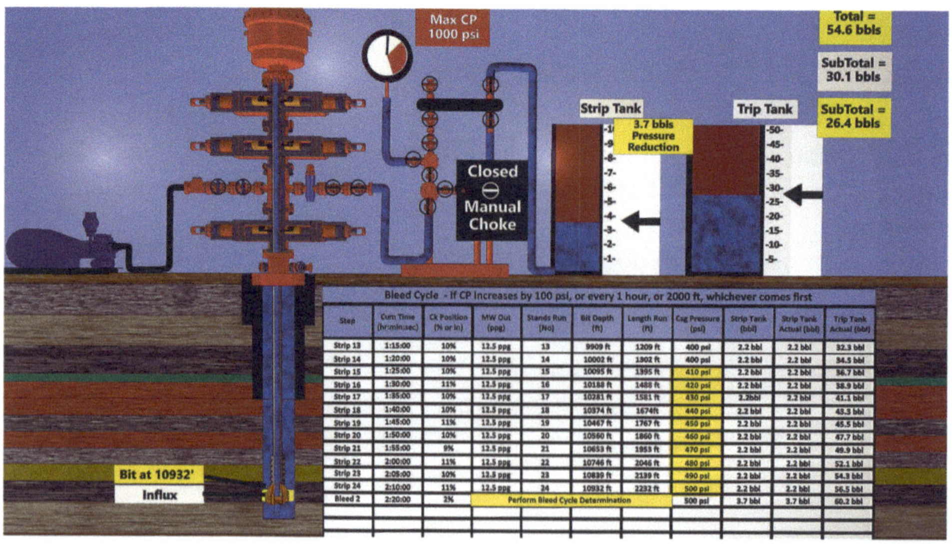

Fig. 3.62 Bleeding 3.7 bbls while holding casing constant (small bleed cycles).

Repeat oil-based strip/connection cycle until TD

With reduced Annular Closing Pressure and leakage established, continue stripping in well until TD is reached. For our example, TD is reached after 29 stands. Utilizing the same tracking system, 2.2 bbls are bled while stripping each stand into the strip tank, then transferred to the trip tank. Upon reaching 29 stands, a total of 71.2 bbls (60 bbls + 2.2 bbls) have been recovered (Fig. 3.63).

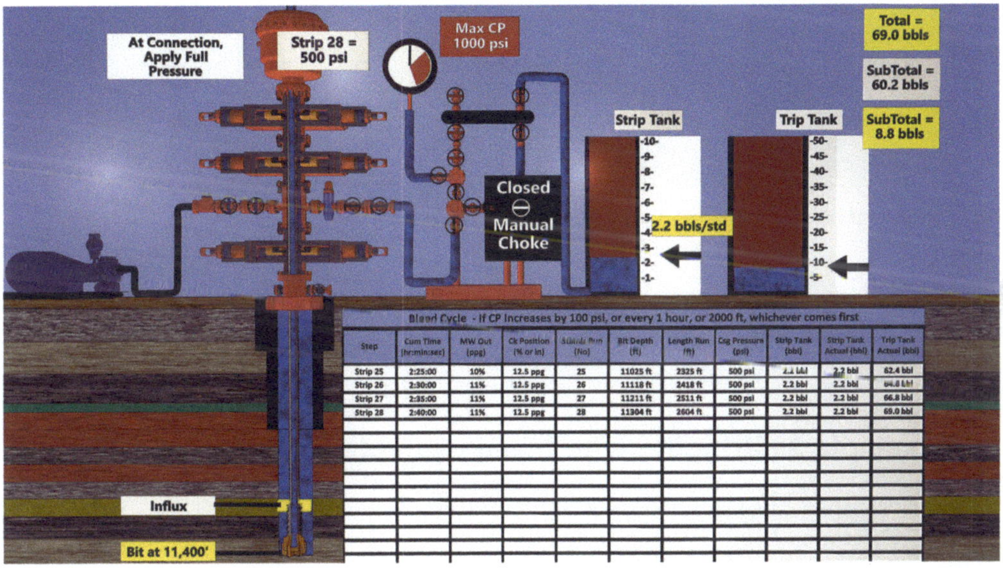

Fig. 3.63 Strip to TD (Stand 27) and shut-in well. Apply full regulated annular closing pressure. Bleed volume of pipe stripped in the hole.

At TD, perform bleed cycle to remove safety margin

Upon reaching TD and before initiating kill operations, the safety margin needs to be removed from the well to ensure accurate SIDPP and SICP when the DP float is pumped. The safety margin is not needed for upcoming well control circulations.

To remove the safety margin (in our example, this is 200 psi), simply bleed back fluid volume equal to the safety margin (for our example, this is 3.7 bbls per 100 psi or 7.4 bbls for 200 psi Safety Margin). The total volume of fluid recovered is 78.6 bbls (71.2 bbls + 7.4 bbls) (Fig. 3.64).

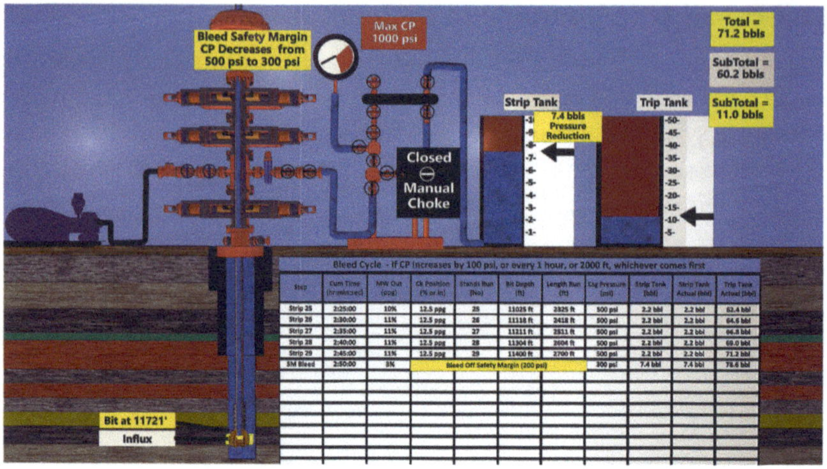

Fig. 3.64 Bleed off safety margin (200 psi).

Preparations are made to bump float and initiate appropriate constant BHP kill method to regain control of the well.

Example 2: Stripping with bleed example (water-based mud)

Water-Based Muds represent a more complicated stripping procedure when compared to Oil-Based Muds. Water-Based muds liquid medium will not allow influx gases to remain in solution and therefore influx rise may cause problems.

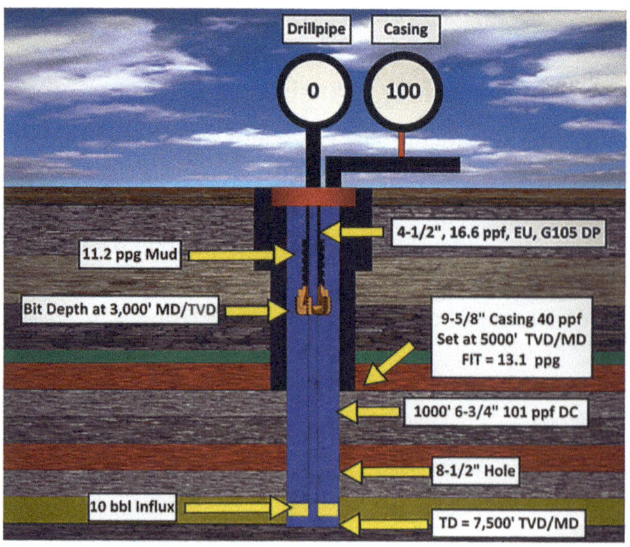

Fig. 3.65 Water-based stripping example well.

48 stands from bottom (Fig. 3.65).
Current Bit Depth = 3000 ft.
TD = 7500′ MD/TVD.
Kick size = 10 bbls.
Float in DP.
SICP = 100 psi.
9–5/8″ 40 ppf (8.755″).
Set at 5000′ TVD, 13.1 ppg shoe test.
8–1/2″ hole.
MW = 11.2 ppg, high viscosity WBM.
DP = 4–1/2″ 16.6 ppf (Internal Capacity = 0.01422 bbl/ft. or 6.6 bbls/5 stds).
Closed-In Disp Cap = 0.0197 bbl/ft.

Studies have shown a gas influx will rise at a rate of 4000 ft./h (this may even be faster in high-angle wells). Before beginning a stripping exercise within water-based fluids, a quick determination should be made on whether the workstring can be stripped to TD before the influx reaches the surface. This determination consists of two parts.

Part 1: Determine how fast the influx will reach surface

a. Gas Migration Distance

$$\text{Gas Migration Distance}_{\text{TVD}_{\text{ft}}} = \frac{\text{Rise in SICP}}{\text{MW}_{\text{ppg}} \times 0.052} \tag{3.63}$$

b. Gas Migration Rate

$$\text{Gas Migration Rate}_{\text{TVD}_{\text{ft}}/\text{min}} = \frac{\text{Gas Migration Distance}_{\text{TVD}_{\text{ft}}}}{\text{Time of Rise}_{\text{min}}} \tag{3.64}$$

For a 7500′ TD well with bit at 3000′ and swap kick on bottom, how much time is needed for the influx to rise to surface unimpeded?

a. Time to Surface

$$\text{Gas Migration Time}_{\text{min}} = \frac{\text{Depth of Rise}}{4000\,\text{ft/h}} \tag{3.65}$$

If the influx is on bottom and neglecting the volume of bubble, what is the time necessary for the influx to rise to surface unimpeded?

$$\text{Gas Migration Time}_{\text{min}} = \frac{7500\,\text{ft}}{4000\,\text{ft/h}} = 1\,\text{h}\,50\,\text{min} \tag{3.66}$$

Part 2: Determine how long it will take to strip to bottom

The average stripping rate is based upon rig performance. For our example, stripping speed is maintained at about 30ft/min (93ft in 3min). Connection time takes 2min with the drillstring not moving. Stripping time would be a total of 93ft in 5min or an average of 18.6ft/min.

a. Average Stripping Rate (ft/min)

$$\text{Average Stripping Rate}_{\text{fpm}} = \left(\frac{93\text{ ft}}{3\text{ min}} + \frac{0\text{ ft(Connection)}}{2\,\text{min}}\right) = \frac{93\text{ ft}}{5\,\text{min}} = 18.6\frac{\text{ft}}{\text{min}} \tag{3.67}$$

b. Strip Time to TD

$$\text{Time to TD}_{\text{h:min}} = \frac{(\text{TD}_{\text{ft}} - \text{Bit Depth}_{\text{ft}})}{\text{Stripping Rate}_{\text{fpm}}} = \text{h : min} \tag{3.68}$$

For a 7500′ TD well with bit at 3000′, how much time is needed to strip to TD.

$$\text{Time to TD}_{\text{min}} = \frac{(7500 - 3000\,\text{ft})}{18.6\frac{\text{ft}}{\text{min}}} = 240\ \text{min or}\ 4\,\text{h} \tag{3.69}$$

Part 3: Comparative analysis

- If Time of Influx to Surface ≥ Time to Strip to TD, forgo stripping operations and initiate volumetric control to allow the influx to expand and reach the surface.
- If Time of Influx to Surface < Time to Strip to TD, stripping operations can commence.

For our example, Time of Influx to Surface (3 h 45 min) < Strip Time to TD (4 h). Conclusion, the influx will reach surface before the bit can be stripped to TD, Therefore, an alternate well control method must be employed (i.e., volumetric control).

Use volumetric control

Use volumetric control to allow influx to expand under controlled environment until influx is at the surface. When the influx reaches the surface, the well is not dead. To maximize safety, the influx must be safely removed. (See Volumetric section for more information.)

Use lube and bleed to remove influx

Use Lube and Bleed Method to safely remove the influx. (See Lube and Bleed Section for more information.)

Develop performance criteria

a) Determine Stripping Application

$$\text{Well Force}_{\text{lbs}} = \text{Largest OD}_{\text{in.}^2} \times 0.7854 \times \text{Wellhead Pressure}_{\text{psi}} \tag{3.70}$$

$$\text{Well Force}_{\text{lbs}} = (6.625)^2 \times 0.7854 \times 100\ \text{psi} = 3500\ \text{lbs} \uparrow \tag{3.71}$$

$$\text{Workstring}_{\text{lbs}} \downarrow = \left(\text{DP Weight}_{\text{ppf}} \times \text{DP Length}_{\text{ft}}\right) + \left(\text{DC Weight}_{\text{ppf}} \times \text{DC Length}_{\text{ft}}\right) \tag{3.72}$$

$$\text{Workstring}_{\text{lbs}} = (16.6\,\text{ppf} \times 6500\,\text{ft}) + (101\,\text{ppf} \times 1000\,\text{ft}) = 208{,}900\,\text{lbs} \downarrow \tag{3.73}$$

$$\text{Workstring}_{\text{lbs}}\ (208{,}900\,\text{lbs}) \downarrow > \text{Well Force}\ (3500\,\text{lbs}) \uparrow,\ \text{Drillstring can be stripped in hole.} \tag{3.74}$$

b) Define Safety Margin

i) Safety Margin is dependent upon shoe strength.

ii) To be established by stripping into closed well and building pressure.

iii) Estimated Safety Margin = 200 psi.
iv) Additive Strip Pressure (1/2 of Safety Margin) = 100 psi.
v) Total Casing Pressure before Stripping is initiated = 200 psi + 100 psi = 300 psi.

c) Determine Strip Volume
i) Strip Volume

$$\text{Strip Volume}\left(\frac{\text{bbls}}{\text{ft}}\right) = \frac{1\text{ ft}}{\text{Closed} - \text{In DP Displacement}_{\text{bbls/ft}}} \tag{3.75}$$

$$\text{Strip Volume}\left(\frac{\text{bbls}}{\text{stand}}\right) = \text{Disp Volume}_{\frac{bbl}{ft}} \times \text{Length of Stand}_{\frac{\text{ft}}{\text{stand}}} \tag{3.76}$$

$$\text{Strip Vol}\left(\frac{\text{bbl}}{\text{stand}}\right) = 0.0197\frac{\text{bbl}}{\text{ft}} \times 93\frac{\text{ft}}{\text{stand}} = 1.8\text{ bbl/std} \tag{3.77}$$

$$\text{Total Strip Vol(bbl)} = 0.0197\frac{\text{bbl}}{\text{ft}} \times (\text{Hole MD}_{\text{ft}} - \text{Bit MD}_{\text{ft}}) \tag{3.78}$$

$$\text{Total Strip Vol (bbl)} = 0.0197 \times (7500' - 3000') = 88\text{ bbls} \tag{3.79}$$

d) Determine Volume and Frequency of Bleed Cycle.
i) *Perform bleed at intervals of one-third of Total Strip Volume or every 1–2 h.*
- For our simulation exercise, 30 ft./min average stripping time is used. For 4500 ft./30 ft./min ~ 2–1/2 h to strip to bottom, therefore bleed cycles limited to two bleed cycles.

ii) $\text{Vol}_{\text{Bleed}}$ (bbls/stand) = ½ Safety Margin.

$$\text{Vol}_{\text{Bleed}}\left(\frac{\text{bbl}}{\text{ft}}\right) = \left(\frac{100\text{ psi}}{(0.052) \times (\text{MW}_{\text{ppg}})}\right) \times \text{Closed In DP Displacement}\left(\frac{\text{bbl}}{\text{ft}}\right) \tag{3.80}$$

$$\text{Vol}_{\text{Bleed}} = 171\text{ ft} \times 0.0197\frac{\text{bbl}}{\text{ft}} = 3.4\text{ bbls where HP is decreased by 100 psi.} \tag{3.81}$$

$$\text{Two bleed cycles have been scheduled} \tag{3.82}$$

iii) Bleed off 3.4 bbls in small increments (When 100 psi of $\text{Vol}_{\text{Bleed}}$ is reached, the influx is allowed to expand). Once bleed volume has been recovered, initiate pumping operations and allow the Back Pressure to rise above previous Aggregate Back Pressure by the Hydrostatic Pressure of Bleed Cycle. (Original SICP + Safety Margin + Back Pressure_1 + Back Pressure_2.)

e) Reinitiate stripping operations by maintain constant casing pressure while maintaining constant displacement volumes.

f) When bit and BHA enters influx, sudden increase in CP can be expected. Increase in pressures should be minor in nature with no corrections needed.

g) At TD, bleed off Safety Margin pressures. Observe stabilized SIDPP and SICP. Initiate an appropriate constant bottom-hole pressure method to kill the well.

Water-based mud example shut-in conditions

While shut-in, evaluation is made. Based on the position of the bit, by time the bit reaches TD, influx will be at or near surface. Decision is made to allow the influx to come to surface in a controlled manner whereupon the influx will be allowed to expand. Once the influx is at the surface, lube and bleed method will be used to remove influx from the well. Once accomplished, the tubulars will be stripped to bottom and the well will be killed.

Shut-in conditions are SICP = 10 bbls, SIDPP = 0 psi (due to float) with a 10 bbl kick (Fig. 3.66).

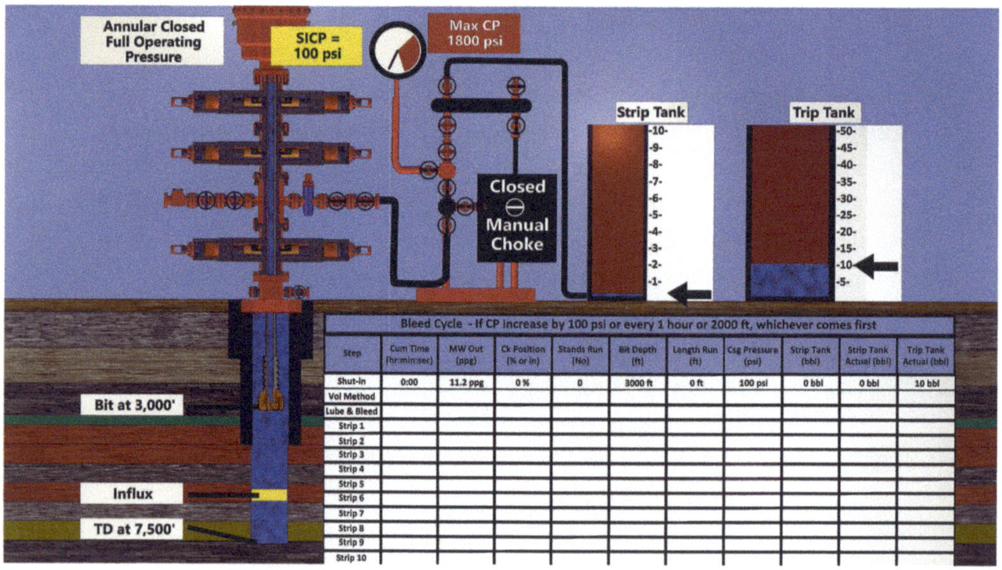

Fig. 3.66 Shut-in conditions.

Using the Volumetric Method, 10 bbls (original kick) + 52 bbls (volumetric bleed) for a total of 62 bbls of fluid removed from well to allow gas to expand. Influx is bled until traces of gas are heard, then shut-in. Casing pressure has increased from 100 to 720 psi (Fig. 3.67).

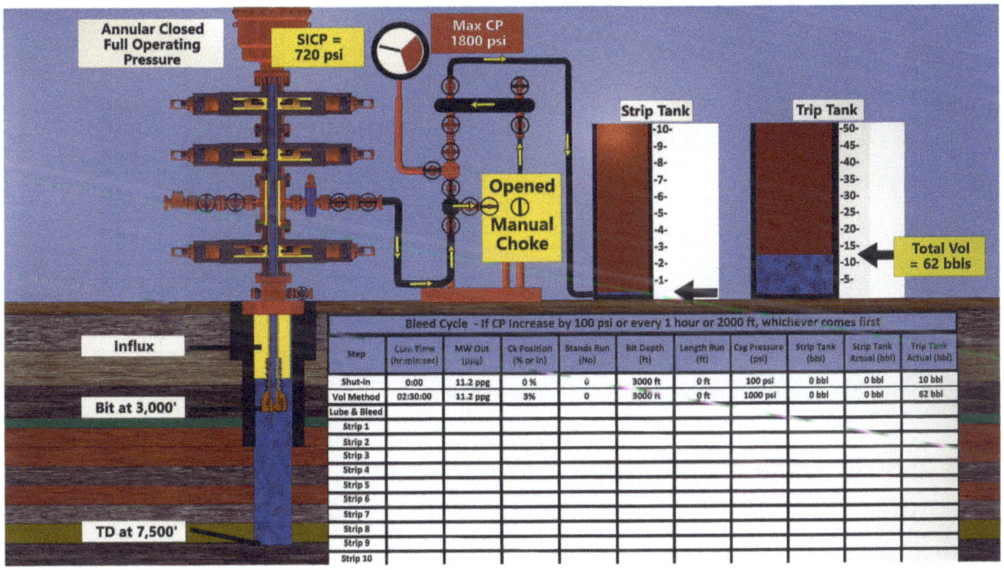

Fig. 3.67 Use volumetric method to get influx to surface.

Step 1: Establish safety margin

Using the Lube and Bleed, KMW is determined to be 11.5 ppg. KMW is pumped into the well while CP is monitored. The fluids are allowed to swap. Then gas is bled off reducing the CP by HP of KMW pumped. This cycle is repeated until gas is evacuated from well. In this example, the CP is reduced to 300 psi after lubricating and bleeding 62 bbls of KMW. During the last lubrication cycle, the CP is increased to 300 psi (Original Shut-In Casing Pressure (100 psi) plus Safety Margin (200 psi) (Fig. 3.68).

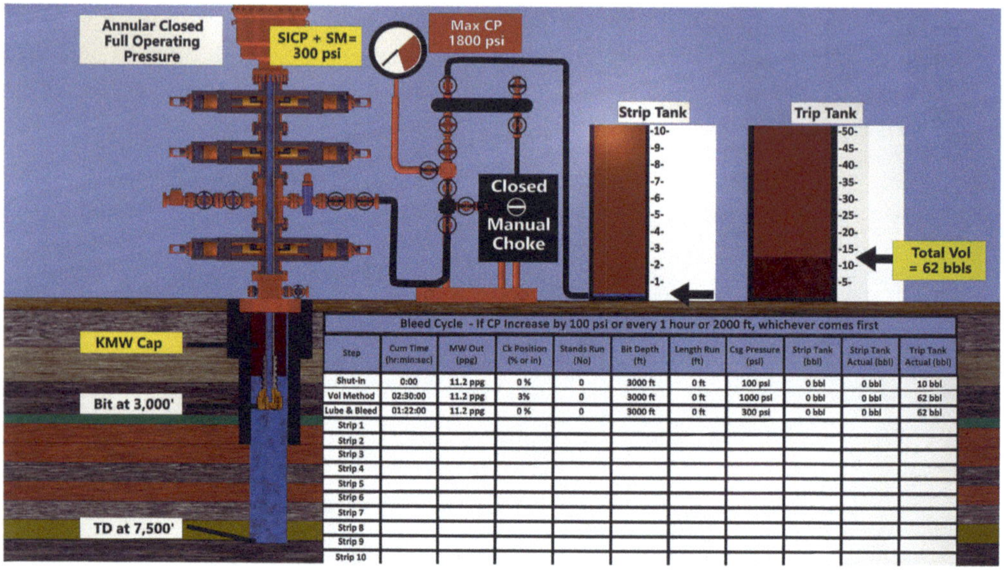

Step	Cum Time (hr:min:sec)	MW Out (ppg)	Ck Position (% or in)	Stands Run (No)	Bit Depth (ft)	Length Run (ft)	Csg Pressure (psi)	Strip Tank (bbl)	Strip Tank Actual (bbl)	Trip Tank Actual (bbl)
Shut-in	0:00	11.2 ppg	0 %	0	3000 ft	0 ft	100 psi	0 bbl	0 bbl	10 bbl
Vol Method	02:30:00	11.2 ppg	3%	0	3000 ft	0 ft	1000 psi	0 bbl	0 bbl	62 bbl
Lube & Bleed	01:22:00	11.2 ppg	0 %	0	3000 ft	0 ft	300 psi	0 bbl	0 bbl	62 bbl
Strip 1										
Strip 2										
Strip 3										
Strip 4										
Strip 5										
Strip 6										
Strip 7										
Strip 8										
Strip 9										
Strip 10										

Fig. 3.68 Use lube and bleed to remove influx. Use pumps to establish safety margin.

Step 2: Strip while monitoring casing pressure

Before stripping is to begin, ensure annular closing pressure is adjusted for leakage and tooljoint pressure ranges are known. Strip tank is to be lined up with the Trip Tank to receive cumulative volumes pumped from the strip tank during connections. Strip tank calculated volumes should be the same volume bleed into strip tank (Fig. 3.69).

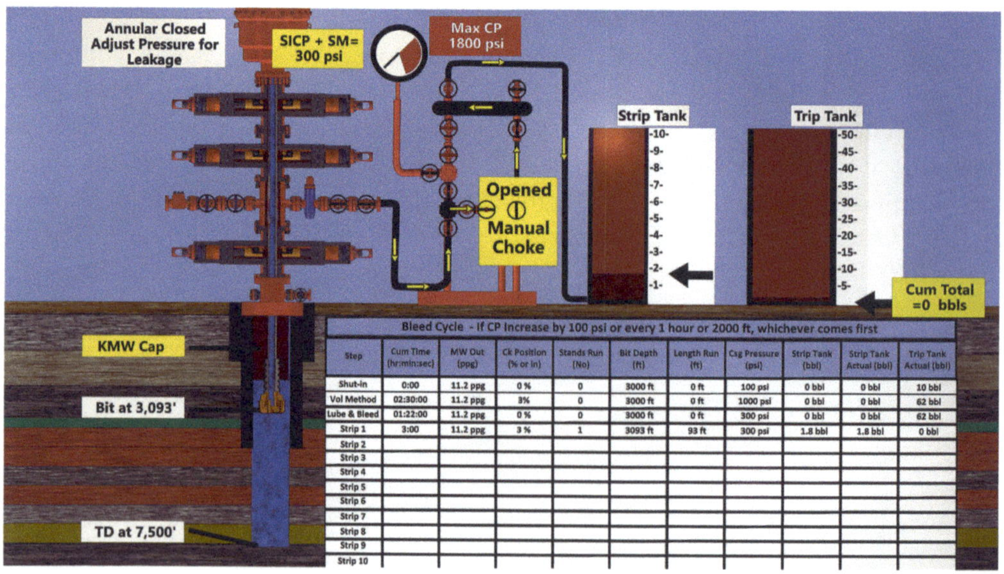

Step	Cum Time (hr:min:sec)	MW Out (ppg)	Ck Position (% or in)	Stands Run (No)	Bit Depth (ft)	Length Run (ft)	Csg Pressure (psi)	Strip Tank (bbl)	Strip Tank Actual (bbl)	Trip Tank Actual (bbl)
Shut-in	0:00	11.2 ppg	0 %	0	3000 ft	0 ft	100 psi	0 bbl	0 bbl	10 bbl
Vol Method	02:30:00	11.2 ppg	3%	0	3000 ft	0 ft	1000 psi	0 bbl	0 bbl	62 bbl
Lube & Bleed	01:22:00	11.2 ppg	0 %	0	3000 ft	0 ft	300 psi	0 bbl	0 bbl	62 bbl
Strip 1	3:00	11.2 ppg	3 %	1	3093 ft	93 ft	300 psi	1.8 bbl	1.8 bbl	0 bbl
Strip 2										
Strip 3										
Strip 4										
Strip 5										
Strip 6										
Strip 7										
Strip 8										
Strip 9										
Strip 10										

Fig. 3.69 Adjust annular pressure for leakage, strip in hole maintaining constant CP with continuously bleeding into strip tank.

Step 3: Connection

When connection point has been reached, stop and shut well-in with full annular closing pressure. Record volumes, pressures, and bit depth. Transfer mud from strip tank to trip tank. Use trip tank for cumulative return volumes and strip tank for individual stand strip volume recovery (Fig. 3.70).

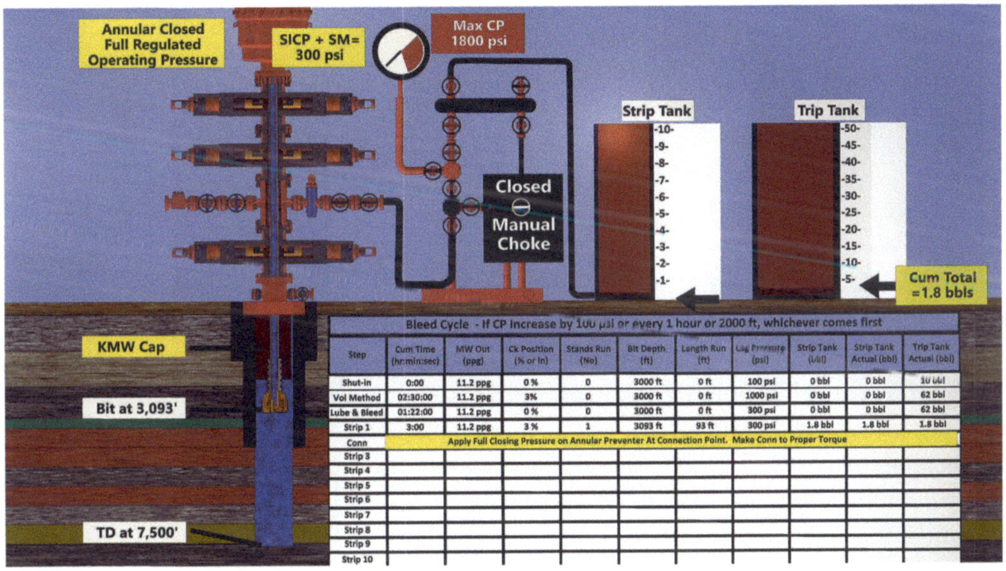

Fig. 3.70 At connection, apply full regulated closing pressure, properly torque connection, transfer mud.

Step 4: Repeat steps 2 and 3 until TD is reached

Continue stripping in well holding casing pressure constant with constant bleeding until TD is reached. Transfer mud from strip tank to trip tank each connection and use the trip tank for cumulative volumes (Fig. 3.71).

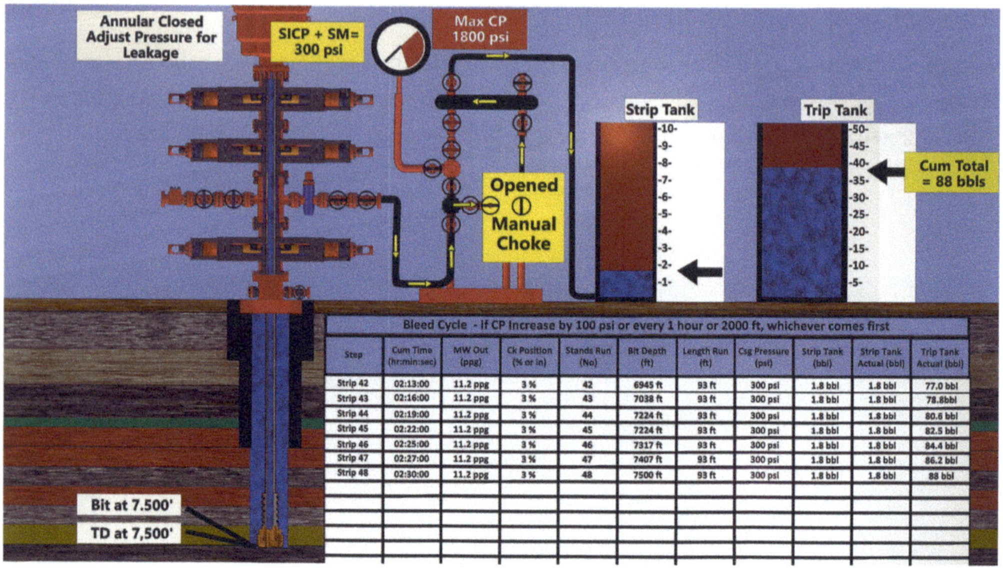

Bleed Cycle - If CP Increase by 100 psi or every 1 hour or 2000 ft, whichever comes first										
Step	Cum Time (hr:min:sec)	MW Out (ppg)	Ck Position (% or in)	Stands Run (No)	Bit Depth (ft)	Length Run (ft)	Csg Pressure (psi)	Strip Tank (bbl)	Strip Tank Actual (bbl)	Trip Tank Actual (bbl)
Strip 42	02:13:00	11.2 ppg	3 %	42	6945 ft	93 ft	300 psi	1.8 bbl	1.8 bbl	77.0 bbl
Strip 43	02:16:00	11.2 ppg	3 %	43	7038 ft	93 ft	300 psi	1.8 bbl	1.8 bbl	78.8bbl
Strip 44	02:19:00	11.2 ppg	3 %	44	7224 ft	93 ft	300 psi	1.8 bbl	1.8 bbl	80.6 bbl
Strip 45	02:22:00	11.2 ppg	3 %	45	7224 ft	93 ft	300 psi	1.8 bbl	1.8 bbl	82.5 bbl
Strip 46	02:25:00	11.2 ppg	3 %	46	7317 ft	93 ft	300 psi	1.8 bbl	1.8 bbl	84.4 bbl
Strip 47	02:27:00	11.2 ppg	3 %	47	7407 ft	93 ft	300 psi	1.8 bbl	1.8 bbl	86.2 bbl
Strip 48	02:30:00	11.2 ppg	3 %	48	7500 ft	93 ft	300 psi	1.8 bbl	1.8 bbl	88 bbl

Fig. 3.71 Strip to TD while holding casing pressure constant with constant bleeding.

After reaching TD holding casing pressure constant with constant bleeding, the well is shut-in. The safety margin is removed by incremental bleeds of return volumes until casing pressure is reduced by 200 psi. The safety margin is not needed for upcoming well control circulations.

Initiate appropriate constant BHP kill method to regain control of the well (Fig. 3.72).

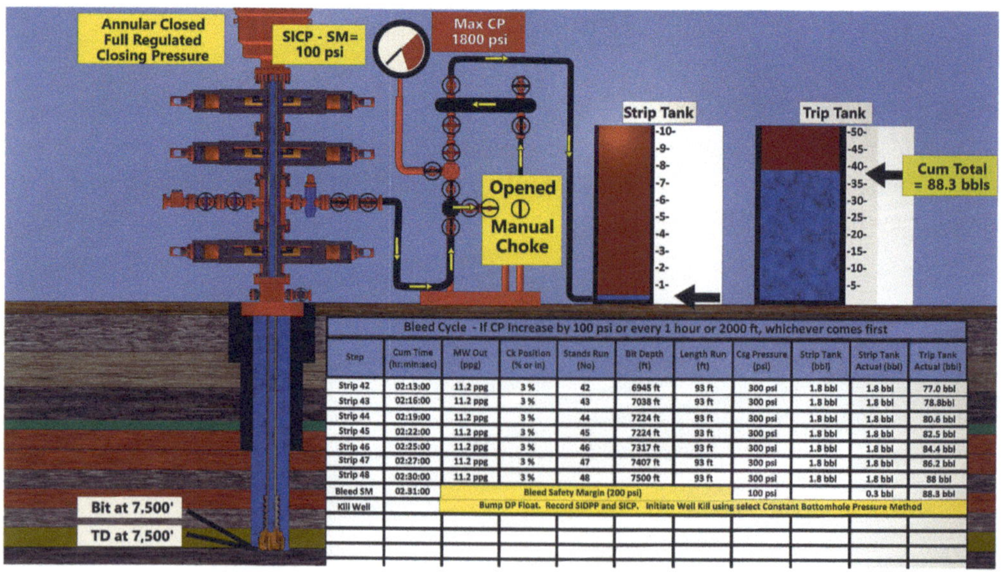

Bleed Cycle - If CP Increase by 100 psi or every 1 hour or 2000 ft, whichever comes first										
Step	Cum Time (hr:min:sec)	MW Out (ppg)	Ck Position (% or in)	Stands Run (No)	Bit Depth (ft)	Length Run (ft)	Csg Pressure (psi)	Strip Tank (bbl)	Strip Tank Actual (bbl)	Trip Tank Actual (bbl)
Strip 42	02:13:00	11.2 ppg	3 %	42	6945 ft	93 ft	300 psi	1.8 bbl	1.8 bbl	77.0 bbl
Strip 43	02:16:00	11.2 ppg	3 %	43	7038 ft	93 ft	300 psi	1.8 bbl	1.8 bbl	78.8bbl
Strip 44	02:19:00	11.2 ppg	3 %	44	7224 ft	93 ft	300 psi	1.8 bbl	1.8 bbl	80.6 bbl
Strip 45	02:22:00	11.2 ppg	3 %	45	7224 ft	93 ft	300 psi	1.8 bbl	1.8 bbl	82.5 bbl
Strip 46	02:25:00	11.2 ppg	3 %	46	7317 ft	93 ft	300 psi	1.8 bbl	1.8 bbl	84.4 bbl
Strip 47	02:27:00	11.2 ppg	3 %	47	7407 ft	93 ft	300 psi	1.8 bbl	1.8 bbl	86.2 bbl
Strip 48	02:30:00	11.2 ppg	3 %	48	7500 ft	93 ft	300 psi	1.8 bbl	1.8 bbl	88 bbl
Bleed SM	02.31:00	Bleed Safety Margin (200 psi)					100 psi		0.3 bbl	88.3 bbl
Kill Well		Bump DP Float. Record SIDPP and SICP. Initiate Well Kill using select Constant Bottomhole Pressure Method								

Fig. 3.72 Bleed off safety margin (200 psi). Initiate killing the well.

Stripping drill (before drilling out casing)

1. Trip drilling assembly until drill bit is ±10 stands above the float collar or at a pre-agreed depth.
   a. For Subsea Operations, ensure drill bit and DP are below the BOPs.
2. Install the Full Opening Safety Valve and close same.
3. Open the choke line valves to a closed choke and close the (upper) annular preventer with full annular closing pressure.
4. Install Top Drive and open the FOSV to simulate recording shut-in drillpipe pressure.
5. Open the choke and circulate until returns are seen in the Strip Tank (Trip Tank). Shut down pumps and close the choke.
6. Pressure up the well to a predetermined value (suggest be between 400 and 500 psi.)
   a. Close the FOSV.
   b. Bleed off above to ensure FOSV is holding, then remove the Top Drive.
   c. Alternatively, you may pressure up the well via the kill line, keeping the full opening safety valve closed. This will eliminate the need for breaking a connection with pressure below the valve.
   d. Record bit depth, pressure, and strip (trip) tank level.
7. Increase in the annulus pressure for the following:
   - Safety Margin (suggest 200 psi)
   - A working pressure increase for gas expansion (suggest 100 psi).
8. Reduce annular closing pressure to a minimum to initiate leakage
   a. Open surge bottle, if fitted.

Step 1: Strip: Establish Safety Margin (200 psi over SICP)
   b. With choke closed, strip pipe until Safety Margin is achieved.

Step 2: Strip: Commence stripping operations
   a. Lower workstring at approximately 30 ft./min.
   b. Choke Operator to monitor and allow the pressure to increase Bleed Increment volume equivalent pressure.
   c. Once this value has been reached, the Choke Operator bleeds off fluid to maintain Bleed Increment value as each stand is lowered.

Step 3: Connection: Stop at connection point, increase to full annular closing pressure, close choke, and set slips. Perform connection.

Note: The Driller to note string weight loss to strip the pipe though the annular and reduced annular pressure. The Driller to note string weight loss for tooljoints to pass through the annular along with increase in closing pressure. This is useful information to have in the case of an actual stripping operation.

Step 4: Strip: Commence stripping two or three more stands in the hole with the choke operator bleeding off fluid to maintain the calculated pressure while stripping.

   a. Utilize several stands to ensure rig crews are proficient at stripping operations.
   b. Record relevant data at the end of each stand and during connections.

9. To discontinue strip drill, bleed off all annulus pressure and open annular preventor.
   a. Reset annular regulator to full closing pressure.

10. Pick up and remove the FOSV. Return systems to drilling alignment.

Manufacturer equipment guidelines

The following series of general manufacturer specifications was developed as examples for what should be investigated. Based on the variety of equipment available on each rig, these general examples are to be used to demonstrate search criteria to be used by specific manufacturer equipment specifications.

General discussion

The recommended hydraulic closing pressure for annular preventers during well control operations is 1500 psi. After the initial seal is obtained, the hydraulic pressure can be reduced to the level necessary to contain well pressure. This minimizes damage to the packing element and extends its effective life. Manufacturers of annular preventers provide charts showing recommended closing pressures for each size preventer based on casing pressure after the well is shut-in. However, the general field rule for surface BOP system is to gradually reduce hydraulic pressure until slight leakage occurs around the pipe. This ensures the packing element in a "relaxed position" for working the drillpipe up and down through the preventer to prevent the pipe form sticking. A small amount of fluid leakage lubricates and cools the packing unit to prolong its life.

Pressure-assisted annular BOPs

Please refer to Manufacturer's specifications for recommended stripping operating pressures (Table 3.5).

Table 3.5 Average closing pressure to establish seal for pressure-assisted annular BOPs.

Pipe O.D. (in.)	7–1/16″ 3M	7–1/16″ 5M	7–1/16″ 10M	7–1/16″ 15M	7–1/16″ 20M	9″ 3M	9″ 5M	9″ 10M	11″ 3M	11″ 5M	11″ 10M	13–5/8″ 3M	13–5/8″ 5M	13–5/8″ 10M	16″ 2M	16″ 3M	16–3/4″ 5M	18″ 2M
6–5/8″	–	–	–	–	–	–	–	–	–	350	–	700	600		350	450	–	500
5″	–	400	–	–	–	–	–	350	450	450	500	800	650	700	400	500	–	550
4–1/2″	350	400	350	2100	2200	400	450	380	450	450	500	900	650	700	500	500	600	600
3–1/2″	400	450	550	2100	2200	500	600	570	550	525	700	1000	700	1200	600	600	650	650
2–7/8″	400	450	750	2100	2200	550	650	700	650	800	800	1100	750	1400	700	700	750	700
2–3/8″	500	500	850	2100	2200	650	750	800	750	900	1100	–	950	1400	800	800	850	740
1.9	600	600	900	–	–	750	850	850	900	–	–	–	1000	1500	900	900	950	850
1.75	700	700	1000	–	–	850	950	1000	950	–	–	–	1000	1500	1000	1000	1050	950
CSO	1000	1000	1150	–	–	1060	1150	1150	1150	1150	1300	1200	1150	2200	1150	1150	1150	1150

The pressures above are to be used as guidelines. Minimum packing unit life will be realized by the use of the lowest closing pressures which will maintain a seal.
Recommended test ppe for maximum packing unit life.
For subsea operations, see the appropriate Operator's Manual for computation of best closing pressures.

Spherical annular preventers

Please refer to Manufacturer's specifications for recommended stripping operating pressures. Most spherical annular preventers open/close with 1500 psi. They operate in three modes (1) closing on stationary pipe or open hole, (2) closing in on casing, and (3) stripping operations (Table 3.6) (Fig. 3.73).

Table 3.6 Example guidelines for closing pressure on casing by spherical preventer.

Spherical size	Working pressure (psi)	Casing size (in.)									
		7″	7-5/8″	8-5/8″	9-5/8″	10-3/4″	11-3/8″	13-3/8″	16″	18-5/8″	20″
20–3/4″	3000	1500	1400	1175	975	790	640	480	300	190	150
21–1/4″	5000	1500	1400	1175	975	790	640	480	300	190	150
21–1/4″	2000	1500	1400	1175	975	790	640	480	300	190	150
18–3/4″	10,000	–	–	–	385	–	–	310	325	–	–
18–3/4″	5000	1500	1400	1175	975	790	640	480	300	–	–
16–3/4″	5000	1500	1400	1175	975	790	640	480	–	–	–
13–5/8″	3000	1500	–	1265	615	415	280	–	–	–	–
13–5/8″	5000	1500	–	1265	615	415	280	–	–	–	–
30″	1000	–	–	–	1000	–	–	1100	–	–	900

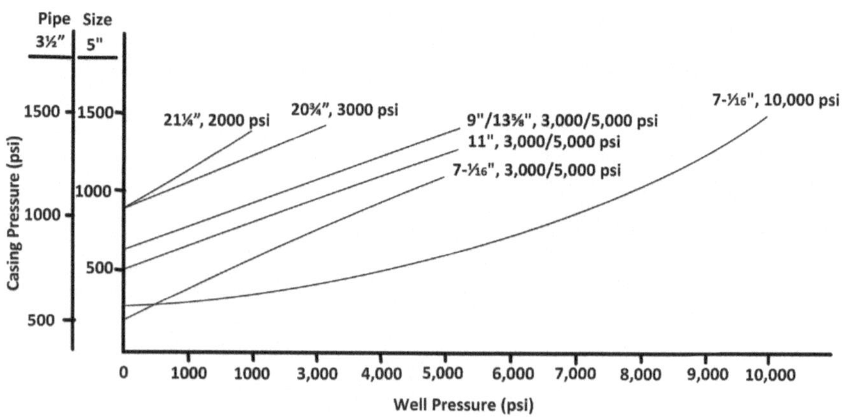

Fig. 3.73 Example guidelines for closing pressures for stripping operations in spherical preventers.

To close spherical on casing, adjust the spherical preventer's hydraulic operating pressure to avoid contact of the sealing element segments with the outer diameter of the casing, if necessary.

Reverse circulating well control method

Well killing techniques are primarily used so sufficient fluid velocities are obtained to circulate out any foreign matter while increasing fluid density to thwart influxes when circulation ceases and minimize time necessary to circulate kill fluids.

Reverse circulating formation fluids and/or influxes on live pressurized wells is/are the most common practice of killing a well. Since reservoir pressures are known and packer fluid contains sufficient density to prevent additional influxes, one reverse circulation will normally kill the well. Drilling wells with open hole are not good reverse circulation candidates.

Initial circulation

Circulating methods are designed to maintain a constant bottom-hole pressure while removing formation fluids (i.e., gas) from the well. As fluid is circulated, the fluid comes in contact with the surface of the drillpipe/drillpipe/tubing, formation and casing, resulting in friction pressure. These friction pressures in combination with hydrostatic pressure of the fluid and pump pressures will yield a bottom-hole pressure which must be at least equivalent to the formation pressure. In our equations, "Friction" means the additive pressure added by friction or "friction pressure."

Normal circulation

DP Side: For normal circulating through drillpipe and up the annulus, the bottom-hole pressure is equal hydrostatic pressure of fluid in drillpipe + SIDPP + friction pressure of fluids traveling down the drillpipe (account for at least 80% of the friction). Friction pressure is the pressure created due to the resistance of the fluid flowing down the drillpipe and out the bit jets. In order to keep BHP, Friction pressure on DP side increases dramatically. Therefore, it is very difficult to manipulate the choke setting and wait for pressure to be transmitted down the annulus and up the drillpipe (Fig. 3.74).

Fig. 3.74 Drillpipe sum of forces when staging pumps up to speed.

Casing Side: If the casing gauge is used to stage pumps up to speed, BHP can be held constant as friction pressure is nominal in the annulus. Bottom-hole pressure is equal to hydrostatic pressure of the annulus plus SICP plus friction pressure. When pumps are brought up to speed, the choke is slowly opened while SICP is held constant. Therefore, the friction pressure is automatically compensated and bottom-hole pressure is kept constant. Mathematically, it is represented by small friction pressure going to zero as choke is opened (Fig. 3.75).

$$\overline{HP}_{Annulus} + \overline{SICP} + \text{Friction}^{\nearrow 0} = \overline{BHP}$$

Fig. 3.75 Annulus drillpipe sum of forces when staging pumps up to speed.

The total equation for both drillpipe and annulus is represented below as (Fig. 3.76):

$$\overline{HP}_{Drillpipe} + \overline{SIDPP} + \text{Friction}\uparrow = \overline{BHP} = \overline{HP}_{Annulus} + \overline{SICP} + \text{Friction}^{\nearrow 0}$$

Fig. 3.76 Total sum of forces when staging pumps up to speed.

Since annular friction pressure can be considered to be negligible, the BHP calculation can be simplified to circulating casing pressure plus annular hydrostatic pressure.

Initiation of circulation of driller's method

In order to keep BHP constant when initiating circulation, if the casing pressure is kept constant, then BHP will remain constant (Fig. 3.77).

$$\overline{HP}_{Annulus} + \overline{SICP} = \overline{BHP}_{Normal\ Circulation}$$

Fig. 3.77 Annular forces needed to keep BHP constant when circulating.

First circulation of driller's method (SIDPP = SICP, well is not dead)

$$\overline{HP}_{Drillpipe} + \overline{\text{Friction}}_{Circulating} = \overline{BHP}_{Normal\ Circulation}$$

Fig. 3.78 Drillpipe forces needed to keep BHP constant when circulating.

Once circulation has been established and pumps are up to speed, read and record the initial DP circulating pressure and maintain this constant until influx is circulated to surface (annular volume + 50%) (Fig. 3.78). Once the influx has been circulated to surface, the well can closed-in while holding casing pressure constant.

Reverse circulation

Comparatively reverse circulation will yield higher bottom-hole pressures as friction pressure resulting from fluid traveling up the drillpipe/tubing is greater due to faster fluid movement. Fluid travels faster, as it flows at the same pump pressure, but within a smaller cross-sectional area (tubing vs. annulus). Fluid will be traveling 4–5 times faster within the drillpipe/tubing resulting in pressure losses of 16–25 times of normal circulation.

With reverse circulating, bottom hole pressure is equal to the additive surface drillpipe/tubing pressure, the drillpipe/tubing hydrostatic pressure and the drillpipe/tubing friction pressure (Fig. 3.79).

$$\overline{\text{SIDPP}} + \overline{\text{HP}}_{\text{Drillpipe}} + \overline{\text{Friction}}_{\text{Drillpipe}} = \overline{\text{BHP}}_{\text{Reverse Circulation}}$$

Fig. 3.79 Static drillpipe forces for reverse circulating.

Bottom-hole pressure is equal to the surface pressure of the casing plus the casing hydrostatic pressure minus the annular fiiction pressure (Fig. 3.80).

$$\overline{\text{SIDPP}} + \overline{\text{HP}}_{\text{Drillpipe}} + \overline{\text{Friction}}_{\text{Annular}} = \overline{\text{BHP}}_{\text{Reverse Circulation}}$$

Fig. 3.80 Static annular forces reverse circulating.

For Reverse Circulation, in order to keep BHP constant when initiating circulation, the casing pressure is kept constant plus a small additional amount of pressure to offset reduction due to Friction Pressure of annulus. Normally, since 10,000 psi pressure gauges read in 100 psi increments, we will derive the following (Fig. 3.81).

$$\overline{\text{HP}}_{\text{Drillpipe}} + \overline{\text{SIDPP}} + \text{Friction}\uparrow = \overline{\text{BHP}} = \overline{\text{HP}}_{\text{Annulus}} + \overline{\text{SICP}} + 100\ \text{psi} + \text{Friction}^{\uparrow\emptyset}$$

Fig. 3.81 Totalized forces when initiating reverse circulating.

Constant bottom-hole pressure

To maintain constant bottom-hole pressure during the circulation, simply maintain constant surface pressure on side without formation fluids and/or influx. If the surface pressure and the hydrostatic pressure stay the same and if the pump rate is kept constant, then bottom-hole pressure must stay constant as friction pressures will not change.

Reverse circulating procedure

The following is a step-by-step procedure for the Reverse Circulating Method of well control.

First circulation

- Bring the pump up to speed holding casing pressure constant at the shut-in casing pressure plus a reasonable safety factor (recommend 100 psi or any reading of 50 psi increments). Manipulate the choke in order to keep casing pressure constant varying pump speed slowly. Always crack open the choke first to ensure choke is operable, before engaging pumps.
- Maintain the initial circulating pressure on the casing until the influx is removed.
- Once the influx is removed, shut down the pump holding casing pressure constant.
  - NOTE: At this point, the shut-in casing and shut-in drillpipe/tubing pressures should be the same. If kill fluid is in the well, theoretically these surface pressures should be 0 psi. If the well is not dead at this point, prepare the required kill weight fluid necessary to control the well. Ensure calculations include temperature correction factor.

Second circulation, if required

- Bring the pump up to speed holding casing constant by regulating the choke on the drillpipe/tubing side.
- Read and record the initial circulating drillpipe/tubing pressure.
- Maintain constant drillpipe/tubing pressure until kill weight fluid fills the annulus.
- Read and record the final circulating casing pressure.
- Maintain the final circulating casing pressure constant until **kill weight fluid** completely displaces the drillpipe/tubing back to surface.
- Shut down the pump holding casing pressure constant.
  - NOTE: The additive safety factor can be reduced at this time, through small multiple bleed offs. Monitor casing pressure, in between bleed increments, to ensure another influx has not occurred.
  - NOTE: Ideally, the well should be dead (i.e., SICP = SITP = 0 psi) when the kill weight fluid completely displaces the tubing.

Reverse circulating considerations

In order to perform safe operation, reverse circulation of a kick requires a review of procedures to ensure proper reactions on the choke are made.

Frictional effects

When reverse circulating, bottom-hole pressure is a function of hydrostatic pressure, surface pressure, and the friction pressure inside the tubing. If choke manipulations do not account for this additional friction pressure, the sum of surface pressure and annular pressure will be less than bottom-hole pressure, which may lead to an influx entering the wellbore.

For Reverse Circulation, in order to keep bottom-hole pressure constant when initiating circulation, the casing pressure is kept constant plus a small additional amount of pressure to offset reduction due to Friction Pressure of annulus.

- NOTE: A large safety factor may be needed when circulating around large OD tools, such as a packer, to compensate for the additional annular friction.

Parameters

If a reverse circulating kill is being attempted on a well where produced gas is in the annulus, the gas migration rate must be overcome by annular fluid velocity. Gas migration rates have been experimentally determined to be between 1000′ and 4000 ft./h. Therefore, when reverse circulating gas kicks, the pump rate used to pump completion/workover fluid down the annulus must exceed the rate of gas migration upward.

- NOTE: Warning, since the gas migration rate determines the minimum circulation rate, it may not be possible to keep the drillpipe/tubing friction from causing the well to go overbalanced and increasing the potential for losing circulation. Given the following information, calculate the required pump rate (in SPM) to effectively reverse circulate this well.

Example 1

- Drillpipe 2–7/8″ 6.5# N-80
- Casing 7″ 26# N-80
- Pump Output = 0.0343 bbls/stroke
- Worst Case Gas Migration Rate = 4000 ft./h
- Fluid Weight (FW) = 10 ppg
- Annular Capacity Factor = 0.0302 bbls/ft.

Solution:

1. Convert the gas migration rate to ft./min

$$\text{Gas Migration Rate}_{\text{ft}/\text{min}} = \frac{\text{Gas Migration Distance}_{\text{ft/hr}}}{\text{Time of Rise}_{\text{min/hr}}} = \frac{4000\text{ft/h}}{60\,\text{min}/\text{h}} = 66.67\text{ft}/\text{min} \tag{3.83}$$

2. Determine the minimum required pump rate in bbl/min.

$$\text{Minimum Pump Rate}_{\text{bpm}} = \left(0.032\frac{\text{bbl}}{\text{ft}}\right)(66.67\,\text{ft}/\text{min}) = 2.01\,\text{bpm} \tag{3.84}$$

3. Determine the Minimum Required Pump Speed

$$\text{Minimum Pump Speed}_{\text{spm}} = \left(\frac{2.01\,\text{bbl}}{\text{min}}\right) \times \left(\frac{\text{stroke}}{0.0343\,\text{bbl}}\right) = 59\,\text{spm} \tag{3.85}$$

Therefore, the minimum required pump speed to efficiently reverse circulate the gas kick is at least 59 strokes per minute.

Poor condition of wellhead and/or production casing

Many wells are reverse circulated due to the poor condition of either the wellhead or the casing at surface. Unfortunately, since higher pressures are induced by larger friction pressures as fluid goes from annulus to drillpipe/tubing, higher pressures within the influx will occur. This means if the influx is gas, gas expansion will not readily occur resulting in higher pressure when gas reaches the surface.

Fortunately, the burst rating of drillpipe/tubing is normally much greater than the burst rating of casing. The condition of tubulars, both casing and drillpipe/tubing, is an important consideration during this type of well control operation.

In order to maximize safety of personnel, surface equipment must be competently fastened, secured, and thoroughly TESTED. Our authors suggest all personnel not required for this operation be evacuated to safe area and operational personnel be located as far away as practical from pressurized circulation lines.

The **advantages** of a reverse circulating kill are:

- The influx or contaminant is contained in the workstring.
- Lower casing pressures when compared to normal circulation.
- Higher burst rating of drillpipe/tubing compared to burst of casing.
- Unique Well Situations where Reverse Circulation **should** be used:
  - When the influx is already in the workstring or drillpipe/tubing, i.e., killing a producing well or taking a kick while washing out or deepening with reverse circulation.
  - When the influx needs to be contained in the drillpipe/tubing or removed quickly, such as DSTs, H_2S, CO_2, etc.
  - When wellhead or casing integrity is suspect, such as a workover on an older well.
- Quick removal of the influx when compared to normal circulation (less volume needed).

The **disadvantages** of a reverse circulating kill are:

- Excessive surface pressures in the drillpipe/tubing due to increased kick height of gas within the tubing.
- Higher pump rates must be used to overcome the migration rate of any gas which is in the annulus.
- Annular friction must be accounted within circulation; otherwise BHP may below Formation Pressure.
- Unique Well Situations Where Reverse Circulation **should not** be used.
  - When the drillpipe/tubing string has a float or restrictive circulation device.
  - When the surface equipment does not give the flexibility for reverse circulating, or cannot safely contain the expected surface pressures.
  - When the influx is already in the annulus, especially if it is gas.
  - When there are jets in a lowest circulation port (i.e., bit), or some other small restriction in the tubing.

CHAPTER FOUR

Well control using specialized equipment

Commentary

Specialized operations require customized equipment and procedures to ensure constant well control. These specialized operations include operations such as wireline, coiled tubing, snubbing, and managed pressure operations. Each specialized equipment operation reviews equipment limitations and most commonly used forms of well control. As these operations represent the cutting edge of technology improvements, individual equipment, and methodologies may change over time as newer, more sophisticated tools are introduced. Regardless of operations, individual components work in harmony to create multiple barriers for the protection of personnel and environment. Proper installation, testing, and maintenance of these specialized services must be performed with precision and thoroughly documented.

Wireline operations well control methods

Well control will be maintained by using well control methods previously discussed, but augmented with specialized equipment and procedures. The use of specialized equipment will vary due to pressure regimes and locations, but in general, offer tools necessary to perform work on wells in which traditional equipment spreads are not economically or operationally viable. Most specialized equipment employs personnel specifically trained in operating their service equipment in harsh environments. The following methodologies are offered for general knowledge of systems and procedures. Individual service company operating procedures may deviate, therefore, please refer to service company provided information for the most specific well control method to be employed.

REFER TO SERVICE COMPANY PROVIDED PROCEDURES FOR SAFE OPERATION OF SPECIALIZED EQUIPMENT AND WELL CONTROL METHODS

Universal Well Control
https://doi.org/10.1016/B978-0-323-90584-8.00004-6

Wireline operations commentary

Well control operations requiring wireline interfacing are normally limited to shutting off flow from the reservoir by installing an isolation plug in a nipple placed within the tubing string. Wireline operations require a small crew and equipment complement. These units are highly portable, easy to install, and test. These units are primarily used for installation of isolation plugs, gas lift mandrels, foreign material, and recovery (sand bailing). The common primary well control equipment employed with solid wireline (slickline) work consists of a pressurized lubricator, hydraulic stuffing box, and BOP to ensure effective well control. To control a well under pressure, the lubricator and BOPs are installed and pressure tested. The appropriate isolation plug is affixed to the wireline and placed within the lubricator. Pressures are equalized and the tool is lowered into the appropriate landing nipple contained within the tubing.

These plug isolation systems are prone to failure due to scaling, formation fines, asphaltenes, and paraffin build up. If a tool becomes "stuck," the wireline can only be worked by pulling and releasing tension. If the tool cannot be recovered, the wireline is pulled until the catch affixed to the tool is sheared. The tool head can be recovered along with the wireline. Unfortunately, due to wear of the wireline, the wireline may fail before the tool head is sheared. In this case, both wireline and tool head may be lost into the well.

Wireline well control operations

Well control operations used with Wireline Operations include bull heading and circulating using 1st circulation of the Driller's Method. Since wireline operations are normally provided by a specific service provider, please refer to specific service provider literature for exact well control procedures.

Solid wireline (commonly called slickline)

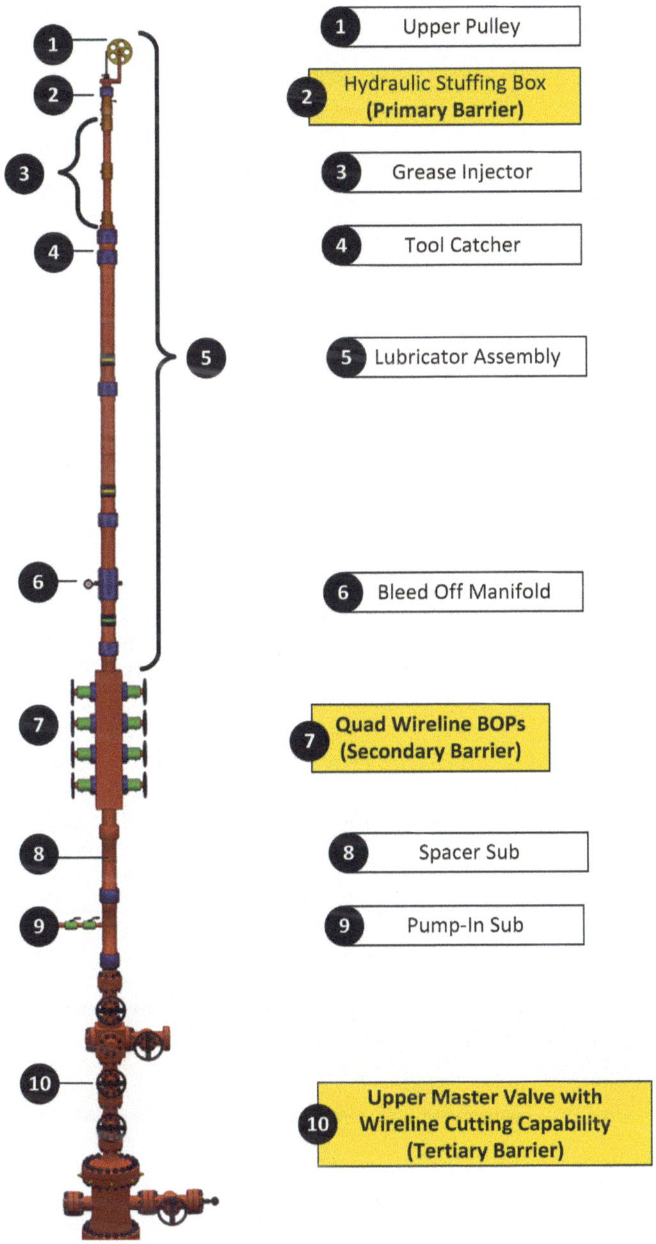

Fig. 4.1 Wireline/slickline assembly.

High-strength steel wireline was developed to reduce size and weight of hoisting/pressure containment equipment (Fig. 4.1). This design reduces overall weight of tool assembly, utilizes smaller sheaves, minimizes the size of the reel, and provides a smaller footprint (i.e., cross-sectional area) for operating in pressurized environments. The most common diameter sizes for this single strand of high-strength steel lines are 0.066, 0.072, 0.082, 0.092, 0.105, 0.108, and 0.125 in. The line can be spooled onto reels in excess of 25,000 ft., if needed. The ultimate yield strength of the slickline is 230,000–240,000 psi. For service in H_2S environments, Type 316 stainless steel is suggested for use due to its resistive properties to corrosiveness of H_2S.

Slickline can be used for:

- Running and pulling gas lift or chemical injection equipment.
- Running and pulling downhole safety valves.
- Bottom-hole pressure, temperature, and fluid sampling.
- Running temperature and pressure surveys.
- Depth measurement.
- Fishing for lost objects and debris.
- Running and pulling flow control devices.
- Opening and closing circulation devices.
- Checking the inside of the tubing for debris, waxes, scale, corrosion, etc.
- Cleaning the inside of the tubing and the completion components.

Breaking strength

Breaking strength refers to a condition where the tensile yield strength of the slickline will be exceeded and may cause failure and separation (Table 4.1).

Table 4.1 Well-measuring wire specifications.
Well-measuring wire specifications (solid wire)

Nominal diameter (in.)	0.066	0.072	0.082	0.092	0.105	0.108
Tolerance on dia (in.)	±0.001	±0.001	±0.001	±0.001	±0.001	±0.001
Breaking strength (lbs)						
Minimum	811	961	1239	1547	1996	2109
Maximum	984	1166	1504	1877	2421	2560
Elongation in 10 in. (%)						
Minimum	1½	1½	1½	1½	1½	1½
Torsions, minimum number of twists in 8 in.	32	29	26	23	20	19

WHEN USED, THE WIRE SHOULD NOT BE WORKED PAST ITS PLASTIC LIMIT APPROXIMATELY 50% OF THE BREAKING STRENGTH.

Coated wireline

Coated wireline was developed for use in tubing that contains an internal coating to prevent corrosion and wear. Service life for coated wireline is shortened due to applied external coating that detrimentally affects bend and yield resistance.

Slickline handling procedures

To maximize the service life of the slickline, the following handling procedures should be implemented:

Properly transferring the measuring line from the ship:

- DO NOT exceed elastic limit.
- During retrieval operations, clean wellbore fluids from slickline before spooling.
- For frequent, repetitive operations, cut, and replace worn sections as needed.
- Do not grip slickline with handling tools with "teeth." These gripping teeth will cause nicking or gouging that may cause the slickline to fail under load conditions.
- Eliminate kinks. If line becomes kinked, replace before usage.
- Properly transfer slickline from transportation spool to slickline reel.

Braided line

Braided wireline (sometimes called stranded line) uses multiple strands of wire that are spun and braided into a single wireline rope. Braiding is used to add strength and increase wear resistance. Braided line may contain one or more electrical transmission wires that are referred to as "Electric Line" or e-line. Braided Line that contains no wound electric cables are called sand lines and are used in heavy-duty operations. Braided line has much greater breaking strength and can be found in the following common sizes, 1/8 in. (0.125″), 9/64 in. (0.141), 5/32 in. (0.156), 3/16 in. (0.187), ¼ in. (0.025), and 5/16 in. (0.312).

Electric line

Braded Line with interwoven electric cables is used to convey and transmit electrical signals from the tools to surface receivers. Electric Lines are commonly found sizes of 3/16″, 7/32″, 5/16″, 3/8″, or 7/16″. Smaller sizes are used for through tubing measurements, while larger sizes are used for open hole logging services. The largest Electric Lines may contain up to 7 electrical cables and can be found in sizes of 15/32″ or 17/32″.

The main uses of electric line are:

- Formation logging and pressure information
- Perforations
- Setting packers and bridge plugs

Braided line running procedures vary from slickline as pressure control may be required around the braded line. Because the braided line is made of a multiple wire strands woven around electric cables, pressurized well fluids may pass around the braids and leak. A hydraulic stuffing box and blowout preventer are used to provide necessary sealing elements and provide a grease seal to mitigate any leakages.

Wireline equipment

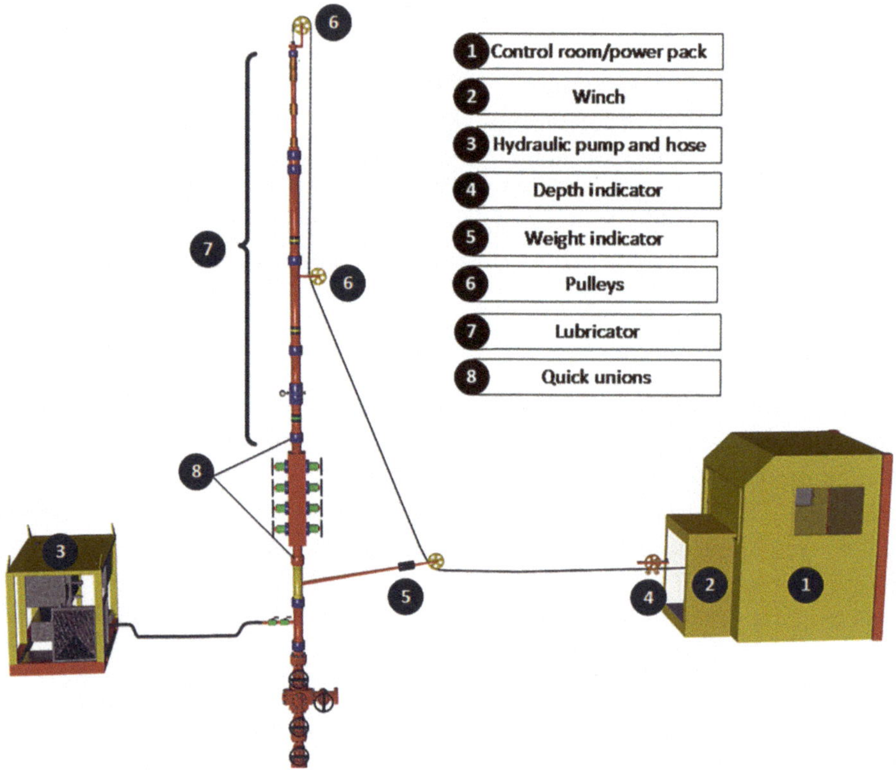

Fig. 4.2 Wireline/slickline assembly.

Wireline equipment complements may vary due to type and level of service (Fig. 4.2). In general terms, wireline equipment may be separated into the following key components:

1. Power Packs
2. Winch
   a. Reel Systems
   b. Depth Indicator (Odometer)
   c. Weight Indicators
   d. Operator's Cabin
   e. Floor Block (Pulleys or Sheaves)
   f. Gin Poles or Masts
3. Well Equipment
   a. Floor Blocks or Pulleys
   b. Stuffing Boxes
   c. Lubricators
   d. Quick Unions
   e. Wireline Valves (Pressure Control Equipment)
   f. Line Wipers

Power packs

Power packs are the portable power plants used to provide hydraulic and electrical power for functioning the equipment complement. Power packs are unitized and can be deployed separately or mounted on trucks/trailers (as needed). The most common types of power packs employ diesel engines as the main power source and drive one or more hydraulic and electrical generators.

A well control equipment control unit could require up to 15 different hydraulic supply and control lines. To ensure that the status, open/closed, of each device is known and that there is no confusion between devices, often a standalone hydraulic control unit is used. An umbilical hose bundle runs between the skid and the well pressure control equipment.

Positioning of the power pack must be established to operate in only areas designated as safe, in accordance to the following:

- Zone 0: Which a flammable atmosphere is continuously present or present for long periods
  - (More than 1000 h per year).
- Zone 1: Which a flammable atmosphere is likely to occur in normal operation
  - (About 10–1000 h per year).
- Zone 2: Which a flammable atmosphere is not likely to occur in normal operation.
  - Will exist only for a short period (<10 h per year).

Winch (reel)

Winch system employs multiple components whereupon the wireline is stored and power reeled with additional equipment necessary to measure line spool, weight, minimize bending, and perform wiping or cleaning of the line. Located in front of an operator's cabin, the powered reel is located in line with the well. Hydraulic power is applied to reel motors and braking systems with operating controls located in the operator's cabin. Multiple reels may be present, depending on application, and can easily reach deployment speeds of 4000 ft./min.

Braking systems are necessary to slow and stop the reel. The braking systems must be reliable over a wide range of temperature and weight ranges. Braking systems can employ one or more of the following designs:

- Band brake where the rotating drum with shoes expand and contact the inside surfaces of reel drum. They use friction for braking.
- Drum brake where a set of brake shoes contact the inner surfaces of rotating drum connected to the reel hub.
- Disc brake where brake pads are hydraulically pressed again rotary plate connected to the reel hub.
- Parking brake where when the reel has stopped, the parking brake can be energized to stop any unwanted reel movement.

Operator console/cabin

The operator's cabin contains instruments and controls necessary to independently operate each piece of equipment within the wireline equipment complement. The cabin is normally positive pressurized and contains sound and vibration insulation. Electric and hydraulic controls allow the operator to vary needed operating parameters to ensure a safe and reliable, environmentally sensitive operation.

Emergency shut down devices (ESDS)

Emergency Shut Down Devices (ESDS) are contained within the operator's control panel. The ESDS systems monitor water and oil pressure/temperature, fuel shut-off device, exhaust temperature, and overspeed limits.

ESDS SYSTEMS CAN BE USED TO SHUT DOWN ENGINES WHICH BREATHE WELLBORE GASES AND KEEP RUNNING, EVEN WHEN THE FUEL SUPPLY IS SHUT-OFF.

Depth indicator (odometer)

One of the most important components within the complement of Wireline Equipment is a rugged and accurate depth indicator device, sometimes referred to as the odometer (Fig. 4.3). These indicators are needed to determine precise tool depths for the tools run into the well.

The minimum dimensions of the depth indicator wheels are dependent upon size, weight, and grade of wire used. The following formula can be used to estimate minimum wheel diameter to minimize bending fatigue.

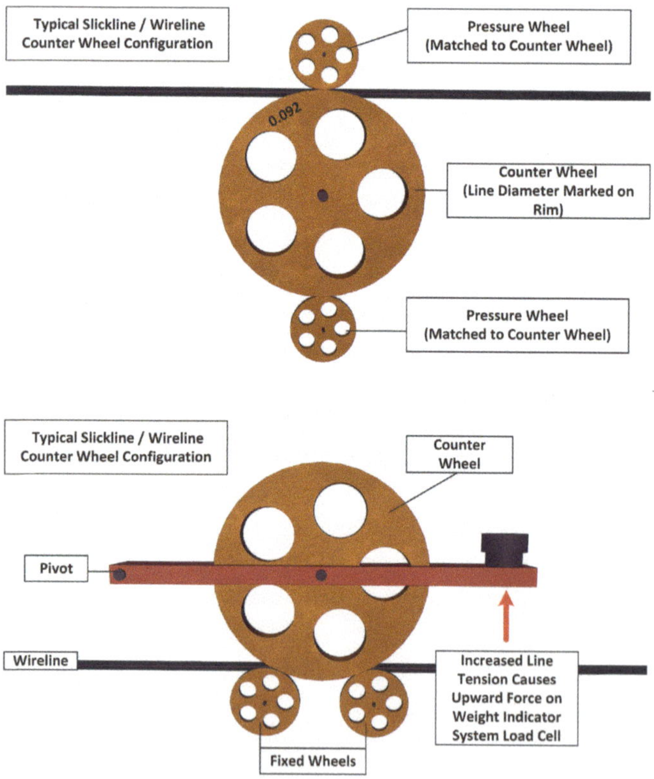

Fig. 4.3 Depth Indicators.

$$\textbf{Wheel Diameter}_{\textbf{in}} = \frac{\textbf{Constant} \times \textbf{Wire Diameter}_{\textbf{in}}}{\textbf{Tensile Strength}_{\textbf{psi}} \times \textbf{106}} \quad (158)$$

where Wheel Diameter_{in} = Wheel Diameter (in.), Constant = Constant of Proportionality, Wire Diameter_{in} = Diameter of Wire (in.), and Tensile Strength_{psi} = Tensile strength of steel wire.

Weight indicators

In order to determine the maximum safe operating load, use of a weight indicator is required. These weight indicators are calibrated in standard weight units and measure the total load on the wireline. The weight indicator is affixed to the wireline with a sensing cell attached to the production tree. The pressure differential (or electrical potential) is calibrated, measured, and displayed on gauge housed on the operator's panel.

Weight indicators are manufactured in three types:

- Mechanical indicator—Normally used for 0.072″ wireline.
- Hydraulic indicator—Uses a load cell, gauge and hydraulic hose. Weight variations are transmitted from the load cell to the operator's panel by the hydraulic hose Due to simplicity, ruggedness and repeatability, hydraulic indicators are widely used.
- Electric indicator—Like the hydraulic indicator, the electrical indicator uses an electronic load cell where deflections within a potentiometer connected to the wireline is transmitter to the operator's console and displayed in a calibrated gauge. These indicators require on-going maintenance to ensure accuracy.

Floor blocks (pulleys or sheaves)

Wireline exiting the reel is normally directed through a series of floor blocks (horizontal to vertical position) and allow the slickline to be routed to the snuffing box. Floor blocks are essential in managing spooling and unspooling for wire onto the reel without kinks. The floor block is normally connected to the load cell of the weight indicator and is used to help determine accurate weight loads.

Lifting equipment

During wireline work requiring use of tall high-pressure equipment complement (i.e. lubricator and BOPs), the removing and replacing of this equipment onto the well becomes problematic. Various forms of lifting equipment can be used to perform these installation and removal services safely. Lifting equipment may include use of stiff legs with block and tackle system (gin pole), electric hoist or hydraulic crane. Specific lifting equipment will be determined by wireline equipment complements, well location, and pressure rating.

GIN POLE AND A BLOCK AND TACKLE MAY BE USED AS A BACKUP, IN THE EVENT OTHER MEANS OF LIFTING ARE NOT AVAILABLE.

Wireline pressure control equipment

Wireline pressure control equipment varies by functionality and company specific designs.

Stuffing box

Wireline Stuffing Box is the primary barrier during slickline operations and is used during work on wells with pressure. The Stuffing Box consists of a packing chamber with an external nut or hydraulic chamber. The Stuffing Box provides a swivel bracket and sheave that guide the wireline through the packing chamber. The radius of the Stuffing Box sheave is interchangeable to accept larger sheaves for larger wireline sizes.

The packing is energized and forms an effective seal around the wireline and preventing well pressures from escaping. The packing is energized by tightening a nut on a "manual" Stuffing Box or by apply hydraulic pressure to a "Hydraulic" Stuffing Box. The Hydraulic Stuffing Box can be remotely operated and is generally used. The pressure necessary to affect the seal will be variable and is also minimized. Excessive forces can prematurely wear the gland packing or cause the packing to become so tight the wireline cannot be moved.

Stuffing boxes contain an integrated plunger (blow out plug or internal BOP) that incorporates passage of the wireline. If the wireline becomes separated and ejected from the well, the integrated plunger is forced up against the bottom of the packing chamber and seals off against well pressure (Figs. 4.4 and 4.5).

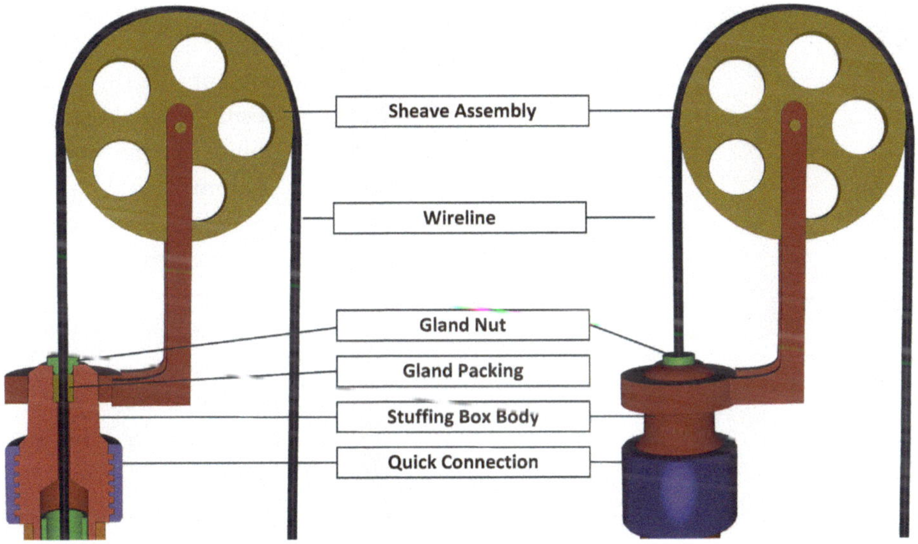

Fig. 4.4 Manual stuffing box.

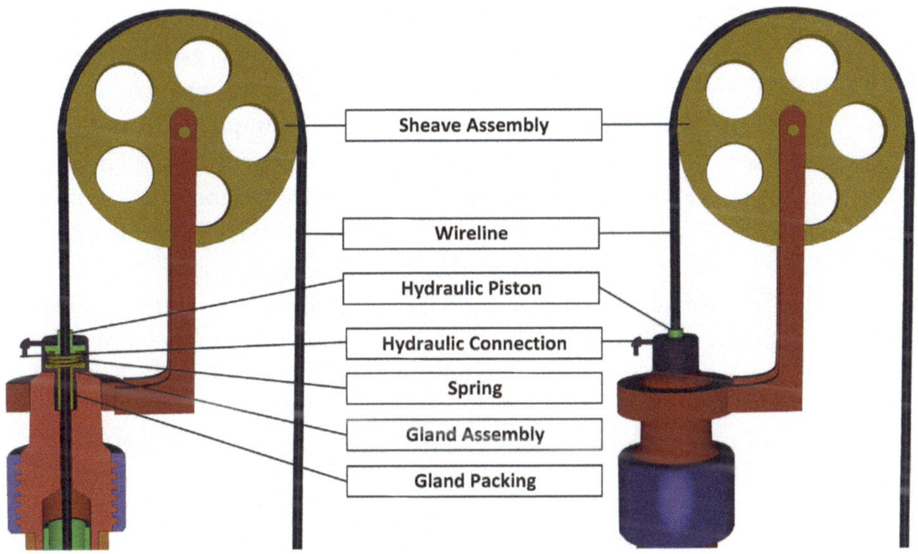

Fig. 4.5 Hydraulic stuffing box.

Grease injector head

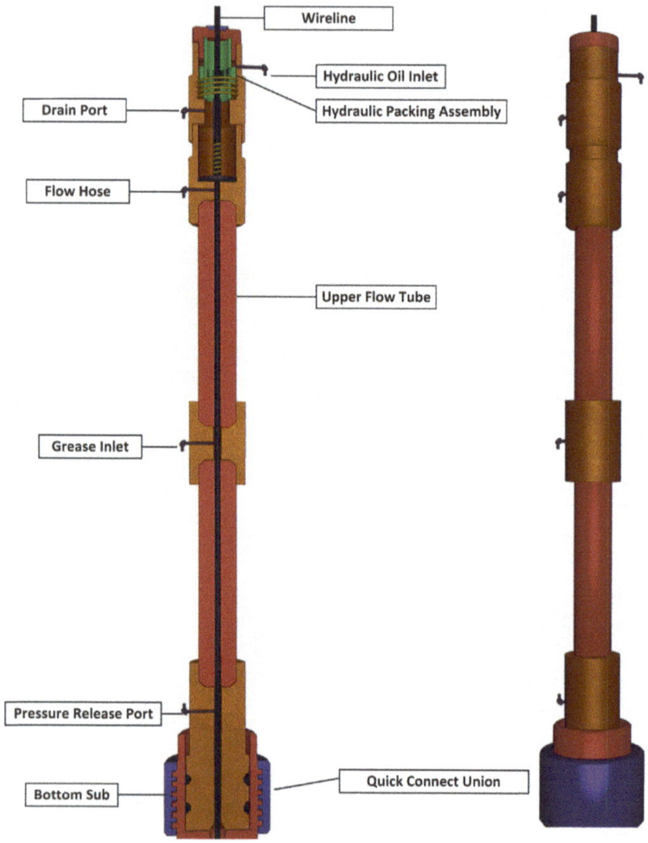

Fig. 4.6 Grease injector head assembly.

The Grease Injector Head provides a method of hydraulically injecting grease to ensure an effective seal of well pressure around wireline (Fig. 4.6). The grease injector head is normally positioned at the uppermost point of pressure control equipment and comprises of two assemblies, the stuffing box and flow-tube assembly.

The stuffing box is designed to provide sealing of stationary wireline through means of a pack-off elastomer energized by hydraulic pressure. The flow tube is located below the stuffing box and is available in two designs. Solid flow tube consists of a simple straight section flow tube with grease injector that must provide a seal while supporting the assembly. Under extreme loads, the solid flow tube may deflect and cause leakage. The concentric flow tube consists of the inner flow tube sized to fit tightly around the wireline and an external sleeve used to support the entire assembly. Under extreme loads, the external sleeve may deflect, while the inner sleeve maintains an effective seal.

To form an effective seal, a viscous grease is hydraulically injected into the flow tube at pressures greater than well pressure. The viscous grease fills the small annular space between the inner wall of the flow tube and the wireline, and forms a liquid seal that contains well fluids while allowing movement.

The number of flow tubes and flow tube sleeves depends on the well pressure:

2 Flow Tubes	0–2000 psi
3 Flow Tubes	2000–10,000 psi
4 Flow Tubes	10,000–15,000 psi

Grease supply pump

The grease supply pump is usually air driven typically requiring between 100 and 150 psi air pressure to function properly. An insufficient air supply can have serious consequences in case of grease seal problems. For example, the pumps use a ratio system from the air pressure supplied to the output pressure. A 40:1 ratio pump will supply 6000 psi grease pressure at 150 psi input, but only 4000 psi at 100 psi input.

There is a nonreturn valve in the grease injection line to protect the grease injection skid from wellbore pressure or fluids in the event of a grease supply line failure. A back-up pump should be rigged up on high-pressure wells.

Tool trap

Used as a safety device, tool traps or tool catchers are designed to prevent the wireline tool string from falling back into the well and also prevent the tools from falling onto personnel once the lubricator has been disconnected. The tool catcher is also designed to prevent the wireline tool from falling back into the well when the wireline is pulled to breakage point in the rope socket (stuck tools) and before the swab valve can be closed.

Designed as a simple flapper-type mechanism, the tool trap allows the wireline tool to pass upwards. The flapper closes once the wireline tool as passed by the flapper. The flapper may be energized and closed manually or by hydraulic means. Most tool traps employ an external device that indicates the position of the flapper and when the wireline tool has passed the flapper alerting necessary closure.

Hydraulic tool catcher

As a preferred safety device, the Hydraulic Tool Catcher also catches the tool string once it has been pulled into the lubricator and acts as a stop (Figs. 4.7 and 4.8). The Hydraulic Tool Catcher is located below Grease Injector Head/Stuffing Box. This suboperates using a collet that locates and catches the fishing neck of the roper socket.

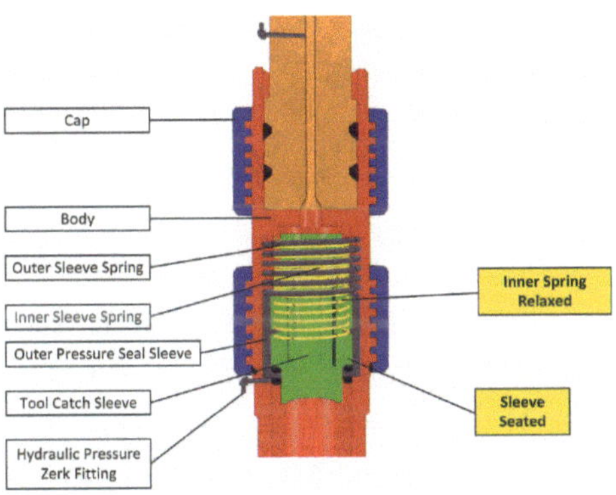

Fig. 4.7 Empty hydraulic tool catcher.

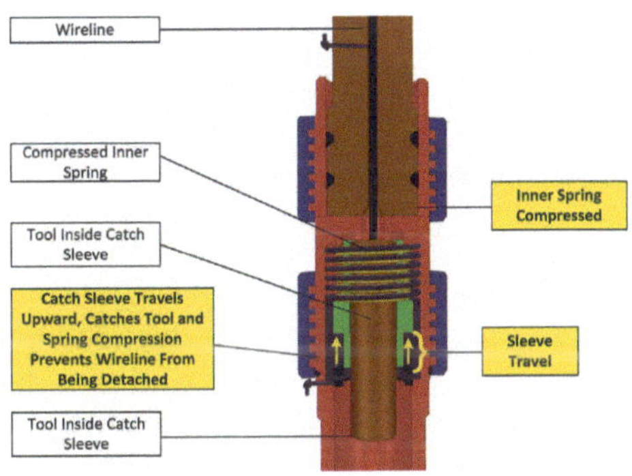

Fig. 4.8 Hydraulic tool catcher with caught wireline tool.

Used as a safety device, the lubricator with tool can be pressure bled and disconnected and the Tool Catcher prevents the tool string from falling out of the lubricator and injuring personnel. The hydraulic tool catcher is the spring catch/hydraulic release type and is therefore permanently in the catch position. To release the tool string, hydraulic pressure must be applied to move the collect from catch to release position. With close tolerances, the collet is designed to catch only the size of fishing neck on wireline. If a separate fishing neck is to be used, the collet must be replaced with collet matched to new fishing neck.

Most hydraulic tool catchers employ features that allow the internal parts of the tool catcher to be loaded from the top and around the wireline. This feature prevents having to cut the wireline and rebuild the rope socket each time the tool catcher is added or removed from the pressure control equipment.

Hydraulic tool catchers can be outfitted with a ball-check assembly and hydraulically operated vent valve. Ball checks are used for pressure containment while vent valves are used for pressure equalization. When the vent valve is hydraulically operated, test fluid can be circulated prior to closing for a pressure test. Without this feature, the grease seal will have to be opened to release trapped air.

Line wipers

When retrieving the wireline from the well, fluid adheres to the wireline. This fluid will become dislodged, drip or is thrown off floor blocks and pulleys. This creates an environmental spill with possible safety concerns. Line wipers are used to clean the line at or near the wellhead. One type of line wiper is housed within an internally split neoprene rubber plug. The plug has a threaded bolt adjustment that compresses the rubber plug unit it envelops the line and strips it of fluid. This line wiper has a hinge pin to anchor it to the floor block. Another type of line wiper operates similarly to a stuffing box, where an expandable gland cleans the line. Oil from the wiper should be collected in a suitable container.

Lubricator

The lubricator consists of a number of tubular sections assembled together with a stuffing box to pack-off wireline on top and BOPs below (Fig. 4.9). The tubular sections are assembled on the ground using a series of quick unions. After make-up, the tubular sections are raised from the ground into the vertical position above the wireline BOPs. The size and length of the assembly will be determined by the diameter of tools to be run into the well along with their length. The lubricator system must have a working pressure rating equal to or greater than the tubing in the well (Fig. 4.10).

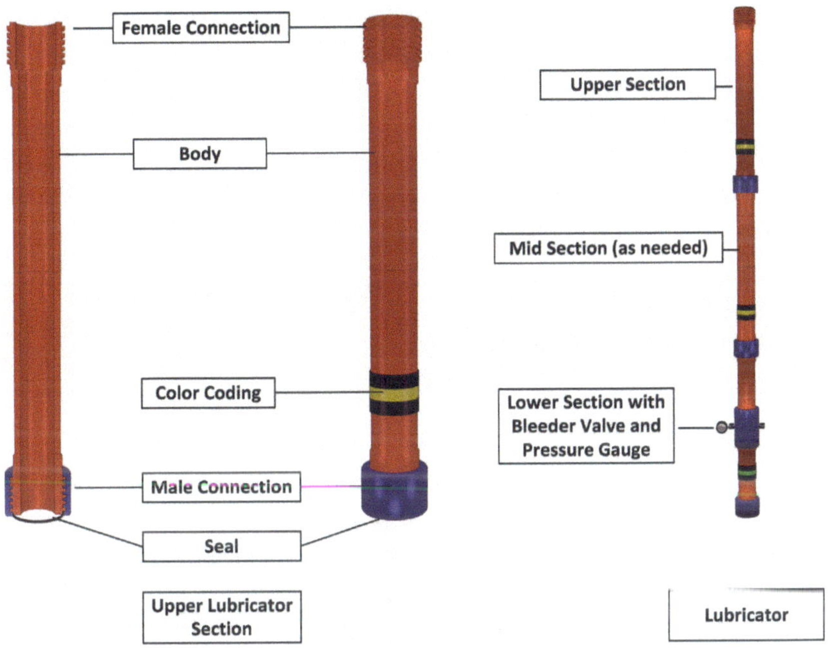

Fig. 4.9 Upper section of lubricator and total lubricator section (without stuffing box assembly).

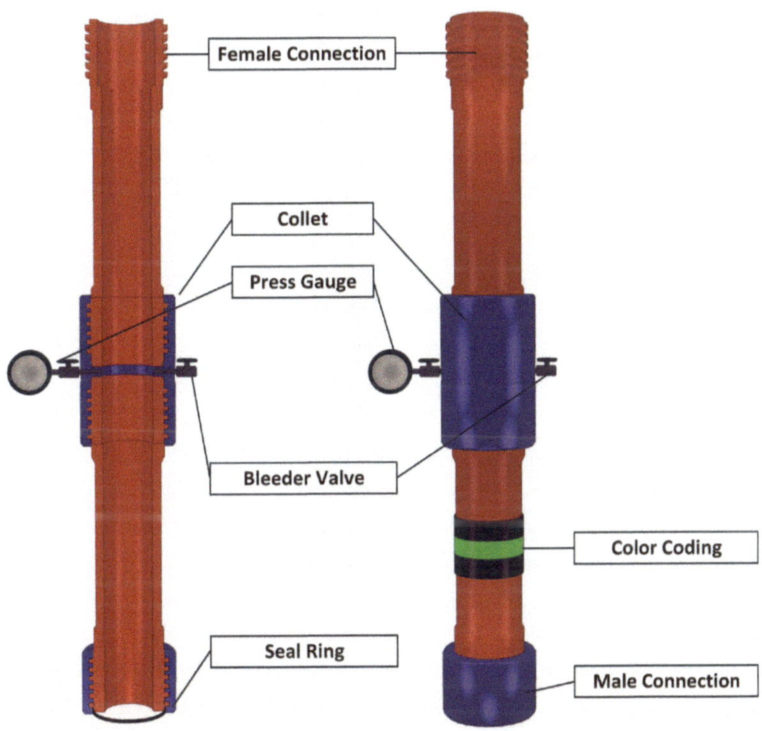

Fig. 4.10 Lower section of lubricator.

The lubricator normally contains multiple diameter sections. The lower section will contain a larger diameter tube, as it must contain the tool that is either being pulled or run. The upper section will only have to contain the smaller diameter tool strings. The lowermost tube is outfitted with a needle valve for bleeding trapped pressure above the BOP or well before breaking off the lubricator to change out tool strings. The bleed valve is outfitted.

A second needle valve may be employed at the bottom lubricator section to provide additional barrier protection and ensure a position for the pressure gauge.

LUBRICATORS ARE TO PRESSURE TESTED TO THE MAXIMUM ANTICIPATED WELL PRESSURE, BUT NOT TO EXCEED THE MANUFACTURER'S RECOMMENDED WORKING PRESSURE

Quick unions

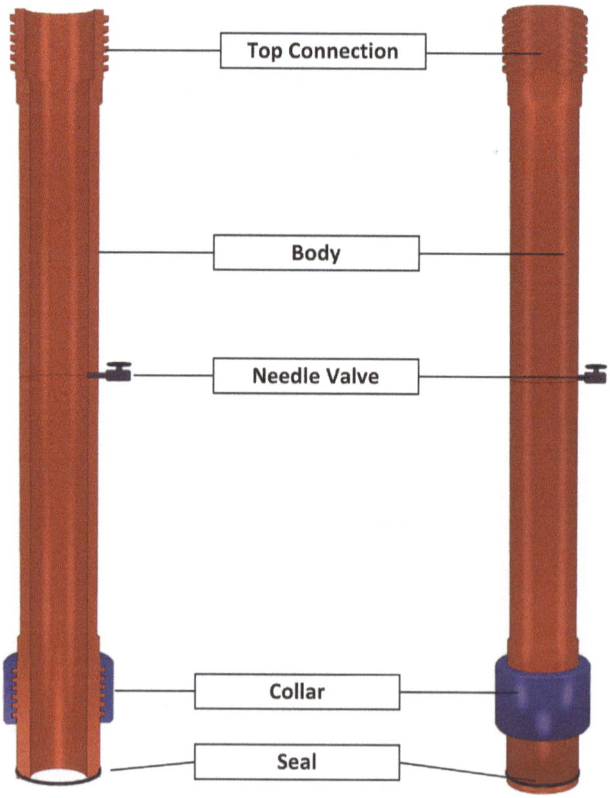

Fig. 4.11 Quick unions.

Quick Unions are connections that are fastened to each end of the lubricator tube and provide means to energize an "O" ring when fastened to hold well pressure (Fig. 4.11). Normally, Quick Unions incorporate an Acme type square threads for fast make-up/break-out and provide means for installing the O-ring for sealing capabilities.

QUICK UNIONS ARE TO BE HAND TIGHTENED ONLY
O RING MUST BE INSPECTED FOR DAMAGE PRIOR TO MAKING A CONNECTION

BOP (Wireline valve)

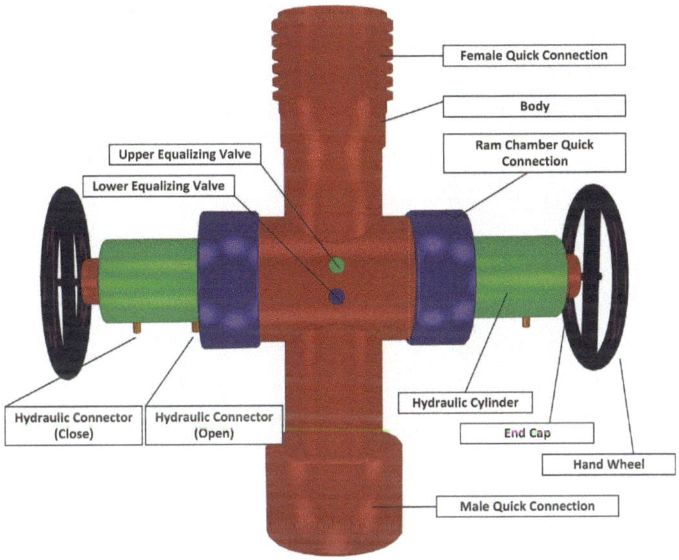

Fig. 4.12 Wireline hydraulic BOP (wireline valve).

Wireline Blowout Preventer (Wireline Valve) is the secondary barrier during slickline operations and is used to perform work on wells with pressure (Fig. 4.12). These blowout preventers are installed to contain wellbore pressure by sealing around wireline or open hole. Installed in a set, these BOPs are mounted below the lubricator and onto the wellhead of tree using quick connections.

Sealing the wireline is accomplished by hydraulically or manually pressing a set of rams together and forcing resilient packing around the wireline to form a seal. Once closed, the pressurized lubricator, can effectively bleed off well pressure via the bleed valve. Once pressures have been normalized, the lubricator can be opened and tools may be removed and/or replaced. With new tools, the lubricator is reenergized with wellbore pressures, and the rams are opened. Once the rams are opened, the tool can now be lowered into the well.

TEST THE WIRELINE VALVE TO THE MAXIMUM ANTICIPATED WELL PRESSURE, BUT DO NOT EXCEED THE MANUFACTURER'S RECOMMENDED WORKING PRESSURE

Manual BOPs

The Manually Operated valves are functioned by rotating two handles, located on the end of pistons, to force internal rams to seal around the wireline. Manual BOPs are primarily used in low pressure, non-H_2S environments.

Hydraulic BOPs

Hydraulically operated BOPs function by using a hydraulic pump connected to the BOP by a high-pressure hose. Care should be taken to minimize the hydraulic closing pressure as to not damage the sealing surfaces. As an emergency back-up, these rams can be closed manually, but will require hydraulic pressure to open.

The rams are designed with elastomeric (rubber) seals on the wireline and upper section of the BOP cavity. Most rams are designed to utilize wellbore pressure to help push the ram upward to assist in forming the cavity seal as well as the wireline seal. Since the rams move slightly upward, pressures above and below the ram must be equalized before the ram can be opened (Figs. 4.13 and 4.14).

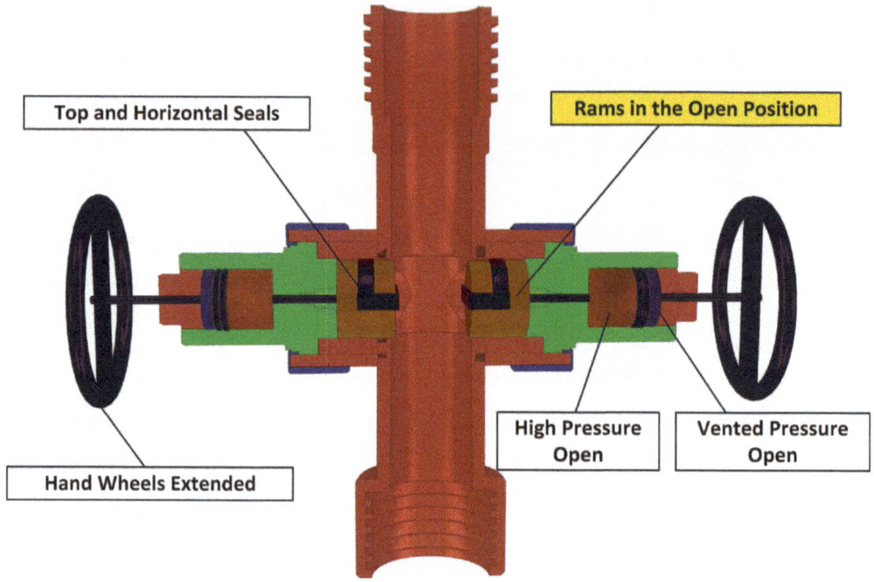

Fig. 4.13 Wireline hydraulic BOP in OPEN position.

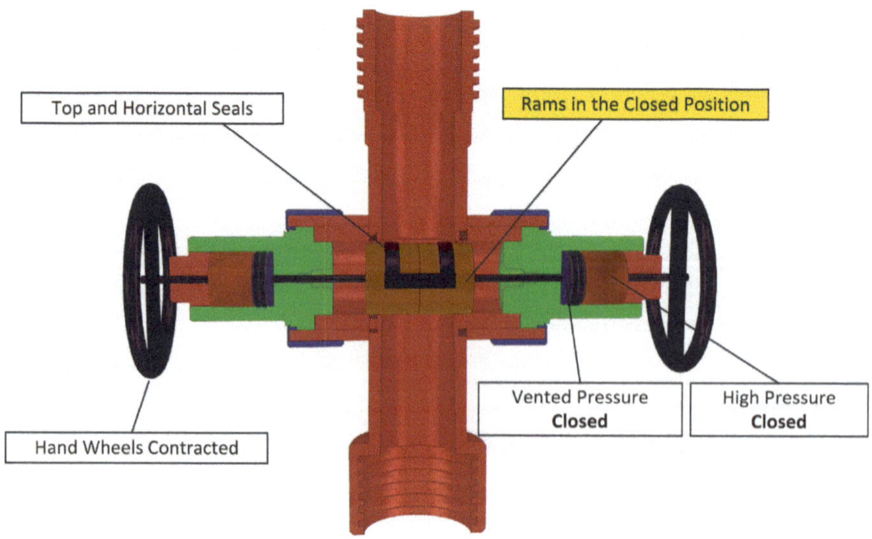

Fig. 4.14 Wireline hydraulic BOP in CLOSED Position.

WIRELINE BOPS ARE DESIGNED TO SEAL ON STATIC WIRELINE CABLE
ALL CABLE MOVEMENT MUST BE STOPPED BEFORE CLOSING BOPS

Multiple rams

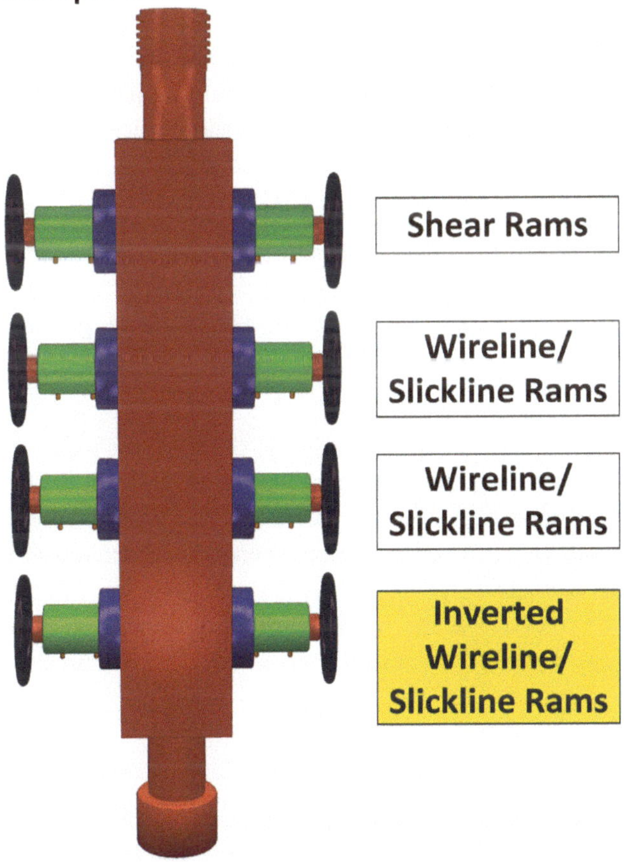

Fig. 4.15 Quad wireline/slickline BOPs.

The advantage of using multiple sets of BOPs will be the number of alternate sealing arrangements that can be selected, as they provided back-up to the primary BOP (Fig. 4.15). Quad or dual rams allow for the lower set of rams to be inverted. This feature allows the quad rams to trap a cavity full of grease between them at higher pressures and prevents this grease from being ejected or pumped down hole. The inverted feature and grease sealing capabilities are suggested for use in gas wells, as gas may migrate through gaps created in braided wireline. By filling the cavity at a pressure higher than wellbore pressure, the injected grease will fill the voids in the braided wireline and displace gas and preventing gas from being ejected into the atmosphere.

Manual wireline cutting

When tools become stuck in the well, the wireline is normally cut as close to the top of the tool string as possible. Since Wireline BOPs do not include a shearing function, the Wireline BOPs are closed and seal around the wireline. The lubricator pressures are bled down and the tools are broken out the opened lubricator. Once complete, a manual cutter is affixed to the wireline at the rope socket when the BOP has been opened. The wireline is then cut and fishing begins.

Shear BOPs

During planning operations for high-pressure and/or high-angle wells, shear valves may be selected for use. This ram type BOP is outfitted with a shearing/sealing shear that can cut wireline or selected components of the bottom-hole assembly. Considerations for shear rams include:

- Due to high-pressure, well condition, nature of the operation or tool configuration, an extra pressure barrier is needed.
- The lubricator isn't of sufficient length to contain the entire work tool string. If wellhead valves were to leak, the ability to isolate the lubricator would be jeopardized.
- If a downhole safety valve has been permanently opened using a sliding sleeve, when the wireline is removed from the well, no secondary barrier exists.

Fishing BOPs

When a routine Wireline operation turns into a fishing job, an additional set of BOPs may be required as well as an additional lubricator to accommodate the wide range, size, and diameter of tools necessary to perform fishing operations.

Equalizing valves

To ensure effective means of compensating for pressures across multiple BOPs, the BOPs are outfitted with an equalizing valve and piping across the two sets of ram preventers. Before closure, the pressure retaining device (i.e., lubricator) is opened, pressure must be equalized across the rams to prevent damage.

BOPs are fitted with equalizing valves across the rams. Before any closed, pressure retaining device is opened, pressure must be equalized across the rams to prevent damage.

Triple BOPs

For work scopes requiring two types of wireline (i.e., slickline and braided line), Triple BOPs are used. These rams can be outfitted with slickline rams on top and braided line rams in the lower two positions. Slickline BOPs hold pressure from below only. Wellbore pressure acts to keep the BOP closed and maintains the seal around the wire.

Shooting nipples

When a conventional lubricator is not feasible and there is only a small risk that the well may become live, logging may be done through an assembly called a "shooting nipple." It is usually made up of a stuffing box mounted on top of a special joint of suitably rated casing.

The shooting nipple has a pump-in sub near the top and a welded flange at the lower end. The shooting nipple is lowered through the preventers and landed on top of the closed blind-shear rams. A set of rams or the annular preventer is then closed to seal on the outside of the casing. Once the rams or annular preventer is closed, the shooting nipple is pulled up until the flange on the bottom of the shooting-nipple pulls against the bottom of the rams or annular preventer.

The shooting nipple is chained or cabled down in the rotary table. It is marked at the rotary table so that the mark can be referenced with respect to the position of the blind/shear rams, should they need be utilized. The assembly is pressure tested as per test procedure. The preferred shooting nipple assembly installation is flanging it to the top of the annular preventer or a set of rams.

THE PREFERRED SHOOTING NIPPLE ASSEMBLY INSTALLATION IS WHEN THE ASSEMBLY IS FASTENED BY FLANGE TO THE TOP OF THE ANNULAR OR PIPE RAMS

Closing a properly sized set of rams around the shooting nipple or flanging to the top of the annular is the safest solution. If only an annular is available, pressure build-up below the wireline assembly may push the shooting nipple assembly up against the annular element causing damage and could jeopardize the sealing capability.

Shooting nipples only have a pack-off element to seal the cable. In a live well situation, the shooting nipple is usually used only to close on a stationary cable to contain pressure. If the tools are pulled from the hole under pressure and the pack-off fails, containment of wellbore fluids will be lost and a dangerous situation can occur. If the wireline assembly is longer than the distance from the pack-off to the top of the blind rams, then the blind rams cannot be closed under the tools. In this case the pressure must be dissipated.

Unanticipated live well pressures may be dissipated by bullheading or by flow diversion allowing the cable and wireline tool to be retrieved.

Smaller diameter slim-line tools are should be run with standard mono-cable or similar diameter coaxial if logs have to be run are run in a pressured well, thus allowing the use of conventional pressure control equipment (Figs. 4.16–4.19).

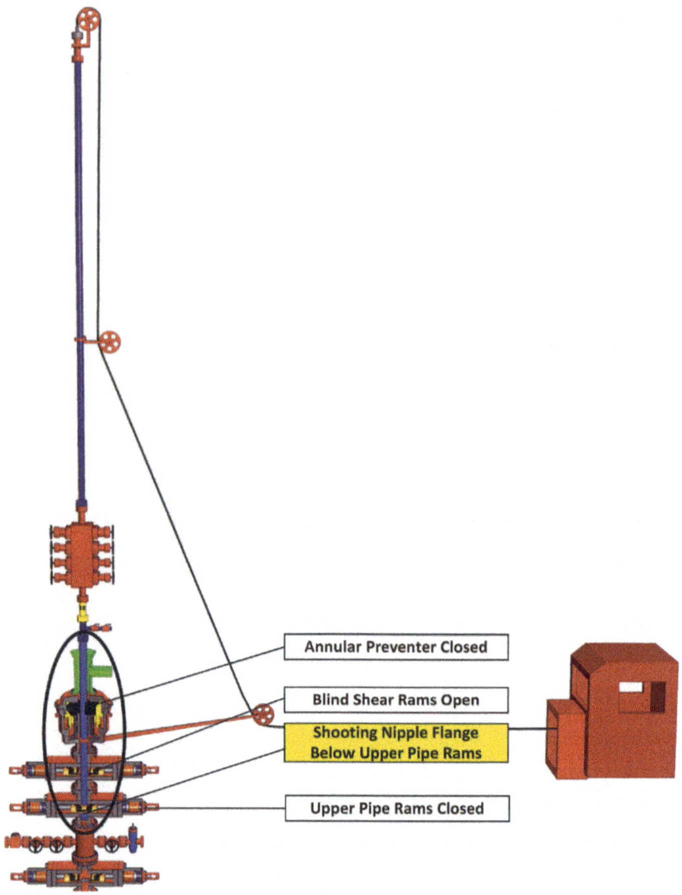

Fig. 4.16 Shooting nipple affixed below set of upper pipe rams.

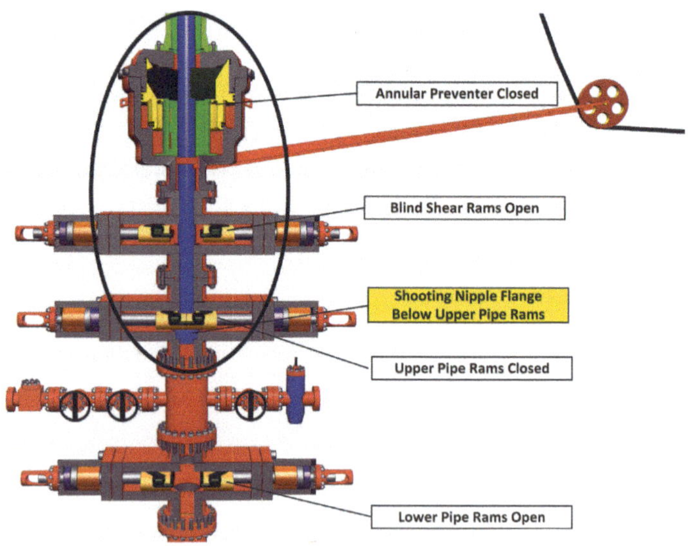

Fig. 4.17 Close-up of shooting nipple installed in BOPs.

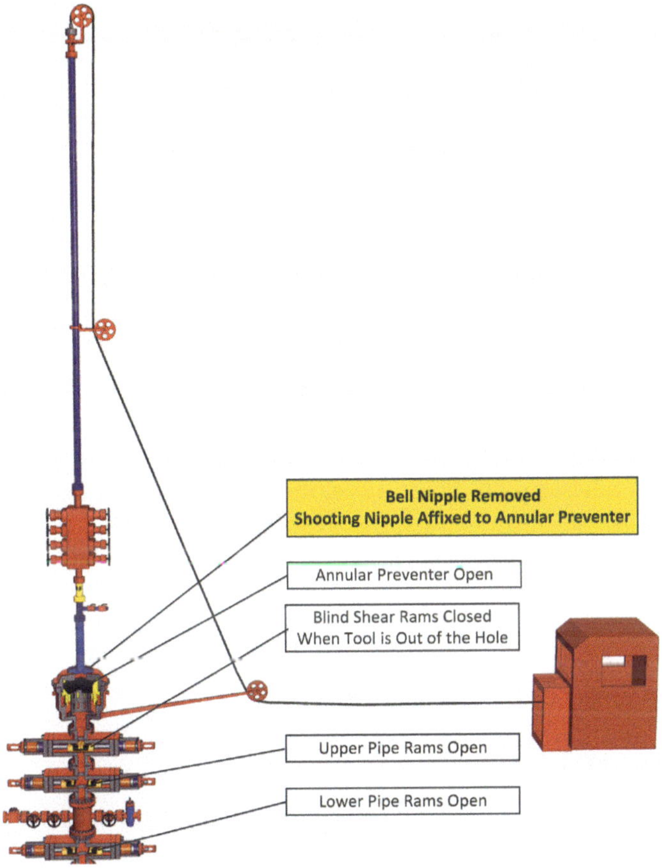

Fig. 4.18 Shooting nipple affixed to annular preventer.

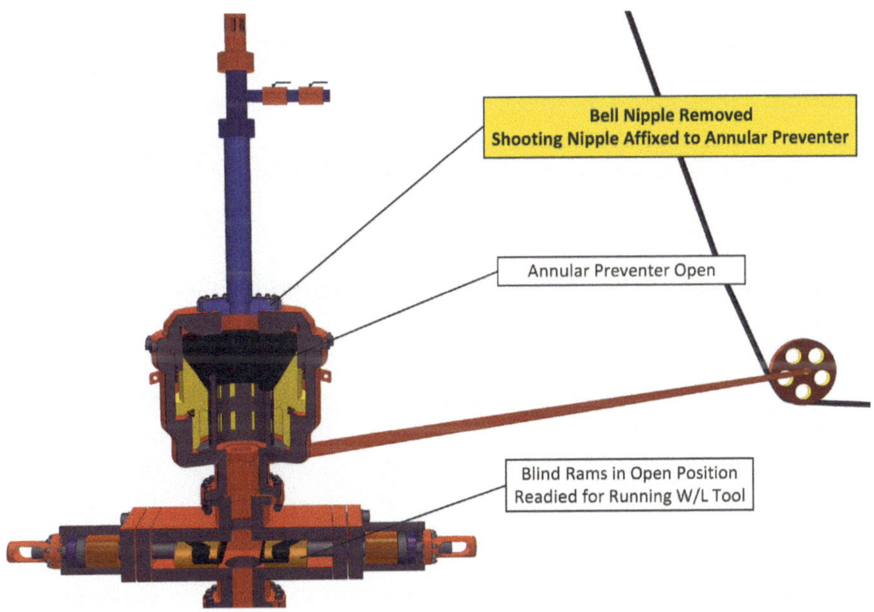

Fig. 4.19 Close-up of shooting nipple affixed to annular preventer.

Coiled tubing well control method

Coiled tubing commentary

Coiled tubing special services can be used in well control by establishing the ability to place KMW within plugged tubulars. Primarily used to place KMW within a plugged tubular system, the Coiled Tubing can be outfitted with a variety of specialized drills that can be used to drill out fines, paraffin, and other blockages. Mills can also be used to drill out scale and corroded tubing plugs. Coiled Tubing can also be used to place acids used to dissolve plugs and scale.

Since Coiled Tubing is reel dispensed, the pipe movement is largely restricted to vertical movement (push and pull). Hybrid systems do exist that imparts rotational abilities, but such systems are not readily available, worldwide. Requiring a small crew complement as well as mobile systems, coiled tubing can be used in a wide variety of well control applications.

Coiled tubing well control operations

Well control operations used with coiled tubing include bullheading and circulating using the Driller's or Wait and Weight Methods. Since coiled tubing operations and equipment will vary between specific service providers, please refer to specific service provider literature for exact well control equipment and procedures.

Wells are normally constructed in two phases. Phase 1 of the drilling operations is to drill the wellbore, install, set, and cement casings to formation rock surfaces. The cement is used to permanently affix the casing to the formation and providing a means of ensuring complete zonal isolation. Phase 2 of the construction process is called the completion phase. When a well is completed, a sacrificial tubular called tubing is installed within the well casing to provide means of transporting formation fluids (i.e. oil, gas, and water) to the surface. The tubing is considered to be sacrificial, as it can be replaced over the life to the well to ensure maximum formation drainage. As such, tubing is designed and selected to enable the performance of wireline operations or small tubing-operations through-the tubing over the life of the well.

Coiled tubing well intervention services offer a means for pumping and circulating fluids within existing wellbore tubulars, even with the wells under pressure. Coiled tubing (CT) is a specialized service that offers a fast, low-cost alternative to well control when compared to traditional hydraulic workover operations. With relative low cost and fast mobilization, coiled tubing is selected for use on a variety of wells with multiple surface and downhole problems.

Introduction to coiled tubing

Coiled Tubing (CT) is a single, long length of pipe that has been continually milled and spooled onto a take-up reel during the manufacturing process. The coiled tubing weight, grade, and size will yield the maximum weight and length. Normal road restrictions for height and weight of truck transported freight and/or crane lifting requirements, along with weight, grade, and maximum weight will ultimately determine the maximum length manufactured. Commonly used CT steels have yield strengths ranging from 55,000 to 120,000 psi and are manufactured in sizes ranging from 0.75″ to 4″ in diameter. Maximum manufactured lengths can approach 30,000 ft. Most coiled tubing reels are in the 20,000 ft. ranges and weigh in around 30 tons.

In order to minimize the weight of the reel, coiled tubing is manufactured from sections of differing wall thickness. The thickest diameter of tubing will be located at the inner end of the reel (the top of the well) where the tensile forces are the greatest. Coiled tubing can be manufactured with up to six sections of varying wall thicknesses forming one "tapered" string of Coiled Tubing (called a 5-step taper).

With elimination of making or breaking connections used in sectioned tubular operations (i.e. Hydraulic Workover Operations), the coiled tubing can be deployed and retrieved quickly and reduces time needed to perform specialized service. Depending on wellbore conditions, running speeds of up to 250 ft./min. Can be achieved. With manufacture of higher yield-strength coiled tubing grades, pressure service limits have continually increased. To overcome frictional and gravitational forces acting against coiled tubing deployment, coiled tubing tractors may be used to pull the coiled tubing into position.

Coiled tubing services incorporate a series of specialized equipment to isolate wellbore pressures while providing means of allowing tubing to enter the well. What separates coiled tubing well control from sectioned tubular workover operations is the requirement to unspool the tubing from the reel, forcing the tubing through a hydraulic-drive injector assembly (gooseneck and motorized drive chains), straighten the tubing and forcing it into the well. The tubing will then pass through a stripper assembly, into a quad ram pass (small set of BOPs), flow tee, well valve(s), and into the well. Operations are performed using hydraulically pressures from a separate power plant combination of prime movers, pumps, and valves providing pressurized hydraulic fluid through a series of high-pressure hoses.

Coiled tubing is used in a variety of application such as:

- Nitrogen lifting wells where HP prevent well flow.
- Velocity string installation/retrievals that include installing small bore strings inside the completion to aid wellbore flows in older wells.
- Fill clean-up operations after workovers, etc.

- Spotting acid at the perforations (either to clean the perforations or remove formation skin damage).
- Fishing operations.
- Cleaning out sumps, rat-holes.
- Spotting cement plugs.
- Through-tubing services such as straddle packers and bridge plugs.
- Use of stiff wireline (Coiled Tubing with logging cable inside).
- Side tracking and drilling small diameter holes.
- Drilling/milling operations inside pipe.
- Opening/closing sliding sleeves, etc.
- ESP deployment.
- Memory PLT logging.
- Fracturing.
- High-pressure Abatement.
- Underbalanced formation drilling.
- Sand jetting perforations.
- Sand cutting tubulars.
- Scale/paraffin removal.

Specialized services also include coiled tubing outfitted internally with electric logging cable and/or two small-bore stainless-steel lines (like DHSV control lines). This type of coiled tubing is used to independently control specialized downhole tools. Multiple strings within the coiled tubing can be used to pump different fluids individually through in the string.

Coiled tubing well control methods

Well control begins with measuring and understanding wellbore pressures, including all annuli, to determine the density of the KMW. A thorough understanding of installed downhole completion equipment and potential problems are necessary to develop an appropriate kill procedure. Capacity and volume calculations are determined. If wellbore pressure exists, the forces necessary to insert coiled tubing into the well will be determined.

Common fluid circulation methods used in well control are the "Driller's" method and the "Wait and Weight" method. These are both constant bottom-hole circulating methods that allow a desired overbalance to the formation. Both methods prevent additional influx of reservoir fluids to the wellbore.

Coiled tubing can be used to conduct kill operations on most existing production system tubulars. The well kill operation is routinely performed by pumping heavier density fluids down the coiled tubing string while entering the desired tubular with formation fluids flowing up the annular space between the coiled tubing and tubing string. Coiled tubing is uniquely designed for deployment and retrieval throughout the well kill operation without interrupting pumping services. This continuous pumping capability provides a great level of control for kill fluid placement and minimizes formation damage. The continuous pumping capabilities allow normal kill techniques to be used such as the Driller's or Wait and Weight well control methods.

Driller's method

The Driller's method is a fluid circulation practice used to displace an influx from the wellbore. It can be used to circulate at a constant rate while running the CT in the hole to the kill depth or once the CT is on bottom. If the circulated fluid density is not of kill weight density then the well will not be dead and the formation pressure will not be balanced by the hydrostatic pressure of the fluid. If the fluid density of the pumped fluid is of high enough density to kill the well then there should be no surface pressure at the end of circulating out the influx.

After the first circulation of the Driller's method is completed and the influx removed, the surface pressure plus the hydrostatic pressure of the fluid will be used to determine the bottom-hole formation pressure and thus the density required for the kill weight mud can be calculated. A second circulation with kill weight fluid is required to balance formation pressure. This second circulation is described below under the Wait and Weight method.

Initiating the Driller's Method: Position the end of the CT at the proper kill depth. Record the shut-in coiled tubing pressure (SICTP) and the shut-in annular pressure (SIAP). Bring the pump up to circulating rate while holding the SIAP constant by manipulating the choke. Once the rate is established, the pump pressure or circulating pressure is held constant until all the influx has been removed from the wellbore and no formation fluid is observed at the surface. Shut down pump and close choke. Record shut-in pressures.

Wait and weight method

The Wait and Weight method is a circulation technique in which the formation pressure is balanced by circulating kill weight mud to displace a lighter fluid. This can be used to circulate out formation fluids or used in the second circulation of the Driller's method. This method is conducted with the bottom of the coiled tubing at the planned kill depth. After the kill weight mud is circulated, the pressure on the coiled tubing and annulus should be zero.

Initiating the Wait and Weight Method: Position the end of the CT at the proper kill depth. Record the shut-in coiled tubing pressure (SICTP) and the shut-in annular pressure (SIAP). Using the SICTP calculate the required kill mud density. If the CT was run with BPvVs, open BPVs to check SICTP. Weight up the required volume of mud in the pits. An LCM pill can be premixed if it is likely to be needed.

Complete the Wait and Weight kill sheet to obtain rates and expected pressures.

Slowly bring pump up to speed to the predetermined kill speed while maintaining the SIAP by choke manipulation. Maintain constant pump rate throughout the procedure. When the pump rate has been established the pre-kill pressure should be recorded. This is the pressure required to pump the lighter fluid through all the tubing on the reel. Once the kill weight mud reaches the tubing guide arch, the pump pressure is recorded as the initial circulating pressure (ICP).

As the kill weight mud travels down the CT in the well, the pump pressure should decrease due to an increase in hydrostatic pressure of the mud. Follow the calculated displacement schedule until the kill mud reaches the bottom of the CT. The pump pressure will be at the final circulating pressure (FCP). Hold that pressure constant until all the light fluid or influx is out of the hole.

Plugged tubulars

In most cases, coiled tubing is used when circulation of the wellbore fluids is not possible by conventional means. The CT can be used to wash out plugged tubulars such as tubing/drillpipe. Depending on future operations, kill weight fluid may or may not be used.

Bullhead method

Bullheading can be accomplished down the CT or CT annulus. If a conduit down the CT cannot be established (such as collapsed or buckled CT), fluid may be pumped through the kill line or flow tee. Using high-pressure pumps, a heavier density kill fluid is pumped at a sufficient rate and pressure to displace fluids within the tubing back through perforations and into the producing zone. The rate and pressures used will be established to minimize potential of formation damage.

Additional well control methods may be used depending on conditions.

- Volumetric (rarely used as gas will have already migrated by the time the CT is rigged up)
- Lubricate and bleed (can be used to reduce surface pressure prior to running in hole with CT)

A detailed discussion of each of these methods is available in this well control manual. In each case, the CT well control method is the same as for jointed pipe operations with four conditions:

- Frictional pressure loss in CT is considerably higher than for drillpipe or typical jointed work strings due to the small ID. This means that an accurate hydraulics simulator is must for planning and executing a CT well control operation.
- For CT operations, the length of tubing on the reel can be much greater than the depth to the end of the tubing. The mud must be pumped through the entire length

of the CT and affects the time required to circulate a fluid to a certain point in the well. In a shallow well with water-based fluids, the influx may migrate to the surface before kill weight fluid exits the end of the CT.

- Dual flapper-type check valves in the BHA isolate CT pressure sensor from reading the shut-in BHP just like a BPV in a drillstring. The CT operator must take steps to open the check valve(s) long enough to make a reading.
- CT operations generally do not have adequate fluid mixing capacity to rapidly mix large volumes of kill weight fluid.

Three methods of opening the check valves so that the CT operator can accurately measure the shut-in BHP are:

- Start pumping slowly while keeping the annulus choke close. Record the CT pressure (reel inlet) at the time when the CT pressure stabilizes or the WHP starts to rise. This will be the shut-in BHP.
- Ramp the pump rate up to the proposed kill rate while maintaining constant WHP. When the pump is up to the kill rate speed, record the CT pressure (reel inlet). This value is the initial circulating pressure (ICP). Subtract the slow kill rate pressure from the ICP to get the shut-in BHP.
- Slowly pump fluid with the annulus choke closed until the WHP increases approximately 100 psi. Shut off the pump and slowly bleed off the WHP through the CT reel. Record the CT pressure (reel inlet) at the time the WHP stabilizes or reaches the initial pressure. This will be the shut-in BHP.

PRESSURE MUST BE MAINTAINED INSIDE THE CT STRING ABOVE THE CT COLLAPSE RATING WHILE IT IS IN USE

Design of coiled tubing string

The design of a coiled tubing string should incorporate all the load conditions that may be placed on the string to ensure success of the job. The wellbore configuration, hydraulically applied loads, wellbore conditions, expected overpull, size of the string, yield strength, and wall thickness of the CT have to be identified and accounted for in the CT string design.

Tensile loads generated by pumping fluids down the CT string have to be accounted for. High viscosity fluids will generate a larger applied tensile load than low viscosity fluids.

When entering a well with surface pressure, there are two main factors to overcome. One is to be sure that the coiled tubing does not collapse by applying pressure inside it equal to the wellhead pressure. The second is to not buckle the CT by applying too large of a compressive load. This load is a combination of forces due to wellhead pressure, friction in the stripper element and other forces that must be overcome to insert the coiled tubing into the wellbore. An antibuckling guide can be used to provide additional lateral support (Figs. 4.20 and 4.21).

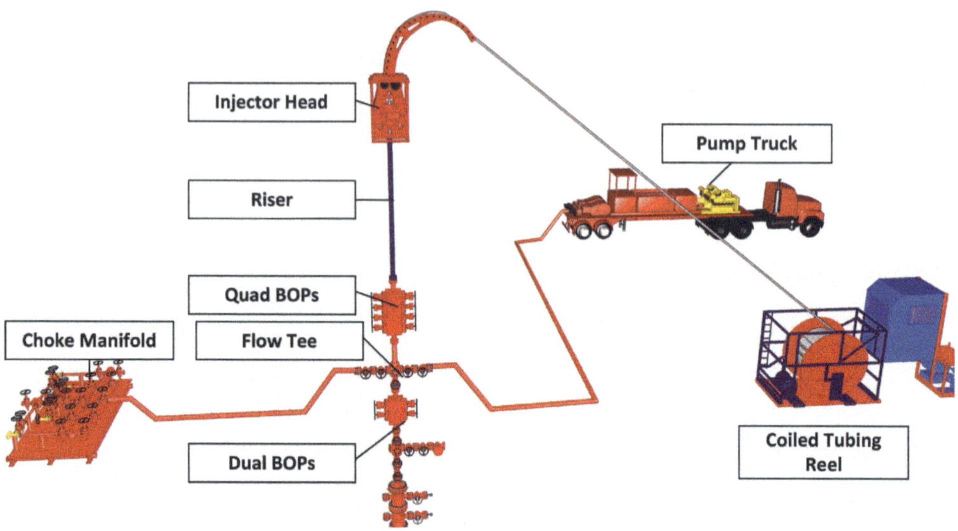

Fig. 4.20 Land-based coiled tubing installation (no crane shown).

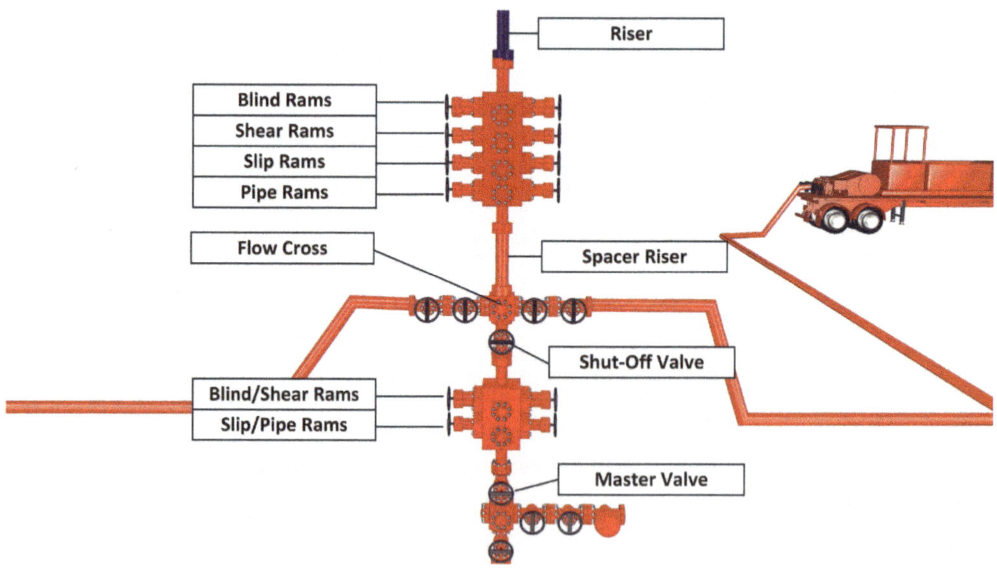

Fig. 4.21 BOP configuration.

Floating operations

Floating rigs present a challenge for coiled tubing work due to heave of the rig. The rig up for the unit must have several additional features to enable coiled tubing to be used effectively. The injector should be secured in a lift frame that houses the injector and CT BOP stack. The frame is normally suspended in the derrick by the traveling block utilizing a lift sub. Heave compensation equipment must be used to reduce the movement of the pipe while it is employed downhole.

The lift frame also should be designed to restrict the motion of the injector within the frame. One method to accomplish this is by using a track and guide system. The injector must be secured within the frame to provide proper alignment of the injector and the riser stack.

There should be work platforms attached to the frame for personnel to use during rig up/rig down.

The stripper should be equipped with a hydraulic releasing, pressure sealing connection.

All riser connections should be flanges having the same bore as the stack.

Coiled tubing specialized equipment

Coiled tubing unit consists of a complete set of specialized equipment that can be used to feed continuous length tubing in a live well while providing means to pump and kill the well. The coiled tubing unit is comprised of four basic elements.

1. Services Power Reel—used for storage, transport, and provide means for pumping service access.
2. Hydraulic Power Pack—consisting of power plant, pumps, manifold, and valves necessary to direct hydraulic fluid for operation of the coiled tubing equipment.
3. Injector Head—used to provide surface driving force for running/retrieving coiled tubing.
4. Quad Blowout Preventer System—used to provide pressure control during operations
5. Control Cabin—used to direct and operate individual coiled tubing equipment for safe, reliable operations.

Coiled tubing services power reel

The coiled tubing services power reel consists of a hydraulically driven reel and support frame (Fig. 4.22). The coiled tubing rests on the core diameter that provides two important functions. Initially, the core diameter allows the coiled tubing to be spun onto the reel in a continuous motion. The interior of the core diameter provides structure that allows the end of the coiled tubing to be interconnected to a pump manifold. This connection passes through the core diameter and affixes to a rotational bearing manifold (called a swivel) that is interconnected to a pumping manifold. For more complex operations, service reels may use an electric swivel that can be used in support of stiff wireline operations. This interconnected "swivel" service allows surface pumping units to be pump fluids into the inside diameter of the coiled tubing.

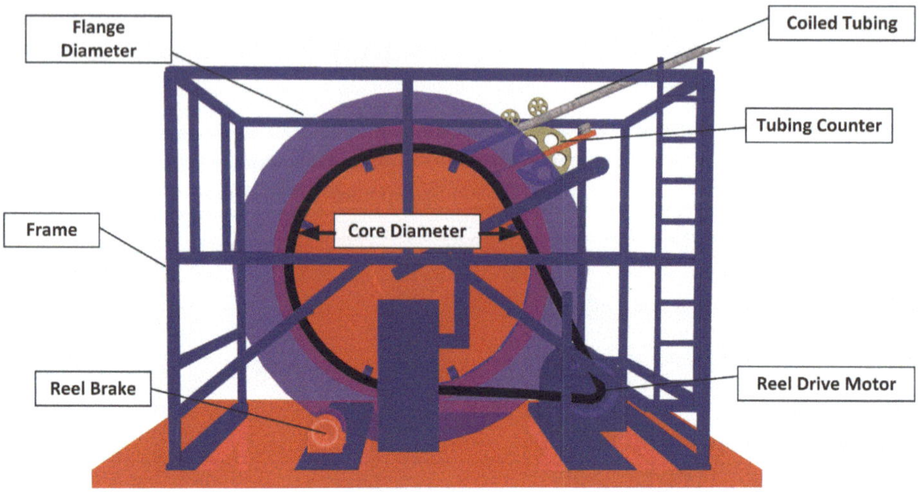

Fig. 4.22 Power reel.

THE CORE RADIUS SHOULD BE AT LEAST TWENTY (20) TIMES THE SPECIFIED COILED TUBING DIAMETER

The coiled tubing services reel incorporates a primary and secondary reel braking system used to slow and stop the reel. To provide rotational motion, a reel drive motor that utilizes a drive belt (or chain) to rotate the reel. As the coil of tubing exits the reel, the tubing will pass through a levelwind guide with tubing lubricator (called levelwind assembly). The guide allows the coil to be side-to-side unspooled in a controlled manner and preventing unwanted lateral travel. Each levelwind assembly contains a mechanical counter used to measure the number of feet of tubing deployed from the reel.

Coiled tubing service life

Every coiled tubing reel has a service life that is largely determined by the services performed to date. The service life of the tubing may be affected by the types of fluids pumped through the tubing. These may include acids, corrosive fluids or exposure to corrosive environments downhole such as H_2S or CO_2. Sand-laden abrasive fluids or cement slurries may erode the tubing due to frictional scouring of the tubing steel.

During operations, the tubing is straightened and forced into the well, subjecting the coil steel to multiple stresses that will affect the life of the tubing. These stresses may include:

- Tension (hanging weight) stresses occurring from the point of suspension (chains) to the bottom of the tubing string.
  - If tension stresses exceed tensile strength of tubing, the tubing may part (usually just below the injector).

- Burst stresses from pressure applied inside the tubing.
- Collapse stresses from pressure applied from outside the tubing.
- Buckling stresses due to high-angle deviation, including horizontal sections.

Buckling of tubing will occur in highly deviated or horizontal wellbores and severely impacts the service life of the tubing. Buckling forces are initiated when the natural radius of curvature is exceeded by the tubing as it is run into the well. The radius of curvature is determined by the natural bend of the tubing as it is stored on the reel. Normally, most radius of curvature of the tubing varies from 13 ft. for 3/4 in. to 42 ft. for 2–3/8-in. tubing.

As the tubing is placed in tension, the radius of curvature is eliminated. As the tubing is lowered into the deviated well, the lower part of the string will experience compressional stresses, the radius of curvature may be exceeded resulting in buckling.

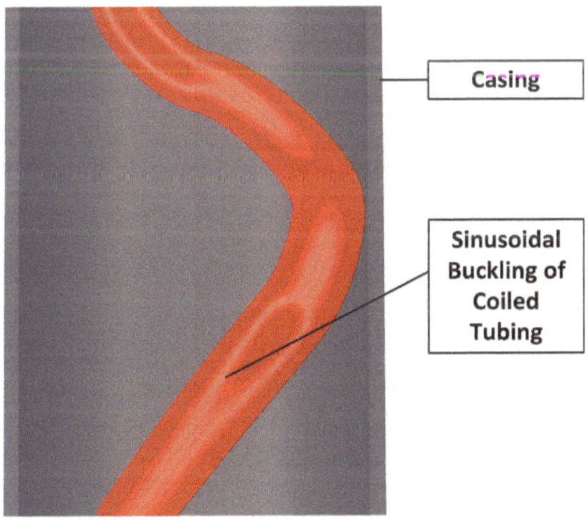

Fig. 4.23 Sinusoidal buckling.

Sinusoidal buckling refers to the resultant stress caused by the force of pushing the tubing into the deviated or horizontal well. These buckling stresses will increase as the tubing is forced into the deviated portions of the well. With sufficient force, the tubing will deform into a sinusoidal configuration. The resulting deformation will be dependent on the dimensions of the tubing and fixed well parameters (Fig. 4.23).

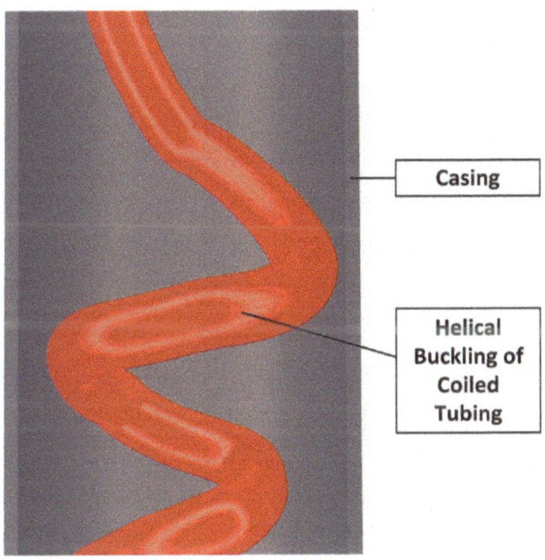

Fig. 4.24 Helical buckling.

For deeper wells, increases in the length of coiled tubing will establish parameters that cause helical buckling. With helical buckling, the tubing is permanently deformed and will not allow the tubing to be pulled from the well. Helical buckling is referred to as "Lock-up" and the tubing is permanently deformed and generates ball of tubing. Helical Lock-up occurs when the frictional forces between coiled tubing OD and well tubular (i.e. casing or tubing) exceeds the pushing force. When these frictional forces exceed the pushing force, coiled tubing will no longer be pushed into the deviated section of the well. If additional coiled tubing is forces into the well, failure of the tubing will occur (Fig. 4.24).

Cycle limitations

Most companies have an established limit the tubing life (~80 cycles), where a cycle is defined when a given length of pipe has been spooled off the drum, forced, and retrieved from the well and spooled back onto the drum. While the service lift is monitored by number of cycles, environment of fluids and well deviations, the size, grade, and wall thickness will determine the maximum service life of the tubing.

Cycles will vary greatly depending on the size, grade, and wall thickness. For example, a 2–3/8″ reel may be limited to only 40 cycles. In order to maximize service lives of coiled tubing, the cycles are weighted for excessive load, high pressure and abrasive fluids (factors commonly used are 2, 3, 4, and 5). Repeated cycling by forcing/retrieval of the same section of tubing may greatly reduce the service life of the tubing. If this occurs, this section of tubing may be removed from the reel and the remaining tubing would be welded together.

To ensure the maximum service life of coiled tubing, the operator records information per job for each length of tubing used. This information may be collected manually or by using computerized systems using real-time data acquisition. The computerized systems allow for predictive modeling such as life expectancy and percentages of success at getting tubing to bottom of a highly deviated well.

Pumping limits

Excessive or high-pressure pumping services, along with abrasive fluids, greatly limit the service life of coiled tubing. Pressures losses can be significant at high pump rates due to the small diameter and overall length of coiled tubing. Information is available outlining safe operational pressure drops for given size, weight, and grade and fluid at varying pump rates. For example, the following limits are offered for review:

- Maximum pumping pressure while the coiled tubing is moving = 4000 psi.
- Maximum pumping pressure with stationary coiled tubing = 5000 psi.
- Maximum differential pressure (collapse pressure) of coiled tubing = 1500 psi.

Ovality

The physical deformity of the tube radius is referred to as "Ovality." Ovality measures the degree of "out of round" of the physical coiled tube. The percentage of ovality can be determined by randomly measuring the tubing internal and external diameters using the following equation. These measurements can be performed using a simple outer and inner set of machinist calipers. The goal is to have the resultant % of Ovality as low as possible. From normal manufacturing processes, expect an initial coiled tubing ovality of ~1%–2%. Ovality affects the ability of a ram BOP to provide a positive seal around the pipe and greatly affect the collapse resistance.

$$\%\mathbf{Ovality} = \frac{(\mathbf{Max\ OD} - \mathbf{Min\ ID})}{\mathbf{Design\ OD}} \tag{159}$$

where

% Ovality = Percentage of Ovality

Max OD = Maximum Outer Diameter of Tube (in.)

Min ID = Minimum Inner Diameter of Tube (in.)

Design OD = Outer Diameter of Tube at Original Conditions (in.)

Hydraulic power pack

The Hydraulic Power Pack consists of the power plant (engine), transmissions, pumps, fluid tanks, manifolds, valves, pneumatic, and electrical systems. The Hydraulic Power Pack will vary for a range of services and coiled tubing injector sizes. The power pack contains a pneumatic (air) supply system capable for starting the engine and provides auxiliary operational requirements. To provide adequate environmental protection, the containerized system includes a series of pans to contain all fluids from the unit and prevent unwanted site spills.

The Hydraulic Power Pack provides all the hydraulic fluid under pressure for safe individual equipment operation. The Pack integrates manifolds, valves, pressure lines, and gauges with individual hydraulic equipment systems. These systems include pumps, hydraulic tanks, crane, injectors, reel, and BOPs. For portability, the Hydraulic Power Pack is mounted normally on a skid or crane trailer.

Injector head

The coiled tubing injector head provides the means of pushing and pulling the continuous coiled tubing in and out of the well. Using a series of traction systems, the continuous tubing is gripped and straightened as it is forced into the well, even if the well is under pressure. This gripping system is adjusted to maintain a tubing injection rate that matches a variety of well conditions. Finally, the system is designed to safety pull the tubing from the well over a variety of well conditions.

Although the methodology for gripping, injection, and retrieving continuous tubing from a well, our industry supports various independent operator equipment proprietary designs. Most designs employ the use of two diametrically opposed sprocket-driven traction chains using counter-rotating hydraulic motors. These systems employ interlocking gripper blocks mounted between links of extremely heavy-duty chain. The gripper blocks can be manufactured with constant diameter sized to the exact specifications of the constant OD tubing. For taper tubing strings, variable diameter gripper block is used where a "V" shape groove is machined into the block. This provides sufficient clearances as the tapered tubing OD varies. Once assembled, the traction chains operate using frictional forces provided by close tolerance chain driven block moving in one direction (Figs. 4.25–4.27).

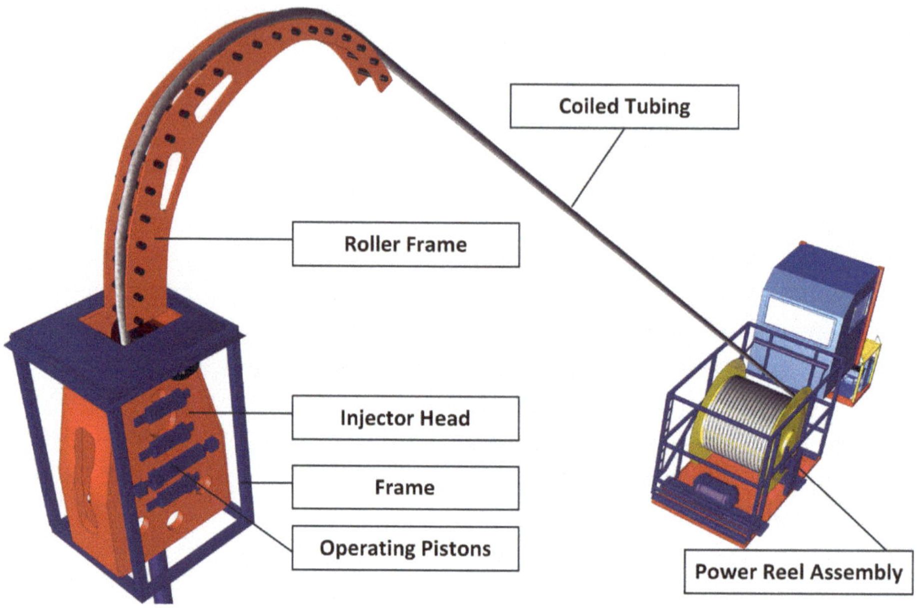

Fig. 4.25 Injector head assembly.

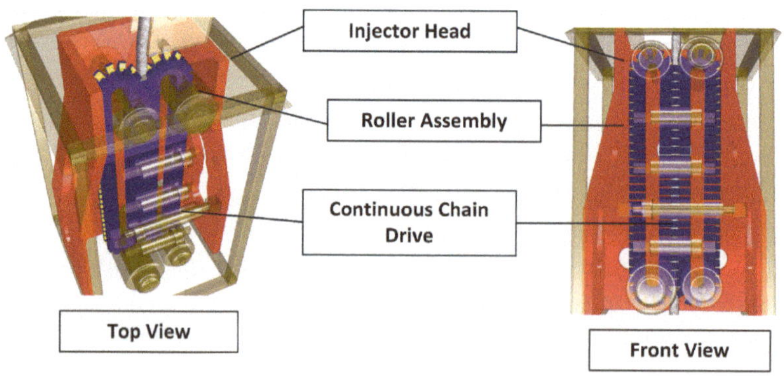

Fig. 4.26 Roller assembly.

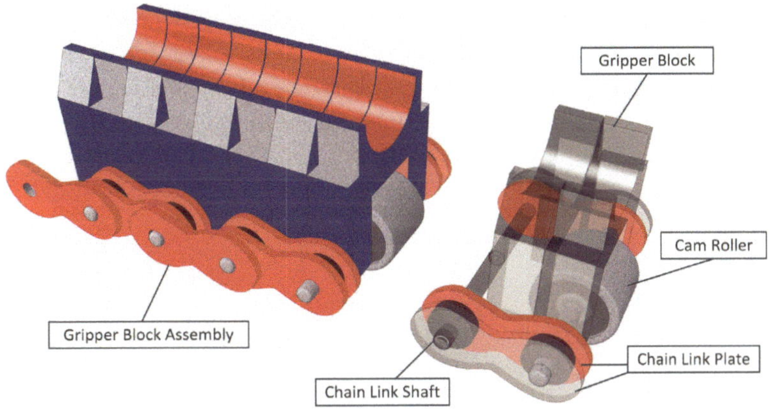

Fig. 4.27 Gripper block assembly.

Traction forces are employed using a series of hydraulic cylinders that provide opposing pressure to the rotating chain blocks. These opposing forces will continually "squeeze" the rotating chain block against the tubing. These forces are continually monitored by the operator maintain sufficient velocities of tubing entering/exiting the well without damaging the tubing. All systems incorporate an emergency pressure source that may be utilized in the event of a primary hydraulic system failure.

Tubing arch guide

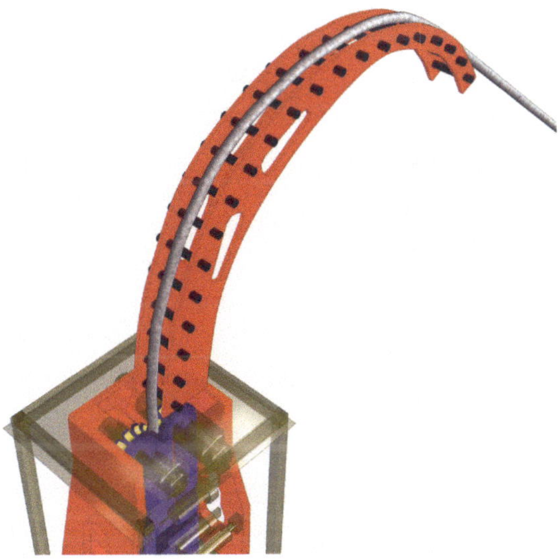

Fig. 4.28 Tubing arch guide.

To assist in spooling of the continuous tubing, a tubing arch guide is used to provide means to support tubing while allowing a sufficient bending radius of the tubing to 90 degree before entering the injector head (Fig. 4.28). The bending radius is specifically designed to prevent premature collapse of the tube from bending forces. Once again, with proprietary designs, individual Tubing Arch Guide equipment will vary between manufacturers. Overall, the Tubing Arch Guide will employ a series of rollers, pads, and secondary rollers that minimizes frictional forces during tubing movement.

RADIUS OF THE TUBING ARCH GUIDE SHOULD BE AT LEAST 30 TIMES THE SPECIFIED OD OF TUBING

As the tubing is spooled out and in, the coiled tubing undergoes a series of bending stresses. The Tubing Arch Guides minimizes these stresses, but does not eliminate them. With maximum bending forces occurring at the base of the Tubing Arch Guide, these systems must be designed to withstand the weight and velocity of reeling the tubing back onto the reel.

Stripper

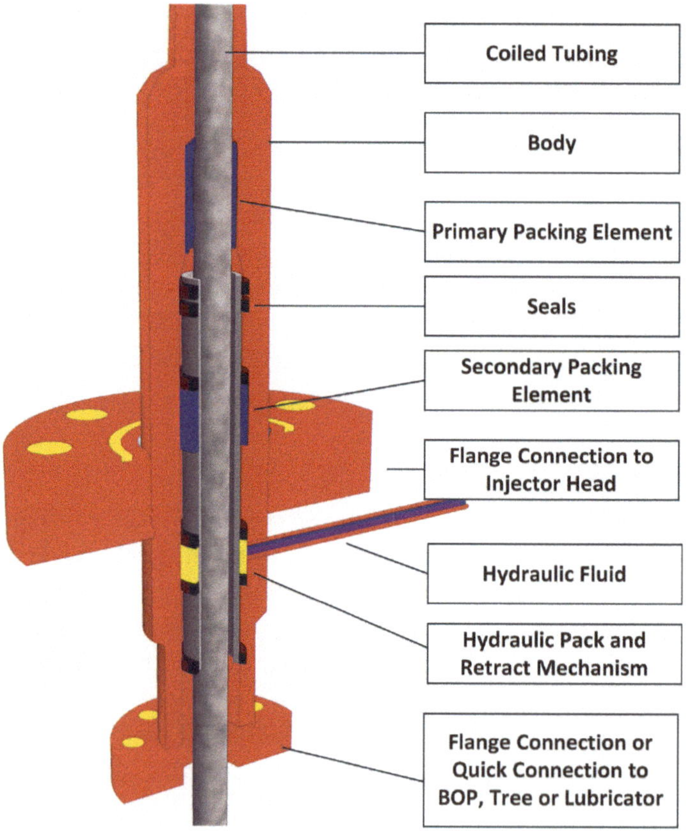

Fig. 4.29 Conventional stripper.

Located below the Injector head, the stripper allows the Injector head to be fastened to the riser or Quad BOPs (Fig. 4.29). Strippers are primary barriers to wellhead pressure. The packing elements are designed to be serviced from above. The energizing stripper employs pressure to lower bushing and compresses packing in an upward direction. Energizing stripper is called "packing." Designed with well-assisted closure where well pressures acting upward assist in closing the element. Strippers are available in 5000 psi, 10,000, and 15,000 psi working pressure.

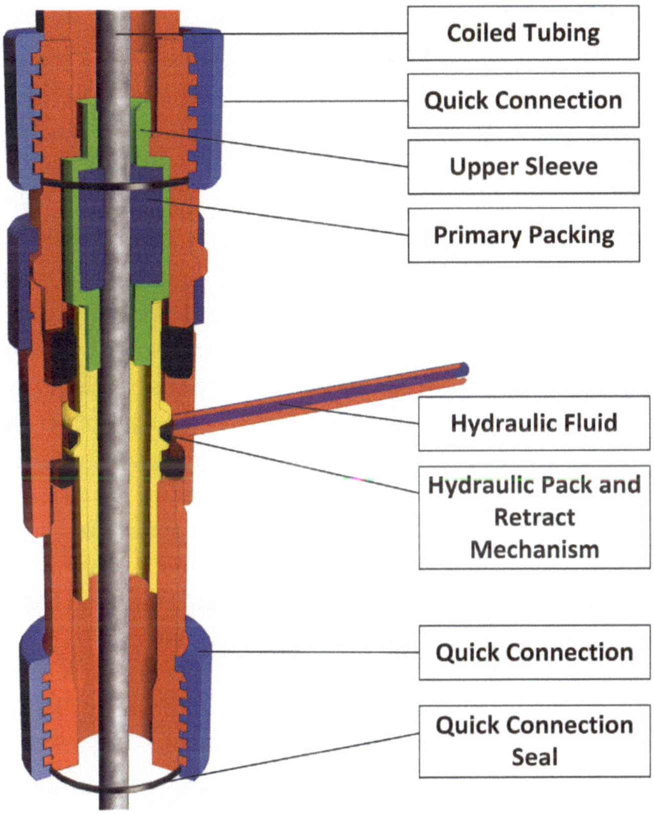

Fig. 4.30 Tandem stripper.

The Tandem Stripper is used in conjunction with a stripper as a back-up in case of wear or failure. Industry recommends the use of a tandem stripper packer because it has two packing elements that can be activated individually. The tandem stripper extends the life of the sealing elements. The inserts used may include urethane, nitrile, or viton elastomer depending on service requirements. Split inserts can be replaced with coiled tubing in service. These tandem stripper designs include (Fig. 4.30).

1) Energizer
2) Packer insert
3) Nonextrusion ring

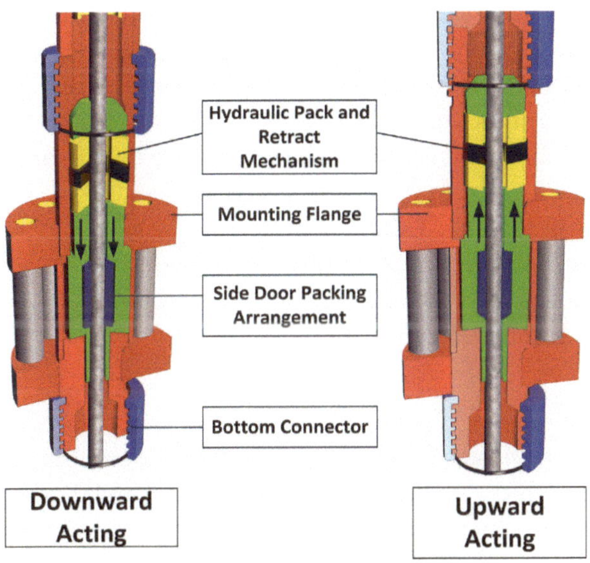

Fig. 4.31 Side-door stripper.

The Side Door Stripper provides easier access to the packing element by utilization of a "side door" (Fig. 4.31). This feature eliminates the need to access the stripper from above.

Quad blowout preventer equipment

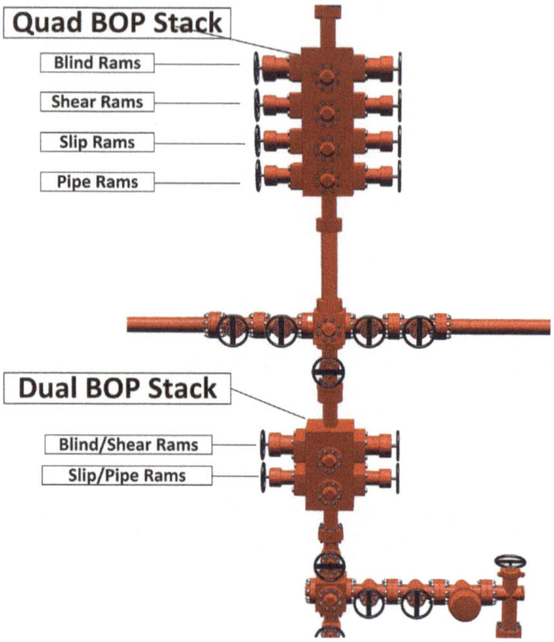

Fig. 4.32 Quad blowout preventer.

The Quad Blowout Preventer Equipment describes a four-ram blowout preventer configuration designed to provide positive sealing against wellbore pressures and secure the well during emergencies (Fig. 4.32). The Quad BOP equipment allows work to be performed under pressure, while tubing is still within the well. Due to size and operating pressure, the Quad BOP incorporates separate blind seal rams and shear rams. Accordingly, these BOPs are considered to be the "secondary" well control barrier in coiled tubing operations (fluid being the primary barrier). The Quad BOPs are normally sized from 3–1/16″ to 7–1/16″ with working pressures from 5000 to 15,000 psi. These BOPs are hydraulically operated at 1500 psi and incorporate manual locks. Once functioned, the BOPs can be manually locked to ensure the BOPs will not prematurely open if hydraulic pressure were to fail.

According to industry standards, Quad BOPs are to be arranged starting from top as (1) Blind Ram, (2) Shear Ram, (3) Slip Ram, and (4) Pipe Ram.

1) Blind Rams—Serve to provide complete sealing of the well when no tubing is present through stack.
2) Shear Rams—Serve to provide sufficient force to shear tubing within the stack.
3) Slip Ram—Serve to hold the tubing in place for shearing and other operations.
4) Pipe Ram—Serve to seal around tubing and isolate wellbore pressure below rams.

Quad BOP Special Features:

- Provide manual override, in case of hydraulic failure
- May be outfitted with a ported type connection for downhole injection
- May be outfitted with Type H union, Type B union, flanged connections
- May include optional ram(s) configuration

Pumping tee

Located immediately below the Quad BOP stack, the pumping tee provides means to circulate fluids from the wellbore through the choke manifold for recover and disposal (Fig. 4.33). The pumping tee also provides means of pumping fluid into to the well via the tubing annulus.

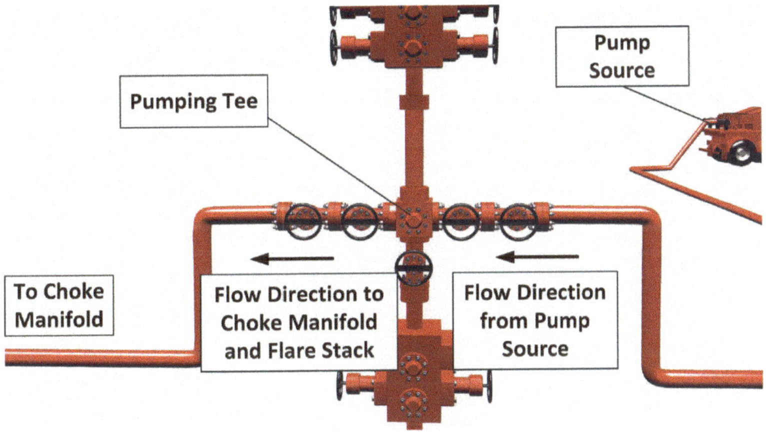

Fig. 4.33 Pumping tee.

Wellhead assembly

The wellhead assembly is a series of valves and wellheads used to support the tubing and provide means for pressure isolation as well as well intervention services (Fig. 4.34). For most coiled tubing operations, the forces generated by pulling and pushing of the coiled tubing, as well as equipment weights are transferred to the multiple casing strings of the well through the wellhead.

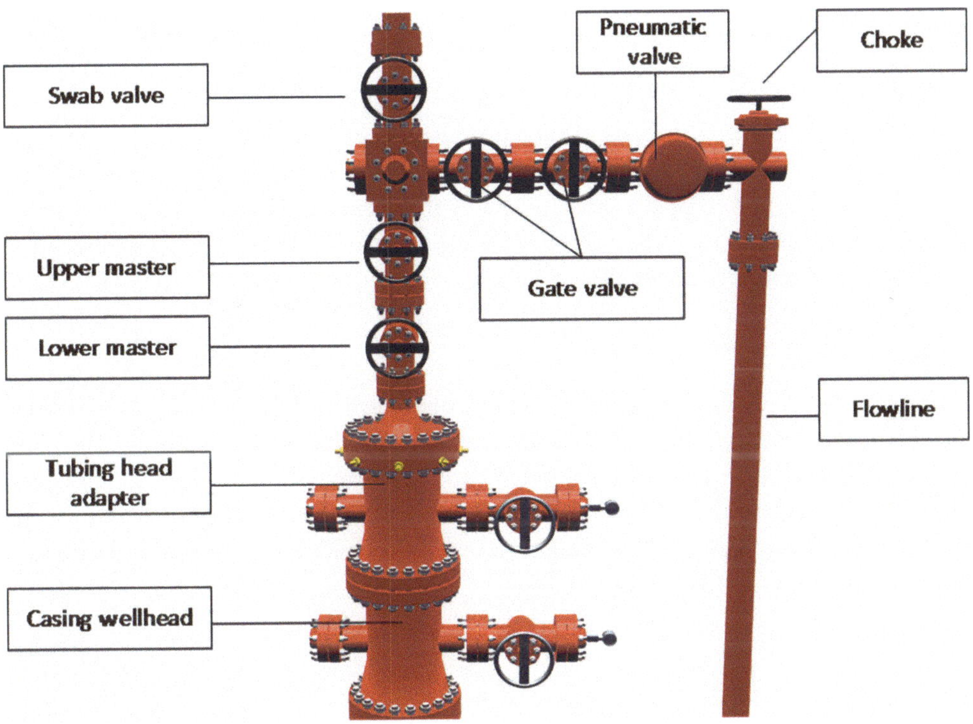

Fig. 4.34 Wellhead assembly.

Coiled tubing support structure

Depending on complexities of the coiled tubing operation, a support structure may be utilized to provide support of the injector head and minimize damage to the coiled tubing by minimizing bending loads (Fig. 4.35). Support Structures are used in order to minimize bending stresses occurring when pulling tubing from well. Heavy-duty support structures can be used to assist in transferring loads from the injector head, through the structure to the ground.

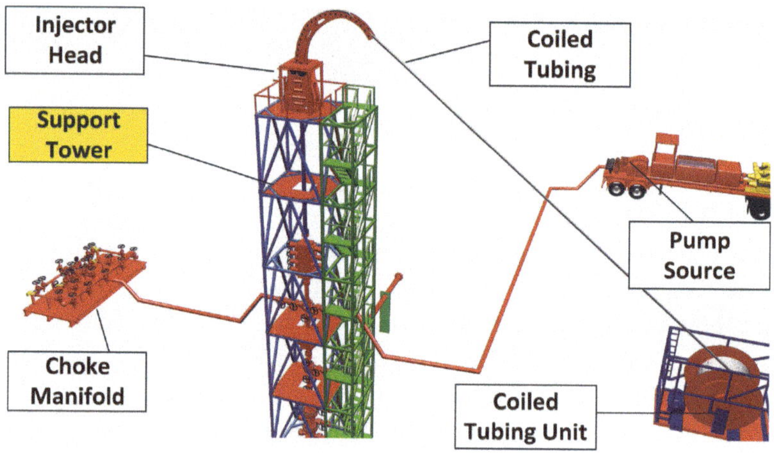

Fig. 4.35 Support tower with climbing tower.

For smaller operations not requiring a heavy-duty support structure, an intermediate sized support structure can be used. For this type of operation, the injector head/tubing combination is supported by the crane with weight transferred to the wellhead. The support structure assists in the direct support of the injector head during operations, minimizing loads on the support crane.

Telescopic legs

Telescopic legs are used to provide leveling of the overall structure (Fig. 4.36). As operational heights may vary, the telescopic legs along with footing pads, allow for level support of the structure and provide sufficient ground clearance. In order to maintain the stability of the structure, each leg sets on an independent surface footpad. Once footpads are in place and the legs are extended, a series of turnbuckles are tightened to provide anchoring support.

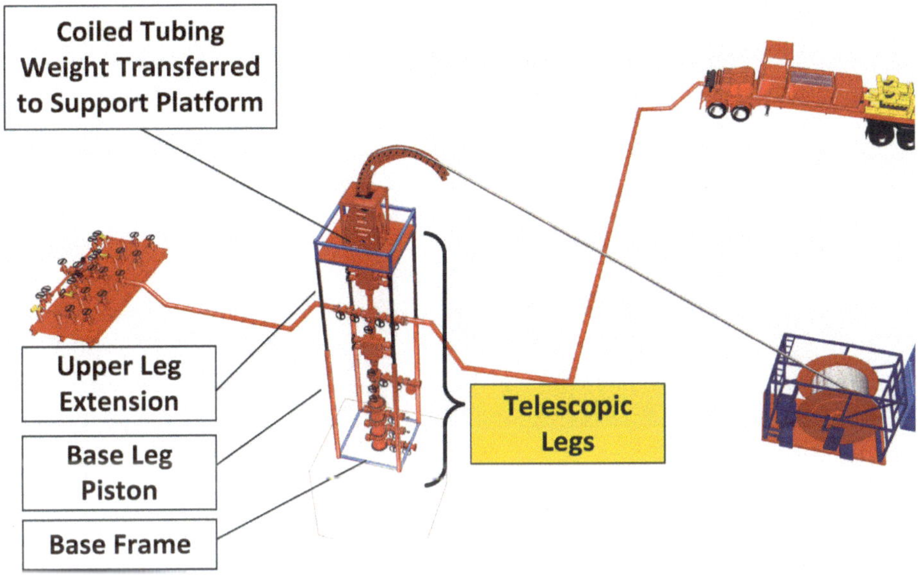

Fig. 4.36 Telescopic legs.

Elevated frame

For offshore operations, injector head support may employ hydraulic elevating frame structure (Fig. 4.37). Once extended in place, the legs are pinned and secured in place. The base of the frame assists in distributing weight of the injector head and tubing to floor.

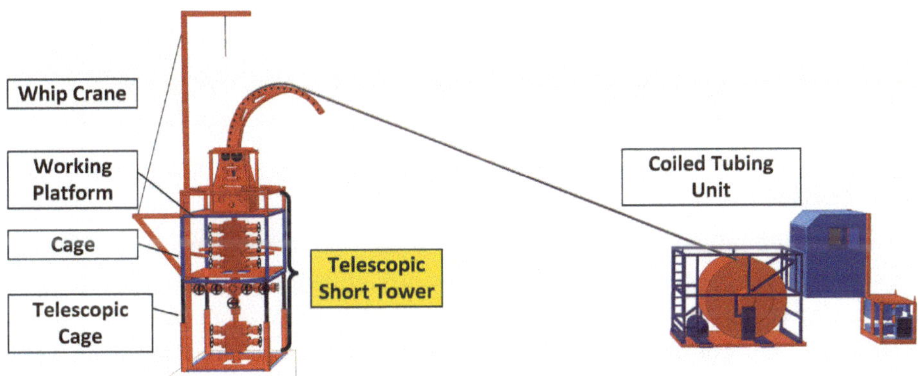

Fig. 4.37 Elevated frame.

Control cabin

Fig. 4.38 Control room.

To control operations, a control cabin is provided to ensure the operator can oversee, direct, and operate every function of the coiled tubing unit (Fig. 4.38). The control cabin is outfitted with an Emergency Shut Down device that can immediate secure the well and equipment during an emergency.

The control cabin is located immediately behind the powered services reel and in-line with the well. Various coiled tubing units have features that allow the control cab to be elevated to ensure clear unobstructed view of the well during operations.

Weight indicator system

Varying weight of the tubing during running operations must be continually measured and tracked to ensure gripping block closing pressures can be adjusted. A system of weight indicators is employed to assist in this determination by determining tensile loads above and below the injector. The weight indicator located above the injector determines maximum tensile loads. The weight indicator below the injector head measures the thrust forces as tubing is forced into the well, referred to as negative weight.

THE THRUST FORCE MUST NOT EXCEED THE WEIGHT OF THE CHAIN DRIVE ASSEMBLY

Control console

Wellhead and pumping pressure gauges, injector chain traction pressure, injector speed, engine throttle, reel controls, reel hydraulic pressure, injector pressure, injector hydraulic pressure, stripper pressure, adjustment controls, Quad BOP stack pressure, Quad ram controls (Fig. 4.39). For operation, the control panel includes an injector brake assembly, load indicator, and emergency kill switches. These controls are normally affixed to a computer for monitoring, recording, and remote evaluation.

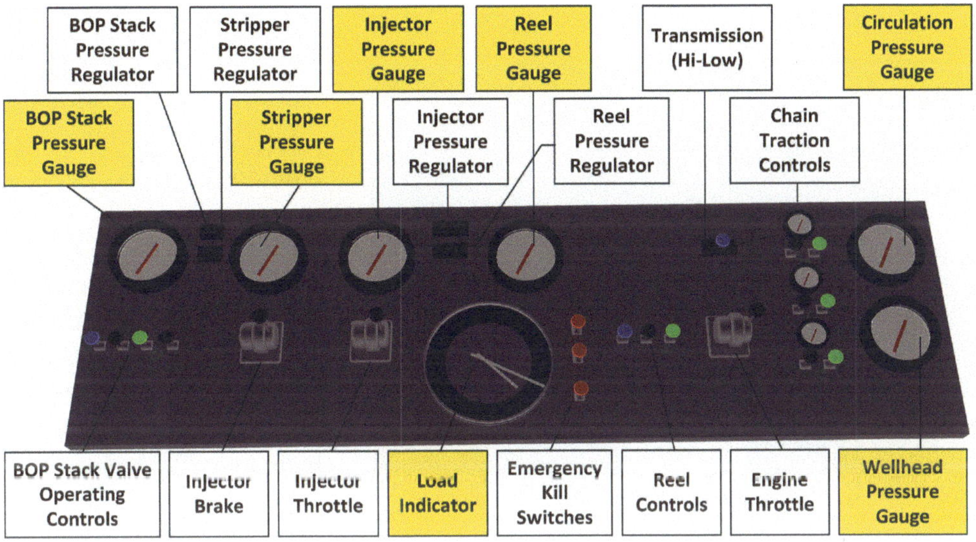

Fig. 4.39 Control room panel.

Stripper console

The Stripper Console includes operation for dual equipment action. The console can be manipulated to "pack" and "retract" hydraulically (Fig. 4.40). Pressures can be monitored to ensure the stripper hydraulic system pressure is maintain within the rating of stripper (most systems are designed for 5000 psi). The stripper console and all equipment MUST be Function tested after initial installation to ensure the proper operation of equipment and ensure installation of correct connections is completed.

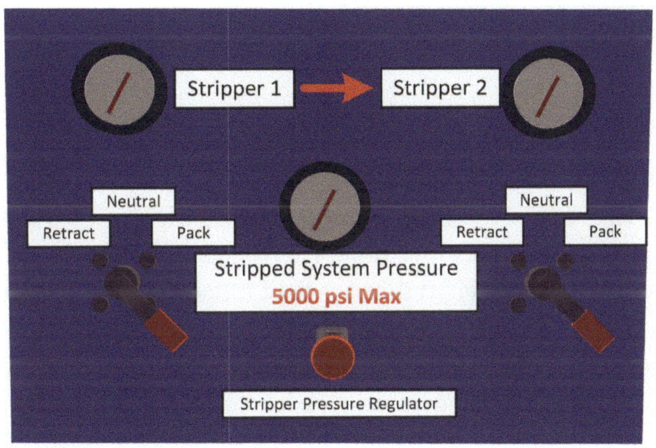

Fig. 4.40 Stripper controls.

BOP console

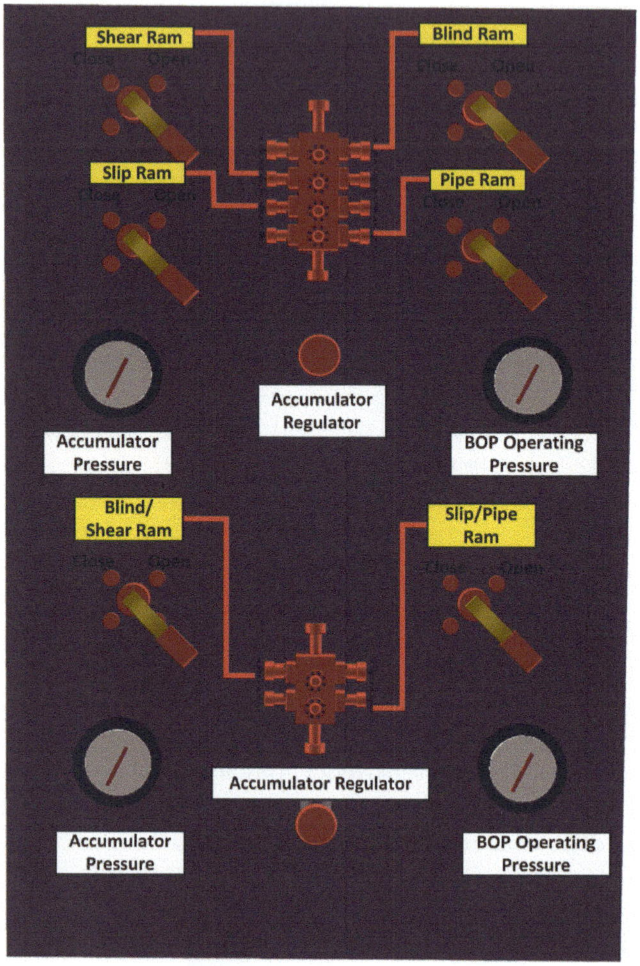

Fig. 4.41 BOP control panel.

The BOP console may be separate of the Control Console (Fig. 4.41). The BOP Console has been designed to require at least two (2) controls be shifted to actuate a ram. This dual control requirement has been employed to aid in the prevention of "accidental" operation. The pressurized hydraulic fluid has been stored within an accumulator bank, which will allow multiple BOP operations without need of powering by pumps.

Coiled tubing connections

Coiled tubing services utilize a variety of connection for the purposes of isolating pressure and transferring tension, compression, and torsional loads from tools onto the coiled tubing. Most connections are designed for easy field installation, maintenance, and reuse.

External slip type

An "external slip" utilizes a slip or grapple-type load placed on the outside diameter (OD) of the tube body. The slips are designed with sharp "spiraled" teeth which "bite" and secure to the coiled tubing. Pressure integrity of for external slips is maintained through a series of elastomeric (rubber) sealing elements that extrude onto the outer diameter of the tubing to provide a seal. To ensure an effective seal, the coiled tubing outer diameter surface must be obstruction free. Any obstructions, such as rust, scale, hydrocarbon residue should be removed for engaging the external slip.

Internal slip type

The internal slip utilizes a slip or grapple-type load to be placed within the inside diameter (ID) of the tube body. The slips are designed with sharp "spiraled" teeth which "bite" and secure to the internal wall of the coiled tubing. The tool connection mechanically wedges into the coiled tubing during make-up. Pressure integrity of for external slips is maintained through a series of elastomeric (rubber) sealing elements that extrude onto the inner wall of the tubing to provide a seal. For an effective seal, the internal coiled tubing manufacturing weld seam must be removed and prepared including grinding, smoothing, and buffing.

SLIP TYPE CONNECTIONS REQUIRE NON-OVAL TUBING FOR EFFECTIVE OPERATION

Swaging tools

For oval tubing, a swaging tool may be used to return the tube body to an "original round" condition. The swaging tool is constructed with a hollow-core hydraulic jack for pulling a short swaging cone up through a longer split skirt expander. The diameters of the swaging cone and skirt expander are selected based on the size, weight, and grade of the coiled tubing body outer diameter and just a few thousandths larger than desired. After swaging the relaxed coiled tubing returns to original round condition.

External dimple type

The external dimply type connection is affixed to the tube body using a series of mechanical screws. The mechanical screws are fastened and compress the steel wall of the tubing. When these forces exceed the yield strength of the steel, a series of "dimples" will appear on the tubing. Variations include template designs. Pressure integrity of for external slips is maintained through a series of elastomeric (rubber) sealing elements that extrude onto the outer wall of the tubing to provide a seal.

Internal dimple type

The internal dimple type connection is affixed to the tube body through a series of mechanical indentations into recesses on an internal mandrel insert. When these forces exceed the yield strength of the steel, a series of "dimples" will appear on the tubing inner diameter wall and provide a means for load transfer that secures the connection to coiled tubing. Pressure integrity of for external slips is maintained through a series of elastomeric (rubber) sealing elements that extrude onto the inner wall of the tubing to provide a seal.

Roll-on type

The roll-on connection is designed to employ a machined insert mandrel that is fitted to the inside of the coiled tubing. The mandrel is machines with multiple circular recesses and the connection is established when tube is mechanically deformed into the machined recesses on the mandrel. Pressure integrity of for external slips is maintained through a series of elastomeric (rubber) sealing elements that extrude onto the inner wall of the tubing to provide a seal.

Weld-on connectors

Weld-on connection is normally limited to the largest coiled tubing diameter where the previous types of connectors are limited and not applicable. Limitations may include excessively large outer diameter, reduction in loads or torque rating, oscillating, and vibratory loads or a restricted internal diameter that greatly reduces the size of pump-down darts or ball. When properly designed and welded, the weld on connections will have the same torque and yield ratings as the body of coiled tubing.

Fig. 4.42 Weld-on connectors.

For proper welding, a gradually tapered insertion neck is used to provide support for preventing concentration of bending loads at a single point where the coiled tubing joins the weld-on connector (Fig. 4.42). The use of a straight wall section eliminates bending forces to the coiled tubing heat-affected zone.

Cooling zones are used to limit the heat from effecting surrounding steel from welding operations. As designed, the weld on connector's straight-wall section outer diameter will exceed the inner diameter of the coiled tubing by several thousand of an inch. The weld-on connection is therefore inserted after the coiled tubing is preheated in preparation for welding operations.

Full bore two-way valve

Developed for subsea completions, full-bore two-way sealing master valves are being installed in a variety of well designs. Designed to hold pressure from both directions, the full-bore two-way valve eliminates the need to install a back-pressure valves to pressure test the BOP stack in coiled tubing and snubbing.

Coiled tubing failure points

Excessive forces used on both internal and external sections of the coiled tubing, along with bending forces may lead to shortened service life of the reel. Observation of marks on the tube can be used to monitor the life expectancy of the tube. Over time, the wear and tear of the forces described below may lead to premature failure of the tube.

Injector gripping forces

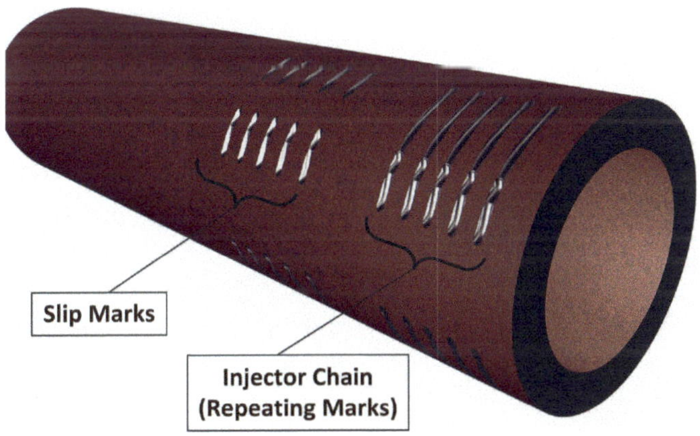

Fig. 4.43 External marks on tubing.

As the injector grips and pushes the tube into and out of the well, the injector chain grippers may damage the tube (Fig. 4.43). To minimize damage, the gripper blocks diameter must be sized to the correct weight, grade, and size of the coiled tubing. The forces used to for gripping or chain tension should be adjusted to minimize damage.

For deeper operations, the coiled tubing will encounter severe tensional loading due to tubing hanging weight. Mechanical misalignments of chain blocks will also exasperate the problems of pipe scaring. Pipe scarring will promote corrosion and weaken the pipe. To offset these problems, the hang-off weight of the coiled tubing should be accomplished using correct slip rams. Slip rams are designed to make continuous contact with the tubing.

Cracks

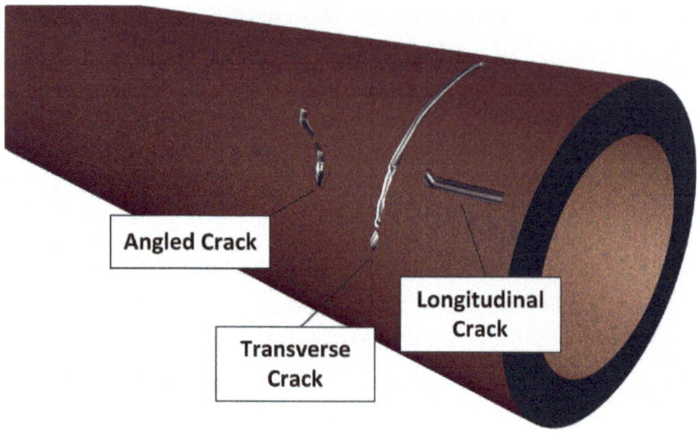

Fig. 4.44 Types of cracks.

Three types of cracks occur in a string of coiled tubing: Angled cracks, transverse cracks, and longitudinal cracks (Fig. 4.44). Barely noticeable at first, cracks will propagate overtime and can lead to premature tubing failure.

Pitting

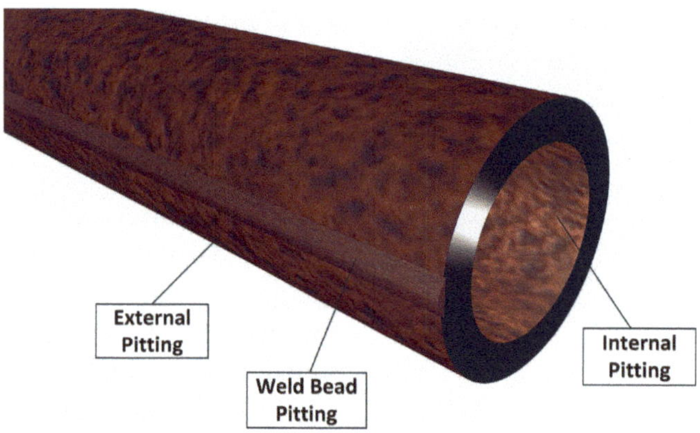

Fig. 4.45 Pitting.

Pitting of the tubing is usually associated corrosion and can occur both externally and/or internally (Fig. 4.45). External pitting may be caused by direct contact to corrosive wellbore fluids, acidic treatments, atmospheric corrosion or a combination of these forces.

Internal pitting is related to improper inhibition of treatment fluids or improper flushing of the coiled tubing after a corrosive fluid job. The internal weld bead may act as an electrolytic corrosive metallic cell allowing accelerated corrosion to take place.

Bending

When the coiled tubing is unreeled, the tubing undergoes bending stresses as the tubing is forced to and from the well (Fig. 4.46). For each tubing cycles (into/out of well), the tubing will experience six significant bending events. Each bending event generates a multitude of bending stresses. Multiple operating cycles will shorten the service life of the tubing. Most operators maintain strict records and set maximum cycles for operations. These cycle records may include simple grading for the job types including pressure and corrosive fluid histories.

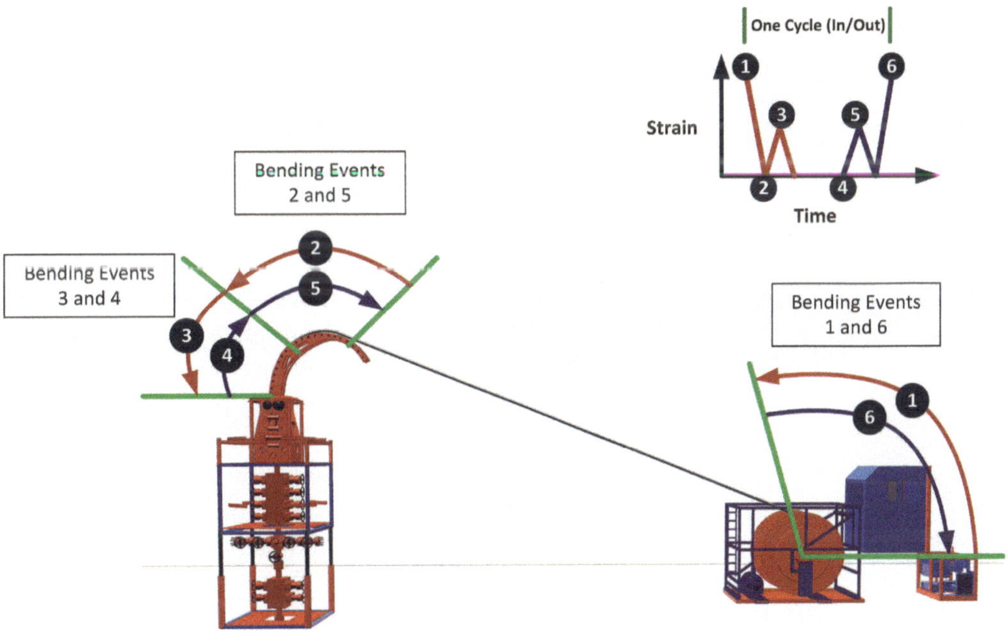

Fig. 4.46 Six bending events per cycle (trip in/out of well).

Abrasion

Fig. 4.47 Localized abrasion.

Abrasion of tubing wall decreases the well thickness and decreases the tubing performance criteria (Fig. 4.47). Sharp scratches, slip teeth marks, and surface gouges will accelerate tubing corrosion. Localized abrasions may develop from contact with well tubulars and may be exasperated by sinusoidal or helical buckling characteristics.

Longitudinal scratches may be an indication of impacting sharp edges within well (i.e., corroded liner tops, etc.). They may also be caused by improperly aligned injector equipment or misaligned spooling and handling equipment. Abrasion may also be cause by careless handling, inadequate maintenance or poor operating equipment.

Ballooning/necking

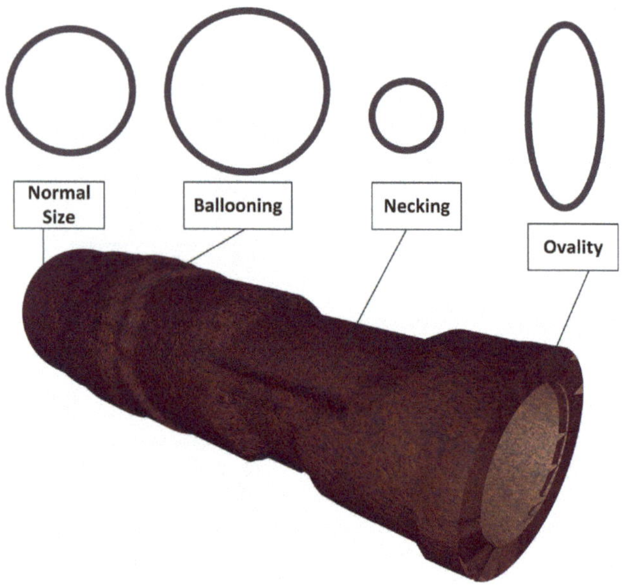

Fig. 4.48 Ballooning and necking.

Ballooning or necking of the tube stock will be caused by a misapplied external force such as poor handling equipment, misaligned injector chain, and pressure control forces such as poorly operated stripper or BOP equipment (Fig. 4.48).

Necking results from stresses applied during tensile loading such as long, hanging tubing in deep wells. Ballooning results from cycling tubing under pressure. Multiple pressure cycles may cause internal stress fatigue allowing the tube to expand.

Coiled tubing advantages

- Run and retrieved while circulating continuously
- Work under pressure
- No connections to make
- Saves time over conventional rig
- Workovers can be accomplished without killing the well thus reducing formation damage
- No need to remove and replace existing tubulars
- Well control operations simplified
- No connections on coiled tubing enables larger a larger tubing size to be run versus pipe with connections.

Coiled tubing disadvantages

- Limited choice of sizes
- Pipe cannot be rotated at the surface (downhole tools available to help overcome this)
- Long lengths of pipe on a spool translate into high pump pressures and low rates
- Bending fatigue due to bending and straightening requires pipe to be cut off at regular intervals and the strength, bust and collapse rating are reduced
- Heavy pipe spools may require large crane for offshore operations
- Wellbore geometry can affect the depth to which the CT can be run depending on its tensile strength

Kicks with pipe to bottom

If the well should kick during operations while on bottom, standard well control techniques will work. Shut-in pressures must be recorded, then control of the well regained by using constant bottom-hole pressure circulating techniques. Often the fluid does not have to be weighted up because the fluid in the well is already a sufficient density. If this is the case, then the Driller's method should be used.

Well kicks with pipe off bottom

The coiled tubing unit's ability to circulate while moving the tubing is ideally suited to handle unforeseen kicks either out or in the hole at the work depth. Well kill options when a well "kicks" with the pipe off bottom are based on how much tubing is in the well and what the shut-in pressures are obtained. Once on bottom, normal kill procedures can be employed. If the pipe is nearly out of the well options include stripping to bottom, or pulling out of the well. If the decision is made to pull out of the well, the blind rams must be closed as soon as the lowermost tool is pulled above it.

Well kicks while out of the hole

In the case where no pipe is in the hole, coiled tubing may not be a viable option since pipe must be forced into the hole under pressure. The force to snub against is likely to buckle coiled tubing. A hydraulic workover unit (snubbing unit) may be required to get pipe back in the hole. Another option may be to set a wireline bridge plug in the casing, bleed off pressure above it and then run drillpipe to the plug. The plug would then have to be drilled out under pressure.

Uses of the coil tubing and nitrogen combination

Treatments involving the use of both coiled tubing and nitrogen can be broadly divided under three subheadings; gas lifting, acidizing, and sand clean outs.

Gas lifting

May be required:

- After recompletion of a producing well or completion of a new well when in the reservoir pressure is insufficient to overcome the hydrostatic pressure of the workover or completion fluid.
- Break out of gas from oil in wells with a high GOR during a period of shut-in. On reopening the well, the gas cap can be dissipated leaving a column of dead oil of sufficient gravity to kill the well.
- After an acid job to encourage clean up.

To lift the oil, coiled tubing would be run into the hole to below the static fluid level and then nitrogen pumped until flow to surface is established. There is no rule as to how deep the column should be gas lifted and how fast the nitrogen should be pumped. However, a rate of 500–1000 scf/min is reasonable, but lift curves can be generated if required before the job.

Acidizing

Two types of acid treatment may require the use of nitrogen to aid clean-up. The nitrogen can be used as gas lift after the job or may be pumped in combination with the acid flush (nitrified acid) in order to lighten the effective hydrostatic pressure on the formation thus providing in situ gas lifting.

- Acid washing is used to remove carbonate scale, detritus, and clay build up from downhole equipment. Occasionally an acid wash on perforations is undertaken to reduce formation skin damage. The acid is circulated to the bottom and the area to be treated allowed to soak. To aid clean up, the acid may need to be nitrified, or nitrogen gas lift may be required.
- Acid squeezing requires the injection of acid into the formation via the perforations. The coiled tubing is run into the production tubing and the acid is pumped from surface through the coiled tubing and once it is in contact with the formation, the well is shut-in and the acid is squeezed into the formation. The acid may be nitrified, usually 300–350 scf/bbl, to aid clean up. Deep gas lifting with nitrogen through the coiled tubing may be necessary on the more reluctant wells.

Sand clean-outs

These can range from small jobs to remove fine insoluble material or much larger operations to remove large quantities of sand from tubing, clear sanded up perforations or clean up a well after a frac job has screened out.

The normal procedure is to pump alternating pills of a viscous fluid (for example HEC) and nitrogen. The viscous fluid provides the solids carrying capacity and the nitrogen when pumped at high pressure provides the fluid column velocity. The coiled tubing is steadily lowered through the sand body until it has been cleared.

Snubbing well control method

Snubbing commentary

Snubbing operations allow a methodology for introducing pipe into a pressurized well by use of hydraulic jacks. Unlike Coiled Tubing, snubbing operations allow rotation as well as vertical movement. Snubbing operations are also designed to operate in high-pressure applications. Due to the small equipment and crew complement, these systems are highly suitable to a variety of well control situations without the time-consuming expense for installation of a portable drilling or workover rig. Snubbing operations are highly adaptable and can be varied in length profiles, BOP complements, and pressure regimes.

Using specialized equipment, snubbing operations allow three critical systems to be employed (1) circulating, (2) hoisting, and (3) rotating. The ability to rotate and hoist tubulars along with the ability to circulate allows complex interference operations to be used.

Snubbing well control operations

Snubbing well control operations are more conventional when compared to wireline and coiled tubing operations, as a conventional circulating system with pumps, tanks, and choke manifolds are used. Depending on the service, a variety of well control methods can be employed from Driller's Method, Wait and Weight, Bullheading, Lubrication, and Bleed methods. As pipe is snubbed into the well, wellbore fluids can be directed to a trip/strip tank to measure fluid displaced from well. To accurately maintain well control, strip/trip measurements are recorded within a strip/trip book.

If conditions exist, a low-volume high pressure pump source can be used to provide back-pressure while snubbing/stripping into the well and constant wellbore pressures can be maintained or regulated by use of a choke. By simply regulating the backpressure, fluids exiting the wellbore are vented through the MGS. Gas vapor from the MGS is vented, while degassed fluid is vented back to the circulating system for processing.

Since snubbing operations and equipment complement will be determined by well complexity and severity of service, the equipment will vary between specific service providers, please refer to specific service provider literature for exact well control procedures.

Snubbing can be defined as the act of forcing a pipe or tubular into a well against wellbore pressure (Fig. 4.49). To safely accomplish snubbing operations on live wells, our industry uses specialized equipment and techniques designed to apply the necessary forces to force tubulars into a live well while safely containing and controlling wellbore pressure and fluids.

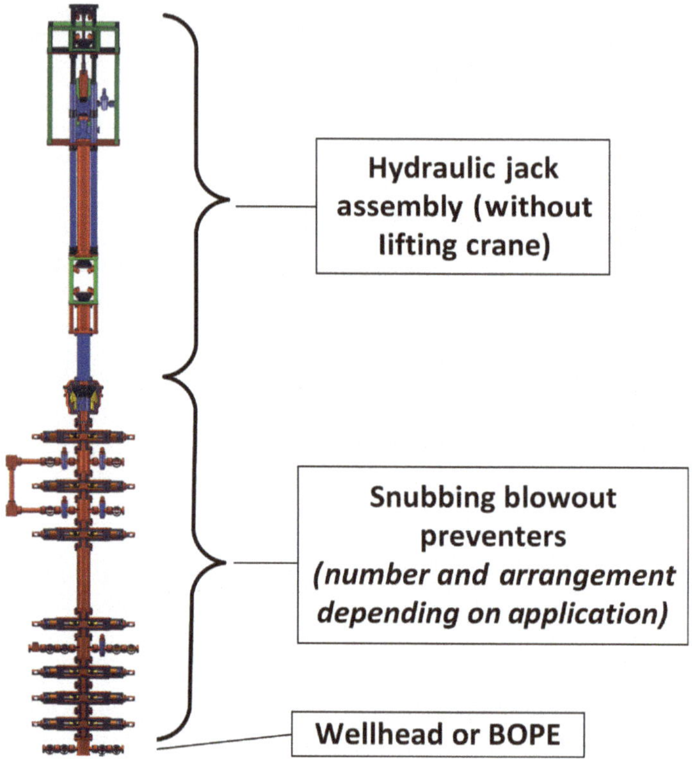

Fig. 4.49 Snubbing stack-up diagram.

To perform live well intervention operations, both coiled tubing and hydraulic workover units (HWO or HWU) can be employed. HWO are normally referred to as "snubbing units" and can perform under higher pressures and more difficult conditions than coiled tubing.

Another type of snubbing unit called a Rig Assist snubbing unit uses the rig hoisting equipment to push or pull pipe from the well under pressure. It is not as versatile as a HWO.

Hydraulic Workover Units differ from coiled tubing and conventional workover rigs, by providing means for ram-to-ram snubbing operations using hydraulic slip sets to latch and force the work string into a live well. HWO equipment has been designed for fit-for-purpose and easy rig-up. Specific HWO equipment sets can be described for their specific designed task including conventional snubbing, rig assist snubbing, workover units, fishing units, rack and pinion units, snubbing units with substructures, long stroke snubbing units, and external frame snubbing units. Snubbing units can be arranged to engulf the existing well's production tree, eliminating the need to remove the tree. Since snubbing units are more robust, they can be used on live wells with or without existing tubing strings.

Remedial operations

As most wells are designed with tubing acting as a sacrificial conduit. This tubular is designed to safely allow the flow of productive hydrocarbons from a reservoir to the surface. Tubing contains profiled nipples that allow for installation of testing, measuring, and flow devices. Tubing is normally isolated from the cemented casing strings by means of an isolation packer. This ensures that corrosive formation fluids only enter the tubing. Therefore, tubing is designed not only to provide a conduit for formation fluids, but sustain any type of wear and tear from this production. Over time, tubing may become clogged with debris/sediment/paraffin or completely fail if the tubing is compromised by corrosion (holes). When this occurs, the wells may be shut-in until such a time that it becomes economical to repair them.

The repairs of each well will be reviewed for safety, efficiency, cost, and complexity. These remedial operations may explore possible problems such as:

1. Repairing mechanical failure
2. Isolating excessive water or gas production
3. Remove tubing blockage (sand, paraffin, asphaltenes, etc.)
4. Replacing tubing and completion equipment
5. Recompleting in newer productive zones
6. Providing means to stimulate the existing reservoir

As planned are developed for returning a well to service, a careful study of the pressure environment, risks, and estimations for success are reviewed. The goal will be to safely return the well to service in order to maximize production. Many remedial services can be accomplished using wireline and coiled tubing operations. For more complex remedial operations, snubbing operations may be employed. Regardless of the type of remedial operation employed, the ultimate goal is to minimize damage to the formation while returning the well to service in order to maximize production.

Snubbing operations

Snubbing operations are normally employed on wells still under pressure that require repairs. The types of wells requiring snubbing operation may vary in complexity, but normally will contain a common obstacle, the well is live and under pressure. Snubbing operations are somewhat complex due to the equipment spread and types of operations performed. In order not to remove the production tree, snubbing equipment is specially designed as a tall structure with most of the load transferred to the wellhead assembly and downhole casing.

PRESSURE THRESHOLD FOR MOBILIZING SNUBBING UNITS IS GENERALLY ≥ 2,000 PSI WHP.

The snubbing rig is a tall and narrow structure allowing it to be deployed in a variety of wells, including within some derricks. The snubbing rig consists of a hydraulic ram system that allows gripping and forcing the workstring into a well. The workstring will enter a series of BOP preventers, specifically designed and rigged to allow ram-to-ram operations so that when a tool joint passes between the rams, pressure can be equalized and allow the string to be forced into the pressurized well.

At the top of the structure is the snubbing basket. This serves as a working platform for snubbing operations with controls and mast for transferring workstring pipe one joint at a time. Located below the basket is the hydraulic jacks whose force can be used to push the workstring in and pull the workstring out of the well. The hydraulic jacks are integrated to a set of traveling slips used to move the workstring. A set of stationary slips are used to hold the workstring as the traveling slips are repositioned after completion of the hydraulic stroke.

Snubbing unit equipment

Snubbing Units are configured to meet well and job-specific demands. The following diagram shows most of the components used within a snubbing unit (Figs. 4.50–4.52).

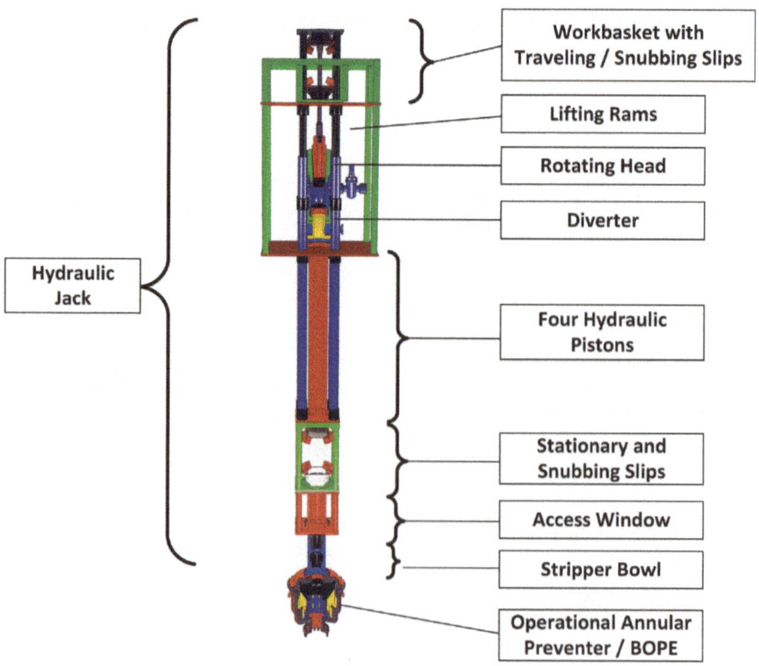

Fig. 4.50 Hydraulic jack (without lifting crane).

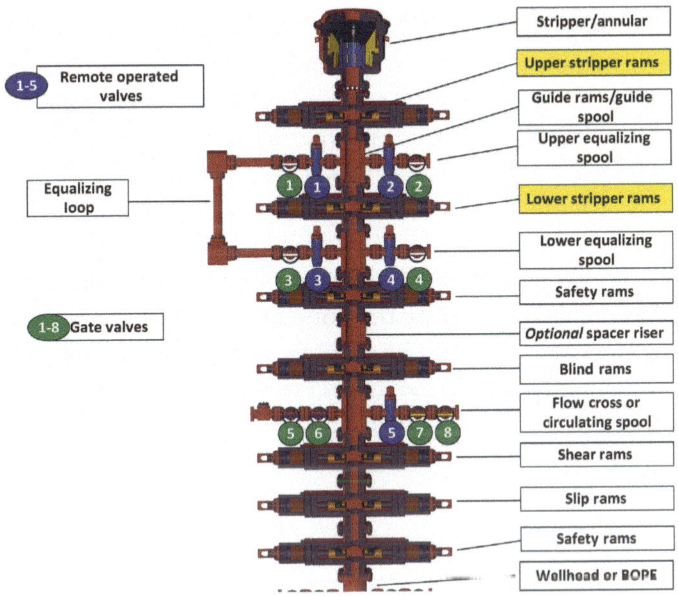

Fig. 4.51 BOP arrangement (varies depending on service needed).

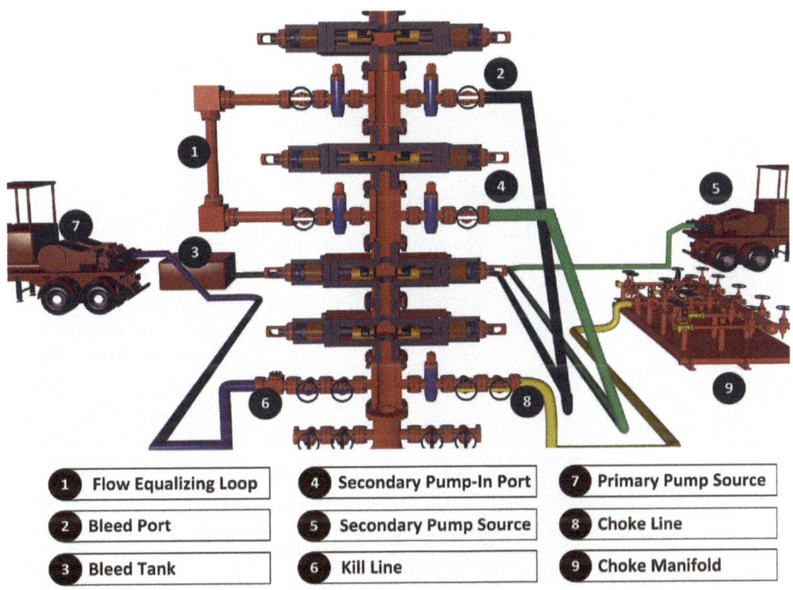

Fig. 4.52 Flow system for snubbing unit.

Stripper bowl assembly

The stripper bowl assembly is a special device primarily used for well control during **low-pressure** operations. The stripper bowl assembly is located directly under the access window. The low-pressure (under 3000 psi) preventer is used to provide pressure seals while tripping tubulars without using stripper rams. Using a stripper bowl assembly instead of stripper rams allows trips to be made more quickly and efficiently without jeopardizing safety.

The working life of the stripper elastomer (rubber) is dependent on pressure service, well fluids, trip speed, lubrication of the rubber, and external pipe roughness. Even though a stripper bowl assembly may be rated to 3000 psi, operating above 1500 psi will normally accelerate deterioration of the stripper rubber. Normal snubbing operations will require changing of the stripper rubber(s) during a job.

The stripper bowl assembly may be used as a wiper rubber to prevent debris from falling into the hole. Depending on the type and model of stripper, the stripper bowl assembly may be wellbore energized.

PRESSURE MAY BE TRAPPED BELOW THE STRIPPER RUBBER

Stripper bowls are available with both single elements and dual elements. Since stripper rubbers are made from different materials, the application will determine which of the different types should be used (Fig. 4.53).

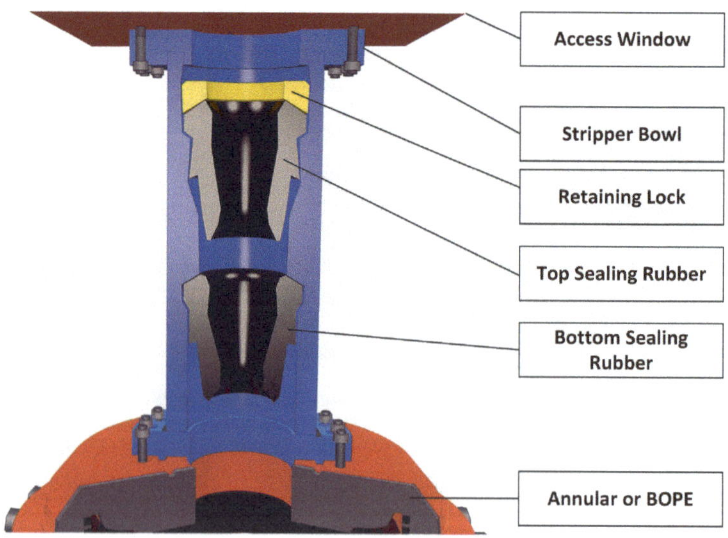

Fig. 4.53 Dual stripper bowl.

Natural rubber

Natural rubber is widely used because of its excellent physical properties and ability to perform in dynamic and demanding drilling environments. It is most often used with water-based mud applications. It often swells and becomes unserviceable when used in oil-based mud applications. Natural rubber has a continuous service temperature up to 225 °F.

Polyurethane

This stripper rubber material has excellent chemical resistance to most oil-based drilling fluids. There is also a low temperature version that is available in this material. Continuous service temperature for polyurethane is up to 180 °F.

Nitrile

Nitrile has good oil resistance and better temperature resistance than the natural rubber. There are several trade-offs. Nitrile does not seal as well against fluid leakage and is less resistant to long-term wear. Continuous service temperature for the Nitrile is up to 250 ° F. Companies are developing low-pressure and a high-pressure nitrile sealing element.

Butyl

Butyl is a high temperature material used in geothermal applications. Continuous service temperature is up to 250 °F.

Annular preventers

Depending on the application, an annular BOP may be used below the snubbing BOPs. It is identical to a standard annular bag-type BOP used in drilling operations although normally of a smaller size. The pressure rating is specified according to the anticipated wellhead pressure. Stripping or snubbing through an annular preventer should be limited to no more than 1000 psi wellhead pressure.

The annular BOP is used when the normal ram-type BOPs cannot seal around something due to its enlarged size, such as a side pocket mandrel, blast joint, slip joint, etc.

When used for snubbing operations, normal practice will have a small (1 gal) accumulator bottle in the closing line hydraulic circuit, situated close to the preventer. This allows the set-up of the tools to be stripped through while maintaining a steady hydraulic closing pressure.

ANNULAR PREVENTERS CAN ONLY BE USED ON WELLS WITH ≤ 2500 PSI
DO NOT USE ANNULAR PREVENTERS WITH SCREW-ON CAP
DUE TO CLOSE PROXIMITY OF CREWS IF ANNULAR LEAKS.

Ram to ram snubbing operations

Since the snubbing unit uses individual sections of workstring, the workstring will contain upsets for each workstring collar. The BOP rams are outfitted with specially designed elastomers that are long-lasting wear-resistant sealing elements. These elastomers will allow thousands of feet of stripping operations to occur and provide sealing capabilities in ranges up to 15,000 psi.

The BOPs are also designed to mitigate entry of the tool joint into a pressurized environment by providing means for pressure isolation and then pressure normalization. This is accomplished using the following steps:

1) With the bottom set of rams closed and wellbore pressures contained, the workstring is snubbed into the well until the tool joint is located immediately above the lower ram.
2) The upper ram is then closed.
3) Wellbore pressure is equalized in the space between the rams housing the tool joint.
4) Once equalized, the lower rams are opened and the workstring is snubbed into the well.

This process is repeated as additional joints of workstring is added. When the workstring is to be pulled from the well, the process is simply reversed. The key for successful snubbing operations is the pressure equalization between the upper and lower pipe ram cavity and the length of service of the ram elastomers (Figs. 4.54–4.57).

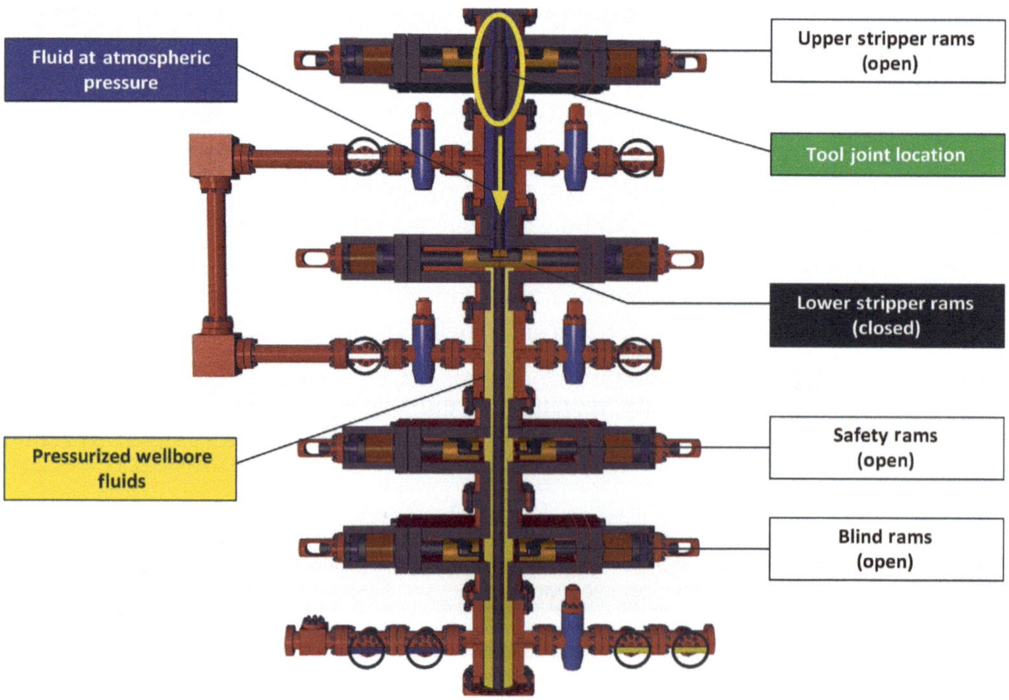

Fig. 4.54 Tooljoint is lowered into position with lower stripper ram closed.

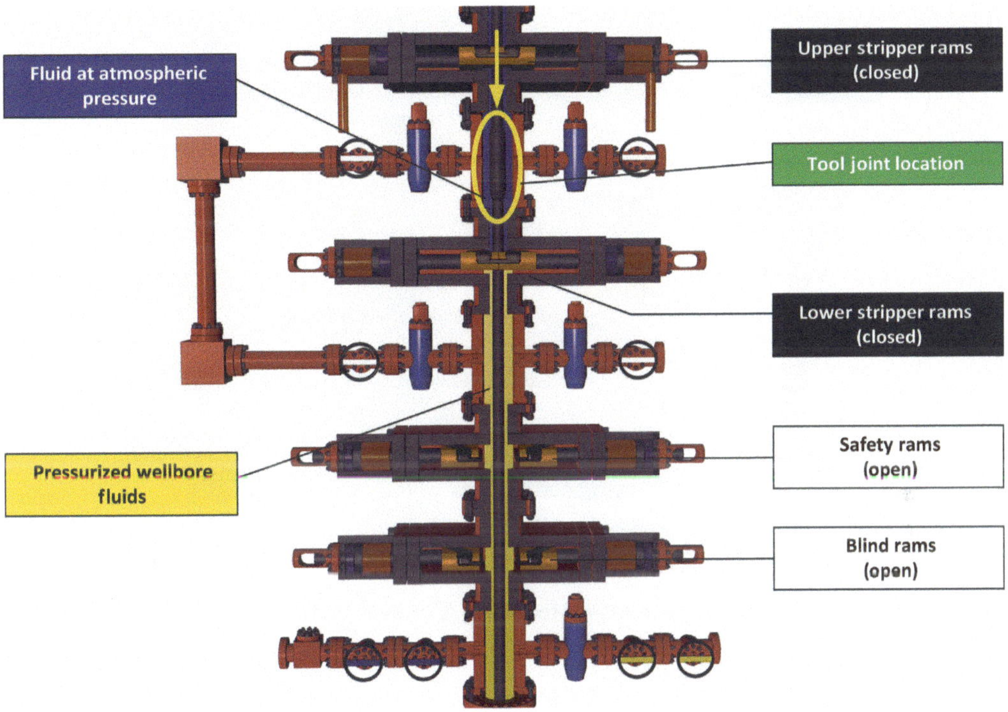

Fig. 4.55 Tooljoint is lowered into position with upper and lower stripper ram closed.

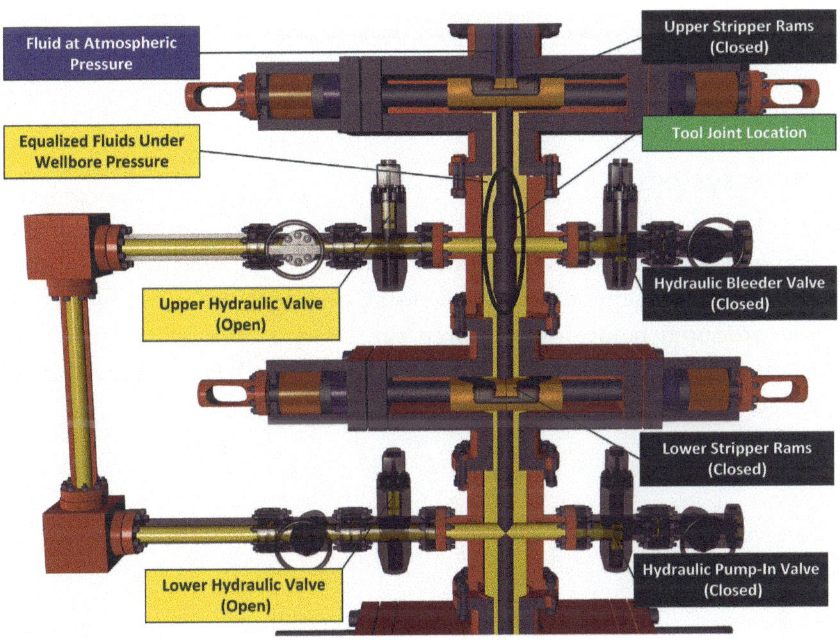

Fig. 4.56 With tooljoint in position, wellbore pressure is equalized between upper and lower stripper rams.

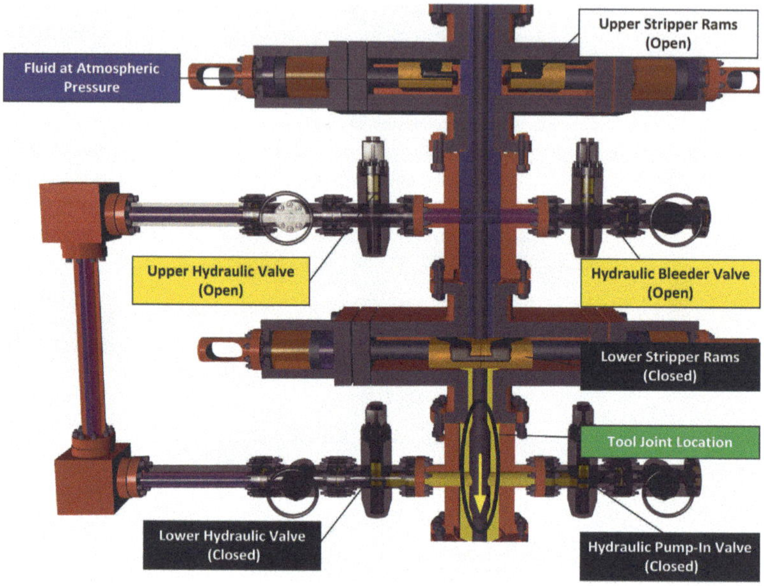

Fig. 4.57 Lower stripper ram is opened and toolstring is lowered into well.

NORMAL STRIPPING RAM TO RAM TAKE LESS THAN 10 ADDITIONAL SECONDS PER JOINT COMPARED TO DEAD WELL OR RUNNING ON AN ANNULAR.

Slip interlock system

The Slip Interlock system was developed to prevent accidental simultaneous opening or partial opening of both slip bowls. This safety feature prevents accidental dropping of the workstring and utilizes a series of sensors to ensure one slip bowl is completely closed and hydraulically pressured before other paired slip bowl is allowed to open. To ensure proper operation, the pressure and slip position sensors must be activated before the interlock is allowed to open (Fig. 4.58).

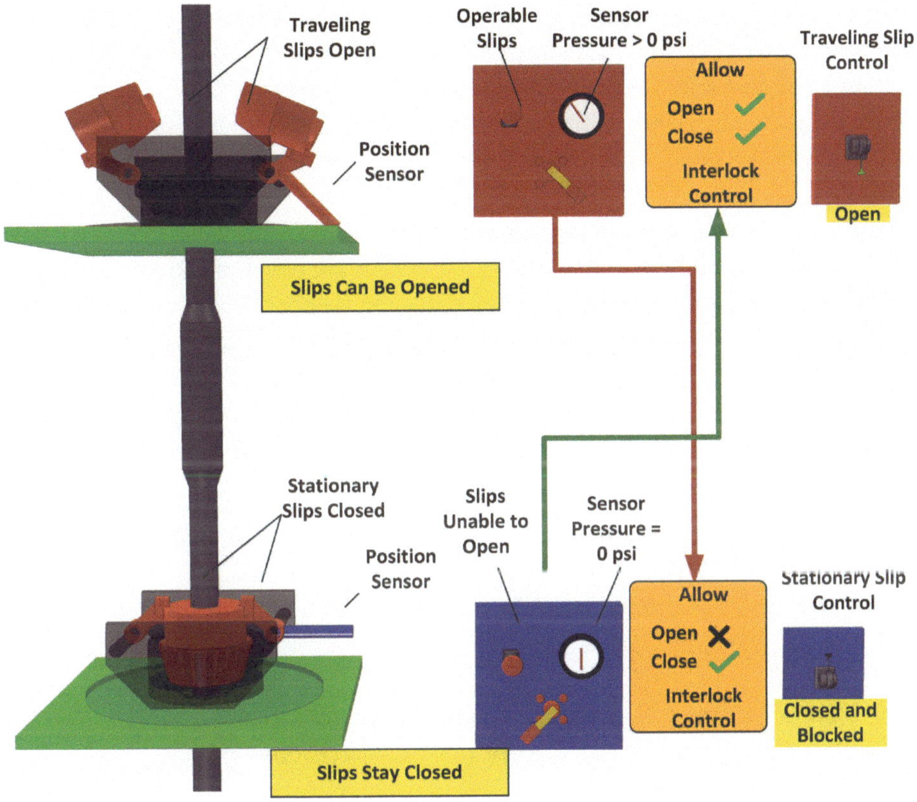

Fig. 4.58 Slip interlock function diagram.

The slip interlock function system provides protection against:

- Operator error
- False bites
- Slip bowl malfunctions
- Slip bowl pressure loss
- Tool joint closure

For standalone units, these snubbing units are normally outfitted with four (4) slip bowls per unit. One pair of snubbing slip bowls and one pair of heavy slips. This means, a pair of slip interlock systems will be needed, as one set is used by the snubbing slip bowl pair (inverted slip bowls) and the other is used by the heavy slip bowl pair.

When sufficient well pressure can "push" the workstring from the well (commonly called "pipe light operations"), an active pair of snubbing slip bowls (traveling and stationary) must be outfitted to ensure one snubbing bowl remains closed on the pipe at all times. When this condition exists, the heavy slip bowls remain in the open position as the slip interlock system has been overridden.

When the workstring has sufficient weight to allow the workstring to fall into the well without having to push the workstring (commonly called pipe heavy operations), the heavy slips (traveling and stationary) are outfitted to ensure one heavy slip bowl remains closed on the pipe at all times. When pipe heavy conditions exist, the snubbing slip bowls remain in the open position as the slip interlock system has been overridden for this condition.

The interlock system does allow manual overrides, allowing the operator to close any or all of the four (4) slip bowls at any time. Overrides are necessary to lower the BHA in the well (at the beginning of the job), at balance point where functional snubbing slips are to be replaced by heavy slips (or vice versa) and end of job when removing BHA from the well.

Barriers

A barrier is any device that prevents the flow of wellbore fluids. Barriers include fluids of sufficient density that prevent formation fluids from entering the well such as brines, seawater, and muds. Barriers are also classified by function and include the following:

a. Closeable Barriers—External and Internal with the ability to open and close, e.g., BOP, safety valves.
b. Mechanical Barriers—Closes off the flow path by sealing the External & Internal.
c. Combination Barriers is a mixture of fluid and mechanical.

Snubbing Primary Barriers: Used **during** normal operations

1. Pressure containment fluid (fluid is well that prevents flow)
2. Stripping Stack (i.e., rams) and BPV: Forms the active barrier while tripping

Snubbing Secondary Barriers: Used **in support** of normal operations or as a contingency (i.e., Safety Ram, additional BOPs safety valve, and plug and nipples)

3. Safety rams (annulus) and TIW valve (inside tubing barrier)

Snubbing Tertiary Barriers: Used in emergency

4. Shear seal
5. Master valve that can cut wireline.

Annular stripping

Annular stripping can be used on wells of low pressure and pipe heavy operations. Since the annular preventer contains an elastomeric (rubber) element and is energized by large piston acting upward to close the annular preventer, the elastomer is deformed and closes around "any type" of tubular in the wellbore. Workstrings with tooljoints can be lowered through the annular preventer by reducing the annular closing pressure low enough to passage of the tooljoint. Once the tooljoint has passed through the annular preventer, the closing pressure is increased to ensure an effective seal around the body of the workstring.

RULE OF THUMB
MAXIMUM PRESSURE FOR STRIPPING PIPE THROUGH AN ANNULAR IS EQUAL TO 40% OF THE MAXIMUM STATIC PRESSURE RATING DRY, *OR* IS EQUAL TO 60% OF THE MAXIMUM STATIC PRESSURE RATING DRY, IF THE PIPE IS LUBRICATED AS IT IS BEING STRIPPED THROUGH THE ANNULAR.

Pressure considerations

Coiled tubing units are normally reserved for well intervention where the tubing remains fixed and the coiled tubing enters the tubing, even when the well is under pressure. Snubbing operations will be confined to Hydraulic Workover Units.

Well control procedures

Normal circulating and noncirculating well control methods can be used in snubbing operations. Constant bottom-hole pressure must be maintained to prevent excessive wellbore pressures and formation breakdown while running pipe into the well. Contin uous manipulation of the choke is required to keep pressures within preplanned calculations. If the workstring is run into a closed well under pressure, buckling, and/or collapse of the workstring can occur.

Snubbing force

Pipe light or pipe heavy conditions can be determined by understanding how force and buoyancy are applied to the workstring. Snubbing forces are the forces necessary to push the workstring into the well. The maximum available snubbing force can be estimated as follows:

150K SNUBBING UNIT HAS APPROXIMATE SNUBBING FORCE OF 75K (1/2 OF THE LIFT)
225K SNUBBING UNIT HAS APPROXIMATE SNUBBING FORCE OF 110K.
PLEASE SEE SPECIFIC MANUFACTURER FOR MOST ACCURATE CAPABILITIES

The maximum snubbing force occurs just as operations begin, as there is not sufficient length of workstring to aid in running the workstring into the well (pipe light conditions). The maximum snubbing force can be determined by the following equation:

$$\textbf{Snubbing Force}_{\textbf{lbs}} = \left\{\textbf{Well Pressure}_{\textbf{psi}} \textbf{ x } (\textbf{DP or DC OD}_{\textbf{in}})^2 \textbf{ x } \textbf{0.7854}\right\} + \textbf{Friction Force}_{\textbf{lb}} \quad (160)$$

where

Snubbing $\text{Force}_{\text{lbs}}$ = Snubbing Force Needed (lbs)
Well $\text{Pressure}_{\text{psi}}$ = Well Pressure (psi)
DP or DC OD_{in} = Drillpipe or Drillcollar Outer Diameter (in.)

0.7854 = conversion constant for area of circle ($\pi/4$)
Friction $Force_{lb}$ = Friction Measured through pressure control elements (lbs)

RULE OF THUMB
FRICTION FORCE ESTIMATION = 20% OF (WELL PRESSURE X (DP OR DC $OD_{in})^2$ X 0.7854)
CONFIRM FRICTION FORCE VIA THE WEIGHT INDICATOR (SLACKOFF WEIGHT VERSES PICKUP WEIGHT)

Example: Assume in this snubbing example that there is 8500 psi of surface pressure and 3000 lbs. of friction necessary to move 2 7/8″ pipe through the BOPs. The formula would be:

$$\textbf{Snubbing Force}_{\textbf{lbs}} = \{8500 \times (2.8752)^2 \times 0.7854\} + 3,000 = 58,183\ \textbf{lbs} \quad (161)$$

This means 58,183 lbs. of snubbing force will have to be overcome in order to push/lower the workstring into the well.

Buckling complications

Buckling occurs when the pipe being snubbed is disfigured, fatigued, and possibly fails during the operation.

Two types of buckling can occur (1) Major axis buckling and (2) Local buckling.

Major axis buckling

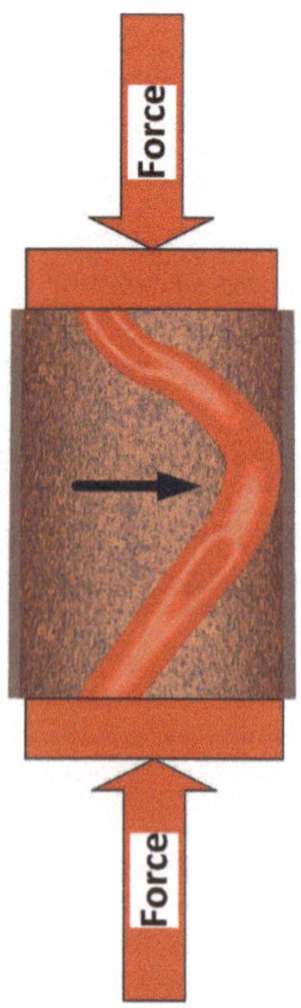

Fig. 4.59 Major axis buckling.

Major axis buckling occurs when the smaller diameter work strings encounter forces from opposite directions cause the workstring tube to bow or flex. If forces are great enough, the protruding tube may become permanently deformed (Fig. 4.59).

Local buckling

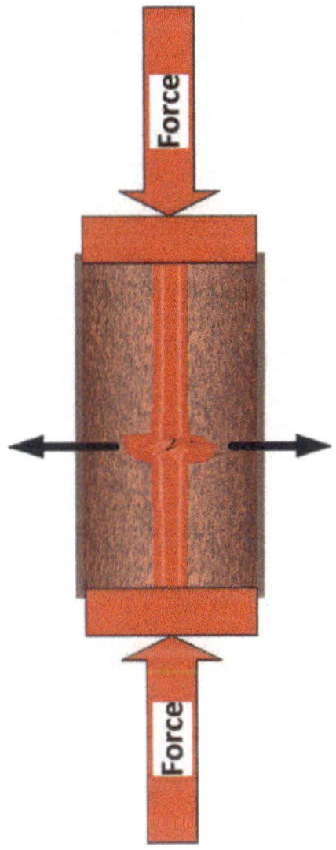

Fig. 4.60 Local buckling.

Local buckling may occur when smaller diameter work strings encounter fast acting or sharp forces from opposite directions that cause deformation over a small area. The tube "balloons" or diameter slightly increases, usually not noticeable to the naked eye. Local buckling more often with larger diameter workstrings (Fig. 4.60).

Snubbing hydraulics

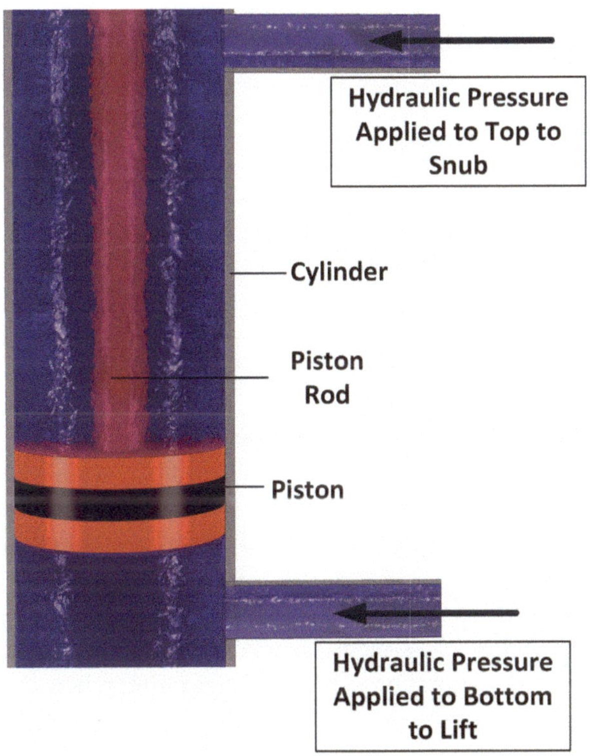

Fig. 4.61 Example of snubbing hydraulics.

The force required to push or pull the workstring out of a pressurized well is provided by the hydraulic cylinders, rams, and power pack (Fig. 4.61). In order to move the workstring, after connection is made via the traveling slips, hydraulic pressure is applied to the top of the hydraulic piston contained within the hydraulic cylinder. This forces the piston downward and snubs the workstring into the pressurized well. Conversely, to lift of pull the workstring, hydraulic pressurized fluid is applied to the lower section of the hydraulic cylinder forcing the piston upward.

Once the snubbing force has been calculated, the hydraulic cylinder operating pressure can be determined in a two-step process.

Step 1: Determine Well Force

Well force is the force acting upward against the largest diameter portion of the workstring to be snubbing into the well.

$$\textbf{Well Force}_{\textbf{lbs}} = \textbf{Largest OD}_{\textbf{in}^2}\ \textbf{x.7854 x Wellhead Pressure}_{\textbf{psi}} \tag{162}$$

where

Well $Force_{lbs}$ = Well Force Acting Upward (lbs)

Largest OD (in.) = Largest Outer Diameter within Workstring (in.)

0.7854 = Conversion Constant (π/4)

Wellhead $Pressure_{psi}$ = Shut-in Well Pressure at Tree (psi)

Step 2: Determine Hydraulic Cylinder Pressure (psi)

Hydraulic cylinder pressure is the operating pressure of the hydraulic cylinder/rod combination necessary to overcome the well force generated by well pressure acting over the cross-sectional area of the largest diameter section of the workstring.

$$\textbf{Hydraulic Cylinder Pressure}_{\textbf{psi}} = \frac{\left(\dfrac{\textbf{Well Force}_{\textbf{lbs}}}{\textbf{Cylinder ID}^2 - \textbf{Piston Rod OD}^2}\right) \textbf{x}\ \ \textbf{.7854}}{\textbf{Number of Cylinders}} \tag{163}$$

where

Hydraulic Cylinder $Pressure_{psi}$ = Hydraulic Pressure to applied to Cylinder (psi)

Well $Force_{lbs}$ = Well Force Acting Upward (lbs)

Cylinder ID = Hydraulic Cylinder Internal Diameter (in.)

Piston Rod OD = Hydraulic Piston Rod Diameter (in.)

0.7854 = Conversion Constant (π/4)

Number of Cylinders = Number of Cylinders (1, 2, or 4)

Friction force is the force is required to overcome the frictional resistance of the elastomers of the blowout preventers under pressure, while the pipe is pushed into the well. The frictional force will depend upon the well pressure, the Drillpipe/Drill Collar size and the roughness of the tube and tooljoint. Because the frictional force always acts to oppose motion, frictional force acts upwards when snubbing into a well. Additional friction forces may result from drag between the workstring and the wall of the cased or open hole as the pipe is pushed into the well.

The snubbing force will change as the workstring is pushed into the well based upon the buoyed weight of the workstring and its physical cross-sectional area. As the workstring is continually pushed into the well, the workstring will approach the balance point, where the buoyed weight of the workstring is sufficient to transition from snubbing force into a stripping force. This transition point can be determined by the following equations:

Determine Buoyancy Factor

$$\mathbf{BF_{dimensionless}} = \frac{\mathbf{65.4 - MW_{ppg}}}{\mathbf{65.4}} \quad (164)$$

where

$BF_{dimensionless}$ = Buoyancy Factor (dimensionless)
65.4 = density of steel (ppg)
MW_{ppg} = Density of Well Fluid (ppg)

STEEL HAS A DENSITY OF 490 LB/FT^3 OR 65.5 LB/GAL.

Determine DC Buoyed Weight

$$\mathbf{DC\ Buoyed\ Weight_{lbs} = DC\ Weight_{lbs/ft} \times DC\ Length_{ft} \times Buoyancy\ Factor} \quad (165)$$

where

$DC\ Buoyed\ Weight_{lbs}$ = Drill Collar Buoyed Weight (lbs)
$DC\ Weight_{lbs/ft}$ = Drill Collar Weight (lbs/ft.)
$DC\ Length_{ft}$ = Drill Collar Length (ft.)
Buoyancy Factor = Buoyancy Factor (dimensionless)

Balance point

Balance point is the point where the weight of the pipe roughly equals the well pressure acting over the cross-sectional area of the workstring. This point is also called the "snub point or transition point."

Balance point while snubbing into dry gas

For determining the balance point while snubbing into dry gas will require two-steps.

Step 1: Determine Well Force

Well force is the force acting upward against the largest diameter portion of the workstring to be snubbing into the well.

$$\mathbf{Well\ Force_{lbs} = Largest\ OD^2 \times .7854 \times Wellhead\ Pressure_{psi}} \quad (166)$$

where

$Well\ Force_{lbs}$ = Well Force Acting Upward (lbs)
$Largest\ OD_{in}$ = Largest Outer Diameter within Workstring (in.)
0.7854 = Conversion Constant ($\pi/4$)

Wellhead Pressure$_{psi}$ = Shut-in Well Pressure at Tree (psi)

Step 2: Determine the Balance Point in Dry Gas

$$\textbf{BalancePoint}_{ft} = \frac{\textbf{Well Force}_{lbs}}{\left(42_{gal/bbl}\ \textbf{x Pipe Capacity}_{bbl/ft}\ \textbf{x Fluid Density}_{ppg}\right) + \textbf{Pipe Weight}_{lbs/ft}} \quad (167)$$

where

Balance Point$_{ft}$ = Balance Point in Dry Gast (ft.)
Well Force$_{lbs}$ = Well Force Acting Upward (lbs)
Pipe Capacity$_{bbl/ft}$ = Pipe Capacity (bbls/ft.)
Fluid Density$_{ppg}$ = Fluid Density (ppg)
Pipe Weight$_{lbs/ft}$ = Pipe Weight in Air Plus Fluid Weight Inside Workstring (lbs/ft.)

PIPE WEIGHT CONSIST OF STEEL WEIGHT IN AIR PLUS FLUID WEIGHT INSIDE PIPE.

Example Snubbing Info:

Tubing Size = 2 3/8″ OD, 4.7 ppf
Capacity = 0.00387 bbls/ft.
Packer Fluid Weight = 11.7 ppg
Shut-in Annulus Pressure = 950 psi. Determine Balance Point in Dry Gas.

$$\text{Well Force} = (2.375) \times (2.375) \times 0.7854 \times 950\,\text{psi}$$
$$= 4208\,\text{lbs.}$$
$$\text{Balance Point in Dry Gas} = 4208 \div [(42 \times 0.00387 \times 11.7) + 4.7]$$
$$= 4208 \div (1.90 + 4.7) = \text{Approx.}638\,\text{ft.}$$

Balance point while snubbing into fluid

For determining balance point while snubbing into a pressurized fluid, a separate two-step calculation is needed. With plugged end environments (tubing stung into packers, etc.), buoyancy factors are not considered.

Step 1: Determine Well Force

Well force is the force acting upward against the largest diameter portion of the workstring to be snubbing into the well.

$$\textbf{Well Force}_{lbs} = \textbf{Largest OD}^2\ \textbf{x .7854 x Wellhead Pressure}_{psi} \quad (168)$$

where

Well Force$_{lbs}$ = Well Force Acting Upward (lbs)
Largest OD$_{in}$ = Largest Outer Diameter within Workstring (in.)
0.7854 = Conversion Constant (π/4)
Wellhead Pressure$_{psi}$ = Shut-in Well Pressure at Tree (psi).

Step 2: Determine the Balance Point in Fluid

$$\mathbf{BP_{ft}} = \frac{\mathbf{Well\ Force_{lbs}}}{\left[(\mathbf{Pipe\ OD\ Cap_{bbl/ft}} \times \mathbf{42_{gal/bbl}} \times \mathbf{Csg\ Fld_{ppg}}) + (\mathbf{Pipe\ ID\ Cap_{bbl/ft}} \times \mathbf{42_{gal/bbl}} \times \mathbf{Tbg\ Fld_{ppg}})\right]} \quad (169)$$

where

Balance $Point_{ft}$ = Balance Point in Fluid (ft.)
Well $Force_{lbs}$ = Well Force Acting Upward (lbs)
Pipe $Capacity_{bbl/ft}$ = Pipe Capacity (bbls/ft)
Fluid $Density_{ppg}$ = Fluid Density (ppg)
Pipe $Weight_{lbs/ft}$ = Pipe Weight in Air Plus Fluid Weight Inside Workstring (lbs/ft)

PIPE WEIGHT CONSIST OF STEEL WEIGHT IN AIR LESS FLUID WEIGHT OF THE O.D. DISPLACEMENT PLUS FLUID WEIGHT INSIDE PIPE

Example balance point in fluid

Tubing Size = 2 3/8" OD, 4.7 ppf
OD Capacity = 0.00549 bbls/ft.
ID Capacity = 0.00387 bbls/ft.
Packer Fluid Weight = 11.7 ppg
Tubing Fluid used = 11.7 ppg
With Shut-in Annulus Pressure = 950 psi
Well Force = 4209 lbs

$$4209 \div (4.7 - (0.00548 \times 42 \times 11.7) + (0.00387 \times 42 \times 11.7)].$$
$$4209 \div (4.7 - 2.69 + 1.90)$$
$$4209 \div 3.91 = 1077\,\text{ft}.$$

The equations used above can be used over a variety of considerations. The following results table demonstrates the variances within the results based upon fluid inside the tubing and run into gas and fluid respectively (Table 4.2).

Table 4.2 Summary table of results.
Table of results

Workstring Run Empty Into Gas	**=**	**895 ft**
Workstring Filled With Fluid and Run Into Gas	=	638 ft
Workstring Run Empty Into Fluid	=	2097 ft
Workstring Filled With Fluid and Run Into Fluid	=	1077 psi

Collapse resistance

As the workstring is snubbed into a pressurized well, the resulting piston force generated by the downward movement of the workstring in a closed system may cause the workstring to collapse. Especially, if the workstring is run dry (no fluid added) into the well. The pressure on the well may have to be maintained in order to prevent packers from unseating or prevent sealing elements from being pumped out of the polished bore receptacle.

Collapse resistance of the workstring will be affected by tension, pressure (internal and external), wall loss (internal and external), and torque.

This example demonstrates the effect tension has on the design collapse of a potential workstring. From these calculations, the collapse resistance falls from 104,000 to 33,000 lbs or at 68% reduction (Fig. 4.62).

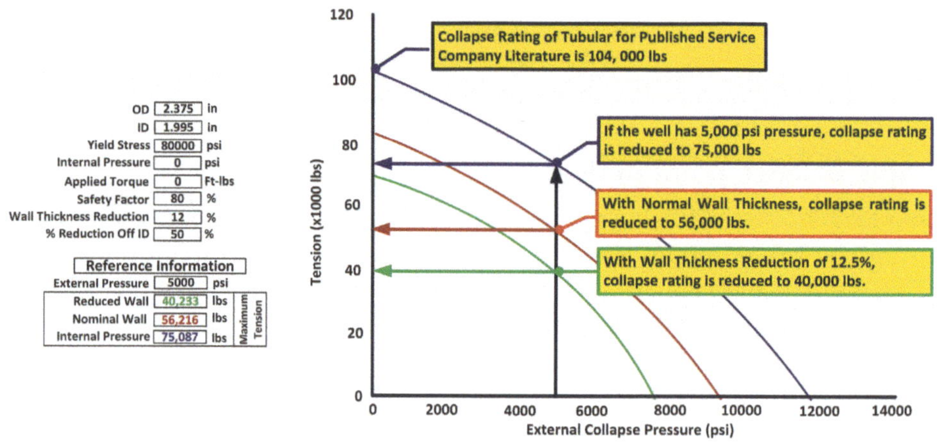

Fig. 4.62 Example of snubbing hydraulics worksheet.

COLLAPSE RESISTANCE OF THE WORKSTRING TENSION, DUE TO PRESSURE (INTERNAL AND EXTERNAL), WALL LOSS (INTERNAL AND EXTERNAL) AND TORQUE MUST BE DETERMINED BEFORE EACH JOB BEGINS.

Managed pressure drilling (MPD) well control method

Managed pressure drilling commentary

Additional specialized services include managed pressure drilling equipment applications. MPD uses a separate circulating system used in conjunction with conventional drilling circulating equipment applications. With the addition of a rotating circulating device, back pressure can be added to the circulating system during connections when the drilling circulation system is shut down. By using constant ECD, the bottom-hole pressure will not deviate and potentially allow formation fluids to enter the wellbore.

With constant ECD, lower mud weights may be used and allow drilling of troublesome formations by minimizing problems. The use of lower mud weights may result in the ability to reach deeper horizons. With deeper penetrations, the need of additional casing strings may be minimized or eliminated.

Managed Pressure Drilling equipment can be used to minimize influx size when a kick is encountered. The use of Coriolis metering can detect kicks as low as 1 bbl. MPD allows for quick reaction for well control, by adjusting circulating back pressure to counter unexpected abnormal pressure. By immediately adding pressure, the kick size may be minimized and allow introduction of KMW quickly and easily through normal circulation. This quick reaction may reduce the need of shutting-in well and circulating using conventional kill techniques.

Managed pressure drilling overview

Managed Pressure Drilling (MPD) is a closed-loop drilling circulation system that minimizes drilling fluid density by imparting and holding a circulating "back pressure" on the well. This is accomplished by installing a rotating circulating head used to perform pressure isolation of the drillstring and diverting annular flow through a manifold system. The manifold system acts to connect a separate fluid pump and remotely actuated hydraulic choke to work in concert to provide and maintain a constant back pressure. The choke operates as a throttling device used to keep back pressures at constant limits, while the separate pump acts to maintain constant back pressure during connections, logging operations or any other operation in which the primary circulating pumps are off line.

The bottom-hole pressure is manipulated by adjusting the choke with no need to vary the mud density or primary pump rate. By minimizing the overbalance, drilling rates will increase, chances for differential sticking are reduced, formation damage is greatly reduced and lost returns may be prevented. By maintaining constant bottom-hole pressures, ballooning effects within specific formations are greatly reduced. Managed pressure drilling allows the ability to extend the depths between casing setting points and may eliminate several casing strings. This system allows drilling within narrow kick tolerance and depleted zones.

In comparison to conventional drilling practices, which are open to the atmosphere, bottom-hole pressure is controlled by manipulating the mud weight and the frictional annular pressure. In conventional drilling, bottom-hole pressure is manipulated by altering the mud density or the pump speed output. For this system to work, a fluid column must extend from the formation to surface. As formations become more complex with lost circulation zones, depleted zones, and narrow kick tolerances, conventional drilling ability to maintain safe operations may become compromised.

Managed Pressure Drilling (MPD) is defined as "*a drilling process used to more precisely control the annular pressure profile throughout the wellbore.*" The objectives of MPD are "*to manage the annular hydraulic pressure profile, avoid drilling problems that contribute to reservoir damage and mitigate problems that might cause loss of the hole.*" The decision to utilize MPD will be based upon applying most appropriate technology to minimize/avoid continuous influx of formation fluids to the surface during drilling activities. Any influx incidental to the MPD drilling operation will be safely contained using an appropriate process.

MANAGED PRESSURE DRILLING IS NOT UNDERBALANCED DRILLING

Variations of managed pressure drilling

Managed Pressure Drilling can be adapted for a variety of well designs over vast geographical environments. Since MPD represents an additive drilling cost, preplanning and selecting the appropriate downhole environment will ensure success in both well control and well costs.

Constant bottom-hole pressure

Constant Bottom-Hole Pressure uses a methodology within MPD whereby bottom-hole pressure is kept constant during connections to compensate the loss of annular friction losses pressure when mud pumps are off.

Pressurized mud cap drilling

Pressurized Mud Cap variation of MPD is determined when drilling operations are executed with no returns to surface. The annulus fluid column, assisted by surface pressure, is maintained above a formation that is capable of accepting fluid and cuttings. A sacrificial fluid with cuttings is accepted by the loss circulation zone. Useful for cases of severe loss circulation that preclude the use of conventional wellbore construction techniques.

PMC uses fluid pumped simultaneously down drillstring and annulus. The annular fluid is referred to as the "mud cap." Cuttings, pump fluids, and any produced fluids are pumped into the formation at the loss zone. This system does not attempt to prevent or cure lost circulation. Used in areas where lost circulation and kicks are prevalent and where it is unsafe to allow formation fluids to flow to surface (i.e., formation fluids with H_2S).

Dual gradient drilling

Dual Gradient Drilling employs the creation of multiple pressure gradients, within select sections of the annulus, to manage the annular pressure profile. Dual gradient drilling includes use of fluids of varying density, dedicated pumps, and subsequent needed combinations.

HSE and performance drilling

HSE and Performance Drilling employs the adoption of MPD tools or processes in wells that can be normally drilled in a conventional way, but employ MPD to overcome concerns for health, safety, environmental, or performance improvement considerations.

Low head drilling

Low Head Drilling is a specialized drilling procedure that employs underbalanced techniques in order to maintain a reduced hydrostatic pressure on the formation. The Low Head Drilling technique used on MPD operations is not intended to create a hydrostatic pressure below the pore pressure, but rather reduce the hydrostatic pressure below the single-phase liquid hydrostatic gradient (i.e., foam drilling).

Subsea mud lift drilling

Subsea Mud Lift Drilling encompasses use of pump on seafloor (or submerged) to pump mud returns to surface and replacing MW with seawater densities (Fig. 4.63). All returns are diverted at the seafloor, then pumped to surface in a separate line for conditioning (Figs. 4.64 and 4.65).

Fig. 4.63 Subsea mud lift drilling.

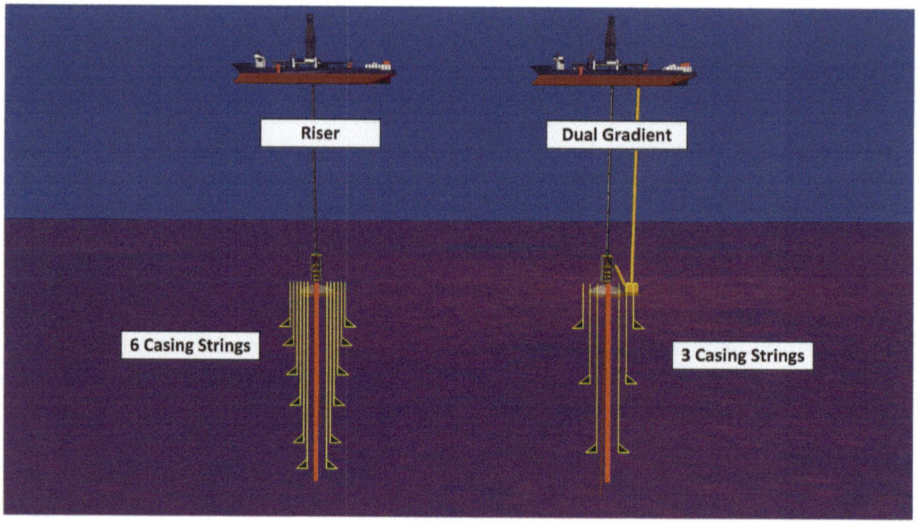

Fig. 4.64 Dual gradient compared to riser drilling.

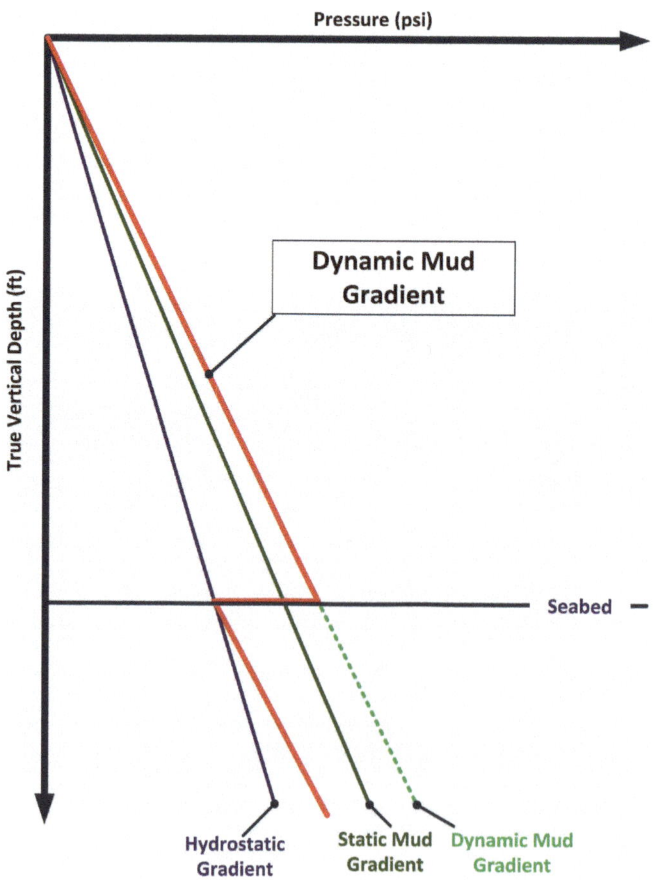

Fig. 4.65 Dynamic mud gradient within dual gradient drilling (subsea mud lift drilling).

MPD IADC classification codes

The International Association of Drilling Contractors (IADC) has established an MPD/UBO Classification System based on three digits, each digit indicating:

1. Risk Level Classification (1 through 5) (Table 4.3)

Table 4.3 MPD risk level classifications.

Risk level classification
1. Well incapable of natural flow to surface. The well is "inherently stable" and is low risk from a well control point of view
2. Well capable of natural flow to surface, but enabling conventional well kill methods and limited consequences in case of catastrophic equipment failure
3. Geothermal and nonhydrocarbon production. Maximum shut-in pressures less than Rotating Control Device (RCD) operating pressure ratings. Catastrophic failure has immediate serious consequences
4. Hydrocarbon production. Maximum shut-in pressures less than RCD operating pressure ratings. Catastrophic failure has immediate serious consequences
5. Maximum projected surface pressures exceed RCD operating pressure ratings, but are below BOP stack ratings. Catastrophic failure has immediate serious consequences

2. Application Category (A, B, or C) (Table 4.4)

Table 4.4 MPD application categories.

Application category
A—Managed Pressure Drilling (MPD)
B—Underbalance Operations (UBO)
C—Mud Cap Drilling (MCD)

3. Fluid System Classification (Level 1 through 5) (Table 4.5)

Table 4.5 MPD fluid system classification.

Fluid system classification

Level	Fluid system	Specific gravity (Unitless)	MWE (ppg)
1	Gas	0.00–0.02	0.0–0.2
2	Mud	0.02–0.07	0.2–0.6
3	Foam	0.07–0.6	0.6–5.0
4	Gasified liquid	0.55–0.9	4.5–7.5
5	Liquid	>0.8	>6.9

When employing an MPD system which uses a full liquid column from TD to the surface, the MASP will depend on the drilling fluid weight, the formation pressure (BHP) and the fracture pressure of the casing/liner shoe. The example below illustrates the surface pressure window when drilling with a fluid which has less hydrostatic pressure than the formation pressure.

Example

A well is being drilled from 9.000′ to 11,000′ utilizing MPD techniques. The formation pressure is 13.0 ppg and the fracture pressure is 16.0 ppg. A 12.0 ppg fluid will drill this interval and maintain a balanced system with surface pressure. The rotating control device (RCD), pipework, and MPD choke is rated to 5000 psi.

Calculate the initial static surface pressure required to balance the formation pressure.

$$\mathbf{Surface\ Pressure_{Inital} = TVD_{Interval} \times 0.052 \times \left(BHP\ MWE_{ppg} - MW\ Hole_{ppg}\right)} \quad (170)$$

where

Surface $\text{Pressure}_{\text{Inital}}$ = Maximum static Surface Pressure to balance BHP (psi)
$\text{TVD}_{\text{Interval}}$ = True Vertical Depth for BHP determination (ft)
0.052 = Conversion Constant
BHP MWE_{ppg} = MWE formation pressure at interval TVD (ppg)
MW Hole_{ppg} Current MW in hole (ppg)

$$\mathbf{Surface\ Pressure_{Inital} = 11,000\,ft \times 0.052 \times (13ppg - 12ppg) = 572\,psi} \quad (171)$$

Calculate the maximum static MASP allowed to prevent fracturing the formation at the casing shoe with the current mud weight.

$$\mathbf{MASP_{Frac} = TVD_{Shoe} \times 0.052 \times \left(ShoeFrac_{ppg} - MW\ Hole_{ppg}\right)} \quad (172)$$

where

$\text{MASP}_{\text{Frac}}$ = Maximum static Surface Pressure to fracture casing shoe (psi)
TVD_{Shoe} = True Vertical Depth of casing shoe (ft)
0.052 = Conversion Constant
$\text{ShoeFrac}_{\text{ppg}}$ = MWE formation fracture pressure at casing shoe TVD (ppg)
MW Hole_{ppg} Current MW in hole (ppg)

$$\mathbf{MASP_{Frac} = 9,000\,ft \times 0.052 \times (16\,ppg - 12\,ppg) = 1872\,psi} \quad (173)$$

The window between the lowest need surface back pressure to prevent a kick and the MASP to prevent fracturing the casing shoe in this example is 572 psi to 1,872 psi. The classification for this well would be:

$$\mathbf{Level\ 5, Category\ A, Fluid\ System\ 5} \quad (174)$$

Barriers

In conventional drilling, annular barriers are independent of each other. The primary barrier is fluid density of a constant fluid column from surface to total depth of well. The secondary barrier is the surface (subsea) pressure containment system or blowout preventers.

In Managed Pressure drilling, since both barriers require pressure containment at the surface, the barriers share common wellbore elements below the BOP rams. The primary barrier is surface control equipment that combines fluid density, friction, and surface pressure. The secondary barrier is the surface (subsea) pressure containment system or blowout preventers. The barrier dependency will also occur in any type of live well operations such as snubbing or production well logging.

Planning MPD

Managed Pressure Drilling begins with accurate downhole prediction modeling of formation fracture gradients and pore pressures. This information is gathered from as many sources as possible, no matter the distance from current well and assuming hydraulic connections between wells. Corrections for pressures may be needed due to uplift or downward location of purposed well formations. Further assistance can be made with computer modeling such as.

Step 1: Develop accurate Fracture Pressure and Pore Pressure Curve.

Step 2: Select Pivot points for downhole environments.

When plotted between the Fracture Pressure Curve and Pore Pressure Curve, the pressure window for Underbalanced Mud (with and without pumps on) forms a cross point above the TD, this means the set pivot point is located their intersection (top) (Fig. 4.66).

NOTE: For a Top Pivot Point, the Fracture Pressure and Pore Pressure windows must diverge with depth.

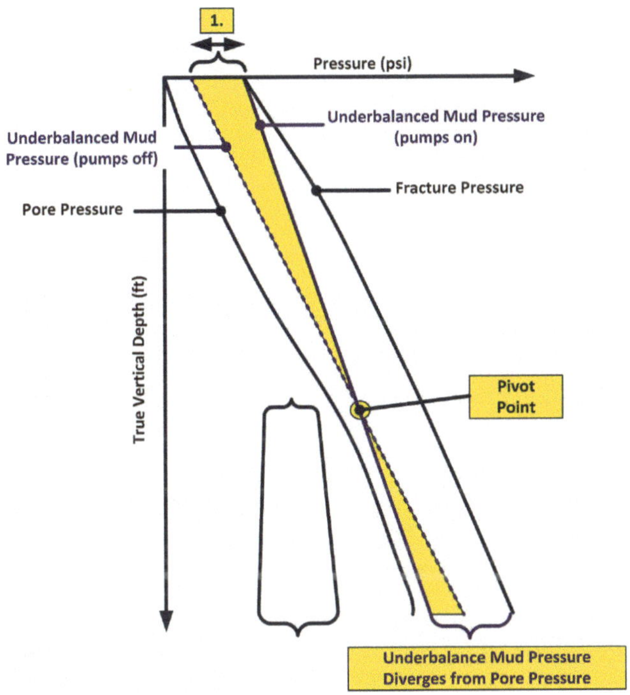

Fig. 4.66 Pressure manipulation at surface to maintain a fixed pressure at pivot point (pressure window diverges).

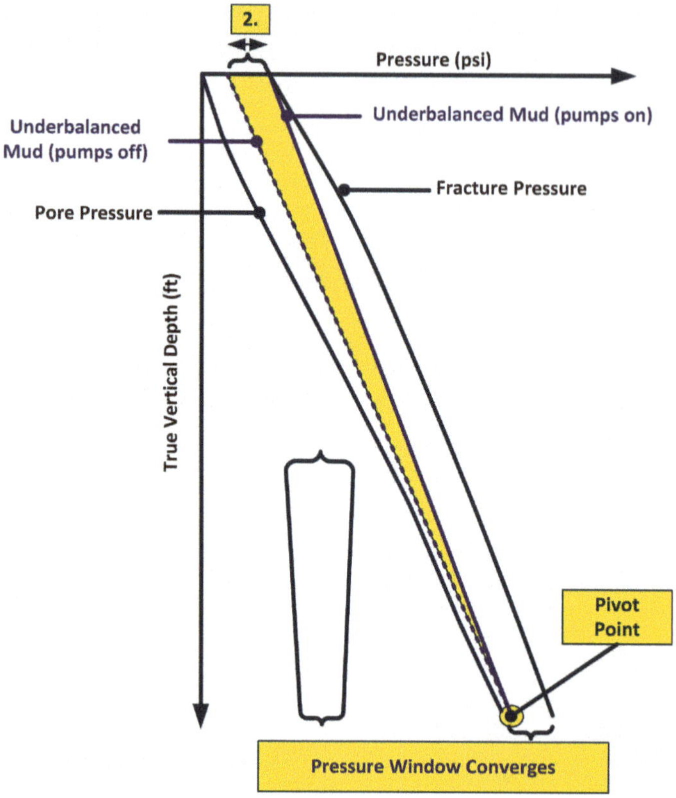

Fig. 4.67 Pressure manipulation at surface to maintain a fixed pressure at pivot point (pressure window converges).

When plotted between the Fracture Pressure Curve and Pore Pressure Curve, the pressure window for Underbalanced Mud (with and without pumps on) forms a cross point at TD, this means the set pivot point at bottom.

NOTE: For a Bottom Pivot Point, the Fracture Pressure and Pore Pressure windows must converge with depth (Fig. 4.67).

Step 3: Determine Surface Pressure Limits.

Normally, there is a set point for protecting surface equipment (normally RCD limit). The RCD has two limits, dynamic and static.

- The surface pressure limit is normally set to the lower dynamic limit.
- In the event of an influx, if the dynamic limit is exceeded, the pipe movement can be stopped.
- Beware for automatic chokes, the system cannot recognize this and the choke will begin to open when the preset limit is reached.
  - **a)** Planned Drilling back pressure (keep as small as possible, 0 psi/0 ppg EMW).
  - **b)** Planned Connection back pressure: PSI or MWE of anticipated Equivalent Circulating Density (psi/MWE).
  - **c)** Back-pressure Limit: PSI or MWE of Maximum Operating Limit of Rotating Control Device.

Step 4: Verify if Surface Pressure (pumps on and off) are within limits.

$$\mathbf{BHP_{MPD} = HP_{Mud} + \Delta P_{Friction} + \Delta P_{Surf\ Pump}} \quad (175)$$

where

BHP_{MPD} = MPD Bottom-Hole Pressure (psi)

SP_{psi} = Surface Pressure (psi)

HP_{Mud} = Hydrostatic Pressure of Mud (psi)

$\Delta P_{Friction}$ = Friction Pressure due to Circulation (psi)

$\Delta P_{Surf\ Pump}$ = Change in Pressure due to Surface Secondary Pump (psi)

Example: 10,000′ TVD hole with 9.5 ppg MW. DP friction = 1800 psi, Annular Friction = 500 psi. SP pumps on = 400 psi, SP pumps off = 900 psi.

Determine BHP with pumps on:

$$\mathbf{BHP_{MPD} = (0.052)(9.5\,ppg)(10,000\,ft) + 500\,psi + 400\,psi = 5,840\,psi} \quad (176)$$

Determine BHP with pumps off (Fig. 4.68):

$$\mathbf{BHP_{MPD} = (0.052)(9.5\,ppg)(10,000\,ft) + 900\,psi = 5,840\,psi} \quad (177)$$

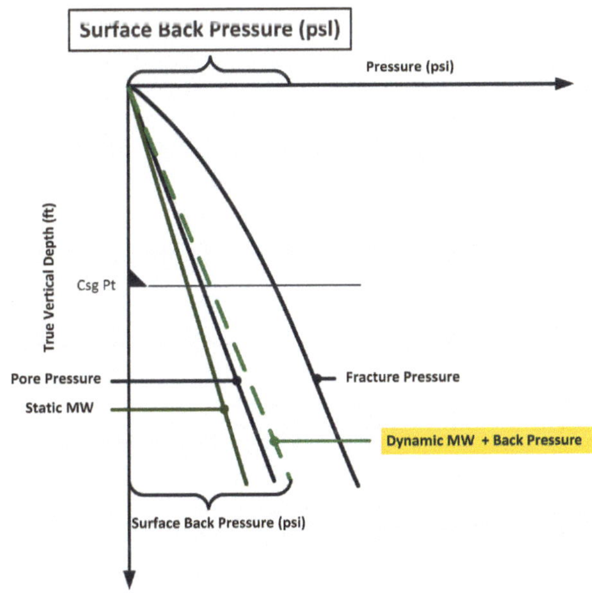

Fig. 4.68 Graphical representation of surface back-pressure window.

Step 5: Verify Injection Pressure Constraints.

Step 6: Revisit as needed.

Connections

The following is an example of how surface secondary pumps maintain back pressure while primary pumps are shut off in order to make connections. The goal is to maintain the ECD at all times during active drilling mode. The process is the same for any operations where the primary pumps are shut down such as:

- Logging
- BOP
- NPT
- R/U for cementing, etc. (Fig. 4.69).

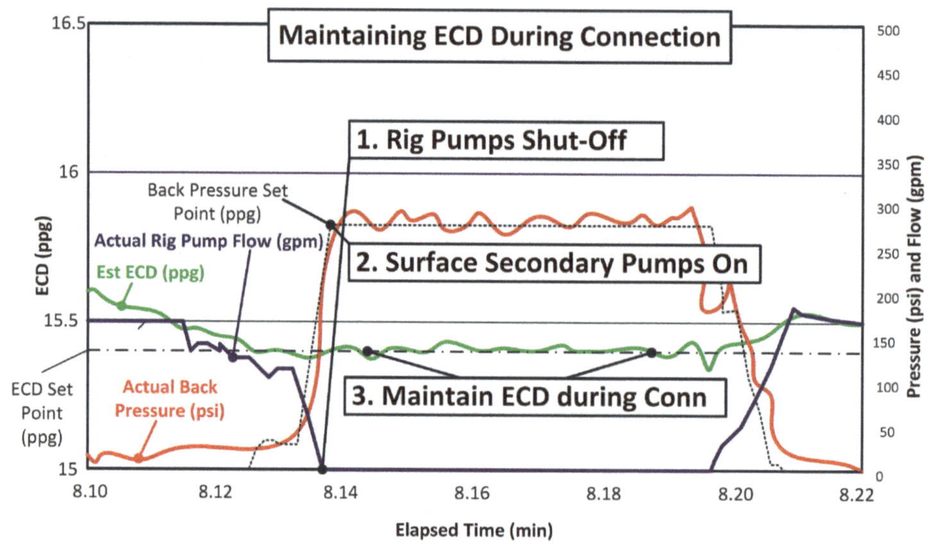

Fig. 4.69 Graphical example of maintaining ECD during connection.

General connection procedure

1) Driller alerts the MPD operator of planned connection operation.
2) The MPD operator initiates the back-pressure pump. The MPD controlling software system will automatically adjust the MPD choke to maintain a constant BHP.
3) The Driller "slowly" brings the rig mud pumps to stop. The MPD controller will automatically adjust the choke to maintain constant BHP by increasing the surface pressure to compensate for the loss of circulating friction.

RIG PUMPS ARE TO BE STARTED AND STOPPED SLOWLY TO ENSURE MPD CONTROLLER CAN AFFECT PROPER CHOKE ADUSTMENTS

4) The Standpipe pressure is bled to zero.
5) The connection is made.
6) The Driller slowly brings the rig mud pump on line to full circulating rate. The MPD controller will automatically adjust the choke to keep BHP constant, decreasing surface pressure as friction pressure increases.
7) Once full circulating rate and pressures are achieved, the surface back-up pump is stopped and the MPD controller will automatically adjust the choke to keep the BHP constant.

Connection schematics

The following equipment schematic demonstrates a conventional drilling flow return system. Starting from the bell nipple, flow stream is directed down the flowline through a header box and into the shale shaker. Once the fluid flows through the solids control system and is treated, the mud returns to the mud pump intake. The mud pumps will pump fluids to the standpipe, down the workstring. Once the fluid exits the workstring and travels up the annulus to the bell nipple, the process begins again (Fig. 4.70).

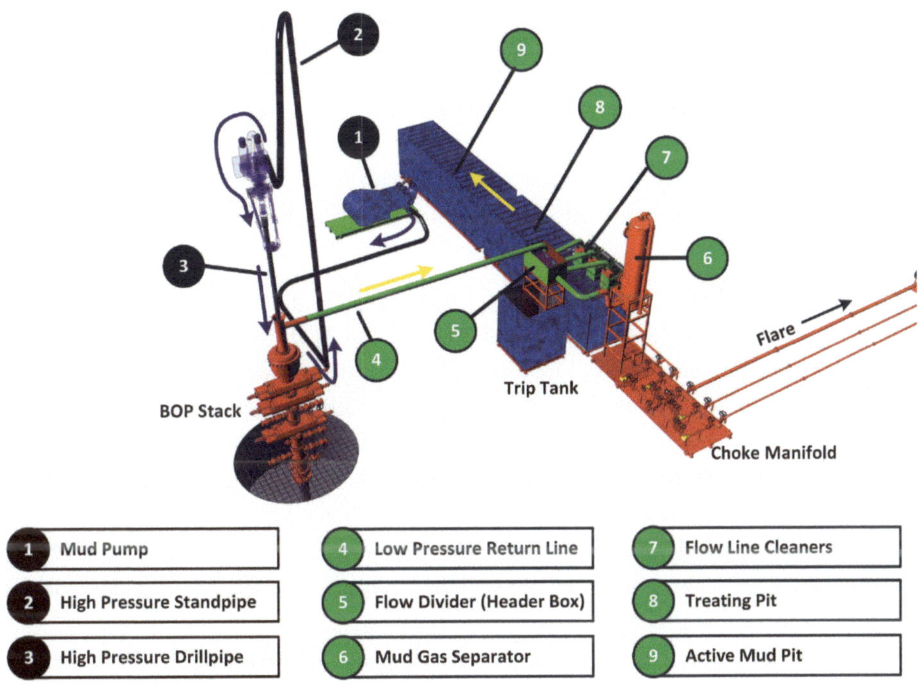

Fig. 4.70 Conventional drilling equipment components.

To begin MPD operations, the Rotating Control Device (RCD), automated choke manifold system, Coriolis flowmeter, Mud/Gas Separator, Flare/vent stack, Surface back-pressure pump, Controller sensors and equipment are installed. This diagram illustrates the beginning of a connection, the back-pressure pump adds mud flow while rig pumps are turned off (Fig. 4.71).

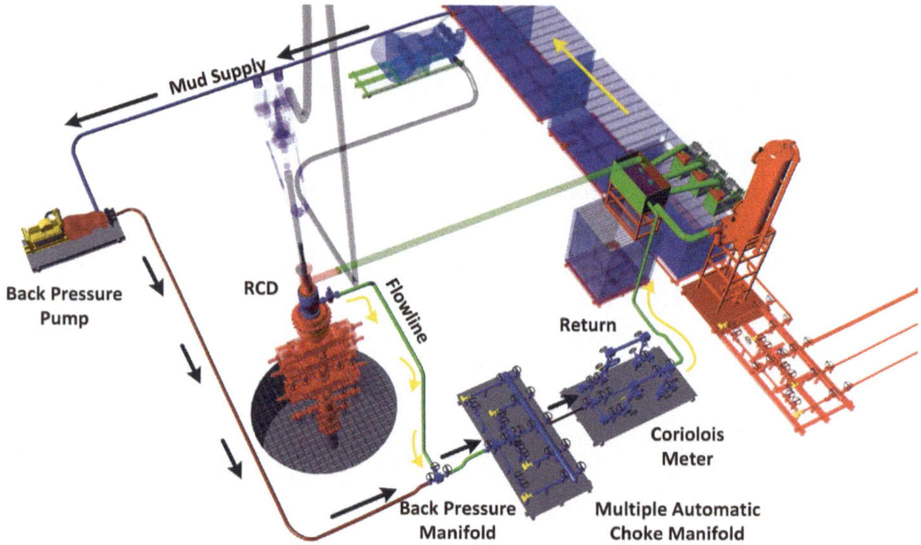

Fig. 4.71 MPD circulating system.

During connections rig pumps are off. The back-pressure pump provides the mud flow through the manifold and into the annulus through the RCD.

Tripping kill strategy

The trip kill strategy will depend on the characteristics of the well formations, MW, and specifications of equipment used. The kill strategy consists of determining if the well is to be killed to conventional overbalanced fluid or not to kill the well and use currently light mud weights. The main reason for drilling with MPD is to ensure the very minimum mud weight is used to prevent formation damage and/or losses. Therefore, the appropriate trip kill strategy is needed.

Considerations for killing the well include determining if objectives of MPD will not be adversely affected by introducing overbalanced fluids to well. For this determination, trip margin fluids must also be considered.

Considerations for not killing the well for trip operations will be dependent upon associated problems that may arise from using overbalanced fluids. For narrow pressure windows, the formations may not withstand higher density fluids, or overbalanced fluids with increase for trip margin.

MPD well control methods

Well control in MPD is more complex when compared to conventional drilling operations. MPD well control begins with a fundamental understanding of the downhole environments to be drilled and accurately determine bottom-hole pressures. By determining accurate bottom-hole pressures and the selection of fluid densities required to drill individual formations, the surface back pressure can be determined.

For conventional drilling operations, the density of the drilling fluid is normally increased and circulated throughout well while simultaneously allowing the influx to expand under controlled conditions to keep bottom-hole pressure above formation pressure (and below fracture pressure).

Dynamic kill

For MPD Well Control, Dynamic Well Control is used. The density of the drilling fluid is not altered, only the back pressure is raised to maintain well control. For dynamic well control to be effective, the kick must be contained as small as possible. Fortunately, MPD allows for quick detection of kicks. Through kick detection, kick volumes are minimized and generates safer well control operations. Using constantly adjusted back pressure, dynamic kill operations can begin immediately allowing faster solutions for well influxes and return to drilling operations quicker when compared to conventional well kill operations.

DYNAMIC WELL CONTROL CAN ONLY BE ACCOMPLISHED ON SMALL SIZED KICKS (< 10 BBLS)

Dynamic Well Control begins by understanding the applied back-pressures system limits for drilling each hole section.

1. Establish Dynamic Limit. This is the operating window between Formation Fracture pressure and Formation Pressure where Hydrostatic Fluid Pressure and additive back pressure is optimized. The difference between the selected HP_{mud} + back pressure and Fracture Pressure will be the Dynamic Limit.
   a. If the kick is outside the Dynamic Limit, more than likely, a conversion to conventional well control will have to occur.
2. Establish appropriate Shut-In and Secure Well Procedure. The Drillpipe float makes it possible to directly read SIDPP.
   a. Pressure up pipe in small increments.
   b. Monitor Surface Pump Pressure (SPP) and Casing Pressure (CP) for indication of the drillpipe float opening.
   c. The opening will show as a small pressure increase in CP.
   d. Once valve is opened, shut down pumps.
   e. Surface Pump Pressure (SPP) after shut down should be equal to SIDPP.

3. Calculate BHP from SIDPP.
4. Determine if Surface Pressure will exceed Surface Limit.
5. Determine if well will be killed via MPD or by Conventional Well Control.
   a. Targeting narrow pressure gradient windows can amplify subsurface thermal effects.
   b. Barite sag may further impact drilling window.
   c. Influx indicators establish maximum limits to be able to perform Dynamic Well Control in MPD. Three possible limits exist (use the lowest).
      i. Ability to reach mass balance before significant gas expansion occurs (usually less the 10 bbls).
      ii. Risk of exceeding fracture pressure at the shoe or other weak point (usually between 8 and 50 bbls).
      iii. Risk of exceeding pressure or flow limits of surface equipment (usually between 25 and 70 bbls).
   d. Decision to be based on case-by-case specifics as one solution does not "fit" every MPD operation.

NO ESTABLISHED INDUSTRY OR GOVERNMENTAL PROCEDURE EXISTS FOR SETTING LIMITS

Using the MPD Controller, the following screen prints demonstrate how a kick is recognized and appropriate actions are taken (Figs. 4.72–4.74).

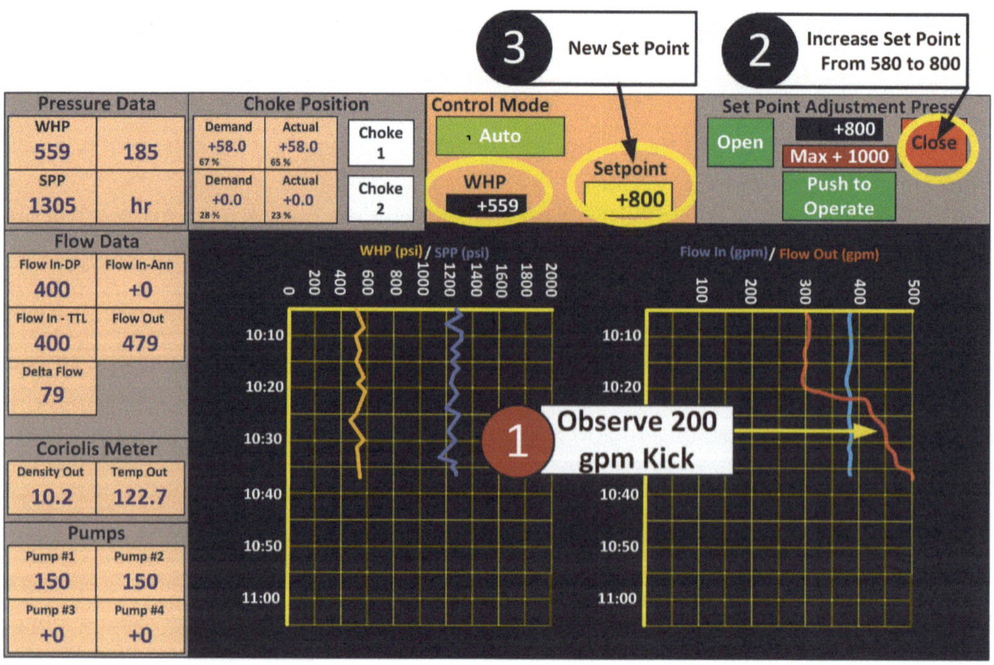

Fig. 4.72 Kick identified (200 gpm), dial in new set point from 550 to 800 psi (not WHP is 559 psi).

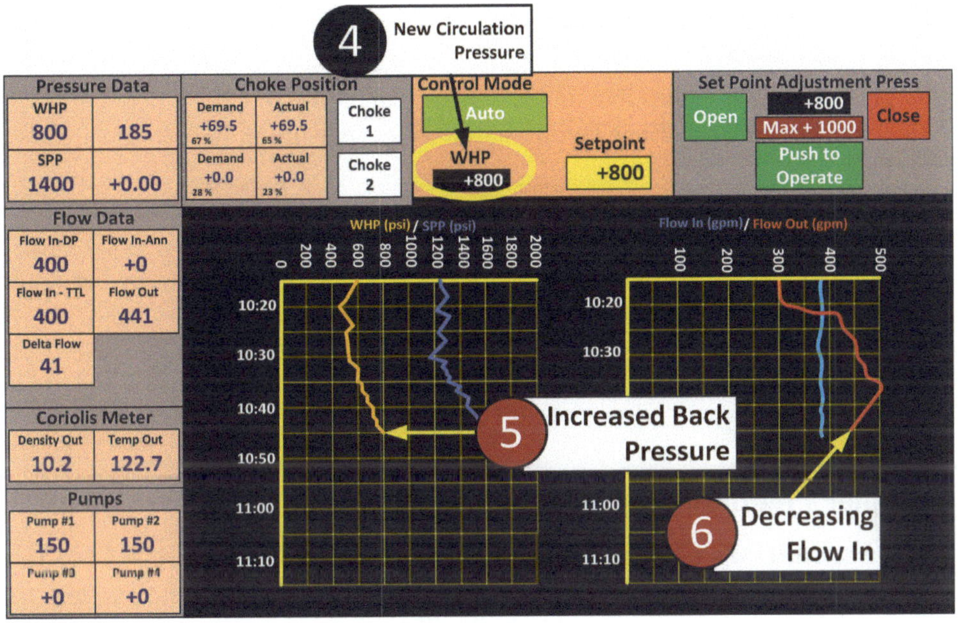

Fig. 4.73 With 800 psi, automatic choke partially closes and system pressure builds to 800 psi. Note pressure dropping.

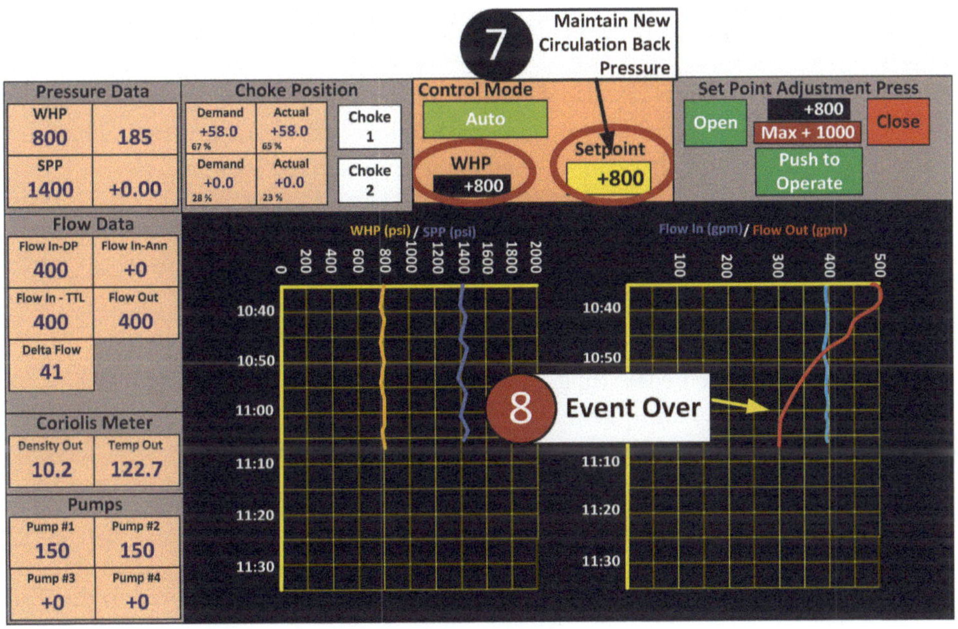

Fig. 4.74 With 800 psi WHP, new back pressure is 800 and event is over.

Limits for dynamic well control

Influx characteristics define maximum limits in which Dynamic Well Control for MPD can be performed. Three possible limits exist (use the lowest):

1. Ability to reach mass balance before significant gas expansion occurs (usually less the 10 bbls).
2. Risk of exceeding fracture pressure at the shoe or other weak point (usually between 8 and 50 bbls).
3. Risk of exceeding pressure or flow limits of surface equipment (usually between 25 and 70 bbls).

Decision to be based on case-by-case specifics as one solution does not "fit" every MPD operation. There are no established procedures to set limits.

Driller's method

During MPD operations with kicks bigger than 10 bbls, the Driller's Method is often used as the primary well control method as the fluid density is not altered. Only the first circulation of the Driller's Method will be necessary in order to remove the influx.

First Circulation of Driller's Method: Bring pumps up to speed (kill rate) while maintain CP constant. Once at speed, read, and record Initial Circulating Pressure (ICP). Maintain ICP until influx has been circulated out and MW in = MW out (Suggest 1–1/2″ hole volumes). **Once the influx has been circulated out, the ICP is now the new back-pressure limit for MPD operations.**

Once the influx has been circulated out and the back pressure is determined to be too high, the mud density may be increased in order to lower the back pressure. The following is a simple calculation that can be used for a revised mud weight.

$$\mathbf{MW_{Revised}} = \left(\frac{\mathbf{ICP_{psi} - Target\ Pressure_{psi}}}{\mathbf{(0.052)\,x\,TVD_{ft}}} \right) + \mathbf{MW_{current}} \tag{178}$$

where

$MW_{Revised}$ = Revised MW (ppg)

ICP_{psi} = Initial Circulating Pressure at Kill Rate (psi)

Target $Pressure_{psi}$ = Target Pressure (psi)

(0.052) = conversion factor

TVD_{ft} = True Vertical Depth (ft.)

$MW_{current}$ = Current Mud Weight is well (ppg)

With the revised MW circulated in place, the back pressure is lowered to the Target Pressure.

Well control matrix

As mandated by governmental agencies, a Well Control Matrix be necessary to safely predetermine how Dynamic Well Control can be used. Dynamic Well Control is the process of increasing the BHP and removing the influx without shutting in the well. This consists of at least a two-step process where the parameters of Kick Characteristics are compared to Operating Pressure characteristics. If outside Dynamic well control constraints, operations become similar to conventional well control (Table 4.6).

Table 4.6 Well control matrix.

Well Control Matrix		Surface Pressure Indicator (See Step 2 Below)			
		At Planned Drilling Back Pressure	At Planned Connection Back Pressure	Planned Back Pressure > Range < Back-Pressure Limit	≥ Back Pressure-Limit
Influx Indicator (See Step 1 Below)	No Influx No Influx	Continue Drilling	Continue Drilling	Increase pump rate, mud weight or both AND reduce surface pressure to planned or contingency levels	Pick up, shut-in and evaluate next action
	Operating Limit	Increase back pressure, pump rate, mud weight or a combination of all	Increase back pressure, pump rate, mud weight or a combination of all	Increase pump rate, mud weight or both AND reduce surface pressure to planned or contingency levels	Pick up, shut-in and evaluate next action
	< Planned Limit	Cease Drilling, Increase back pressure, pump rate, mud weight or combination of all	Cease Drilling, Increase back pressure, pump rate, mud weight or combination of all	Pick up, shut-in and evaluate next action	Pick up, shut-in and evaluate next action
	≥ Planned Limit	Pick up, shut-in and evaluate next action	Pick up, shut-in and evaluate next action	Pick up, shut-in and evaluate next action	Pick up, shut-in and evaluate next action

Step 1: Determine Influx Indicators.

Influx Indicators include rate of kick, duration of kick and volume gains. Of these, only the volume gained is an absolute indicator of the kick. The rate of the kick must be determined by quality equipment with tolerances for accurately determining tolerance.

1) Influx Rate
   - **a)** At the Operating Limit (Low) of <0.1 bbl/min
   - **b)** Less than Planned Limit (Medium) of <1.0 bbl/min

c) Greater than or Equal to Planned Limit (High) of >1.0 bbl/min
2) Volume Duration
a) At the Operating Limit (Low) of <1 min, light influx
b) Less than Planned Limit (Medium) of 1 min > Range < 10 min,
c) Greater than or Equal to Planned Limit (High) > 10 min
3) Volume Gain
a) At the Operating Limit (Low) of <0.5 bbls
b) Less than Planned Limit (Medium) < 1.0 bbls
c) Greater than or Equal to Planned Limit (High) > 1 bbl.

Step 2: Determine Surface Pressure Indicators. Include alarms and leaks that are physical warning alarms or observations specify problems such as (1) ambient hydrocarbon gas detected, (2) Hydrocarbon gas and fluid leak detected, (3) Drilling fluid leak detected, and (4) RCD seal leak.

The following represents an example of Surface Pressure Indicators. Actual information to be provided by case-to-case limits as defined by FP, PP, MW, and back pressure

1) Planned Drilling back pressure
a) 0 psi or 0 ppg Mud Weight
2) Planned Connection back pressure
a) 382 PSI or 0.7 MWE of anticipated Equivalent Circulating Density (psi/MWE)
3) Back-pressure Limit
a) 1092 PSI or 2.0 MWE of Maximum Operating Limit of Rotating Control Device

Measuring equipment to be used must be extremely accurate in order to measure surface pressures to an acceptable tolerance.

Operational considerations

Automated Systems: Caution should be taken when MPD system is in the fully automated mode. Operators must fully understand how the system will react to outside data and limitations of the models.

Safety Systems: Safety systems setup, testing, and operation should be understood by all operators. Limits and alarms should be properly set and/or calibrated.

Set Point for Casing Shoe: One operational problem area for automatic chokes will be casing shoe fracture limit. Casing shoe fracture information (Leak-Off/FIT pressure/EMW including casing size, weight, grade, 80% burst and setting depth), is used as a "set point" for the shoe. Remember, the Automated MPD System will automatically open the choke to ensure well pressures do not exceed casing shoe fracture pressures. Once the influx is above the shoe and the influx is gas, the gas must be allowed to expand casing CP to rise. The automatic feature will open the choke, causing the influx HP to decrease, allowing BHP to be reduced. If BHP < PP, secondary kicks will occur. During well control operations, manual control should be engaged while the automatic choke should be disabled and not automatically open.

Set Point for Surface Equipment: The Rotational Control Device (RCD) operating limit is normally the set point for protecting all surface equipment. The operating limit is set with lower dynamic limit (not static limit). By using the lower limit, during well control and if the dynamic limit is reached, pipe movement can be immediately stopped. If the automatic choke feature is engaged, when the lower dynamic limit is reached, the choke will begin to open decreasing BHP. If BHP < PP, secondary kicks will occur.

Back-Pressure Control: Alternating between standpipe pressures and back pressure must be understood. When RSS or motor system are used, pressure drops will ensure when these systems are moved off bottom. The automated MPD system will compensate for the loss in standpipe pressure by increasing back pressure. Since the pressure drops are contained within the tool, the result of additive back pressure will be an artificial increase in BHP.

Stroke Counter: If the stroke counter were to fail, the automatic control system would respond by identifying a "kick situation" (mass out greater than mass in) and will begin initiating an appropriate kick response.

Mass Flow Coriolis Meter. These meters are used to accurately measure flow from the well. The mass out is converted to volume. The basis for dynamic well control is that flow in equals flow out. The basic mass flow equation is shown below (Fig. 4.75):

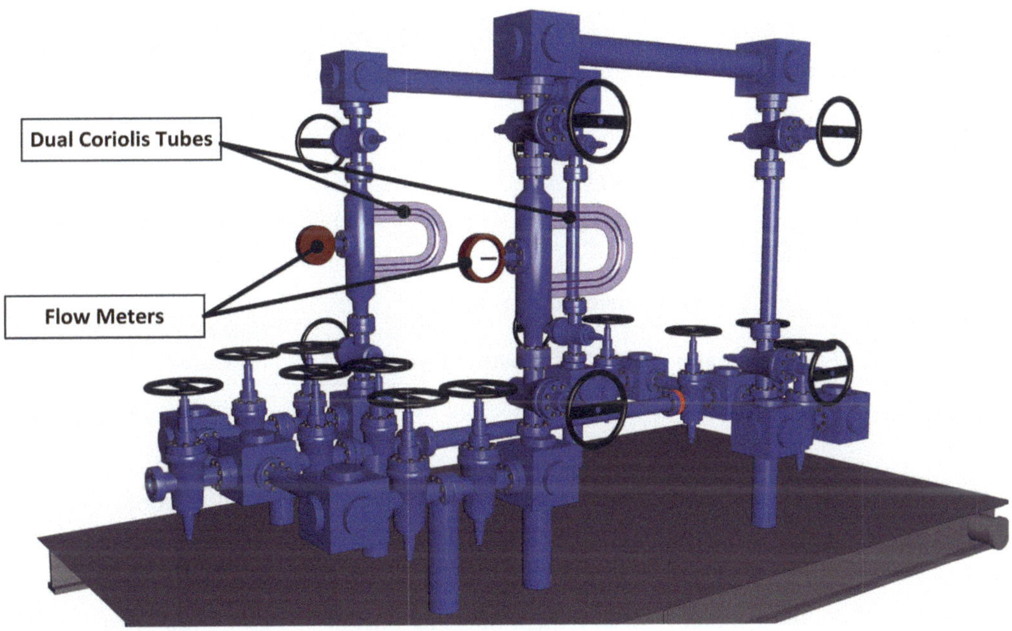

Fig. 4.75 Coriolis flow meter skid (dual flow meters).

Mass Flow Metering: Mass Flow Coriolis meters may suffer from interference caused by extreme vibration. Mass Flow Coriolis meters should be manifolded to ensure back pressure can be added behind the meter to help mitigate vibrations.

Vibrations may be induced from:

- Cavitation from closed chokes
- Pumps with damaged pulsation dampeners
- Deck vibrations
- Mechanical means
- Hydraulic means

$$\mathbf{Q_m} = \left(\frac{\mathbf{K_\mu - I_\mu \omega^2}}{\mathbf{2Kd^2}}\right) \mathbf{x}\, \boldsymbol{\tau} \tag{179}$$

where

Q_m = Mass Flow
K_μ = Shape-dependent factor
I_μ = Inertia of the tube
ω = Vibration frequency
K = Shape dependent factor
d = Width
τ = Time lag

Modeling: All MPD modeling must be updated with most up-to-date MW in use, including cutting loading and temperature effects. The model should be accurate across the entire range of pump rates including complex shut off and isolation.

Risk analysis

A risk analysis must be performed and risk reducing measures applied to reduce risk as low as reasonably practical. The following is an example for suggested equipment standards for risk analysis (Table 4.7).

Table 4.7 Suggested equipment standards for risk analysis.

Element name	Failure scenario	Probability reducing measures	Consequence reducing measures
Wellhead	Leak in flanged connection below BOP	All bolts, studs, and nuts in Wellhead and stack shall be torqued to manufacturer specifications. Inspection for internal wear during installation and removal. Minimize number of flanges used	Monitoring for leaks with gas detection installed on specific connections Kill fluid available for immediate use
Casing	Wearing a hole in the casing during drilling	Measurement of casing thickness prior to and during operations. Magnet in the flowline to measure metal. Conduct wear estimates during operations	Continuously monitor the "B" Annulus pressure Kill fluid available for immediate use

Table 4.7 Suggested equipment standards for risk analysis.—cont'd

Element name	Failure scenario	Probability reducing measures	Consequence reducing measures
Casing Cement	Leak through cement and up the annulus	Pressure testing of cement to formation leak-off/formation integrity pressure. Assessment of cement bonding	Continuously monitor the "B" Annulus pressure Kill fluid available for immediate use
Surface well control equipment	Leak in surface well control equipment	High quality certified equipment, periodic testing, preventative maintenance plan and solutions. Only manufacturers original equipment will be acceptable as replacement parts for use in safety critical equipment	Have competent personnel involved in the operation Kill fluid available for immediate use

Managed pressure drilling equipment

Managed Pressure Drilling equipment complement will vary depending on pressure regimes, well construction techniques, mud systems used and location. Most MPD systems consist of the Rotating Control Device (RCD), automated choke manifold system, Coriolis flowmeter, Mud/Gas Separator, Flare/vent stack, Surface back-pressure pump and an operational "Controller" unit with computerized panel, touchscreen monitor and read out near Driller's console. For floater applications, more equipment will be installed including kick detection systems and MPD Riser joint (consisting of RCD, Riser Annular Preventer, and Flow Spool) (Fig. 4.76).

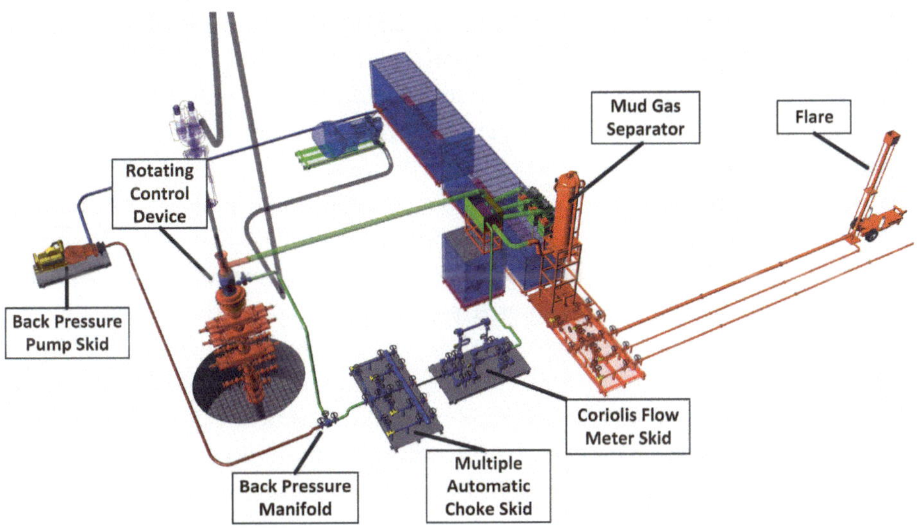

Fig. 4.76 MPD equipment groups.

Rotating control devices (RCD)

Rotating Control Devices (RCD), sometimes referred to as rotating heads, are drill through device with a rotating seal that allows rotation of the drilling string and "containment of pressure" by use of seals or packers that seal against the drill string (Fig. 4.77). These are devices used to form a close-loop circulating system diverting flow to chokes. This device is **not recognized as a barrier** (such as a BOP) as its function is diverting flow.

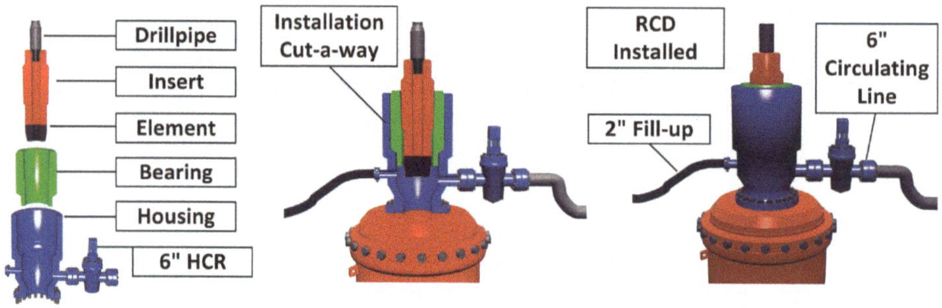

Fig. 4.77 Rotating control device.

ROTATING CONTROL DEVICES ARE NOT HIGH-PRESSURE BLOWOUT PREVENTERS
ONLY NOV PCWD IS AN API CERTIFIED ANNULAR PREVENTER

Installations will vary depending upon application. The illustrations above have been developed for a drilling application. During drilling operations, while tripping in or out of the well, the RCD provides a continuous seal against the drillpipe, tooljoints, and drill collars. The elastomeric element provides a seal against the tubular and the housing. The bearing is designed to provide an effective seal during tubular rotation. Mud, dust, air, gas, steam, or hydrocarbon mists are diverted through the side outlet and into a choke manifold ensuring safety of rig crews on the drill floor.

Each manufacturer has individual performance designs that will allow effective pressure sealing from 500 to 2500 psi applications in sizes up to 20″ diameters. Regardless of design, RCDs are considered to be a secondary sealing equipment and **should not be used as standard drilling well control equipment**. Depending on manufacturer, the RCD should be reviewed to ensure they can contain fluid above the uppermost top seals.

The service life of the RCD will depend upon fluid compatibility, lubrication, condition of drill string components (i.e., hard banding, tong marks), alignment of the drill string through RCD and differential pressures employed. To maximize service life, the

RCD and elastomers should be matched for entire well service conditions such as pressure, temperature, and hydrocarbon compatibility.

RCD specifications

- Static Pressure Ranges: From 3000 to 5000 psi.
- Operating pressures reduce when rotation is employed. The faster the rotation, the lower the operating pressure will be
  - Rotating at 100 RPM, Ranges from maximum of 2500–3000 psi.
  - Rotating at 200 RPM, Ranges from maximum of 1500–2000 psi.
- Height Ranges from 40″ to 69.5″.
- See Manufacturer specifications for more information.

Automated flow (choke) manifold

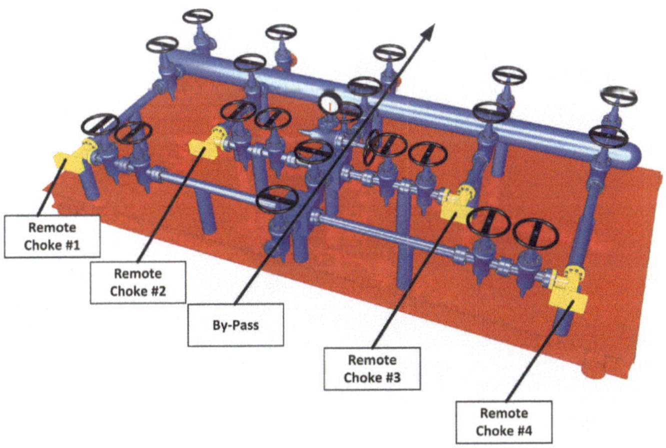

Fig. 4.78 Automated flow (choke) manifold.

Since MPD operations are dependent upon maintaining an accurate back pressure and flow control, all systems use an automated choke manifold system (Fig. 4.78). These systems will integrate quick-acting remotely operated flow valves (chokes), manifolds, valves, and pressure relief valves within their unit skids. For efficient operations, a minimum of two flow valves (chokes) are needed to provide redundancy and allows for isolation of repairs due to plugging and erosion. With the primary flow entering the manifold from the RCD line, a secondary flow line attached to the rig choke manifold is desirable for redundancy. Larger flow rates with less pressure drop and larger cutting sizes, it is desirable to use larger chokes, valves, and pressure lines of 3–1/16″ diameter or greater.

Mass coriolis flowmeter

As MPD operations require accurate pressure and flow information in order to maintain needed back pressure, the flowmeter device will be installed downstream of the automated flow manifold (Fig. 4.79). Most flowmeters used in MPD operations include a Coriolis tube flowmeter. In a Coriolis mass flowmeter, the tube's harmonic defections (swinging) are generated by vibrations within the tube as fluid flows through it. The amount of total deflection (twist) is proportional to the mass flow rate of fluid passing through the tubes. Sensors within the flowmeter measure this twist and convert this into a digital linear scale that is reproduced at the controller's station.

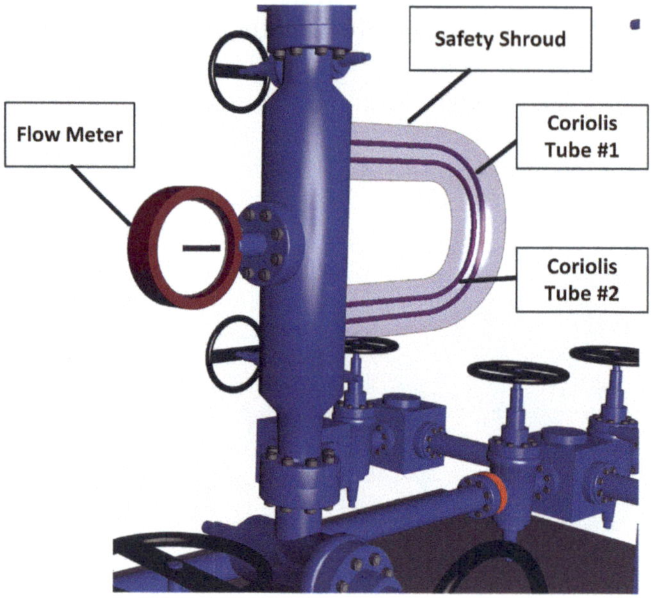

Fig. 4.79 Mass coriolis flowmeter components.

Mud gas separator

Located downstream of the Mass Flowmeter will be the MGS (Fig. 4.80). The MGS may be provided by the MPD service contractor or they may elect to utilize the rig MGS. As the name suggests, the MGS employs a simple technology by using rig mud level at the bottom of the tank to generate a small pressure within the system. Gaseous fluid is pumped into the top of the vessel and is distributed in a cascading effect onto baffle plates. Baffle plate serve an important feature by allowing the fluid to spread into thin sheets, lowering the surface tension of the fluid and allowing the gas to liberate efficiently. Gas rises and exits the top of the vessel through the vent line. The degassed mud falls to the bottom of the vessel and travels back into the mud pits by means of a mud leg piping.

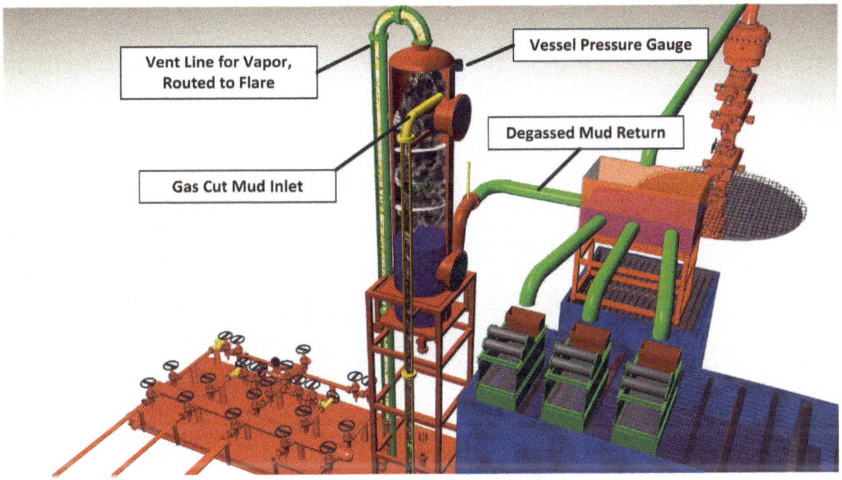

Fig. 4.80 Mud gas separator.

Flare stack

In order to maintain safety, the vented gases may contain flammable hydrocarbon gases (Fig. 4.81). To ensure these gases do not pose risks to worker's health or contaminate the environment, the gases are directed to a flare stack. Flare stacks vary in processing amount by size, pumping diameters, stack height, and flaring head. The gases pass through the vent line at pressures a little above atmosphere pressure, enter the flare chimney, and proceed upward where the flare head initiates a flame. When the gases mix with the atmosphere, the air/fuel combination ignites. Once ignited, the gases are chemically altered inert and pose no danger to personnel or the environment.

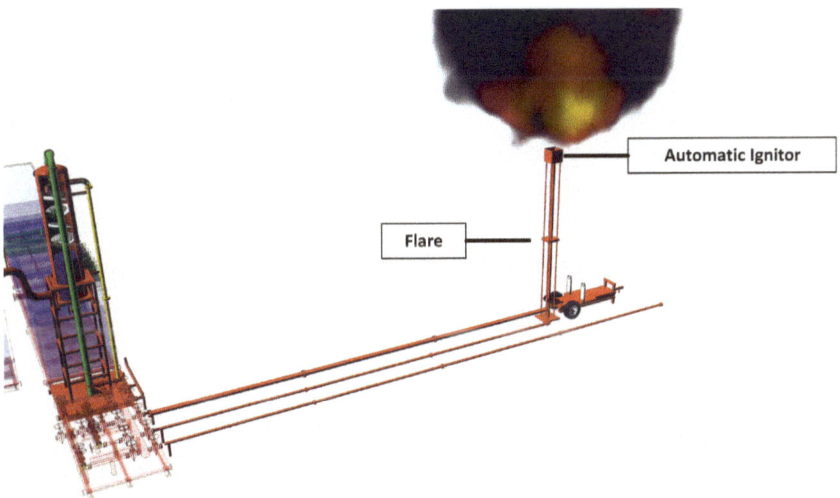

Fig. 4.81 Flare stack.

Surface back-pressure pump

The surface back-pressure pump is normally plumbed from rig mud system and manifolded upstream of the MPD Automated Flow (Choke) Manifold (Fig. 4.82). The surface back-pressure pump consists of an oilfield triplex pump and direct drive motor or an electrical motor system. Alternately, depending on service, the surface back-pressure pump may be driven by an electrical motor system connected to the rig's power system.

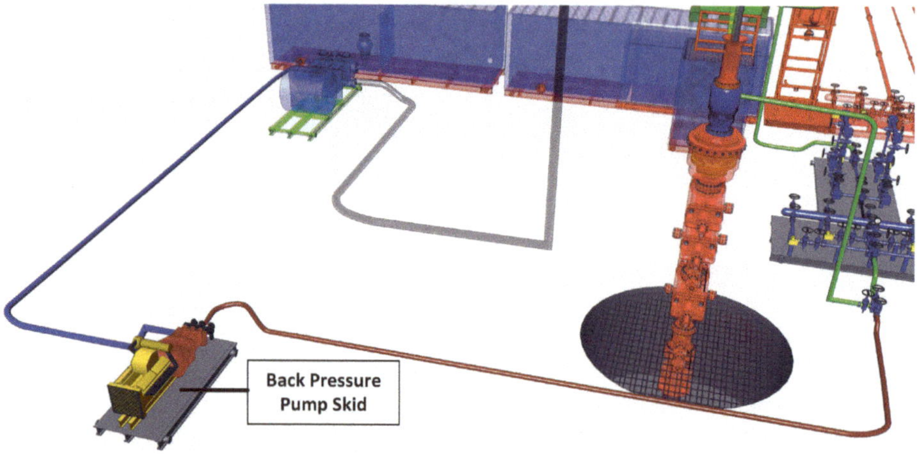

Fig. 4.82 Surface back-pressure pump.

By plumbing the surface back-pressure pump upstream of the Automated Flow (Choke) Manifold, fluid under pressure can be provided when rig pumps are shut down. This pressurized fluid will maintain the back pressure of the system and continue to exit through the Automated Flow (Choke) Manifold, MGS and return degassed fluid to the rig systems return mud pits.

The surface back-pressure pump is controlled via the "Controller's console" located in the Driller's cabin. Limits for operation are set to ensure the pump automatically starts when back pressures start decreasing as rig pumps are slowly brought off line during connections, etc. When rig pumps are slowly brought on line, the surface back-pressure pump is disengaged and stopped.

Controller unit

The Controller unit may be plumbed directly into the Driller's Cabin or may be stand alone with skid trailer or located on the rig floor. All the controls for operation of the RCD, Automated Flow (Choke) Manifold, MassCoriolis Flowmeter and Surface back-pressure pump are plumbed into this workstation. The operator (or controller) will set pressure point limits such as:

- Static RCD Pressure Limits (for connections and tripping)
- Dynamic Pressure Limits (for drilling operations)
- Kick Limits (for drilling and static operations)
- Freeing Stuck Pipe Limits (static operations)
- Lost Circulation (dynamic operations)

The computer interfaces manipulate pressure, temperature information as well as determine fluid compressibility and gas expansion for more accurate BHP determinations. The computer system has built in redundancies with multiple back-up sensors and power supplies.

CHAPTER FIVE

Reference: Surface well control equipment

Commentary

Since Blowout Prevention Equipment acts as an emergency barrier during well control events, rig crew personnel must understand equipment function, testing, and operating parameters. The following chapter reviews surface well control equipment function, specifications, and testing procedures. The following information is provided as a means of understanding well control equipment operating functions. As each rig contains a customized complement of individual manufacturer equipment, rig equipment standards and specifications should be referenced and used for operational knowledge and testing parameters. For operational consideration, this section includes suggestions based upon experiences of the authors.

BOPE used in testing have been developed with the Blind/Shear Ram on top. This arrangement may differ from field operations, as ram placement is at the discretion of the drilling company or the operating company. Based on author experience, the Blind Shear Ram on top is suggested to offer dual barriers when workstring is present. If leakage occurs, this arrangement will allow additional well control equipment to be added (i.e., snubbing unit) by ensuring at least two barriers in well at all times.

Surface well control equipment

Surface well control equipment has been developed to give a "general" understanding of the basic function of each type of blowout preventer equipment and accessories. Conceived as general discussion, individual manufacturer technical manuals are to be addressed for equipment specifications, operating functions, and parameters.

WHEN NIPPLING UP THE PREVENTERS, ENSURE THE RING GROOVE IS CLEAN AND FREE OF OBSTRUCTIONS.
DO NOT USE SOLVENTS TO CLEAN GROOVE.
INSERT CORRECT RING FOR APPROPRIATE MACHINED GROOVE.
UNDER NO CIRCUMSTANCES SHOULD PIPE DOPE BE USED!

Universal Well Control
https://doi.org/10.1016/B978-0-323-90584-8.00007-1

PRIMARY SEALING SURFACES WITHIN RAMS, ANNULAR AND GATE VALVES ARE ELASTOMER (RUBBER)
STEEL REINFORCEMENT AND ENCASEMENT ASSISTS IN PROVIDING SUPPORT AGAINST LEAKS.

Diverters

Diverters are usually low-pressure control systems for redirecting flow away from the rig safely. Diverters are used onshore primarily with surface drilling and before intermediate pipe is set, with expectations of shallow gas or when air drilling. The illustration above demonstrates a drilling application. The hydraulic diver valves are to be sized to match the divert lines. Hydraulic gate valves are preferred due to high forces over larger diameters. The hydraulic valves should be plumbed in a fashion, where the valves are opened first, before the diverter is closed. The upwind divert valve is then closed to direct wellbore fluids only downwind (Fig. 5.1).

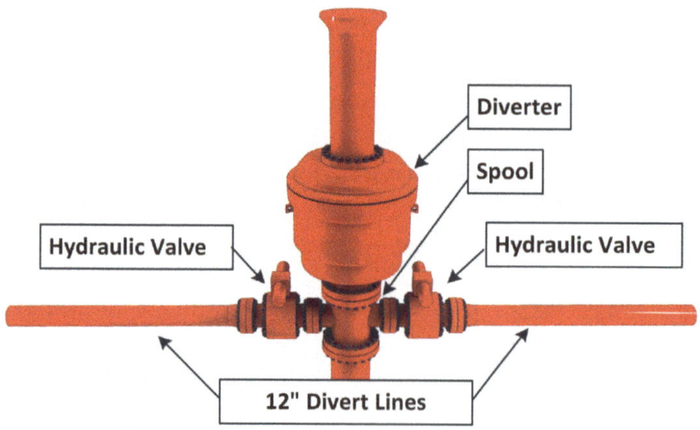

Fig. 5.1 Diverter system.

DIVERTERS ARE NOT WELL CONTROL DEVICES
DIVERTERS SIMPLY PROVIDE A MEANS TO DIVERT WELLBORE FLUIDS AWAY FROM THE RIG FLOOR IN PREPARATION OF EVACUATION.

Rotating control device (rotating heads)

Rotating Control Devices (RCD), sometimes referred to as rotating heads, are devices used to form a close-loop circulating system diverting flow to chokes. This device is not recognized as a barrier (such as a BOP) as its function is diverting flow (Fig. 5.2).

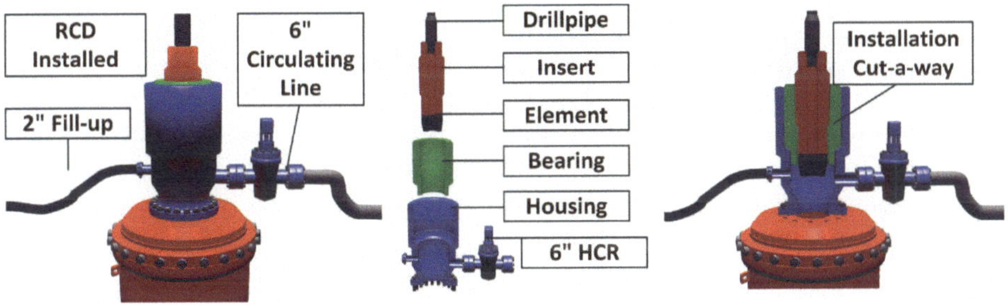

Fig. 5.2 Rotating control device.

Installations will vary depending upon application. The illustrations above have been developed for a drilling application. During drilling operations, while tripping in or out of the well, the RCD provides a continuous seal against the drillpipe, tooljoints, and drill collars. The elastomeric element provides a seal against the tubular and the housing. The bearing is designed to provide an effective seal during tubular rotation. Mud, dust, air, gas, steam, or hydrocarbon mists are diverted through the side outlet and into a choke manifold ensuring safety of rig crews on the drill floor.

Each manufacturer has individual performance designs that will allow effective pressure sealing from 500 psi to 2500 psi applications in sizes up to 20" diameters. Regardless of design, RCDs are considered to be a secondary sealing equipment and should not be used as standard drilling well control equipment.

Managed pressure drilling applications

Two or more outlets are normally installed within the RCD. The larger outlet is used for returns that can be processed through chokes or directed to a Managed Pressure Drilling (MPD) automated choke system. For MPD applications, a separate choke system would be employed allowing conventional circulation to rig supplied choke or gases circulation through the MPD automated choke. The MPD automated choke would be plumbed to the MGS for separation of gas and fluid and safely venting gas away from the rig. If wellbore pressures become excessive, conventional well control preventers would be used following standardized well killing practices.

RCDs are used often in underbalanced or MPD drilling applications in order maintain a constant "back pressure" while drilling the well. Since RCDs have flexible elements used while tripping, the elastomeric elements may require frequent replacement.

ROTATING CONTROL DEVICES ARE NOT WELL CONTROL DEVICES
SIMPLY PROVIDE A MEANS TO DIVERT WELLBORE FLUIDS AWAY FROM THE RIG FLOOR.

Annular preventers

The annular preventer consists of a large "bag-type" elastomer (rubber) preventer element designed to close on, secure, and seal on a variety of wellbore tubulars such as the drillpipe or drillpipe/tubing, drill collars, heavy-weight and the Top Drive stand pipe. Since the annular preventer closes on any tubular, this preventer is normally the primary preventer used to shut-in the well (Fig. 5.3).

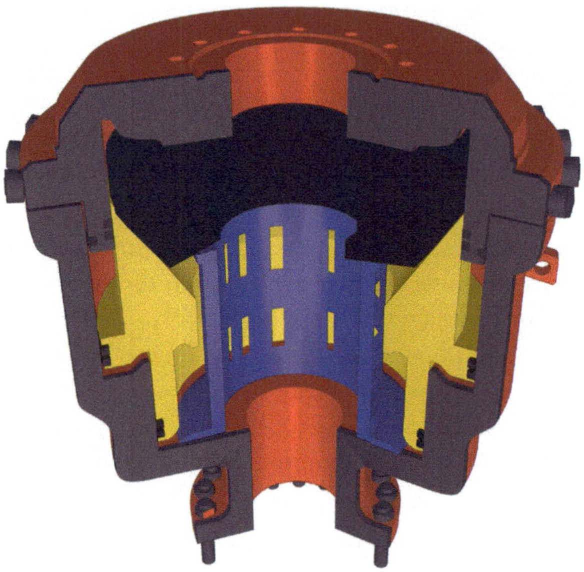

Fig. 5.3 Cut-a-way diagram of annular preventer.

When the annular preventer is transported to location, the manufacturer threaded sealing plugs should be inserted in the closing and opening chamber ports. This will prevent corrosive, foreign material, and debris from entering the closing piston from entering the preventer.

Annular preventers function the same way by using pressurized hydraulic pressure to drive a piston upward, then deforming the element in and around wellbore tubular. Features vary between manufacturers, but generally begin with a closing chamber of greater volume than the opening chamber. This feature allows the preventer to close and seal at lower pressure when compared to the pressure necessary to open the preventer. Some manufacturers direct wellbore fluids underneath a piston area generating a "pressure assisted" closure.

When closed, annular preventers do not employ locking devices. This means, if the closing line "leaks," hydraulic pressure will decrease and may allow the preventer to open prematurely. Therefore, annular preventers must be thoroughly tested to ensure no leakages.

Lifting suggestions

THE PICK-UP EYES ARE DESIGNED TO PICK UP <u>ONLY</u> THE ANNULAR PREVENTER, NOT THE ENTIRE STACK

Transporting suggestions

DO NOT TRANPORT THE ANNULAR PREVENTER UPSIDE DOWN
IF TRANSPORTED UPSIDE DOWN, DAMAGE MAY OCCUR TO PISTON AND ELEMENT

DO NOT USE RAGS AS TRANSPORTATION SEALING PLUGS
USE ONLY APPROPRIATE MANUFACTURE SEALING PLUGS

Closure suggestions

DO NOT CLOSE THE ANNULAR PREVENTER ON OPEN HOLE, EXCEPT IN AN EMERGENCY SITUATION.
CLOSURE ON OPEN HOLE WILL RESULT IN PERMANENT DAMAGE TO THE PREVENTER ELEMENT

ANNULAR PREVENTERS ARE NOT DESIGNED TO CLOSE ON WIRELINE
IF CLOSED ON WIRELINE, REPLACE ANNULAR ELEMENT UPON COMPLETION OF EVENT

Ram preventers

The pipe rams are designed to close only on the full diameter of the pipe or tubing. These rams should be closed when the annular preventer leaks excessively or the Maximum Allowable Surface Pressure will exceed the working limitation of the annular preventer element (Fig. 5.4).

One advantage to using a set of pipe rams over an annular preventer is the ability to close and lock the preventer if hydraulic pressure is lost for any reason.

Fig. 5.4 Cut-a-way diagram of pipe rams.

BEFORE PIPE RAMS ARE CLOSED, ENSURE PIPE IS PROPERLY SPACED OUT TO ENSURE NO TOOL JOINT IS WITHIN THE RAM.

Notice the rams contain a rubberized sealing element located at top of ram and ram face. This ensures an effective pressure seal (Fig. 5.5).

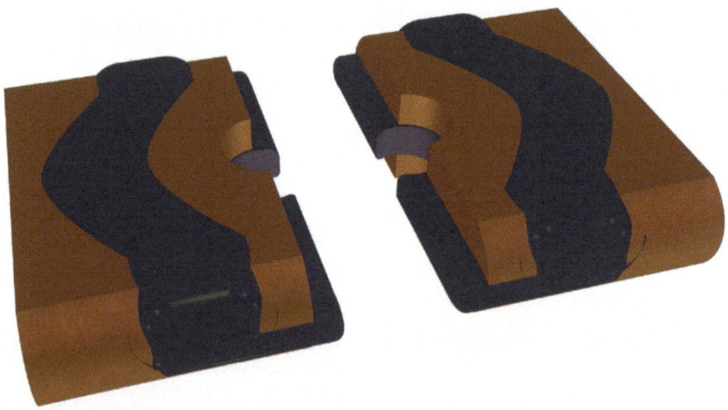

Fig. 5.5 Pipe rams.

DO NOT CLOSE PIPE RAM PREVENTER ON OPEN HOLE OR TOOL JOINT. PERMANENT DAMAGE TO RAM WILL RESULT

RAM PREVENTERS ARE DESIGNED TO HOLD PRESSURE FROM ONE DIRECTION ONLY. ENSURE THE RAMS ARE INSTALLED CORRECTLY WITH ELASTOMER ELEMENT ON TOP SIDE

Blind ram preventers

Fig. 5.6 Blind rams.

The blind rams are designed to close <u>only</u> on open hole. These preventers should be operated and left closed after every trip to ensure objects are not dropped down the hole (Fig. 5.6).

Notice the rams contain an elastomer (rubber) sealing element located at top of ram and ram face. This ensures an effective pressure seal.

BLIND RAMS ARE ONLY TO BE CLOSED ON OPEN HOLE

Full opening safety valves (FOSV)

Full Opening Safety Valve (FOSV) is designed to shut off flow from drillpipe or tubing. This valve is not considered to be well control equipment unless the associated closure wrench is easily **<u>assessable</u>** to everyone on the Rig floor. The FOSV should always be stored in the open position. Prior to installation, verify the FOSV is open. The FOSV is to be "stabbed" prior to closing any preventer. Once installed, close the FOSV **<u>before</u>** closing appropriate preventer (Fig. 5.7).

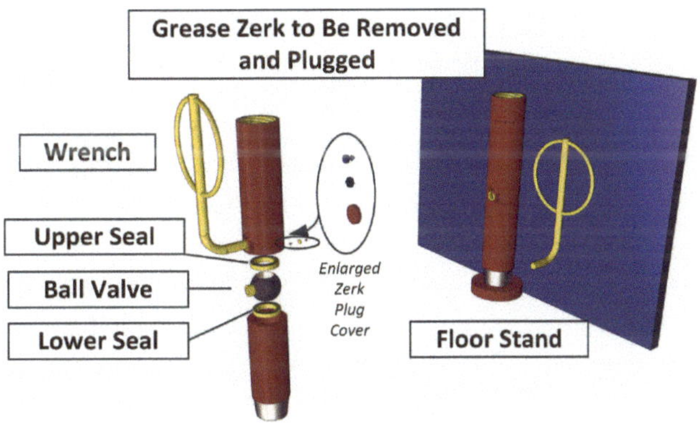

Fig. 5.7 Full opening safety valve (older split body model shown).

Several models of FOSV are offered by various manufacturers including TIW (sometimes called the TIW valve). Most FOSV are manufactured as within a single gate section. Older split gate models can present problems if tongs are misplaced onto the body and torque is applied.

Grease energized sealing rings provide the sealing force to the ball gate. In order to ensure that the FOSV is a pressure containment device, the grease zerk that is added during maintenance must be removed and the manufacturer supplied blanking plug must be reinstalled. If the zerk is left within the body of the FOSV, pressurized fluid may leak out this orifice as a grease zerk is only a very low-pressure check valve.

IF APPLICABLE, THE FOSV MUST HAVE THE GREASE FITTING REMOVED AND THE BULL PLUG INSTALLED TO BE CONSIDERED A SAFETY VALVE.

THE FOSV CLOSURE WRENCH MUST BE LOCATED ON THE RIGFLOOR AND EASILY ASSESSABLE TO CREW MEMBERS

Inside blowout preventers (IBOP)

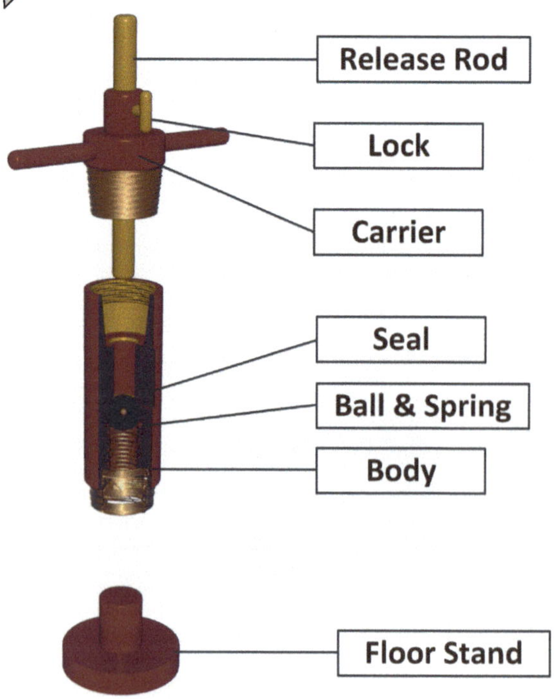

Fig. 5.8 Inside BOP (with ball and seat).

This is a back-pressure type valve (of float valve) that allows stripping or running pipe into the hole without fluid flowing upward through the valve. It can be stabbed and make up on the pipe **only at very low flow rates**. The best method is to stab and close the full-opening safety valve first, then install the inside BOP if the decision is made to go back into the hole (Fig. 5.8).

Three major sealing mechanisms are commonly used (1) dart and seat, (2) ball and seat, and (3) flapper type. The dart and ball seat operate by compressing a spring to allow flow around the seal. The flapper type uses a gate affixed to a spring. With no downward pressure, the gate spring forces the gate closed. When pumping begins, the gate opens.

Each type of IBOP is pressure assisted, as wellbore pressures acting from below help keep the valve closed. Valves are normally outfitted with springs that will require 200 psi to open.

The IBOP is designed to prevent fluid flow from the well. After installation and make up the upper sub is then removed and additional pipe may be added as desired. Unfortunately, they can only be stabbed with very low flow rates coming from the pipe.

The main disadvantage to using the inside BOP with the restrictive internal diameter that will not allow passage of items such as perforation tool in the event the drillstring becomes plugged during well control.

The IBOP is to be readily available for emergencies. The IBOP thread profiles must match the tubulars being run into the well (including crossovers, if needed). The IBOP should be stored on a stand, with the handling carrier installed. The release rod must be positioned to open the IBOP, the rod is locked and readied for emergency service.

Choke manifold

The choke manifold is the primary pressure throttling and control device used when circulating out a kick. The choke manifold consists of valves, spacers, chokes, gauges, and pressure chambers to provide primary and secondary control. Larger choke systems contain two (2) hydraulically controlled adjustable chokes and two (2) manually operated chokes. The piping of the panel is conFig.d to provide at least two (2) upstream gate valves for isolation and repair in an emergency (Figs. 5.9 and 5.10).

Fig. 5.9 Class III choke manifold (1 hydraulic choke + 1 manual choke).

Fig. 5.10 Class IV choke manifold (2 hydraulic chokes + 2 manual chokes)

Regular maintenance must be performed on each gate valve and choke orifice. During drilling and work over operations, the choke manifold valve configuration is to be inspected and verified before starting each tour. All valves should be tagged as locked out or tagged out.

Depending on location, maintenance frequency should be adjusted to incorporate regional weather environments. For the most part, maintenance should include a **daily** flushing of lines to prevent solids settling. This may include back flowing diesel throughout entire manifold to prevent settling, freezing, or contamination. This maintenance should include daily lubrication to the valves and chokes. Pressure gauges accuracy should be verified with readouts from the Driller's Console.

CHOKE MANIFOLD VALVES AND CHOKES SHOULD BE OPERATED AND LUBRICATED DAILY

Gate valves

Gate valves are designed to maintain an effective seal against wellbore pressures. This effective seal is made by forcing the gate against the sealing element within the body of valve. These types of valves must be lubricated **daily** so the sealing element of the stem is energized.

Most valves are designed to hold pressure from one direction. Care should be taken when installing these types of valves to ensure they are installed in the correct position (Figs. 5.11 and 5.12).

Fig. 5.11 Gate valve open.

Fig. 5.12 Gate valve closed.

Ancillary equipment

Kelly drilling system

Kelly drilling systems have provided the backbone of rig design and are still in use. This type of hoisting/rotation system is used mainly in straight holes, low angle holes, and workover operations. The Kelly system comprises of swivel, pipe spinner (optional), square, or hexagonal Kelly and Kelly bushing. The Kelly drilling system incorporates two surface valves called the upper and lower Kelly cocks, one on each end of the Kelly. The Kelly bushing engages the rotary table allowing rotation and torque to be applied to the drillstring. The Kelly provides a means to circulate drilling fluids from the pumps/pits through the Kelly hose, into the swivel and down the Kelly/drillpipe (Fig. 5.13).

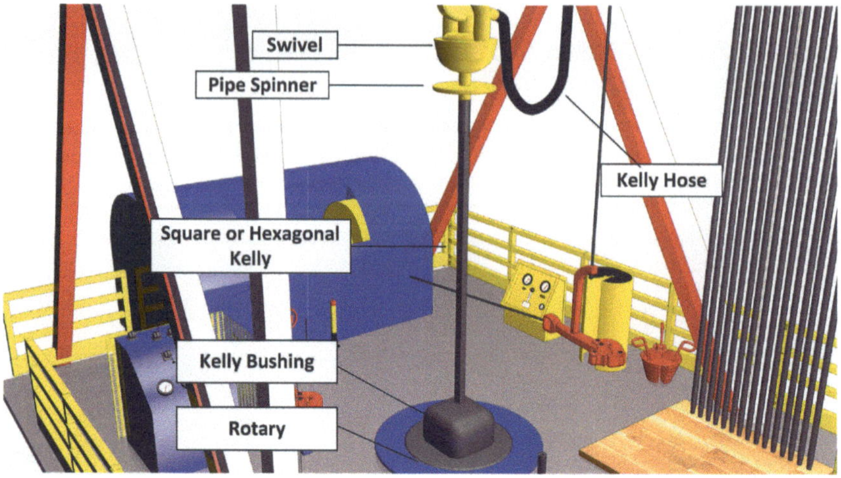

Fig. 5.13 Kelly system.

The upper and lower Kelly cock valves are used to isolate the drillstring and provide secondary back-up in the event the drillpipe float and leak in swivel/Kelly hose during an event. The upper Kelly cock is a manually operated valve. Closure may require harnessing and hoisting personnel to close valve with a wrench.

The lower Kelly cock is a drillpipe connectable valve and operates as a full opening safety valve. Once again, this valve is manually closed with a wrench.

The Kelly system's limiting factor is the inability to provide rotation while pulling out of the hole. For high angle and horizontal wells, back reaming is needed.

UNDERSTAND LOWER KELLY COCK VALVE SPECIFICATIONS
SEVERAL MANUFACTURERS OFFER MODELS WRENCH KEY OPERATING STEMS WHICH MAY LEAK UNDER PRESSURE AND CANNOT BE STRIPPED INTO THE HOLE.

Top-drive system

The Top Drive System provides rotations, hoisting, and circulating capabilities within one unitized piece of equipment. With system integration, the Kelly and rotary have been removed. Two high strength industrial electric motors provide rotation and torque directly to the drillstring. Housed within a carrying structure, the Kelly hose and goose neck allow drilling fluids to be pumped into a central shaft. The motor drives the shaft into a low-to-high speed transmission and allow circulation through an upper and lower IBOP and into the drillstring. Rotational torque and hoisting capabilities are also imparted onto the drillstring. To help guide this massive equipment and keep everything aligned, most Top Drive Systems are hoisted within a track structure mounted to the derrick (Fig. 5.14).

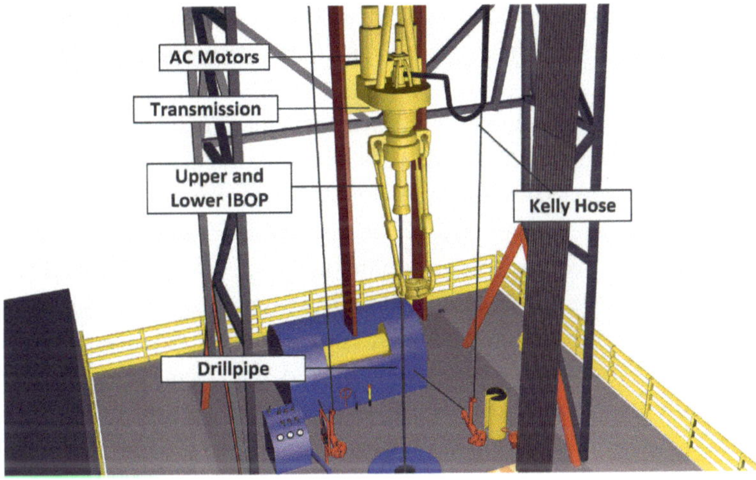

Fig. 5.14 Top drive system.

Blowout preventer review checklist

The pressure rating of all pressure control equipment (BOP, valves, etc.) must be equal to or greater than the Maximum Allowable Surface Pressure. Any deviation from this policy should have prior approval of the Person-in-Charge.

The bore size of the preventer stack must be adequate to allow passage of any tool or tubular to be run in the well.

High-pressure spacers and drilling spools are used within preventer stacks to allow the preventer stack to be properly spaced out and allow for circulating through kill and choke lines. Ram type preventers have circulating outlets located within the preventer under the sealing surface and can be used when there are space limitations within the substructure. The diameters of all preventer side outlets are designed to match choke and kill line diameters.

All ram preventers are being outfitted with operable locking devices. Locking devices must be installed and tested for operation before operations are to commence.

Adequate guy wires and turnbuckles are required to stabilize the preventer stack. These stabilization wires are to be fastened to the substructure in an upward fashion. Guying downward into the substructure could cause undo buckling forces onto the preventer stack, if the rig were to settle.

New metal ring gaskets should be used each time a flange is assembled. Flange grooves are to be well cleaned, dry, and free of any type of lubricant. Most preventer stacks used RX or BX ring gaskets. R type ring gaskets are NOT TO BE USED on individual pressure components.

All preventer packing elements and gaskets will be inspected at the time of installation. Elements and gaskets with defects are to be rejected as defects can lead to failure during operations.

In operating side valves, the inside valves are considered master valves and would normally never be opened or closed when there is pressure unless the outside valve is closed.

Preventer assemblies are to be shopped every 3–5 years for inspection and repair. During the shop inspection, the preventers will be disassembled and inspected for tolerances of wear and tear. If the preventer is found to be out of specification standards or requires welding, all repairs are to be conducted to applicable specification standards. The service company providing this service must be licensed by the manufacturer and have an operating kiln/furnace to allow components to be heated and stress relieved for each type of repair. New elastomers, bolts, and nuts are to be provided during the inspection. The pressure hull is to be tested to applicable standards. After assembly the preventer is to be pressure tested to the maximum operating range of the preventer. Repair reports are to include all pressure charts, replacement parts along with quality assurance and quality control (QA/QC) signatures, date, and facility that performed repairs.

All flange bolts should be checked periodically and tightened or replaced as required.

All testing shall be done with clean water. All preventers and valves should be tested separately where possible.

If desired, pressure control equipment may be tested to its nominal working pressure. This may be done by seating a proper test plug in the bowl and the casing head spool before applying the test. The maximum hang down load for a given test pressure should be checked with the wellhead manufacturer or discussed with the appropriate personnel.

Blowout preventer control manifolds must be clearly labeled to indicate that preventers are operated by the respective control valves. Opening and closing positions for the control valves should be indicated.

Stations are to be installed at a safe distance, at an easily accessible location, preferably near an escape route.

Blind Ram controls shall be equipped with a safety device or shield to prevent accidental closure. The safety shield **shall not prevent blind ram's closure from a remote station.**

Never leave a prevent control in the "blocked" position once operations have commenced.

Ring gaskets

The pressure integrity of all preventers is only insured if the correct ring gasket is installed and tested properly. The ring gasket is the PRIMARY seal on all preventers. If a used, damaged or corroded ring gasket is installed in a corroded or pitted pressure groove, the pressure integrity of the ENTIRE stack is jeopardized. Therefore, it is imperative the

Person-in-Charge personnel verify the pressure grooves to be clean of foreign materials (such as rust, paint, and grease) and the ring gasket is not pitted, corroded, or damaged, before pressure testing the stack.

Ring gaskets come in a variety of sizes, fitting all double-studded adapters, wellheads, and BOPs. The ring gaskets come in the following designations (Fig. 5.15):

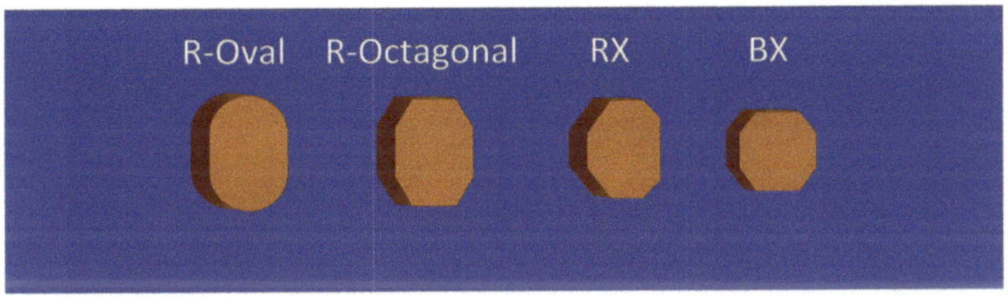

Fig. 5.15 Types of ring gaskets.

Type "R" ring gaskets

These ring gaskets were the first design used for pressure control. These types of ring gaskets can only be used for **gate valves**.

The gasket is permanently deformed when the Type "6B" flange is mated and fastened together. This type of flange does allow flange bodies to meet. The Type "R" ring gasket supports the weight of the flange and flange body. The gasket becomes permanently deformed and should be immediately discarded upon disassembly of preventers (Fig. 5.16).

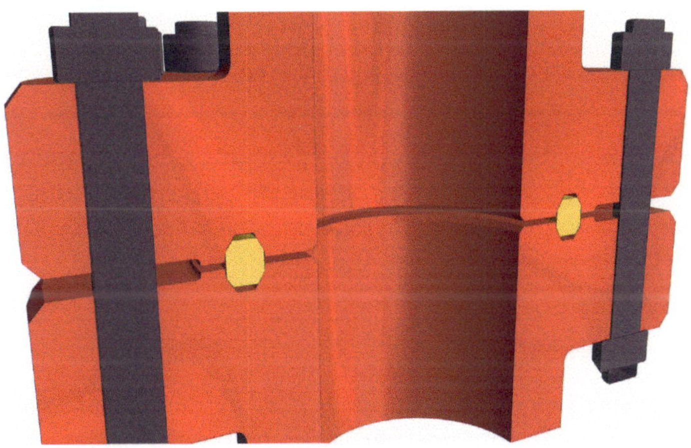

Fig. 5.16 Type R wellhead (ring gasket has all weight).

DO NOT USE A TYPE "R" RING GASKET IN BOP STACK.
ONLY USE TYPE "RX"RING GASKETS FOR TYPE R WELLHEADS AND VALVES

RX and BX ring gaskets were the second and third generation ring gaskets, respectively. They are superior to Type "R" ring gaskets since they are pressure assisted. The Type "RX" ring gasket does not support the flange or the flange bodies for Preventers and Therefore they are not permanently deformed. In an emergency, these types of ring gaskets may be reused (Fig. 5.17).

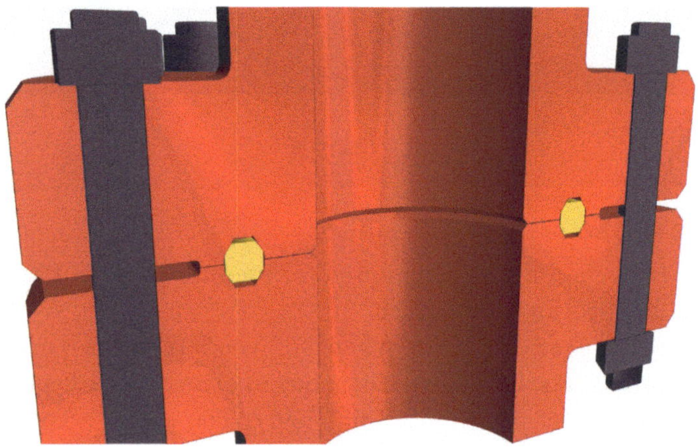

Fig. 5.17 Type BX wellhead (wellhead has all the weight).

RX RING GASKETS ARE USED FOR PRESSURE RATINGS OF UP TO 10,000 PSI.
BX RING GASKETS ARE USED FOR PRESSURE RATINGS FROM 5,000 PSI TO OVER 15,000 PSI.

Caution should be used when reusing any type of ring gasket to ensure pressure integrity of the entire stack. Closely check the ring and ring groove for pitting, corrosion, and foreign materials (such as grease). The ring should be installed in a clean, dry ring groove with no grease. Under no circumstances should pipe dope be used.

Annular preventers

The two general types of annular preventers are the pressure assist type and the spherical type. On the blowout preventer stack, the annular preventer is normally the first preventer used to shut-in the well. Although annular preventers are designed to seal-off on any type of tubular in the wellbore (i.e., drillpipe/tubing, drill collars, tooljoints, Kelly), permanent damage to preventer element will result if closed on **open hole.**

DO NOT CLOSE ANNULAR PREVENTER ON OPEN HOLE
THIS WILL RESULT IN PERMANENT DAMAGE TO PREVENTER ELEMENT.

ANNULAR PREVENTERS ARE NOT DESIGNED TO CLOSE ON WIRELINE
IF CLOSED ON WIRELINE, REPLACE ANNULAR ELEMENT UPON COMPLETION OF EVENT

For the pressure assist type, closure is accomplished by applying hydraulic pressure to raise a contractor piston. As the piston travels upwards, it displaces and deforms a rubber sealing element radically inward that eventually contacts and seals on the outside of any pipe that is in the hole. Compression of the rubber throughout the sealing area assures a seal against any shape (Figs. 5.18–5.20).

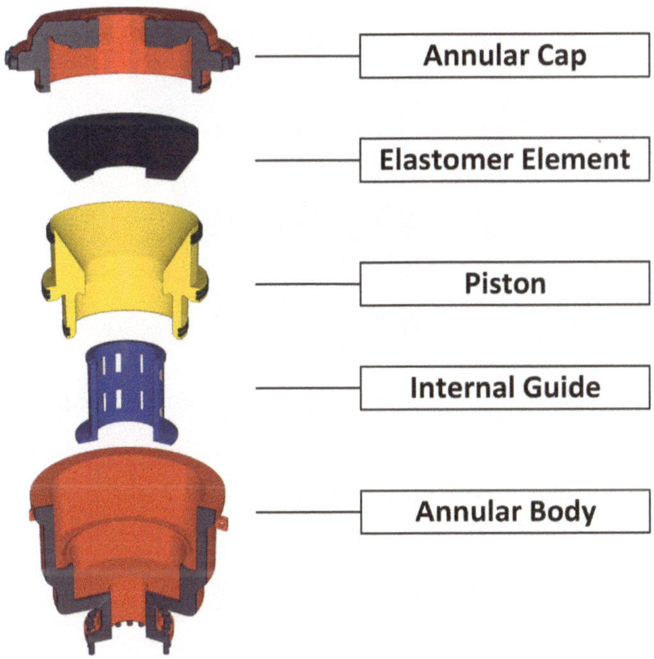

Fig. 5.18 Pressure assisted annular assembly diagram.

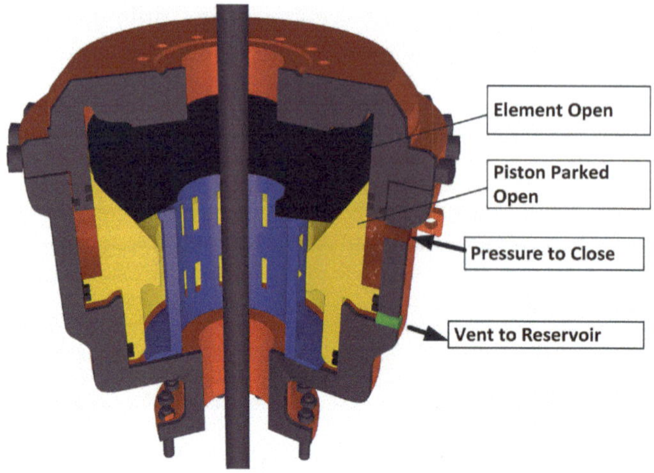

Fig. 5.19 When open, piston parked, and open.

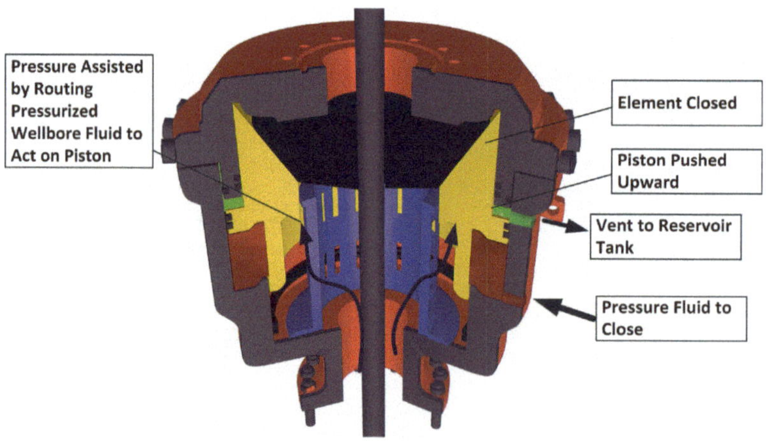

Fig. 5.20 When closed, piston travels upward and elastomer deforms around pipe. wellbore fluid assist with closure.

Pressure assisted annular features

- One-piece rubber sealing element.
- Only two moving parts (piston and packing element).
- Lifting eyes (to be used to lift **ONLY** the annular preventer).
- Screwed cap for field replacement of annular element.
- Element wear can be determined in the field.

Do's and don'ts for pressure assisted annular preventers

Can perform a remaining element life by measuring remaining piston stroke. The piston stroke has a direct correlation to the condition of the packing unit.The condition of the packing unit can be determined by:

(1) Inserting a 5/16" rod into the drilled passage of head.
(2) Measuring the piston stroke on a new packing unit.
(3) Using the appropriate test pipe diameter (usually 4-1/2")
(4) Take subsequent measurements each time the annular BOP is tested.
(5) The remaining life of the packing unit is proportional to amount of piston stroke left using the following formula:

$$\%\text{of Remaining Life} = \left(\frac{\text{Stroke Length in Used Condition inches}}{\text{Stroke Length in New Condition inches}}\right) \tag{5.1}$$

(6) If full stroke of the piston is reached before a seal is obtained, the packing unit should be replaced since any further increases of closing pressure will not cause element to seal off.

The pressures of this chart are to be used as guidelines for initiating an effective seal of the annular pressure while minimizing wear. These pressures were derived for new packing elements under clean test conditions. Maximum packing unit life will be realized by use of the lowest closing pressure that will maintain a seal (Table 5.1).

Table 5.1 Example of average closing pressure table.
Average closing pressure for initial seal off in pressure assisted annular (See Manufacturer Specifications for exact closing pressure information)

Pipe O.D.(in) Rating (psi)	13-5/8" 3M psi	13-5/8" 5M psi	16" 2M psi	16" 3M psi	16¾" 5M psi	18" 2M psi
4½"	900	650	500	500	600	600
3½"	1000	700	600	600	650	650
2-7/8"	1100	750	700	700	750	700
2-3/8"	–	950	800	800	850	750

Suggestions (Do's)

- Always threaded plugs into opening and closing chamber ports when transporting. This will keep foreign objects from entering chamber and contaminating hydraulic fluid and corroding chamber.
- Keep annular preventer element clean and free of obstructions. Wash out annular preventer at end of every trip. Especially after any or all cement jobs, as cement will harden behind element and inhibit or prevent proper operations.

- Keep threaded connections and ring groove located on top of preventer clean and free of obstructions. To accomplish this, the flow nipple or "pitcher" nipple should be installed with a ring gasket and all bolts should be used.

Avoid the following (Don't)

- Do not close annular preventer on open hole, permanent damage to the preventer element, and reduced operating life will result.
- Do not store or transport the annular preventer "Upside down." This will result in permanent damage to the preventer piston, housing, and element.
- Do not lift the entire BOP stack with lift eyes for the annular preventer.
- Do not use pipe dope on the screwed cap threads. This will cause galling and lead to abnormally high torque needed to open preventer.

Spherical annular preventers

The "spherical" preventer derives its name from the semi-circular profile on the inside of the cover. Closing pressure moves the piston upward, and deforms the sealing element upward and radically inward along the profile, until a seal is made against the pipe in the hole (Fig. 5.21).

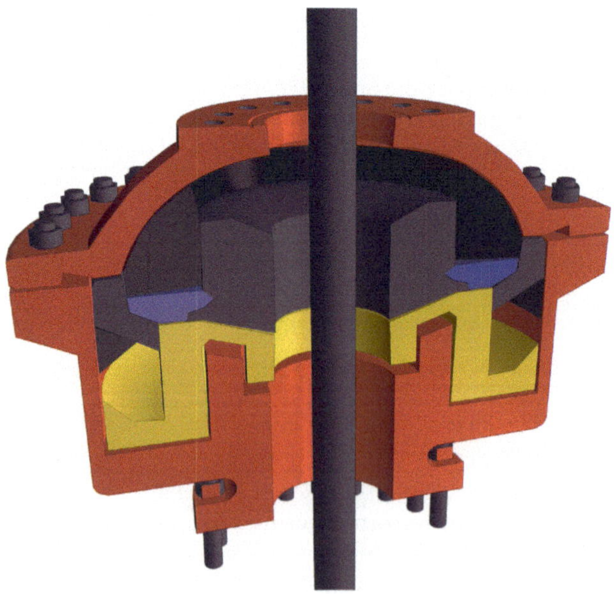

Fig. 5.21 Spherical preventer: when closed, piston travels upward and elastomer deforms around pipe.

Features common to 2000–5000 psi spherical annular preventers

- Closing pressure for all pipe is **1500 psi** (for stripping operations see charts below).
- One-piece rubber sealing element.
- Only two moving parts (piston and packing element).
- Element can be changed with pipe in the hole.
- Well pressure assists in closing of sealing element.
- Lifting eyes (to be used to lift **ONLY** the annular preventer).
- For preventers greater than 13-3/8" and closed on pipe greater than 7-5/8", the closing pressure should be reduced below 1500 psi to prevent pipe deformation. Use the following chart (Tables 5.2–5.4):

Table 5.2 Spherical preventer closing pressures.
Maximum spherical annular preventer closing pressures
To prevent element and pipe damage

		Casing size (In.)									
Spherical size (in)	**Wellbore pressure (psi)**	**7"**	**7-⅝"**	**8-¾"**	**9-5/8"**	**10¾"**	**11¾"**	**13¾"**	**16"**	**18-½"**	**20"**
21¾"	5000	1500	1400	1175	975	790	640	480	300	190	150
21¾"	2000	1500	1400	1175	975	790	640	480	300	190	150
18¾"	5000	1500	1400	1175	975	790	640	480	300	-	-
16¾"	5000	1500	1400	1175	975	790	640	480	-	-	-
13-5/8"	3000 or 5000	1500	1265	890	615	415	280	-	-	-	-
30"	1000	-	-	-	1100	-	-	1100	-	-	900

Table 5.3 Spherical preventer closing pressures for stripping operations.
Recommended closing pressures for stripping operations
20"—2000 psi spherical preventer

Pressure	4½"	3½"	2-7/8"	2-3/8"
0	1000	1100	1200	1300
250	1050	1150	1225	1325
500	1100	1175	1275	1350
750	1150	1225	1325	1375
1000	1200	1275	1375	1400
1250	1275	1300	1400	1425
1500	1325	1350	1425	1450
1750	1350	1375	1450	1475
2000	1375	1400	1475	1500

Continued

Table 5.3 Spherical preventer closing pressures for stripping operations—cont'd
Recommended closing pressures for stripping operations
18¾"—5000 psi spherical preventer

Pressure	4½"	3½"	2-7/8"	2-3/8"
0	675	875	1000	1150
500	725	900	1050	1200
1000	800	1000	1100	1250
1500	850	1050	1150	1300
2000	975	1150	1200	1325
2500	1050	1200	1250	1375
3000	1200	1250	1300	1400
3500	1250	1300	1350	1425
4000	1300	1350	1400	1450
4500	1350	1400	1450	1475
5000	1400	1450	1500	1500

Table 5.4 Spherical preventer closing pressures for stripping operations.
Recommended closing pressures for stripping operations
13-5/8" 5000 psi spherical preventer or
12" 3000 psi spherical preventer

Pressure	4½"	3½"	2-7/8"	2-3/8"
0	650	825	950	1050
500	700	875	1050	1100
1000	750	950	1150	1150
1500	800	1050	1225	1200
2000	900	1100	1250	1250
2500	1050	1175	1300	1300
3000	1200	1200	1350	1325
3500	1250	1300	1375	1375
4000	1300	1350	1400	1400
4500	1400	1400	1450	1450
5000	1500	1450	1500	1500

Spherical sizes not listed above require no pressure adjustments when closing on casing (Table 5.3).

Do's and Don'ts for spherical preventers

Suggestions (Do's)

- Always threaded plugs into opening and closing chamber ports when transporting. This will keep foreign objects from entering chamber and contaminating hydraulic fluid and corroding chamber.

- Keep annular preventer element clean and free of obstructions. Wash out annular preventer at end of every trip. After any and all cement jobs, cement must be thoroughly washed out to prevent excess from hardening behind elements and preventing operation.
- Keep threaded connections and ring groove located on top of preventer clean and free of obstructions. To accomplish this, the flow nipple or "pitcher" nipple should be installed with a ring gasket and all bolts should be used.

Avoid the following (Don't)

- Do not close annular preventer on open hole, permanent damage to the preventer element, and reduced operating life will result.
- Do not store or transport the annular preventer "Upside down." This will result in permanent damage to the preventer piston, housing, and element.
- Do not lift the entire BOP stack with lift eyes for the annular preventer.

Sealing elements

All preventer manufacturers provide sealing elements of different composition that are designed for use in specific wellbore environments (Tables 5.5 and 5.6).

Table 5.5 Pressure assisted elastomer table.
Pressure assisted sealing elements

Element type	Color code	Letter code	Recommended usage
Natural Rubber	Black	R	Water-Based Mud –30 to 225°F
Synthetic (for H_2S)	Red	S	Oil muds with aniline point of 20–190°F
Neoprene	Green	N	Oil muds with temperatures of –30 to 170°F

Table 5.6 Pressure assisted elastomer table.
Spherical sealing elements

Element type	Color code	Letter code	Recommended usage
Natural rubber	Red	1 or 2	Low Temperature with Water-Based Mud –30 to 225°F
Nitrile synthetic (for H_2S)	Blue	5 or 6	Oil muds with aniline point of 20–190°F
Neoprene	Black	3 or 4	Water-based muds with H_2S of 30 to 170°F

ALL RUBBERIZED PREVENTER ELEMENTS MUST BE STORED IN A COOL, DARK LOCATION SINCE RUBBER WILL DETERIORATE WITH ULTRAVIOLET RAYS FROM THE SUN

Ram type preventers

The pipe ram will only seal on the particular pipe size that matches the ram block installed at the time of closure. The sealing elements will be damaged if pipe rams are closed on the open hole. Pipe rams form a very effective seal since the sealing elements are backed-up with heavy steel ram blocks. Most pipe rams can be locked in the closed position and may be able to be operated both mechanically and hydraulically (Fig. 5.22).

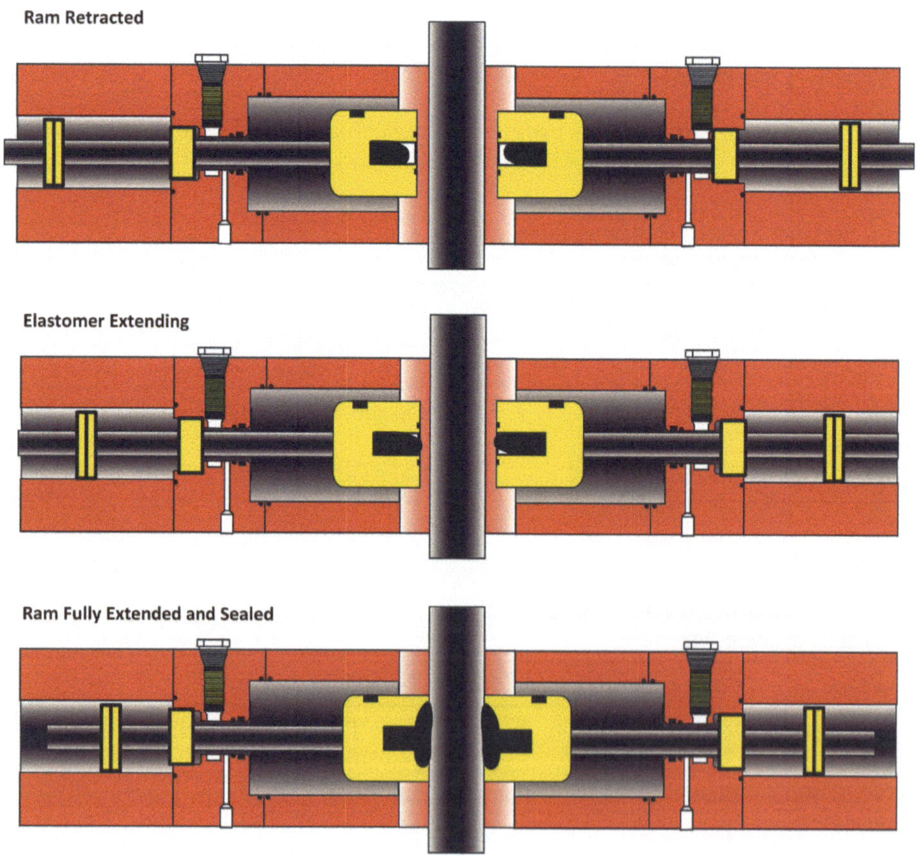

Fig. 5.22 Diagram demonstrating how elastomers form seal.

DO NOT CLOSE RAM PREVENTER ON OPEN HOLE.
THIS WILL RESULT IN PERMANENT DAMAGE TO SEALING ELEMENT AND RAM.

Pipe rams should never be closed on open hole during a function test. The self-feeding action of the ram packers causes the rubber seal to extrude a considerable distance into the wellbore. This extrusion result in permanent damage to the rubberized element and will cause early seal failure.

Many pipe rams are able to support the weight of the entire drillstring if a tooljoint is lowered onto the top of a closed ram (Fig. 5.23).

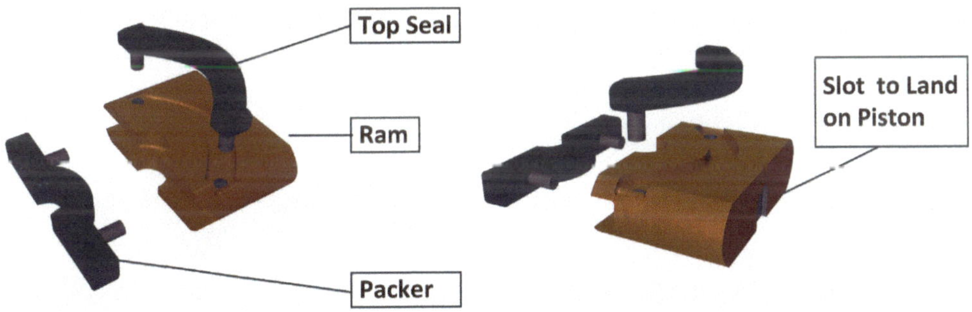

Fig. 5.23 Pipe ram preventer assembly drawing.

Please refer to the schematic. Notice the top sealing element. This element allows an effective seal between ram and the upper section of the ram cavity. When the rams are extended, wellbore fluids will generate an upward force ensuring an effective top seal.

Blind shear rams

Blind Shear rams differ from Blind Rams, as they are designed to shear or cut drillpipe, not drill collars, dril-lpipe collars, or casing. Before shearing can take place, the drillstring must be put in tension before activating the shear (Fig. 5.24).

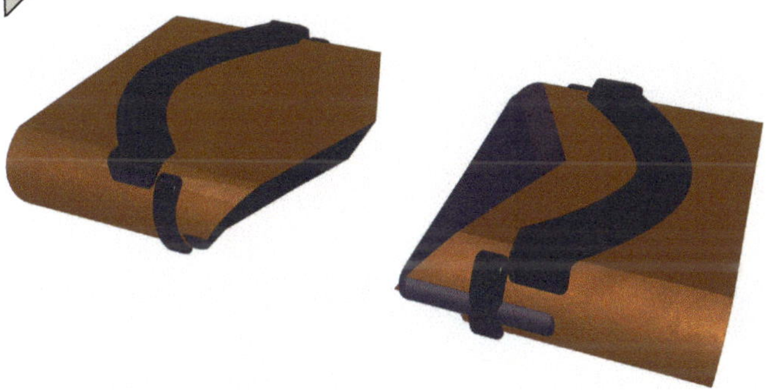

Fig. 5.24 Bladed shear ram.

Both blind shear and blind rams are to be closed at the end of every trip or when the well is free of tubulars to ensure objects are not dropped into the well.

Once closed, the two rams form an effective, one-piece, solid ram preventing flow from the wellbore. These rams also have rubberized elements help form the effective seal.

Pipe rams and blind rams should be installed so the ram doors are positioned over and shield the valves installed on the casing head below. Rams should be installed in such a way they prevent objects from falling on kill lines or circulating lines.

BLIND SHEAR AND BLIND RAMS ARE THE ONLY RAMS DESIGNED TO CLOSE ON OPEN HOLE

Hinged door ram type preventers

Ram preventers are manufactured in multiple pressure sizes and for extreme conditions. The rams basically operate the same way, where high-pressure hydraulic fluid is pumped into a piston attached to the ram and forces the ram closed. Manufacturers differ in how to access the ram and provide two basic designs (1) Hinged Door and (2) Hydraulic Access (Figs. 5.25–5.26).

Fig. 5.25 Hinged door rams from front.

Fig. 5.26 Hinged door rams from back.

Hinged Door preventers have bonnet doors on hinges, which when opened will allow access to be ram cavity to change ram rams (Fig. 5.27).

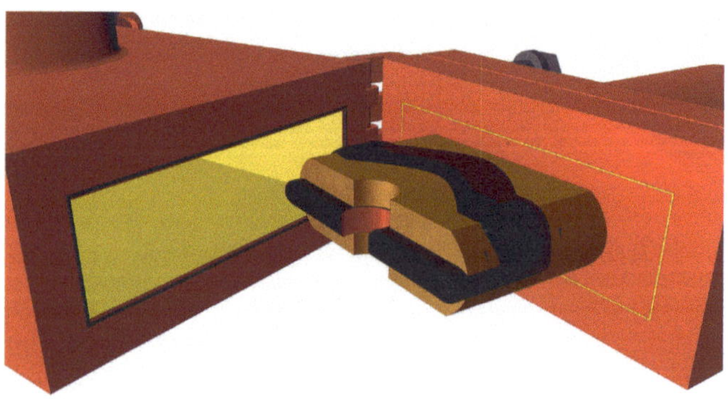

Fig. 5.27 Opened with cavity and bonnet door gasket.

Most hinge door designs incorporate heavy duty hinges needed to support heavy doors and provide routing of the closing and opening hydraulic pressure lines. To ensure proper operation of the closing and opening forces, hinge door seals are equipped with life-long steel reinforced elastomer elements.

General Ram Change Procedure for Hinged Doors

1. Open the rams by applying opening pressure.
2. Disengage all bonnet bolts on both bonnets.
3. Manually swing both bonnets open clear of BOP stack.
4. Close the rams by applying closing pressure.
5. Remove the ram assembly by attaching lifting eyes in the holes provided and lift straight up (Fig. 5.28).

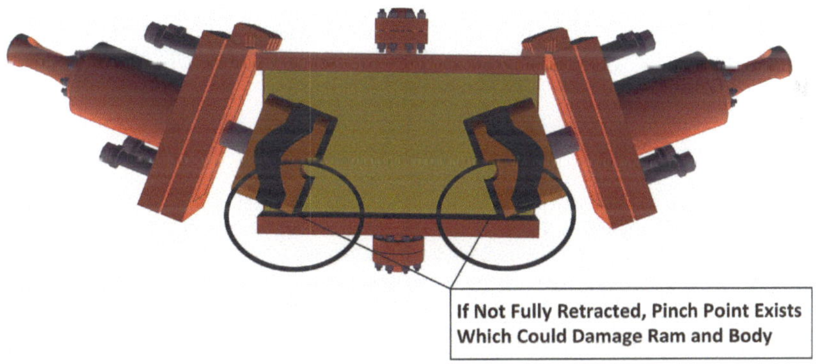

Fig. 5.28 Opened with cavity and bonnet door gasket.

CAUTION
DO NOT USE HYDRAULIC PRESSURE ON CLOSING CHAMBER BEFORE BONNET DOORS ARE OPENED. IF PRESSURE IS APPLIED BEFORE DOORS ARE OPENED THE RAM BLOCKS WILL BIND AGAINST RAM BODY AND RAM PISTON WILL BE DAMAGED. DAMAGE TO RAM PISTON CAN ONLY BE REPAIRED IN SHOP, NOT ON LOCATION.

CAUTION
THESE PREVENTERS CAN INADVERTANTLY BE INSTALLED UPSIDE DOWN.
SINCE THEY ARE DESIGNED TO HOLE PRESSURE FROM ONE DIRECTION, THEY WILL FAIL IF INSTALLED UPSIDE DOWN.
NOTE POSITION OF FLOW OUTLET IS BELOW RAM

Hydraulic access ram preventers

The Hydraulic Access Ram Preventer is a wellbore pressure assisted ram preventer and represents the most manufactured ram type preventer worldwide. All ram preventers manufactured today are equipped for H_2S service (Fig. 5.29).

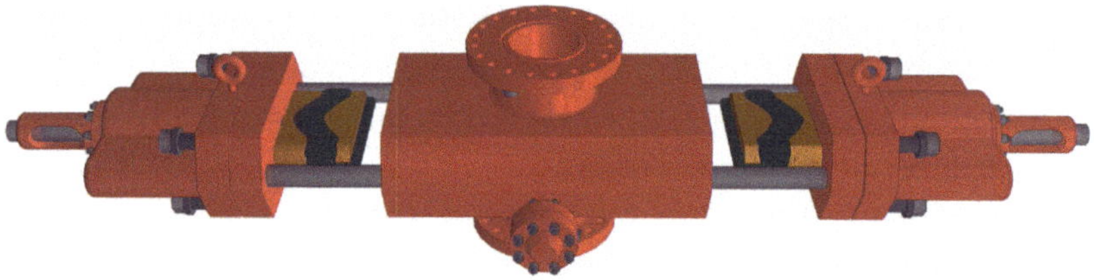

Fig. 5.29 Hydraulic access ram preventer.

The hydraulic access ram preventer has been designed to allow easy access to the ram. After loosening ram bonnet bolts, "closing" hydraulic pressure can be applied to open the bonnet. The end assembly is hydraulically pumped to the open position.

The lifting eyes provided by the manufacture are only to be used to lift the ram bodies for service. These are simple threated lifting eyes and will not support the weight of the ram.

CAUTION
DO NOT USE RAM LIFT EYES TO LIFT PREVENTER.
AS SHOWN IN THE SCHEMATIC ABOVE, THE RAM LIFT EYES ARE STORED ON THE PREVENTER ITSELF. THESE EYES MUST NOT BE USED TO LIFT THE PREVENTER.

CAUTION
THESE PREVENTERS CAN INADVERTANTLY BE INSTALLED UPSIDE DOWN.
SINCE THEY ARE DESIGNED TO HOLE PRESSURE FROM ONE DIRECTION, THEY WILL FAIL IF INSTALLED UPSIDE DOWN.
NOTE POSITION OF FLOW OUTLET IS BELOW RAM

Secondary seals for ram preventers

Manufacturers of ram preventers offer features such as a secondary sealing element for emergency situations. This emergency seal can be activated for the functioning ram rod. It is very important wellbore pressures be isolated from the operating cylinders on all ram preventers. Normally, the closing chamber is provided with a primary lip seal installed in the bonnet through which the operating rod passes (ram rod). If fluid pressure should bypass the primary seal and enter the operating cylinder, it is possible the ram preventer could be forced open. To prevent this occurrence, a series of secondary and tertiary seals are provided that may include:

- Back-up "O" rings
- Plastic packing injection seal (for emergency use only)
- Vent to the atmosphere.

In order for the secondary seal to work, the Weep Hole must be free of obstructions so if a leak develops, wellbore fluid will vent freely to atmosphere. The plastic packing ring will be energized by removing the cover bolt, inserting an Allen wrench and torquing down on plastic set screw. This forces plastic packing through check valve and energizes the tertiary seal (Fig. 5.30).

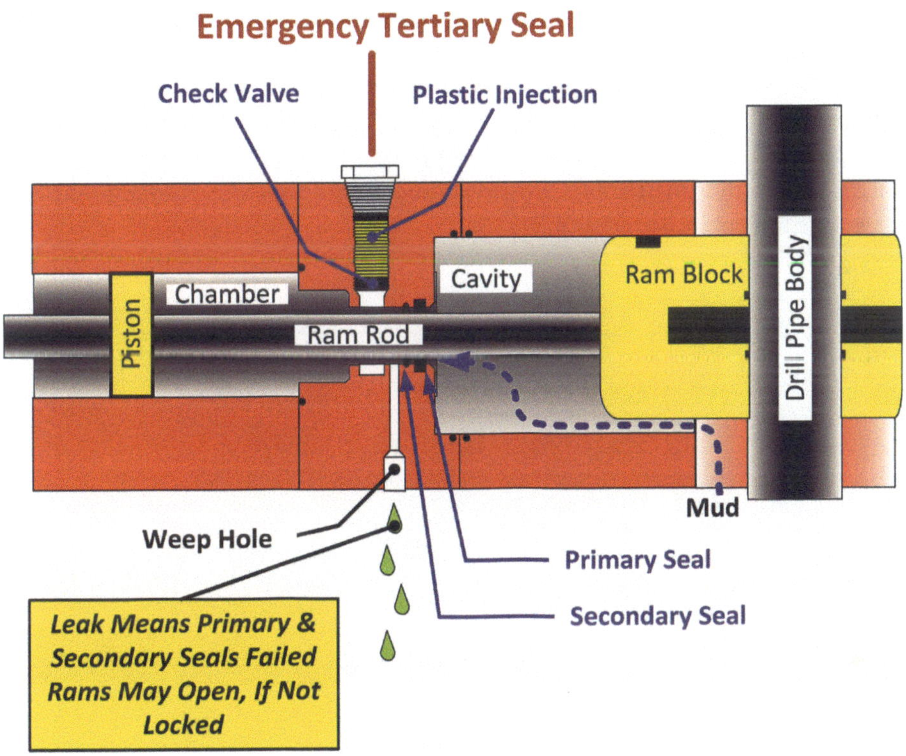

Fig. 5.30 Secondary seal assembly.

THE TERTIARY SEAL IS TO BE USED ONLY IN CASE OF EMERGENCY

If the Weep Hole shows fluid discharge during testing of the stack, the inner seals must be replaced by an authorized manufacturer repairman. **Do not energize plastic tertiary seal unless of extreme emergency.**

BOP closing equipment (accumulators)

BOP Closing Equipment is static pressure storage vessels designed to deliver pressurized hydraulic fluid to the closing chambers of the Blowout Preventer

Equipment **without assistance of any rig power**. Since BOP Closing Equipment "accumulates" pressurized hydraulic fluid, these units are sometimes referred to as Accumulators.

The BOP Closing Equipment must hold fluid under pressure without leaks or excessive pumping and must maintain sufficient reserve capacity to close all preventers and open HCRs with the recharging pumps off. The BOP Closing Equipment has been engineered to provide reliable, hydraulic power automatically requiring certain skills and knowledge to keep the system operating successfully. A firm understanding of the system and its components will enable any crew member to maintain the unit at its best performance level.

Principle of Operation: To provide pressurized hydraulic fluid to the closing chambers of the BOPs in a matter ofs. Ensures a secondary pressure containment on well, regardless of tubulars in use (Fig. 5.31).

Fig. 5.31 Diagram of BOP closing equipment.

Principle of Design: The BOP Closing Equipment may seem confusing when all components are studied simultaneously, but if separated into individual components, the system is easily understood. The designs are so basic, little or no changes have been made since the 1930s.

The basic components may be grouped into (1) Reservoir Fluid System, (2) Pumping System, (3) Accumulator System and (4) Manifold System.

Reservoir fluid system

The Reservoir Fluid System begins with the proper sizing of the atmospheric tank, which is designed to provide storage of fluids equal to twice the total volume of combined accumulator bottle banks. This system provides storage of clean hydraulic fluid at surface or atmospheric pressure. In order for the BOP Equipment to function properly, the hydraulic fluid should be tested periodically to ensure proper lubricity is available. Normally, hydraulic fluid is white in color and should be changed out if discoloring (defining contamination) is apparent. The fluid is normally colored white to ensure hydraulic leaks are easy-to-view (Fig. 5.32).

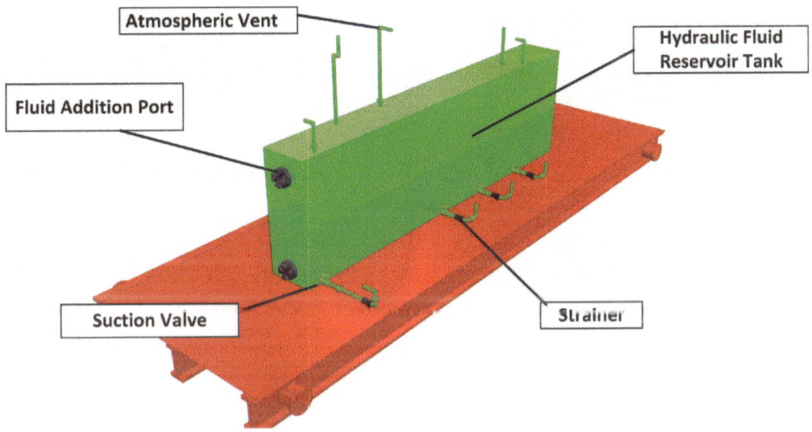

Fig. 5.32 Labeled diagram of reservoir fluid system.

The Reservoir Fluid System provides a means of directing hydraulic fluid from atmospheric tanks hydraulic pumps through a high-pressure line with shut off valve and strainer. Before operations begin, ensure all shut off valves are opened. Strainers are to be cleared and cleaned per manufacturer specification and before changing out discolored hydraulic fluid.

Equipment assurance standards state:

(1) Each closing unit should have a fluid reservoir with a capacity equal to or at least twice the useable fluid capacity of the accumulator system.

(2) If Shearing rams are present, increases to the reservoir volume should be considered.

(3) Fluid is stored at atmospheric pressure.

Quick Check List:

1. Suction valves to be open
2. Atmospheric vent to be clear and functioning
3. Sufficient available volume for 1.5 times the volume necessary to Close BOPs and open HCRs

Pump system

Equipment standards state "two or three independent sources of power should be available on each closing unit. Normally, air, electric, or nitrogen are used. The following information has been developed for a 3000 psi BOP Closing System with 1000 psi precharge pressure (Fig. 5.33).

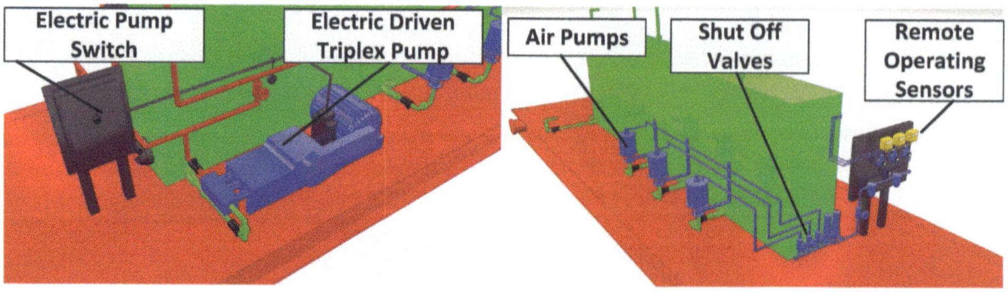

Fig. 5.33 Labeled diagram of pump system.

Work Pump: Air pumps operate extremely efficiently and are used to pressure fluids from atmospheric pressure to 2700psi quickly. Air pumps operate from supplied "regulated rig air" supplied via the hydro-pneumatic pressure switch. This switch automatically stops air operated pumps when pressure reaches 2700psi and initiates pumping when pressure drops approximately 400psi. Most BOP closing systems utilize 60:1 or 40:1 ratio pumps. For a 60:1 ratio pump, 1 psi rig air translates into 60psi usable hydraulic pressure. Since most rig air systems operate at 120psi air pressure, the pumps must be regulated to ensure the air pumps receive about 50psi for a 3000-psi accumulator system. These types of pumps are extremely efficient and provide an "explosion proof" means of maintaining hydraulic pressure.

Polishing Pump: Electrical pumps deliver precise quantities of fluid and provide means to finish reach hydraulic pressure limits from 2700psi to 3000psi without problems. Electrical power is used to drive pumps that contain minimum and maximum pressure switch limitations. These pumps are designed to automatically charge the system and automatically shut down when a predetermined maximum pressure limitation is reached. The electrical pressure switch is used to determine the pressure of the system. This switch automatically stops pumps when accumulator pressure reaches 3000psi and initiates pumping when pressure drops below 2700psi. The pressure range is to verify during an accumulator drill.

FOR SURFACE ACCUMULATORS: WITH THE ACCUMULATOR REMOVED FROM SERVICE, EACH PUMP SYSTEM SHOULD BE CAPABLE OF CLOSING THE ANNULAR PREVENTER ON DRILLPIPE AND OPENING THE REMOTE VALVE AND OBTAIN A MINIMUM OF 200 PSI ABOVE PRECHARGE PRESSURE WITH TWO MINUTES.

THE PUMP SOURCE IS DESIGNED TO RECHARGE THE ENTIRE ACCUMULATOR SYSTEM CAPACITY FROM 0-3000 PSI IN FIFTEEN MINUTES OR LESS.

Quick Check:

1. Source Main Air Supply Valve and Air Pump Isolation valves to be open during operations
2. Air Filter to be purged according to manufacturer specifications or as needed.
3. Reservoir isolation valves are open and strainers cleaned to manufacture specifications

Accumulator system

The Accumulator System is used to store fluid under pressure, until demand, then successfully route hydraulic fluid under pressure to appropriate BOP preventer or HCS as directed by the manifold system. Accumulator valving also contains an accumulator relief valve which is set to rupture at 3300 psi and direct fluid back into the reservoir. Accumulator bottles should be precharged as labeled and normally are precharged to 1000 psi. Unfortunately, over years of service, many accumulator systems no longer contain proper badging on bottles and systems. When this occurs, there is no way to properly ensure this equipment has been manufactured to manufacturer specifications and international standards (Fig. 5.34).

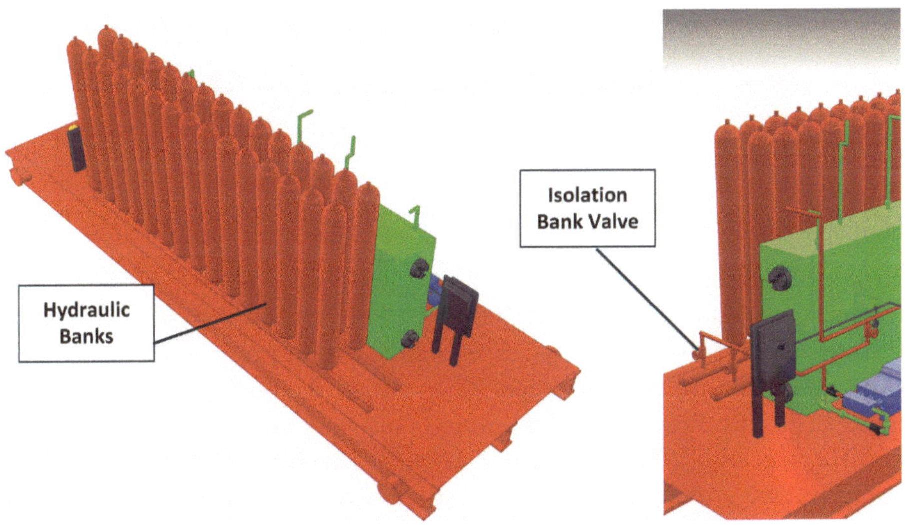

Fig. 5.34 Labeled diagram accumulator bank.

PRECHARGE TO ACCUMULATOR BOTTLES LIMITED TO NITROGEN ONLY
DO NOT USE OXYGEN.

Quick Check:

- Accumulator Bank Isolation Valve—Manually operated, normally open
- Accumulator Bank Bleeder Valve—Normally closed.

Accumulator storage system comprises of hydraulic fluid under pressure contained within a pressure storage device (bottle). The bottle contains nitrogen precharged rubber bladders which are designed to separate nitrogen from the hydraulic fluid. The bottles are precharged so a greater amount of pressurized hydraulic fluid can be readily available than if the bottles were at atmospheric pressure. This nitrogen precharged system allows for smaller, lighter accumulator designs to be used. The accumulator bottle contains a one-way poppet valve allow fluid inside the vessel but not allowing the rubber bladder to extrude (Fig. 5.35).

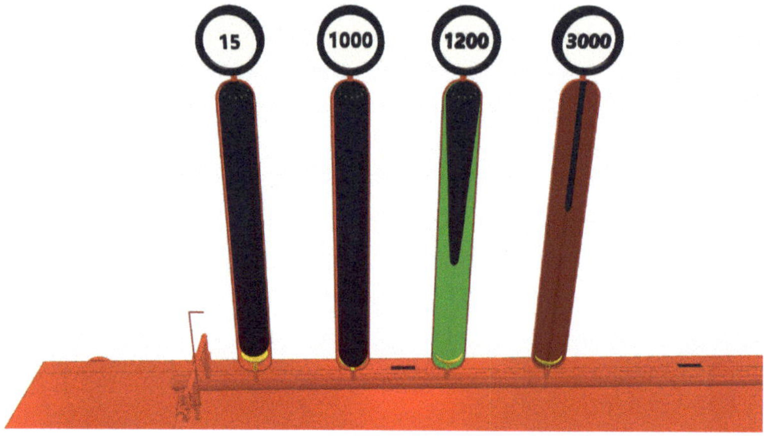

Fig. 5.35 Accumulator bottle with bladder.

Starting from left to right

- Bottle 1 (0 psi): As an example, this 3000-psi accumulator system starts at atmospheric pressure. A rubber bladder is used to separate nitrogen gas from the hydraulic fluid. with no pressure, the bladder simply hangs within the bottle.
- Bottle 2 (1000 psi Precharge): Demonstrates nitrogen has been added to the bladder from the top with 1000 psi sustained energy. The bladder is designed to extend throughout the bottle and close the operating valve located at the bottom of the bottle. This is the "Precharge Pressure" limit and is the point the lower valve is closed. This prevents the bladder for extending to the bladder bank piping system.
- Bottle 3 (1200 psi Lower Limit): Represents the requirement for BOP Closing System's lower operating limit is 200 psi over precharge. The green color represents fluid which contains 200 psi worth of hydraulic pressure.
- Bottle 4 (3000 psi Operating Pressure): Demonstrates the maximum operating limit of the 3000-psi accumulator (represented by red). The fluid stored within this 3000 psi to 1200 psi pressure range is called "usable fluid." Usable fluid represents the pressurized fluid available to operate BOPE.

Accumulator sizing

BOP Closing System requirements will vary depending on authorities overseeing operational area. The most common minimum requirements designate 100% of fluid volume needed to close ALL preventers within the stack plus the volume required to open an HCR valve and have a system operating pressure 200 psi above the minimum recommended precharge pressure remaining on the BOP Closing System with pumps off.

From minimum requirements, standard operational requirements have been developed. For a 3000-psi system with 1000 psi precharge, the standard operational requirements designate 150% of the fluid volume required to close ALL preventers and open an HCR valve and have 1200 psi system pressure remaining on the BOP Closing System with pumps off.

Determination of number of bottles needed

Step 1: Determine Total Closure Volume Gallons Plus Safety Factor. Determine total closure volume (usable volume) fluid needed to open the HCR(s) and close all the BOPs plus 50% Safety Factor. Use Tables 5.11 to 5.19 to determine closing volumes. Track closing ratios identified within these Tables to determine minimum operating pressures.

Step 2: Determine Minimum Operating Pressure. Determine the lowest threshold pressure for each individual equipment function as follows:

For equipment with published Closing Ratios,

$$\text{MinimumOperatingPressure}_{(\text{psi})} = \frac{\text{FullOperatingPressure}_{(\text{psi})}}{\text{ClosingRatio}} \tag{5.2}$$

For equipment without published Closing Ratios,

or

Minimum Operating Pressure = 200 psi above precharge pressure.

Step 3: Determine Usable Fluid Volume. As pressurized hydraulic fluid enters and occupies the internal volume of each bottle, the precharged bladder is compressed. When discharged, the pressurized fluid exits the bottle through a shut off orifice located at the bottom of each bottle. Since the bottle has two limits, minimum operating pressure and maximum operating pressure, only a **portion** of the stored pressurized fluid can be used. Calculate individual equipment usable fluid volume.

$$\text{VolumeUsableFluid}_{\text{gal}} = \left(\frac{\text{PrechargePressure}_{\text{psi}}}{\text{Min Opr Press}_{\text{psi}}} - \frac{\text{PrechargePressure}_{\text{psi}}}{\text{Max.OperatingPress}_{\text{psi}}}\right) \tag{5.3}$$

Example of bottle determination

Determine the number of bottles needed to provide 1.5 times the closing volume for

- 1—13-5/8″ 10,000 psi Hydril GX Annular Preventer
- 1—13-5/8″ 15,000 psi Cameron Type "U" Shear Ram
- 3—13-5/8″ 15,000 psi Cameron Type "U" Shear Ram
- 2–3″, 15,000 psi Cameron Type "F" HCR Valves

Step 1: Determine closure gallons plus safety factor

See Table 5.7.

Table 5.7 Closure volume plus 50% safety factor.

Equipment	Size and pressure rating (psi)	Table	Gallons to close (gal)	Closure volume +50% Safety factor (gal)	Close ratio
1—Hydril GX Annular BOP	13-5/8″ 10,000 psi WP	T25	24.14	36.21	N/A
1—Cameron Type "U" Shear Rams	13-5/8″ 15,000 psi WP	T28	16.40	24.60	6.60
1—Cameron Type "U" Pipe Rams	13-5/8″ 15,000 psi WP	T28	16.40	24.60	10.60
1—Cameron Type "U" Pipe Rams	13-5/8″ 15,000 psi WP	T28	16.40	24.60	10.60
1—Cameron Type "U" Pipe Rams	13-5/8″ 15,000 psi WP	T28	16.40	24.60	10.60
1—Cameron Type "F" HCR Valve	3″, 15,000 psi WP	T31	0.49	0.74	N/A
1—Cameron Type "F" HCR Valve	3″, 15,000 psi WP	T31	0.49	0.74	N/A

Step 2: Determine minimum operating pressure

Determine the lowest threshold pressure for <u>each</u> individual equipment function. For the shear ram, the minimum operation pressure would be determined as follows. Repeat this for each individual equipment.

$$\text{Minimum Operating Pressure for Shear Ram}_{(\text{psi})} = \frac{15,000\ \text{psi}}{6.6} = 2273\ \text{psi} \tag{5.4}$$

Step 3: Determine usable fluid volume

Determine usable volume and number of bottles per individual equipment item from individual minimum operating volume.

$$\text{OperatingThresholdFactor} = \left(\frac{1000\ \text{psi}}{2273\ \text{psi}} - \frac{1000\ \text{psi}}{3000\ \text{psi}}\right) = .1067_{\text{dimensionless}} \tag{5.5}$$

$$\text{UsableVolume}_{10\ \text{gal bottle}} = .1067\, x\, 10_{\text{gal/bottle}} = 1.07\ \text{gal}_{10\ \text{gal bottle}} \tag{5.6}$$

$$10\ \text{GallonBottlesNeeded} = \frac{\text{Closure Vol} + 50\%\text{SafetyFactor}}{\text{UseableVolume per 10 gallonbottle}} = \frac{24.60}{1.07} = 23.06_{10\ \text{gal bottles}}$$

$$\text{RoundUp} = 23.06_{10\ \text{gal bottles}} = 24_{10\ \text{gal bottles}} \tag{5.7}$$

Table 5.8 demonstrates bottles needed per BOPE based upon operating threshold factors.

Table 5.8 Summary table of bottles needed.

Equipment	Size and pressure rating (psi)	Table	Gallons to close (gal)	Closure volume +50% safety factor (gal)	Close ratio	Minimum operating pressure	Operating threshold factor	Useable Vol Per 10 gal bottle	10 gal bottles needed	Rounded up
1—Hydril GX Annular BOP	13-5/8″, 10,000 psi WP	T27	24.14	36.21	N/A	1200	0.5000	5.00	7.24	8
1—Cameron Type “U” Shear Rams	13-5/8″ 15,000 psi WP	T30	16.40	24.60	6.60	2273	0.1067	1.07	23.06	24
1—Cameron Type "U" Pipe Rams	13-5/8″ 15,000 psi WP	T30	16.40	24.60	10.60	1415	0.3733	3.73	6.59	7
1—Cameron Type "U" Pipe Rams	13-5/8″ 15,000 psi WP	T30	16.40	24.60	10.60	1415	0.3733	3.73	6.59	7
1—Cameron Type “U” Pipe Rams	13-5/8″ 15,000 psi WP	T30	16.40	24.60	10.60	1415	0.3733	3.73	6.59	7
1—Cameron Type “F” HCR Valve	3″, 15,000 psi WP	T33	0.49	0.74	N/A	1500	0.3333	3.33	0.22	1
1—Cameron Type “F” HCR Valve	3″, 15,000 psi WP	T33	0.49	0.74	N/A	1500	0.3333	3.33	0.22	1
						Bottles required for usable volume				**55**

Operating manifold system

The operating manifold is a series of valves, regulators and piping which directs properly regulated pressurized fluid to the appropriate BOPE and HCRs for function. The bank of accumulator bottles routes high-pressure hydraulic fluids (3000 psi) to a series of regulators that regulates high hydraulic pressure to an acceptable lower operating pressure required to function the BOPE to manufacturer specifications. Manufacturer closing pressure requirements are set to minimize wear and tear on the BOPE while providing closure within acceptable limits. Additional pressure (3000 psi) is available for emergencies. Reduced pressure requirements for annulars run in the range of 600 to 800 psi, while rams and HCR range around 1500 psi (Fig. 5.36).

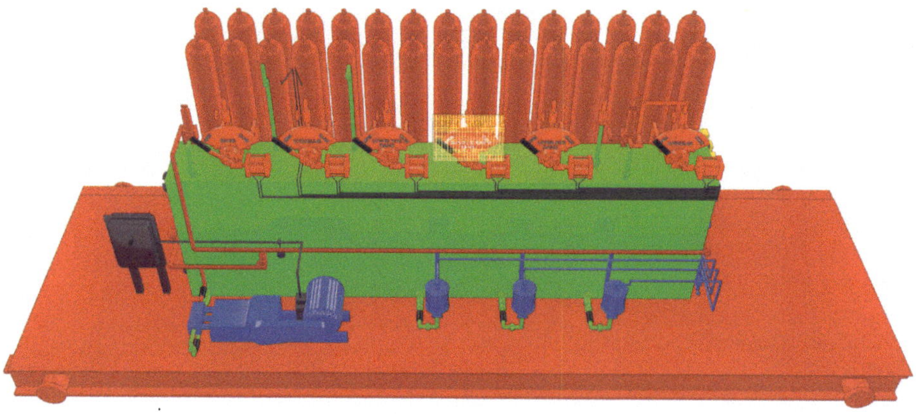

Fig. 5.36 Operating manifold system.

SET REGULATED PRESSURES FOR ANNULAR AND RAMS TO MANUFACTURER BOP SPECIFICATIONS

The cornerstone of the manifold is the series of four-way valves and piping necessary to direct regulated hydraulic pressure to the closing chamber appropriate BOP while providing means of venting the opening chamber fluid to the reservoir tank. The entire hydraulic system is a closed loop system, providing energized hydraulic fluid to the appropriate BOP while also providing means for return fluids from the opening chamber. The hydraulic system is a heavy duty closed loop system (Figs. 5.37 and 5.38).

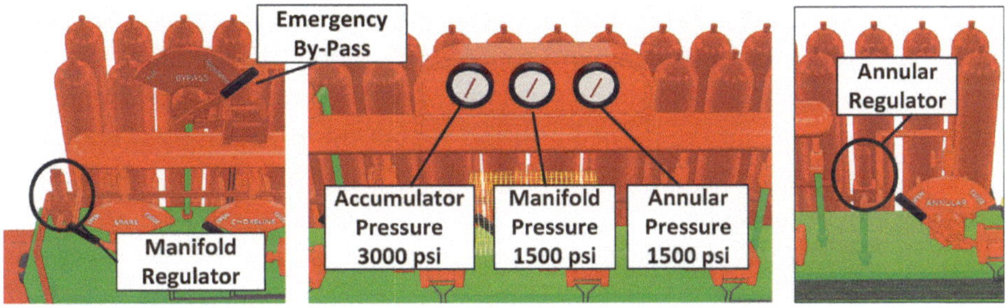

Fig. 5.37 Regulators and pressure gauges.

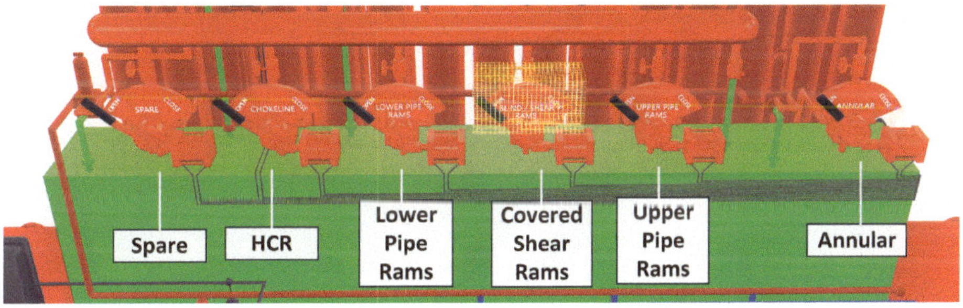

Fig. 5.38 4-Way valves.

The manifold system comprises of individual components which work in unison to help direct pressurized fluids to proper BOPE

1. From the top right, a four-way emergency by-pass valve has been installed to apply unregulated, full operating pressure to BOPE. The by-pass valve is used to provide maximum shearing force to the shearing ram (if applicable) or provide suitable pressure in case of a leak.
2. Manifold gauges should be well light and easy to read
   a. Accumulator Pressure = 3000 psi
   b. Manifold Pressure = 1500 psi (adjust manifold regulator as needed).
   c. Annular Pressure = 600–800 psi (adjust annular regulator as needed)
3. Not all 4-way valves should be either in the Open and Closed Position.
   a. Block position only to be used for repairs and while moving.
   b. Shear Rams should include a cover which must be lifted in order to function
      i. No pins or locks are acceptable.

Open position 4-way valve

When handle is shifted to the open position, regulated pressure hydraulic fluid is routed to the open side of the pistons within the blowout preventer (route shown as yellow arrows in illustration above). The BOP operating piston is forced into the open position. Hydraulic fluid in the closing chamber is routed back to the BOP Closing System's reservoir tank and vented to atmospheric pressure (route shown as black arrows in illustration above). The ram is in the parked position within the BOP cavity (Fig. 5.39).

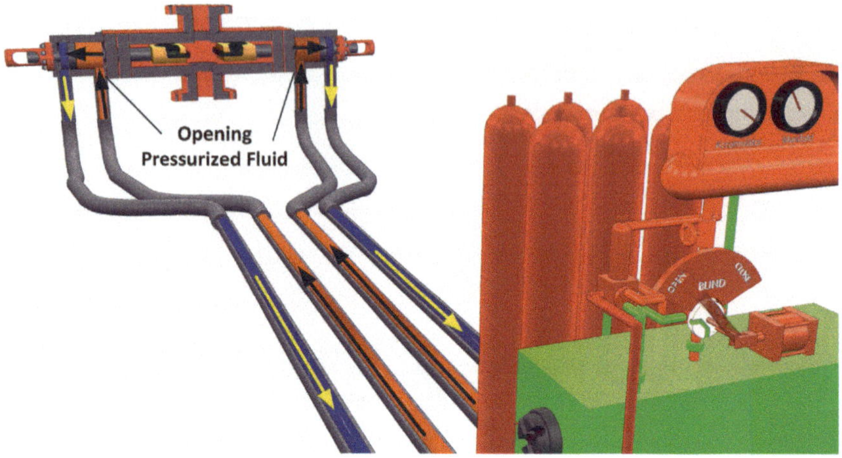

Fig. 5.39 Open function on blind rams.

Closed position 4-way valve

When handle is shifted to the closed position, regulated pressurized hydraulic fluid is routed to the close side of the pistons within the blowout preventer (route shown as yellow arrows in illustration above). The BOP operating piston is forced into the closed position. Hydraulic fluid in the open chamber is routed back to the BOP Closing System's reservoir tank and vented to atmospheric pressure (route shown as black arrows in illustration above). The ram is forced into the wellbore area (Fig. 5.40).

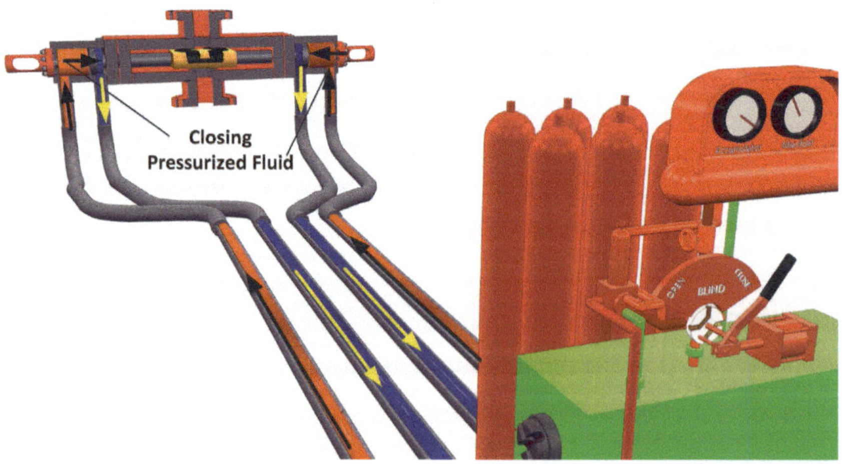

Fig. 5.40 Closed function on blind rams.

Block position 4-way valve

When handle is shifted to the block position, the 4-way valves allow pressure of one set of lines to vent to the reservoir tank while trapping pressure in the other set of lines. If the 4-way valves allow pressure normalization, both sets of lines will be depressurized by providing a leak path. This leak-by does not always occur and may allow-pressure to be trapped on the pressurized side. This results in not knowing if the ram has been fully retracted into the cavity and may allow the ram to extend within the wellbore (Fig. 5.41).

Fig. 5.41 Blocked function on blind rams.

- Select Block after Closing—May cause rams to extend into cavity
- Select Block after Open Position—Rams may travel into the cavity due to vibrations within BOPE caused by operations.

Block is normally selected before rig moves to ensure all lines have been depressurized. During operations, block position should never be selected. If 4-way valve is set in the block position during operations, an investigation should commence as probable leak exists and block was selected to keep pumps from operating non-stop.

DURING OPERATIONS, BLOCK POSITION MUST NEVER BE SELECTED
IF SELECTED, RAM MAY EXTEND INTO THE WELLBORE

Remote panel

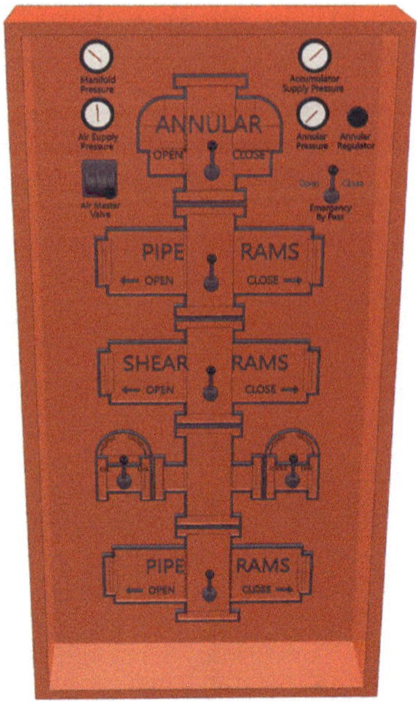

Fig. 5.42 Air actuated remote bop panel (drill floor).

Every rig is outfitted with one or more Remote BOP Closing System panels. These panels are manufactured in various configurations, but operate the same way. These panels use regulated rig air to remotely function each 4-way valve. For operation to occur, the panel's air supply must be energized (by holding lever or depressing a switch). with the lever or switch energized, the appropriate BOP is selected for closure (Figs 5.42–5.44).

The rig air is routed to the BOP Closing System by hard piping, arriving into pressure solenoids, which direct regulated and scrubbed source air to the appropriate 4-Way pneumatic piston. When energized, the piston is extended causing the 4-way handle to function closed.

Regulated air lines are routed through remote sensors to each 4-way valve piston for remote operation.

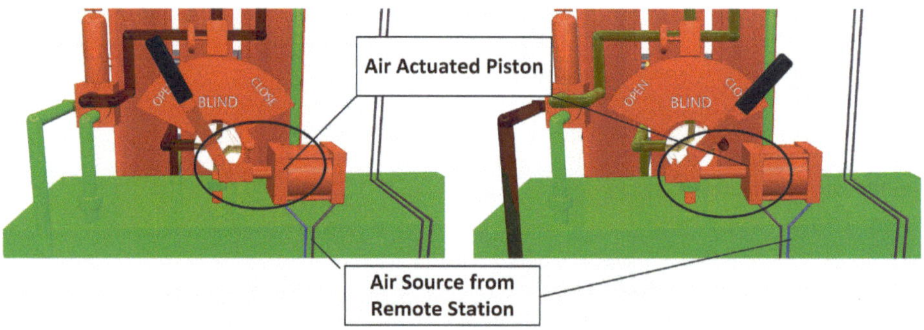

Fig. 5.43 Remote station air actuated 4-way valve pistons.

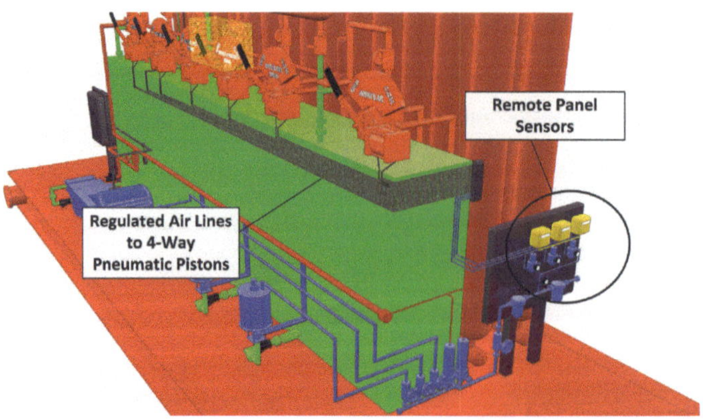

Fig. 5.44 Remote air operated 4-way valve piston.

Accumulator volume requirements

To ensure sufficient accumulator capacity to close all Blowout Preventers during an emergency, the accumulator should contain 1.5 times the volume necessary to close all BOP equipment with minimum of 200 psi above precharge pressure remaining in system. This provides a 50% safety margin providing complete replenishment of the fluid in the "closing chamber" lines at the time preventers are activated.

Six-step quick-check for BOP closing system

1. Ensure the airline is open to the air driven pumps.
2. Ensure electric starter switch is in AUTO position.
3. Accumulator bank isolation valve must be open.
4. Accumulator bank bleeder valve must be closed.
5. No four-way valves should be in the block position.
6. Accumulator, Manifold and Annular pressure gauges read correct pressure.

BOP closing system (accumulator) drill

To be performed alternately at BOP Closing System Unit and Remote BOP Closing System Panel (s).

1. Turn off all BOP Closing System pressurizing sources.
2. Record initial accumulator, manifold and annular pressures.
3. One at a time, close each preventer and record time to close. **DO NOT CLOSE BLIND RAMS**. Where applicable, substitute the opening of one pipe ram to simulate closure of the blind rams.
4. Record final accumulator, manifold and annular pressures.
5. To successfully pass the BOP Closing System drill, each BOP must close within the times listed below (Table 5.9):

Table 5.9 Closure times for BOPE.

Closure times for BOPEs	
Closing Ram BOP	$\leq$30 s
Closing Shear ram	$\leq$30 s
Annular BOP <18-3/4" ID	$\leq$30 s
Annular BOP $\geq$ 18-3/4" ID	$\leq$45 s
Choke and Kill valves	$\leq$Minimum observed ram close response time
Diverter Sequence	$\leq$45 s

The BOP Closing system remaining pressure must be at least (Table 5.10):

Table 5.10 Remaining system pressure after BOP closing system drill.

Precharge press (psi)	Remaining system pressure after BOP closing system drill
750	1290 psi accumulator pressure remaining in a 1000 psi BOP closing system
1000	1385 psi accumulator pressure remaining in a 2000 psi BOP closing system
1000	1500 psi accumulator pressure remaining in a 3000 psi BOP closing system
1500	1610 psi accumulator pressure remaining in a 5000 psi BOP closing system

6. Observe the remaining pressure for at least 5 min to detect any possible ram piston seal leaks.
7. Reopen the BOPs.
8. Vent reservoir tank to atmospheric by removing filling port cover. Open Accumulator Bank Bleeder Valve and drain fluid from accumulator bottles to reservoir tank. Observe accumulator pressure gauge for abrupt bleed-off.
   - Abrupt bleed-off pressure designates the lowest precharge in accumulator bottle system. If this pressure is below 990 psi, the accumulator system should be repaired.
9. Close Accumulator Bank Bleeder Valve and turn pump source on. The entire system should recharge in 15 min or less. The BOP Closing System drill is now concluded.

Sizing volume tables

See Tables 5.11–5.19.

Table 5.11 Sizing volume tables for annular BOPS (Part 1 of 3).
Annular blowout preventers

Model or type	Bore size (in)	Max working press (psi)	Hydraulic operating press (psi)	Gallons secondary (psi)	Gallons to close (gal)	Gallons to open (gal)
			MH Koomey			
N	7-1/16"	3000	1500		2.85	2.24
N	7-1/16"	5000	1500		3.86	3.3
N	7-1/16"	10,000	1500		9.42	7.08
N	7-1/16"	15,000	1500		11.2	7.5
N	11"	3000	1500		7.43	5.54
N	11"	5000	1500		9.81	7
N	11"	10,000	1500		25.1	18.98
N	13-5/8"	3000	1500		11.36	8.94
N	13-5/8"	5000	1500		17.98	14.16
N	13-5/8"	10,000	1500		37.18	26.5
N	21-1/4"	2000	1500		31.05	18.93

Continued

Table 5.11 Sizing volume tables for annular BOPS (Part 1 of 3)—cont'd
Annular blowout preventers

Model or type	Bore size (in)	Max working press (psi)	Hydraulic operating press (psi)	Gallons secondary (psi)	Gallons to close (gal)	Gallons to open (gal)
Hydril (Division of G.E. Oil and Gas)						
GK	7-1/16"	3000	1500		2.85	2.24
GK	7-1/16"	5000	1500		3.86	3.30
GK	7-1/16"	10,000	1500		9.42	7.08
GK	7-1/16"	15,000	1500		11.20	7.50
GK	7-1/16"	20,000	1500		10.90	7.20
GK	9"	3000	1500		4.33	3.41
GK	9"	5000	1500		6.84	5.80
GK	9"	10,000	1500		15.90	11.95
GK	11"	3000	1500		7.43	5.54
GK	11"	5000	1500		9.81	7.98
GK	11"	10,000	1500		25.10	18.97
GK	11"	15,000	1500		26.67	20.45
GK	13-5/8"	3000	1500		11.36	8.94
GK	13-5/8"	5000	1500		17.98	14.16
GK	HL 13-5/8"	5000	1500		17.98	26.50
GK	13-5/8"	5000 (9.75)	1500		34.53	24.66
GK	13-5/8"	10,000 (SH)	1500		37.18	26.50
GK	13-5/8"	10,000 (LH)	1500		37.18	12.59
GK	16-3/4"	2000	1500		17.46	15.80
GK	16-3/4"	3000/5000	1500		28.70	19.93
GX	11"	10,000	3000		17.88	17.88
GX	11"	15,000	3000		24.14	24.14
GX	13-5/8"	10,000	3000		24.14	24.14
GX	13-5/8"	15,000	3000		34.00	34.00
GL	13-5/8"	5000	3000	8.24	19.76	19.76
GL	16-3/4"	5000	3000	16.6	35.30	35.30
GL	18-3/4"	5000	3000	20	44.00	44.00
GL	21-1/4"	5000	3000	29.5	58.00	58.00

Table 5.12 Sizing volume tables for annular BOPs (Part 2 of 3).
Annular blowout preventers

Model or type	Bore size (in)	Max working press (psi)	Hydraulic operating press (psi)	Gallons secondary (psi)	Gallons to close (gal)	Gallons to open(gal)
Hydril (Division of G.E. Oil and Gas)—Continued						
MSP	7-1/16"	2000	1500		2.85	1.98
MSP	9"	2000	1500		4.57	2.95
MSP	11"	2000	1500		7.43	5.23
MSP	20-3/4"	2000	1500		31.05	18.93
MSP	21-1/4"	2000	1500		31.05	18.93
MSP	HL 21-1/4"	2000	1500		31.75	19.25
MSP	29-1/2"	500	1500		60	Vent
MSP	30	1000	1500		87.6	27.8
FSP	28"	2000	1500		90	75
Cameron International Corporation (Division of Schlumberger)						
A	7-1/16"	5000	1500		2.20	1.90
A	7-1/16"	10,000	1500		4.00	3.98
A	11"	5000	1500		7.80	7.56
A	11"	10,000	1500		12.10	10.34
A	13-5/8"	5000	1500		15.50	13.50
A	13-5/8"	10,000	1500		21.50	18.20
A	13-5/8"	5000	1500		33.00	26.40
D	7-1/16"	3000	3000		1.69	1.39
D	7-1/16"	5000	3000		1.69	1.39
D	7-1/16"	10,000	3000		2.94	2.55
D	7-1/16"	15,000	3000		6.94	6.12
D	7-1/16"	20,000	3000		8.38	7.56
D	11"	3000	3000		5.65	4.69
D	11"	5000	3000		5.65	4.69
D	11"	10,000	3000		10.15	9.06
D	11"	15,000	3000		23.50	21.30
D	13-5/8"	3000	3000		12.12	10.34
D	13-5/8"	5000	3000		12.12	10.34
D	13-5/8"	10,000	3000		18.10	16.15
D	16-3/4"	3000	3000		22.30	19.00
D	16-3/4"	5000	3000		22.30	19.00
D	18-3/4"	5000	3000		35.60	29.00
D	18-3/4"	10,000	3000		51.00	45.10
D	20-3/4"	3000	3000		40.50	28.40
D	21-1/4"	2000	3000		40.50	28.40

Table 5.13 Sizing volume tables for annular BOPS (Part 3 of 3).
Annular blowout preventers

Model or type	Bore size (in)	Max working press (psi)	Hydraulic operating press (psi)	Gallons secondary (psi)	Gallons to close (gal)	Gallons to open(gal)
Shaffer (Division of NOV)						
S	7-1/16"	3000	1500		4.57	3.21
S	7-1/16"	5000	1500		4.57	3.21
S	7-1/16"	10,000	1500		17.11	13.95
S	9"	3000	1500		7.23	5.03
S	9"	5000	1500		11.05	8.72
S	11"	3000	1500		11.00	6.78
S	11"	5000	1500		18.67	14.59
S	11"	10,000	1500		30.58	24.67
S	13-5/8"	3000	1500		23.50	14.67
S	13-5/8"	5000	1500		23.58	17.41
S	13-5/8"	10,000	1500		40.16	32.64
S	16-3/4"	5000	1500		33.26	25.61
S	18-3/4"	5000	1500		48.16	37.61
S	18-3/4"	10,000	1500		85.00	66.00
S	21-1/4"	2000	1500		32.59	16.92
S	21-1/4"	5000	1500		61.37	47.76
S	30"	1000	1500		122.00	55.00
Premier pressure control (Regan and Torus)						
Type K	7-1/16"	3000	3000		0.50	
Type K	7-1/16"	3000	3000		0.80	
Type K	7-1/16"	3000	3000		1.60	
Type K	9"	3000	3000		4.10	
Type K	11"	3000	3000		5.70	
Type K	11"	3000	3000		7.60	
Type K	11"	3000	3000		11.10	
Type K	11"	3000	3000		10.30	
Type K	13-5/8"	3000	3000		15.30	
Type K	13-5/8"	3000	3000		19.90	
Type K	16-3/4"	3000	3000		25.00	
Type KFD	16-3/4"	500	1000		3.00	
Type KFD	18-3/4"	500	1000		3.00	
Type KFD	21-1/4"	500	1000		3.00	
Type KFD	21-1/4"	500	1000		3.00	
Type KFDJ	11"	2000	2500		3.00	
Type KFL	13-5/8"	3000	Well + 500		19.50	
Type KFL	13-5/8"	5000	Well + 500		22.00	
Type KFL	13-5/8"	10,000	Well + 500		24.50	
Type KFL	16-3/4"	3000	Well + 500		25.75	
Type KFL	16-3/4"	5000	Well + 500		29.00	
Type KFL	16-3/4"	10,000	Well + 500		31.50	
Type KFL	20-3/4"	2000	Well + 500		28.50	

Continued

Table 5.13 Sizing volume tables for annular BOPS (Part 3 of 3)—cont'd
Annular blowout preventers

Model or type	Bore size (in)	Max working press (psi)	Hydraulic operating press (psi)	Gallons secondary (psi)	Gallons to close (gal)	Gallons to open(gal)
Type KFL	20-3/4"	3000	Well + 500		32.00	
Type KFL	20-3/4"	5000	Well + 500		35.00	
Torus	7-1/16"	3000	3000		4.50	
Torus	7-1/16"	6000	3000		4.50	
Torus	9"	3000	3000		8.25	
Torus	9"	6000	3000		8.25	

Table 5.14 Sizing volume tables for pipe or blind rams (Part 1 of 3).
Ram blowout preventers

Model or type	Bore size (in)	Working pressure max. (psi)	Hydraulic working press (psi)	Closing ratio	Operating ratio	Gallons to close (gal)	Gallons to open (gal)
Cameron International Corporation (Division of Schlumberger)							
QRC	7-1/16"	3000	1500			0.81	0.95
QRC	7-1/16"	5000	1500			0.81	0.95
QRC	9"	3000	1500			2.36	2.70
QRC	9"	5000	1500			2.36	2.70
QRC	11"	5000	1500			2.77	3.18
QRC	11"	10,000	1500			2.77	3.18
QRC	13-5/8"	3000	1500			4.42	5.10
QRC	16-3/4"	10,000	1500			6.00	7.05
QRC	17-3/4	5000	1500			6.00	7.05
U	7-1/16"	3000	3000	6.90:1	2.20:1	1.30	1.30
U	7-1/16"	5000	3000	6.90.1	2.20:1	1.30	1.30
U	7-1/16"	10,000	3000	6.90:1	2.20:1	1.30	1.30
U	11"	3000	3000	7.30:1	2.50:1	3.50	3.40
U	11"	5000	3000	7.30:1	2.50:1	3.50	3.40
U	11"	10,000	3000	7.30:1	2.50:1	3.50	3.40
U	11"	15,000	3000	9.80:1	2.20:1	6.20	6.10
U	13-5/8"	3000	3000	7.00:1	2.30:1	5.80	5.40
U	13-5/8"	5000	3000	7.00:1	2.30:1	5.80	5.40
U	13-5/8"	10,000	3000	7.00:1	2.30:1	5.80	5.40
U	13-5/8"	15,000	3000	10.60:1	3.60:1	10.60	10.40
U	16-3/4"	3000	3000	6.80:1	2.30:1	10.60	9.80
U	16-3/4"	5000	3000	6.80:1	2.30:1	10.60	9.80
U	16-3/4"	10,000	3000	6.80:1	2.30:1	12.40	11.60
U	18-3/4"	10,000	3000	7.40:1	3.70:1	23.10	21.20

Continued

Table 5.14 Sizing volume tables for pipe or blind rams (Part 1 of 3)—cont'd
Ram blowout preventers

Model or type	Bore size (in)	Working pressure max. (psi)	Hydraulic working press (psi)	Closing ratio	Operating ratio	Gallons to close (gal)	Gallons to open (gal)
U	20-3/4"	3000	3000	7.00:1	1.30:1	8.40	7.90
U	21-1/4"	2000	3000	7.00:1	1.30:1	8.40	7.90
U	21-1/4"	5000	3000	6.20:1	4.00:1	29.90	27.20
U	21-1/4"	10,000	3000	7.20:1	4.00:1	26.90	24.50
U	26-3/4"	3000	3000	7.00:1	1.00:1	10.80	10.10
U-Shear	11"	3000	3000	12.00:1	4.80:1	7.60	7.40
U-Shear	11"	5000	3000	12.00:1	4.80:1	7.60	7.40
U-Shear	11"	10,000	3000	12.00:1	4.80:1	7.60	7.40
U-Shear	11"	15,000	3000	15.20:1	3.70:1	9.00	8.90
U-Shear	13-5/8"	3000	3000	10.80:1	4.50:1	10.90	10.50
U-Shear	13-5/8"	5000	3000	10.80:1	4.50:1	10.90	10.50
U-Shear	13-5/8"	10,000	3000	10.80:1	4.50:1	10.90	10.50
U-Shear	13-5/8"	15,000	3000	16.20:1	6.00:1	16.20	16.00
U-Shear	16-3/4"	3000	3000	10.40:1	4.40:1	19.00	18.10
U-Shear	16-3/4"	5000	3000	10.40:1	4.40:1	19.00	18.10
U-Shear	16-3/4"	10,000	3000	10.40:1	4.40:1	19.10	18.20
U-Shear	20-3/4"	3000	3000	10.80:1	1.70:1	14.90	14.30
U-Shear	21-1/4"	2000	3000	10.80:1	1.70:1	14.90	14.30
U-II Pipe	18-3/4"	10,000	3000	6.70:1	2.50:1	24.70	22.30
U-II Pipe	18-3/4"	15,000	3000	9.30:1	3.50:1	34.70	32.30
T-Pipe	13-5/8"	10,000	3000	8.60:1	4.30:1	13.90	12.90
T-Pipe	18-3/4"	15,000	3000	6.70:1	3.10:1	24.30	22.40

Table 5.15 Sizing volume tables for pipe or blind rams (Part 2 of 3).
Ram blowout preventers

Model or type	Bore size (in)	Working pressure max. (psi)	Hydraulic working press (psi)	Closing ratio	Operating ratio	Gallons to close (gal)	Gallons to open (gal)
			MH Koomey				
PL-HD	7-1/16"	3000	1500	7.75:1	2.50:1	1.02	0.96
PL-HD	7-1/16"	5000	1500	7.75:1	2.50:1	1.02	0.96
PL-HD	7-1/16"	10,000	1500	7.75:1	2.50:1	1.02	0.96
PL-PRC	7-1/16"	3000	1500	7.75:1	2.50:1	1.02	0.96
PL-PRC	7-1/16"	5000	1500	7.75:1	2.50:1	1.02	0.96
PL-PRC	7-1/16"	10,000	1500	7.75:1	2.50:1	1.02	0.96

Continued

Table 5.15 Sizing volume tables for pipe or blind rams (Part 2 of 3)—cont'd
Ram blowout preventers

Model or type	Bore size (in)	Working pressure max. (psi)	Hydraulic working press (psi)	Closing ratio	Operating ratio	Gallons to close (gal)	Gallons to open (gal)
PL-PRC	11"	5000	1500	7.75:1	2.50:1	3.05	2.95
PL-PRC	11"	10,000	1500	7.75:1	2.50:1	3.05	2.95
PL-HD	13-5/8"	5000	1500	7.75:1	2.50:1	6.25	5.78
PL-HD	13-5/8"	10,000	1501	7.75:2	2.50:2	6.25	5.78
PL-PRC	13-5/8"	5000	1500	7.75:1	2.50:1	6.25	5.78
PL-PRC	13-5/8"	10,000	1501	7.75:2	2.50:2	6.25	5.78
PL-PRC	21-1/4"	2000	1500	7.75:1	2.50:1	12.65	12.18
PL-PRC	21-1/4"	5000	1500	7.75:1	2.50:1	9.7	9
PB-PRC	7-1/16"	15,000	1500	25.1[a]	25.1[a]	1.1	7
PB-PRC	11"	15,000	1500	25.1[a]	25.1[a]	0.75	3.3
PB-PRC	13-5/8"	5000	1500	25.1[a]	25.1[a]	2.8	2.8
PB-PRC	13-5/8"	10,000	1500	25.1[a]	25.1[a]	2.8	2.8
PB-PRC	13-5/8"	15,000	1500	25.1[a]	25.1[a]	3.54 Pipe 5.30 Shr	3.54 Pipe 5.30 Shr
PB-PRC	18-3/4"	10,000	1500	25.1[a]	25.1[a]	11.5 Pipe 19.30 Shr	11.50 Pipe 19.30 Shr
PB-PRC	18-3/4"	15,000	1500	25.1[a]	25.1[a]	11.5 Pipe 19.3 Shr	11.50 Pipe 19.30 Shr
			Shaffer(Division of NOV)				
SL/w 10" Piston	7-1/16"	10,000/ 15,000	3000	7.11:1	3.37:1	2.72	2.32
	13-5/8"	3000/5000	3000	5.54:1	3.00:1	5.44	4.46
	16-3/4"	50,000	3000	5.54:1	2.03:1	6.07	4.97
SL/w 14" Piston	7-1/16"	10,000/ 15,000	3000	13.94:1	7.14:1	6.00	5.57
	11"	10,000	3000	7.11:1	7.62:1	9.45	7.00
	11"	15,000	3000	7.11:1	2.80:1	9.40	8.10
	13-5/8"	5000	3000	10.85:1	10.02:1	11.00	10.52
	13-5/8"	10,000	3000	7.11:1	4.29:1	10.58	10.52
	13-5/8"	15,000	3000	7.11:1	2.14:1	11.56	10.52
	16-3/4"	5000	3000	10.85:1	5.77:1	11.76	10.67
	16-3/4"	10,000	3000	7.11:1	2.06:1	14.47	12.50
	18-3/4"	10,000	3000	7.11:1	1.83:1	14.55	13.21
	18-3/4"	15,000	3000	10.85:1	1.68:1	14.62	13.33
	21-1/4"	10,000	3000	7.11:1	1.63:1	16.05	13.86

Continued

Table 5.15 Sizing volume tables for pipe or blind rams (Part 2 of 3)—cont'd
Ram blowout preventers

Model or type	Bore size (in)	Working pressure max. (psi)	Hydraulic working press (psi)	Closing ratio	Operating ratio	Gallons to close (gal)	Gallons to open (gal)
LWS w/ Various Pistons ( )	7-1/16" (6.5)	5000	1500	5:45:01	1.93:1	1.45	1.18
	9" (8.5)	5000	1500	5.57:1	3.00:1	2.58	2.27
	11" (6.5)	3000	1500	5.45:1	1>16:1	1.74	1.45
	11" (8.5)	3000	1500	5.57:1	2.09:1	2.98	2.62
	11" (14)	5000	1500	16.00:1	3.41:1	9.50	8.90
	20-3/4" (8.5)	3000	1500	5.57:1	0.78:1	5.07	4.46
	20-3/4" (10)	30,000	1500	8.16:1	1.15:1	7.80	6.86
	20-3/4" (14)	3000	1500	16.00:1	2.21:1	14.50	13.59
	21-1/4" (8.5)	2000	1500	5.57:1	0.78:1	5.07	4.46
	21-1/4" (10)	2000	1500	8.16:1	1.15:1	7.80	6.86
	21-1/4 (14)	2000	1500	16.00:1	2.21:1	14.50	13.59

Effective opening and closing ratios of 25:1 are provided for calculation.
[a]Model PB (Pressure Balanced) Preventers do not oppose wellbore pressure to open or close.

Table 5.16 Sizing volume tables for pipe or blind rams (Part 3 of 3).
Ram blowout preventers

Model or type	Bore size (in)	Working Pressure max. (psi)	Hydraulic working press (psi)	Closing ratio	Operating ratio	Gallons to close (gal)	Gallons to open (gal)
Hydril (Division of G.E. Oil and Gas)							
Manual Lock Pipe Rams	7-1/16"	3000/5000	3000	4.80:1	1.50:1	1.00	0.93
	7-1/16"	10,000	3000	7.70:1	1.70:1	1.90	1.80
	7-1/16"	15,000	3000	7.10:1	6.60:1	3.70	3.40
	9"	3000/5000	3000	4.50:1	2.60:1	1.90	1.90
	11"	3000/5000	3000	6.00:1	2.00:1	3.30	3.20

Continued

Table 5.16 Sizing volume tables for pipe or blind rams (Part 3 of 3)—cont'd
Ram blowout preventers

Model or type	Bore size (in)	Working Pressure max. (psi)	Hydraulic working press (psi)	Closing ratio	Operating ratio	Gallons to close (gal)	Gallons to open (gal)
	11"	10,000	3000	6.90:1	2.40:1	5.20	5.00
	11"	15,000	3000	7.20:1	3.24:1	8.80	8.10
	13-5/8"	3000/5000	3000	4.80:1	2.10:1	5.40	4.80
	13-5/8"	10,000	3000	10.20:1	3.80:1	11.80	11.80
	20-3/4"	3000	3000	4.75:1	0.98:1	8.10	7.20
	21-1/4"	2000	3000	4.75:1	0.98:1	8.10	7.20
	21-1/4"	5000	3000	10.20:1	1.90:1	17.50	16.60
Manual Lock Shear Rams	11"	3000/5000	3000	5.60:1	4.20:1	5.50	5.00
	11"	10,000	3000	11.70:1	4.00:1	8.80	8.20
	11"	15,000	3000	7.20:1	3.24:1	8.80	8.20
	13-5/8"	3000/5000	3000	10.10:1	4.70:1	11.50	11.20
	13-5/8"	10,000	3000	10.20:1	3.80:1	11.80	11.80
	20-3/4"	3000	3000	10.14:1	2.20:1	17.20	16.30
	21-1/4"	2000	3000	10.14:1	2.20:1	17.20	16.30
	21-1/4"	5000	3000	10.20:1	1.90:1	17.50	16.60
Multi Position Lock Pipe Rams	7-1/16"	3000/5000	3000	5.40:1	1.50:1	1.20	0.93
	7-1/16"	10,000	3000	8.20:1	1.70:1	2.00	1.80
	7-1/16"	15,000	3000	7.60:1	6.60:1	3.90	3.40
	11"	10,000	3000	7.60:1	2.40:1	5.70	5.00
	13-5/8"	3000	3000	5.20:1	2.10:1	5.90	4.90
	13-5/8"	5000	3000	5.20:1	2.10:1	5.90	5.20
	13-5/8"	10,000	3000	10.60:1	3.80:1	12.90	11.80
	13-5/8"	15,000	3000	7.74:1	3.56:1	12.60	11.00
	16-3/4"	10,000	3000	10.60:1	2.41:1	15.60	14.10
	18-3/4"	10,000	3000	10.60:1	1.90:1	17.10	15.60
	18-3/4"	15,000	3000	7.27:1	2.15:1	19.40	16.70
	20-3/4"	3000	3000	10.60:1	0.98:1	18.00	16.30
	21-1/4"	2000	3000	10.60:1	0.98:1	18.00	16.30
	21-1/4"	5000	3000	10.60:1	1.90:1	19.30	16.60
Multi Position Lock Shear Rams	11"	3000/5000	3000	6.00:1	4.20:1	6.00	5.00
	11"	10,000	3000	12.40:1	4.00:1	9.30	8.20
	11"	15,000	3000	7.60:1	3.24:1	9.30	8.10
	13-5/8"	3000/5000	3000	10.60:1	4.70:1	12.00	11.20
	13-5/8"	10,000	3000	10.60:1	3.80:1	12.90	11.80
	13-5/8"	15,000	3000	7.74:1	3.56:1	12.60	11.00
	16-3/4"	10,000	3000	10.6-:1	2.40:1	15.60	14.10
	18-3/4"	10,000	3000	10.60:1	1.90:1	17.10	15.60
	18-3/4"	15,000	3000	7.27:1	2.15:1	19.40	16.70
	20-3/4"	3000	3000	10.60:1	2.20:1	18.00	16.30
	21-1/4"	2000	3000	10.60:1	2.20:1	18.00	16.30
	21-1/4"	5000	3000	10.60:1	1.90:1	19.30	16.60

Table 5.17 Sizing volume tables for hydraulically operated valves (Part 1 of 3).
Hydraulically operated valves

Model or type	Line size (in)	Working pressure max. (psi)	Hydraulic opening pressure	Gallons to close (gal)	Gallons to open (gal)
Cameron International Corporation (Division of Schlumberger)					
HCR	4"	3000	1500	0.52	0.61
HCR	4"	5000	1500	0.52	0.61
HCR	6"	3000	1500	1.95	2.25
HCR	6"	5000	1500	1.95	2.25
Type F	2"	960	1500	0.1	0.1
Type F	2"	3000	1500	0.1	0.1
Type F	2"	5000	1500	0.16	0.16
Type F	2"	15,000	1500	0.16	0.16
Type F	2"	960	1500	0.1	0.1
Type F	2"	3000	1500	0.1	0.1
Type F	2"	5000	1500	0.16	0.16
Type F	2"	15,000	1500	0.16	0.16
Type F	2-1/2"	960	1500	0.13	0.13
Type F	2-1/2"	3000	1500	0.13	0.13
Type F	2-1/2"	5000	1500	0.2	0.2
Type F	2-1/2"	15,000	1500	0.2	0.2
Type F	2-1/2"	960	1500	0.2	0.2
Type F	2-1/2"	3000	1500	0.2	0.2
Type F	2-1/2"	15,000	1500	0.4	0.4
Type F	3"	960	1500	0.15	0.15
Type F	3"	2000	1500	0.15	0.15
Type F	3"	3000	1500	0.24	0.24
Type F	3"	5000	1500	0.24	0.24
Type F	3"	10,000	1500	0.28	0.28
Type F	3"	15,000	1500	0.49	0.49
Type F	4"	3000	1500	0.3	0.3
Type F	4"	5000	1500	0.3	0.3
Type F	4"	10,000	1500	0.59	0.59
Type F	6"	3000	1500	0.84	0.84
Type F	6"	5000	1500	0.84	0.84
Type DV	4"	3000	1500	1.1	0.8
Type DV	4"	5000	1500	1.1	0.8
Type DV	6"	3000	1500	3.6	2.1
Type DV	8"	3000	1500	5.6	2.4
Type DV	10"	3000	1500	11.4	5.7
Type DV	10"	5000	1500	11.4	5.7
Type DV	12"	3000	1500	22.7	11.8

Table 5.18 Sizing volume tables for hydraulically operated valves (Part 2 of 3).
Hydraulically operated valves

Model or type	Line size (in)	Working pressure Max. (psi)	Hydraulic opening pressure	Gallons to close (gal)	Gallons to open (gal)
		M^C^Evoy (division of Schlumberger/Cameron)			
AC Valve w/ U-1	2"	2000	2500	0.11	0.13
Hydraulic Operator	2"	3000	2500	0.11	0.13
	2"	5000	2500	0.11	0.13
	2"	10,000	2500	0.20	0.21
	2-1/2"	2000	2500	0.23	0.26
	2-1/2"	3000	2500	0.23	0.26
	2-1/2"	5000	2500	0.23	0.26
	2-1/2"	10,000	2500	0.42	0.45
	3"	2000	2500	0.25	0.30
	3"	3000	2500	0.46	0.51
	3"	5000	2500	0.46	0.51
	4"	2000	2500	0.62	0.69
	4"	3000	2500	0.62	0.69
	4"	5000	2500	0.98	1.04
C Valve with RM-1	2"	5000	1500	0.10	0.11
Actuator	2-1/2"	5000	1500	0.12	0.13
	3"	5000	1500	0.23	0.25
	4"	5000	1500	0.44	0.50
E Valve with RM-1	1-13/16"	10,000	1500	0.08	0.09
Actuator	2-1/2"	10,000	1500	0.16	0.18
	2-9/16"	10,000	1500	0.30	0.33
	3-1/16"	10,000	1500	0.36	0.37
	4-1/16'	10,000	1500	1.00	1.07
EDU& EU w/ U-1	3"	5000	2500	0.47	0.52
Act	3-1/16"	10,000	2500	0.47	0.52
		Shaffer (Division of NOV)			
Flo-Seal	2" Reg	2000	3000	0.20	0.20
Flo-Seal	2"	2000	3000	0.20	0.20
Flo-Seal	2" Reg	3000	3000	0.20	0.20
Flo-Seal	2"	3000	3000	0.20	0.20
Flo-Seal	2" Reg	5000	3000	0.20	0.20
Flo-Seal	2"	5000	3000	0.20	0.20
Flo-Seal	2-1/16"	10,000	3000	0.40	0.40
Flo-Seal	2-1/16"	15,000	3000	0.40	0.40
Flo-Seal	2-1/2"	2000	3000	0.30	0.30
Flo-Seal	2-1/2"	3000	3000	0.30	0.30
Flo-Seal	2-1/2"	5000	3000	0.30	0.30
Flo-Seal	3"	2000	3000	0.30	0.30
Flo-Seal	3"	3000	3000	0.30	0.30

Continued

Table 5.18 Sizing volume tables for hydraulically operated valves (Part 2 of 3)—cont'd
Hydraulically operated valves

Model or type	Line size (in)	Working pressure Max. (psi)	Hydraulic opening pressure	Gallons to close (gal)	Gallons to open (gal)
Flo-Seal	3"	5000	3000	0.30	0.30
Flo-Seal	3-1/16"	10,000	3000	0.60	0.60
Flo-Seal	4"	3000	3000	0.80	0.80
Flo-Seal	4"	5000	3000	0.80	0.80
Flo-Seal	4-1/16"	10,000	3000	0.80	0.80
Flo-Seal	6"	3000	3000	0.80	0.80
Type DB	2-1/16"	5000	3000	0.15	0.20
Type DB	2-1/16"	10,000	3000	0.15	0.20
Type DB	2-1/16"	15,000	3000	0.26	0.29
Type DB	3-1/8"	5000	3000	0.20	0.25
Type DB	3-1/16"	10,000	3000	0.35	0.40
Type DB	3-1/16"	15,000	3000	0.35	0.40
Type DB	4-1/16"	5000	3000	0.35	0.40
Type DB	4-1/16"	10,000	3000	0.45	0.50
Type DB	4-1/16"	15,000	3000	0.45	0.50

Table 5.19 Sizing volume tables for hydraulically operated valves (Part 3 of 3).
Hydraulically Operated Valves

Model or type	Line Size (in)	Working pressure Max. (psi)	Hydraulic opening pressure	Gallons to close (gal)	Gallons to open (gal)
			Shaffer (Division of NOV)		
Flo-Seal with Ram Lock	2" Reg	2000	3000	0.3	0.3
	2"	2000	3000	0.3	0.3
	2" Reg	3000	3000	0.3	0.3
	2"	3000	3000	0.3	0.3
	2" Reg	5000	3000	0.3	0.3
	2"	5000	3000	0.3	0.3
	2-1/16"	10,000	3000	0.4	0.4
	2-1/16"	15,000	3000	0.4	0.4
	2-1/2"	2000	3000	0.3	0.3
	2-1/2"	3000	3000	0.3	0.3
	2-1/2"	5000	3000	0.3	0.3
	3"	2000	3000	0.4	0.4
	3"	3000	3000	0.4	0.4
	3"	5000	3000	0.4	0.4
	3-1/16"	10,000	3000	0.4	0.4
	4"	3000	3000	0.6	0.6
	4"	5000	3000	0.8	0.8
	4-1/16"	10,000	3000	0.8	0.8
	6"	3000	3000		

Mud gas separators (MGS)

Mud Gas Separators (MGS) are designed to provide a physical means to effectively separate mud and gas circulated from well. In addition, MGS are designed to provide means to safely vent gas at atmospheric pressure and return degassed mud to the active mud pits. Depending on mud properties, a vacuum or impingement type degasser, normally located within the mud separating and processing equipment of the mud pits, may remove residual entrained gas exiting the mud gas separator.

With the well shut-in after a kick has occurred and without broaching the casing seat, using various constant bottom-hole kill techniques, the kick can be safely circulated to the surface. To ensure safety when gas gets to surface, the MGS must be of sufficient size control and process the amount of gas, which will be liberated during the transition from pressurized wellbore fluids to atmospheric pressure.

Processing rate

MGS are designed and selected largely based upon processing rates of gas defined in MMSCFPD measurements. The processing rate is listed on the "identification plate" containing serial number, model number, etc. Most MGS used in the oilfield are designed to process about 35 MMSCFPD.

Unfortunately, not every MGS contains processing rates or may lack identification plate. The following is a "rule of thumb" that can be used to estimate processing rates. If the MGS processing rate is not known and the kick is of relatively large size within OBM, a rental MGS may be used as an option to ensure safety during a well control event.

$$Q = 0.008085\, x\, \text{SICP}_{\text{psi}} \times \text{SCR}_{\text{spm}} \times \text{PumpOutput}_{\text{bps}} \times \text{Pump Efficiency}_{\%} \qquad (5.8)$$

where:

Q = Estimated Flow rate (MMSCFPD)
0.008085 = Conversion factor
SICP = SICP at initial conditions (psi)
SCR = Slow Circulating Rate (stk/min)
Pump Output = Pump Output * Efficiency (bbl/stk)
Pump Efficiency = Rated Efficiency (98%)

Example:

With kick shut-in, stabilized SICP = 300 psi
SCR = 40 spm
Pump Output = 0.0998 bbl/stk with 98% efficiency

$$Q = 0.008085 \times 300\,\text{psi} \times 40\,\text{spm} \times 0.0998\,\text{bps} \times 0.93 = 9.5\,\text{MMSCFPD} \qquad (5.9)$$

Using the Comparison Guideline Table below, in order to ***process 9.5 MMSCFPD, the selected MGS of 6' Diameter (72 in)*** will be required. Generally speaking, the larger the diameter of the MGS, the more gas processing will be available (Tables 5.20 and 5.21).

Table 5.20 Comparison table of MGS diameters.

Description	2′ MGS	4′ MGS	6′ MGS	8′ MGS	10′ MGS	12′ MGS
Separator diameter	2 ft	4 ft	6 ft	8 ft	10 ft	12 ft
Feed Inlet Pipe Diameter	8 in	8 in	8 in	8 in	8 in	8 in
Ht. between inlet and gas outlet	1.5 ft	1.5 ft	1.5 ft	1.5 ft	1.5 ft	1.5 ft
Vent Line diameter	12 in	12 in	12 in	12 in	12 in	12 in
Mud Seal Height	10 ft	10 ft	10 ft	10 ft	10 ft	10 ft
Est. Separator Capacity (MMSCFPD)	1.4	5.6	12.6	22.5	35.2	50.7

Table 5.21 Design criteria used for comparison Table 5.21.

- TVD = 10,000 ft
- Original Mud Wt in Well (ppg) = 13 ppg
- Annular Cap (DP/Csg) = 0.0703 bpf
- Slow Circ Rate (spm) = 30 spm
- Slow Circ Rate Press (psi) = 790 psi
- Pump Output (bbl/stk) = 0.0998 bps
- Slow Circ Rate (bbl/min) = 3 bpm
- Vent Line Effective Length (ft) = 200 ft
- Stabilized SIDPP (psi) = 200 psi
- Stabilized SICP (psi) = 300 psi
- Pit Gain (bbls) = 10 bbls
- Kill Mud Weight (ppg) = 13.4 ppg
- Gas Gradient = 0.1 psi/ft

The diameter of the MGS is maintained within regulated overland road transport limitations. Therefore, land MGS designs are normally limited by diameter size and length. Offshore MGS may be constructed in larger diameter within the shipyards.

MGS design considerations

Every Mud Gas Separator incorporates three basic functions. The primary function will be to provide means to efficiently liberate free gas from the fluid stream. The primary function occurs as the fluid stream enters and impacts the wall or impingement plate of the separator (Fig. 5.45).

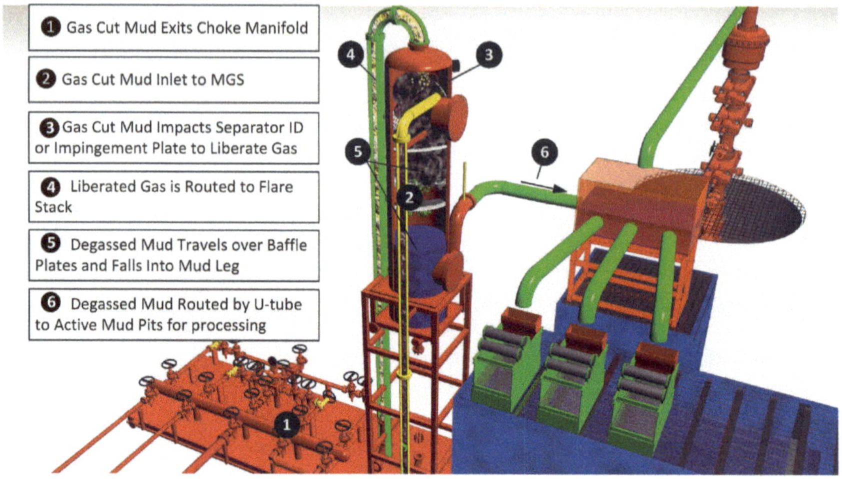

Fig. 5.45 Vertical two-phase separator process.

The secondary function consists of providing a means to lower surface tension of the fluid and allow more efficient liberation of free gas. As the fluid cascades over the perforated baffle plates, the fluid is spread into thin layers that lower the surface tension of the fluid. The fluid cascades much like a water fall from one baffle plate to another, also providing means of liberating free gas.

The final stage is allowing the degassed fluid to fall to the bottom of the MGS. As the fluid volume increases, the fluid is forced from the separator downstream to the mud processing tank/equipment. Separators are designed to maximize the retention time for the fluid to remain in the simulator to maximize gas recovery.

Four step design review

Step 1: Minimum Diameter: The efficiency and effectiveness of gas processing (separation capacity) is largely dependent upon two design criteria. The fixed diameter of the vessel relegates the overall peak gas processing by defining surface area that is employed to define separation ability. As vessel diameter increases, the processing volume also increases dramatically. Minimum diameter limitations will be defined for worst-case analysis of deepest well design with maximum processing envelopes. For drilling applications, minimum diameters should be limited to no less than six feet (72 in.) (Fig. 5.46).

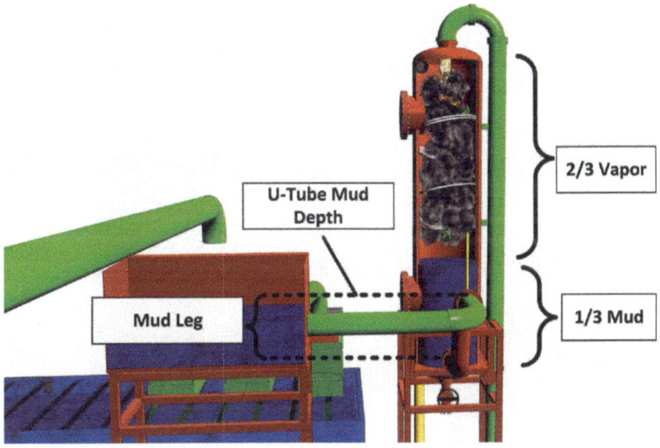

Fig. 5.46 Mud leg should be approximately 1/3rd of MGS height.

Step 2: Minimum Height. The height of the separator will determine "retention time" or the time gas-laden fluids are processed. Once again, the height and retention time should be determined for the worst-case analysis of deepest well design at maximum processing envelopes. Most normal operating designs begin at 18 ft.

Step 3: Friction Pressure of Vent Line. The back pressure that occurs as the gases flow out the vent line will affect the separation capability. In order to minimize this back pressure, vent lines should be sized as large as practical. Furthermore, these vent lines should be constructed as straight as possible (minimize all turns and eliminate any 90 degree turns). 90 degree turns should be replaced by two 45 degree turns with a minimum of 8 ft. between turns.

Step 4: Mud Leg Height. Mud leg height is determined by the distance from source mud (mud pit, flow director, shale shaker flow weir, etc.) and the vertical distance represented within the MGS due to "U" tube effects. The physical vertical location of the MGS and "U-tube" effects of the source mud will determine the active mud level within the MGS (Figs. 5.47–5.50).

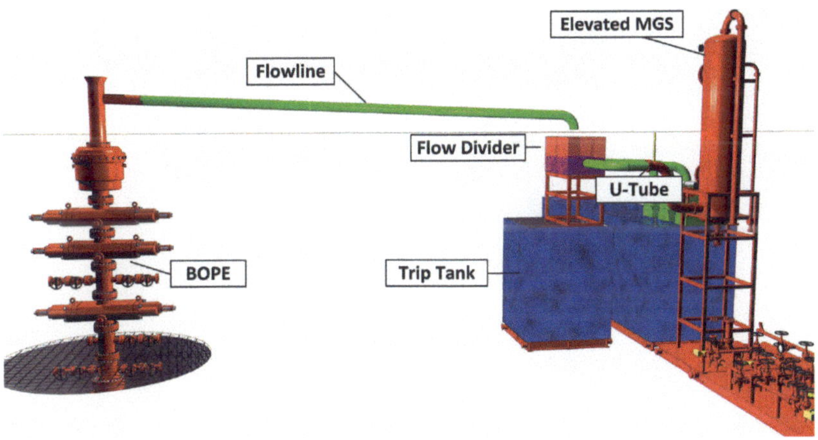

Fig. 5.47 Elevated MGS plumbed to flow divider.

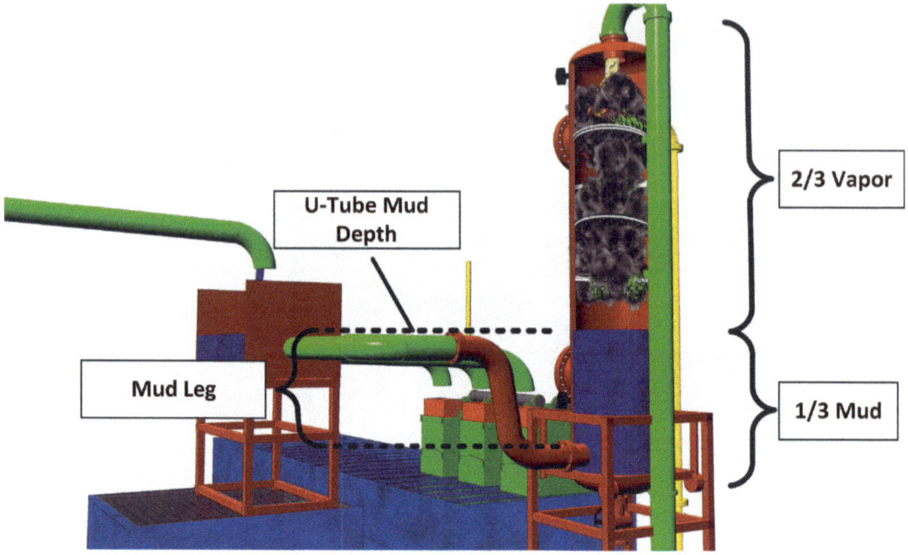

Fig. 5.48 Elevated MPG cutaway diagram.

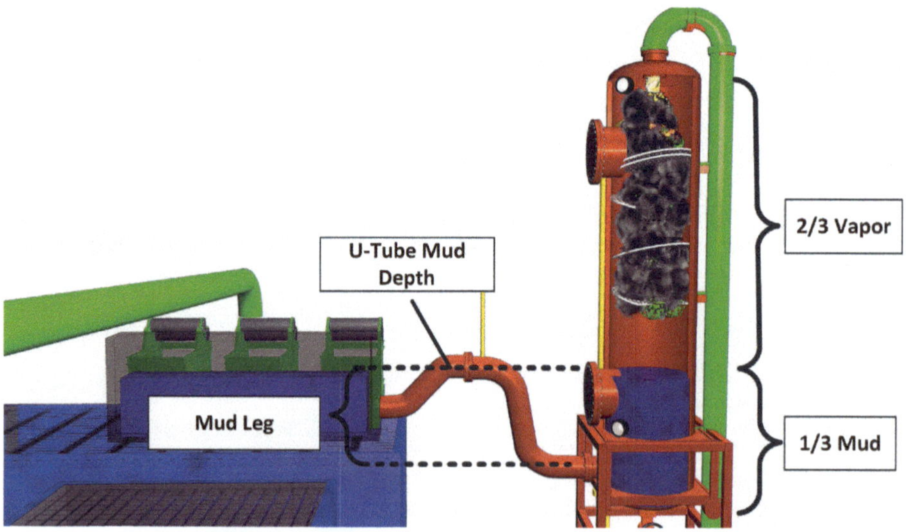

Fig. 5.49 Elevated MGS plumbed to flow divider or flow header.

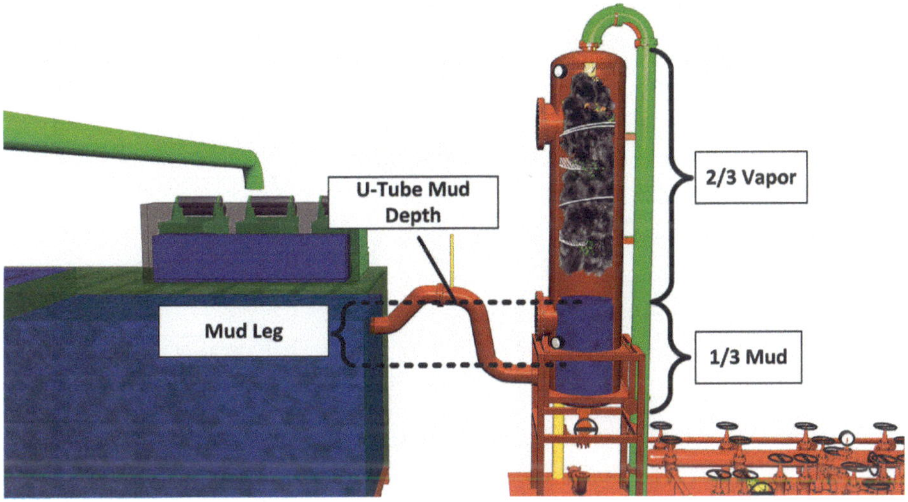

Fig. 5.50 Elevated MGS plumbed to mud pit.

Under no circumstances should the baffle plates be submerged in the processed mud. To prevent submergence of baffle plates within processing mud, operating mud level must remain at least 18" below the lowest baffle plate.

Overall, a simple 1/3rd mud and 2/3rds gas processing volume will achieve these design criteria. The mud leg should not occupy more than 1/3rd of the internal volume of the MGS. This level should be adjusted to ensure no baffle plates are submerged within the operating mud volume. The remaining 2/3rd of the separator volume is dedicated to vapor from the liberation of free gas.

Closed bottom design (most preferred MGS)

Closed Bottom MGS is fully enclosed vessel with a mud return valve and piping system located at the lowest point of the vessel. The mud exiting the vessel enters a mud line and directed back to the pit through an inverted "U"-shaped bend. Fluid level within the Close Bottom Separator may be adjusted by manipulating the length of the U-shaped bend (Fig. 5.51).

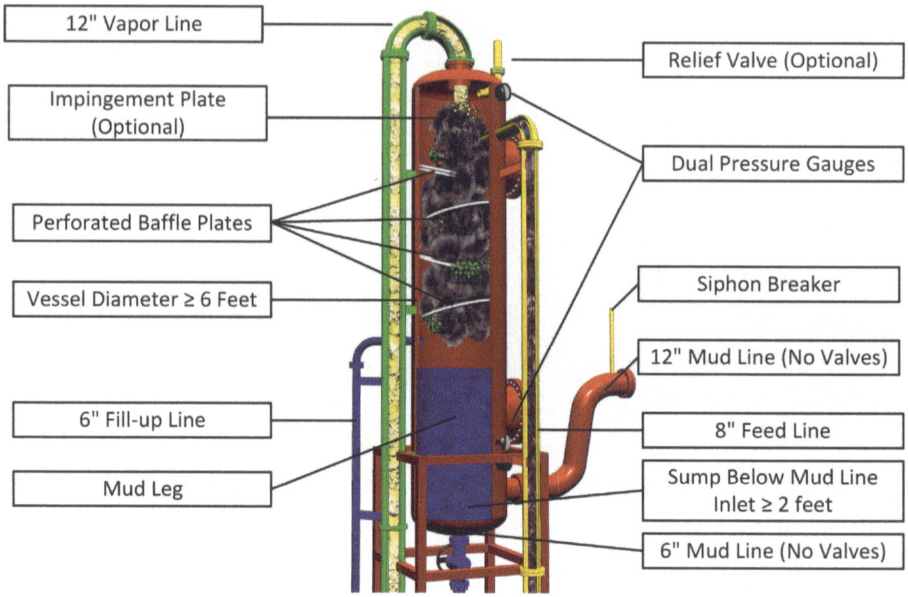

Fig. 5.51 Closed bottom separator features.

CLOSE BOTTOM MUD GAS SEPARATORS ARE MOST PREFERRED TYPE
FOR GREATEST CAPACITY, UTILIZE LARGE DIAMETER SEPARATOR INLET, GAS VENT LINE AND MUD RETURN LINE IMPERATIVE FOR PROPER FUNCTIONING, ALL MANWAYS ARE PROPERLY SEALED.
VENT LINES AND FLARE LINES ARE TO BE AT LEAST 12" DIAMETER

Preparation to perform kill operations

MGS operating mud leg volumes, processing capabilities, inlet fluid composition, rates, and complexities of gases liberated yield pressure variances within the separator during well control events. The mud leg (comprising of the hydrostatic pressure of the mud over the vertical height of the fluid within the separator isn't the mud leg the vertical height of the mud in the separator? If it is higher it will flow out) determines the operating envelop. This pressure is developed based upon the original fluid density contained by the source fluid. As most MGS are stored without fluid, the separator mud leg volume must be established before well control operations begin by adding fluid via the fill-up line.

Blow through

During well control operations, blow through may occur if the MGS is unable to efficiently process the amount of fluid and gases entering/exiting the separator. If the processing fluid rates exceed the MGS processing capability, fluid may enter the vent line resulting in a back pressure being applied to the MGS. As the internal MGS operating pressure increases due to vent line back pressure, the mud leg may be pushed out by gases (called blow through) and gases will be directed to mud pit area. If gases escape into the mud pit area, the chances of an explosion become greatly elevated.

Blow through response

Since blow through is largely dependent upon processing capabilities of the MGS, a reduction in flow rate may be necessary to prevent blow through from occurring. Follow established practices in lowering rates by predetermining new pressures based upon the following equation.

$$\text{New PumpPressure}_{psi} = \left(\frac{\text{New SPM}}{\text{Old SPM}}\right)^2 x \text{ Old PumpPressure}_{psi} \tag{5.10}$$

where:

New Pump Pressure$_{psi}$ = New Pump Pressure (psi)
New SPM = New SCR (spm)
Old SPM = Old SCR (spm)
Old Pump Pressure$_{psi}$ = Old Pump Pressure (psi)

If blow through occurs, shut down operations and shut-in using established well control procedures. The mud pit area should be evacuated until such time operations can resume safely. In preparation to start-up and continue well control operations, the MGSs mud leg must be reestablished by pumping whole mud from the active pit into the MGS. The vent line should be inspected to ensure any blockages

Dual pressure sensors or gauges (most preferred)

In order to prevent blow through from occurring, the MGS operating pressures can be monitored throughout the well kill operation. The most accurate method of monitoring operating pressures is the use of two (or more) pressure gauges or sensors. If pressure gauges are to be used, they should be positioned for easy reading and may require individual external illumination sources for night time observation. For remote observation, the use of pressure sensors connected to pressure gauges on the Driller's Console are preferred (Fig. 5.52).

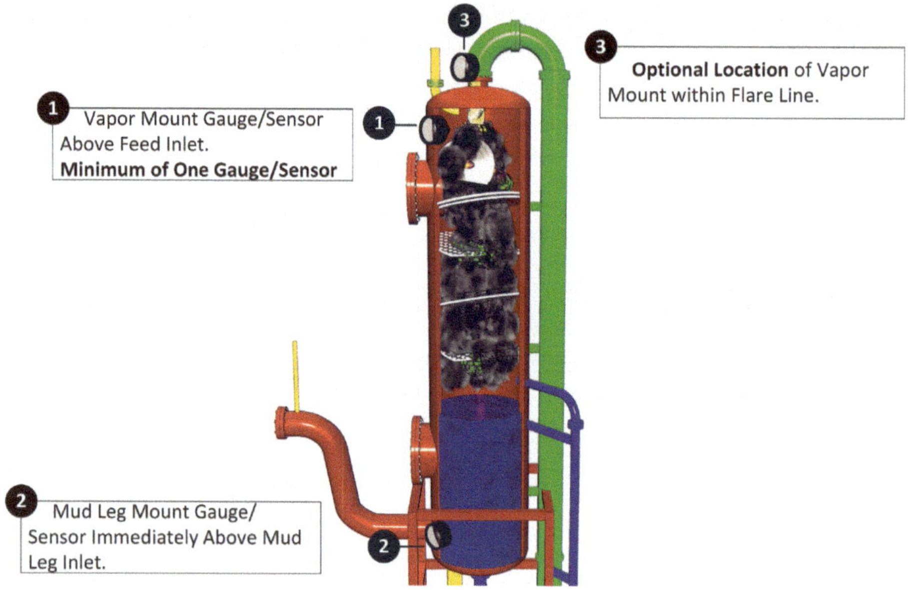

Fig. 5.52 Dual sensor (gauge) locations.

The vent line sensor/pressure gauge is to be mounted above the Inlet Line Feed. The mounting point should be located away from mud "splatter" as the inlet fluids enter the separator. An alternate mounting point is within the vent line near the MGS. This pressure gauge will measure the vent line pressures during operation. If the vent line pressures increase, the probabilities of fluid entering the vent line increase.

The lower sensor/pressure gauge should be mounted above the mud leg inlet. This gauge measures the overall hydrostatic pressure of the fluid column within the simulator. If the hydrostatic pressure decreases, blow through may occur.

Continual observations of these pressure gauges will give an indication of increasing chances of blow through. If Vapor Gauge ~ Mud Leg Gauge, blow through is imminent.

Vapor Gauge (0—10 psi) (Figs. 5.53–5.55).

Fig. 5.53 Initial Conditions.

Fig. 5.54 Peak gas to surface.

Fig. 5.55 Plugging flare line.

Mud Leg Gauge (0—20 psi) (Figs. 5.56–5.58).

Fig. 5.56 HP of mud leg.

Fig. 5.57 HP with degassed mud.

Fig. 5.58 Blow through (Vapor ~ Mud Leg).

One pressure sensors or gauges (good)

For MGS separators containing one pressure sensor, the proper location would be near the vent line or within the vent line. Unfortunately, most single pressure gauges are located in the upper half the MGS separator. Regardless of the location, this gauge should be routinely monitored during well control operations. If the gauge shows increasing pressure, the chances of blow through become elevated. (increasing pressure below shows flare line plugging. We need to differentiate)

Vapor Gauge (0—10 psi) (Figs. 5.59–5.61).

Fig. 5.59 Initial conditions.

Fig. 5.60 Peak gas to surface.

Fig. 5.61 Plugging flare line.

During well control operations, vapor gauge should be continually monitored to identify partial plugging of flare line represented as an increase in pressure. Blow through limits cannot be determined from one gauge. If partial plugging is observed, the well can be shut-in and circulation stopped followed by investigation and mitigation of plugged flare line. As circulation is established, to lower the chances of potential flare line plugging, a lower circulation rate is warranted.

No pressure gauge

Most MGS lack pressure sensors or gauges. Therefore, the operating pressures within the separator are unknown and little foresight is given for blow through conditions. For these simulators, the vent line should be observed for wellbore fluids. With long distances between the MGS and flare pit, such observations may be difficult.

Open bottom design (not preferred)

The Open Bottom MGS design incorporates a wide-open lower section of the MGS submerged within the small mud tank. The mud level control within the MGS is dependent upon the mud contained within the mud tank and the adjusted level of the MGS within the tank. If blow through were to occur, gases would accumulate within the walls of the tank creating a confined space entry situation. Venting of gases heavier than air (such as H_2S) within a confined space, would be problematic. Therefore, this vessel design is not recommended for well control applications (Fig. 5.62).

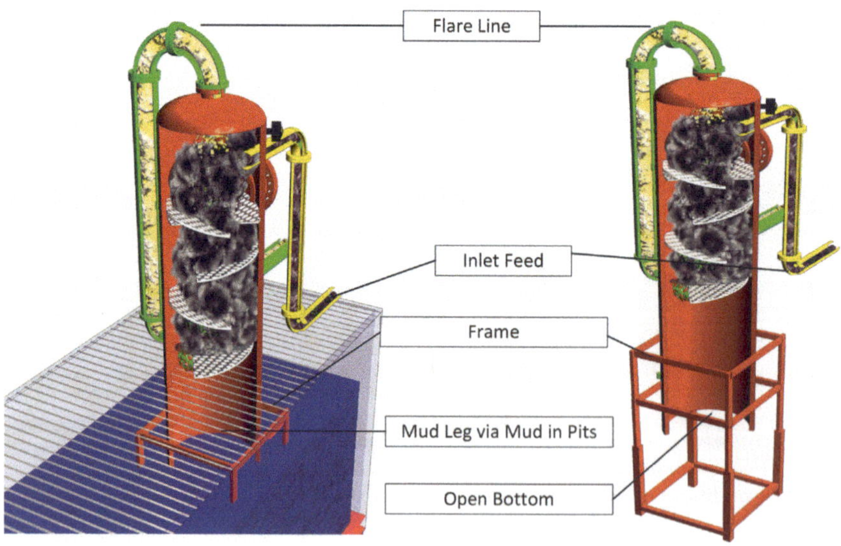

Fig. 5.62 Open bottom separator.

Float type MGS (not preferred)

The Float Type system consists of an automatic fluid leveling system contained within a closed bottom MGS. The float system is comprised of a float within the MGS that floats on top of the degassed mud. The float is connected via pneumatic or manual linkages to an operating valve located on the mud leg (shouldn't this say mud inlet). As the float is raised, the valve is partially closed to reduce the amount of fluid entering the MGS. Conversely, when the float is lowered, the valve is partially opened allowing more fluid to enter the MGS (Fig. 5.63).

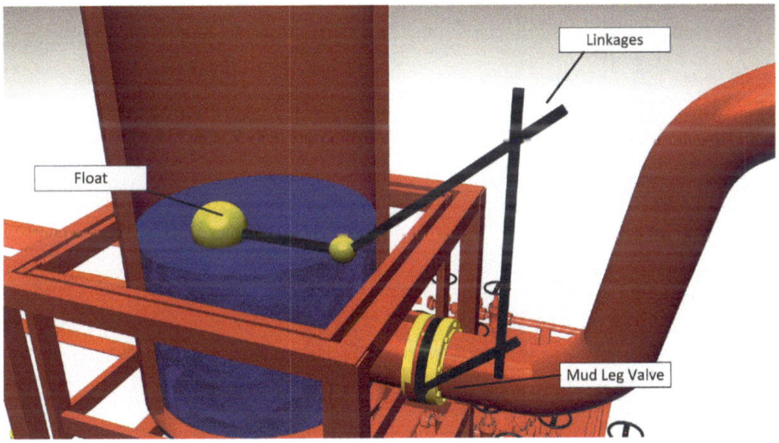

Fig. 5.63 Float type of MGS.

The float system is prong to linkage failure that may result in closure of the throttle valve within the mud leg. If closed, mud levels within the MGS will rise causing overflow through the vent line. The actual float spheres allow accumulations of solid material (dried mud) causing them to become heavier. The mud leg "throttle" valves are also prone to plugging due to solids settling from muds within the mud leg line. The float type system requires continuous maintenance to ensure proper operation. For these reasons, the float type MGS is not preferred for well control operations.

Well/preventer stack classification

Class 1 well

The Class 1 preventer stack is intended for use only on wells where there is no known chance of flow to surface. This is those wells with static fluid levels below ground level (Fig. 5.64).

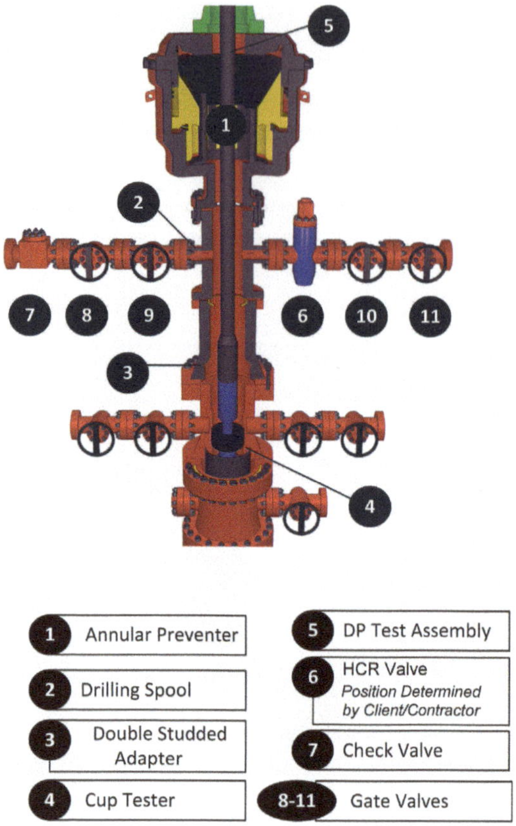

Fig. 5.64 Class 1 BOP stack.

It is composed of a single annular preventer that is capable of closing on almost any surface passed through it and in an emergency situation, might be shut-in on open hole.

Note: According to manufacturer's recommendation, annular preventers may be shut on an "open" hole once. They cannot be closed on open hole for second or third time.

Note: If shut on open hole, the packing element must be changed out prior to use again.

Testing

The BOPE should be tested on every installation or every 21 days, whichever is less.

Maintenance

During normal operations, packing elements should be replaced when they fail a pressure test or when visual inspection indicates the element might be damaged in any manner. Ensure the stack is properly washed out after any cement job to ensure the cement will not solidify behind element rendering the element useless.

THESE MAINTENANCE PRACTICES SHOULD BE USED FOR ALL ANNULARS ON ANY STACK CONFIGURATION

Class 2 wells

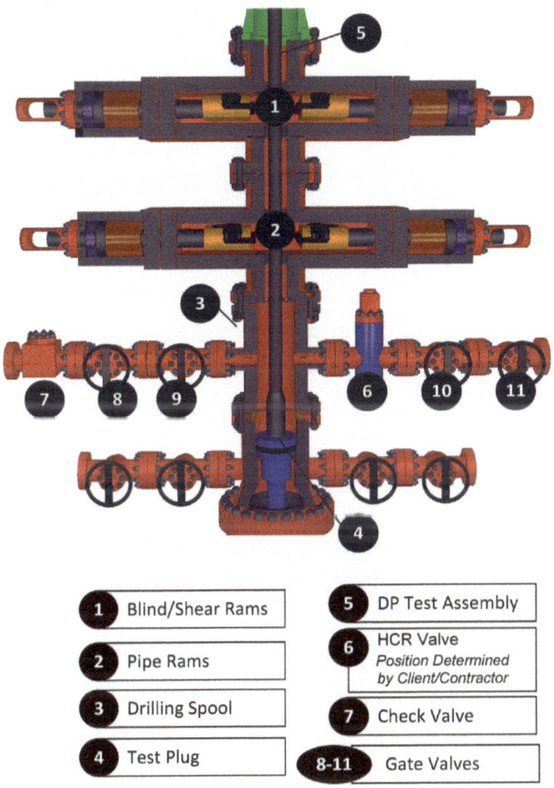

Fig. 5.65 Class 2 BOP stack.

The Class 2 preventer stack is intended for use on any well with a low GOR (i.e., no potential for free-flowing gas) and where maximum anticipated shut-in or pump-in pressure is less than 1000 psi. It is composed of a single blind or blind/shear ram, with or without a spool or adaptor (Fig. 5.65).

This well classification is intended for use on any well with a low GOR (i.e., no potential for free-flowing gas) and where maximum anticipated shut-in or pump-in pressure is less than 1000 psi. It is composed of a single blind or blind/shear ram, with or without a spool or adaptor.

This arrangement can still close on almost any surface passed through it. However, it has the ability to shut on open hole without pipe in the hole or by dropping the pipe) with a blind ram or by shearing the pipe with a blind/shear ram. The outlets on the casing should be utilized for circulating, etc., only if there is no spool or if the outlets on the ram cannot be used.t

Testing

The BOPE should be tested on every installation or every 21 days, whichever is less. If the blind or blind/shear rams cannot be tested on location they should be stump tested every three months.

Maintenance

In high temperature use (i.e., , steam flood) manufacturers recommend changing the **annular** packing element every two months. During normal operations the annular or ram elements should be replaced whenever they fail a pressure test or when visual inspection indicates the elements are damaged in any manner. The blind rams should be shopped immediately if they fail an operator test. In addition, the blind or blind/shear rams should also be shopped every three years for teardown and inspection.

Class 3 wells

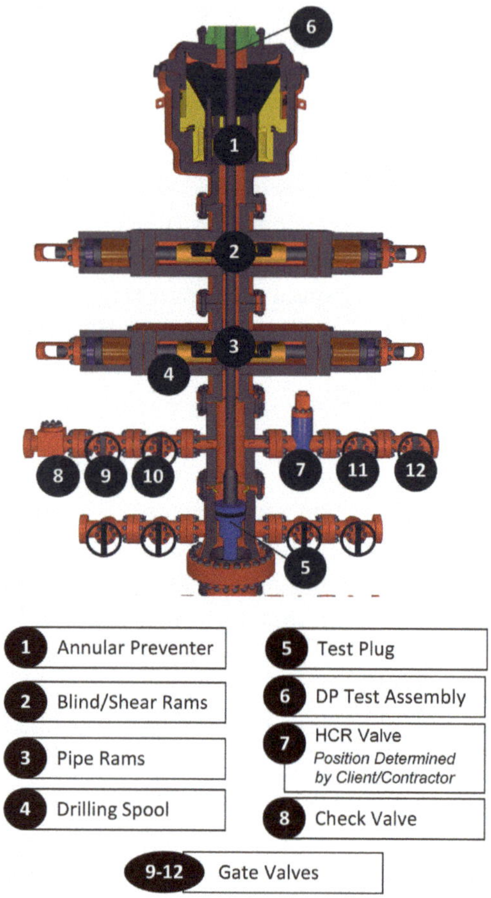

Fig. 5.66 Class 3 BOP stack.

The Class 3 preventer stack is intended for use on any well with the potential for free-flowing gas or where maximum shut-in or pump-in pressures are in excess of 2000 psi. It is composed of a double ram (pipe ram on bottom) and an annular preventer (with or without a spool or adaptor flange) (Fig. 5.66).

This arrangement has the ability to close on most any surface passed through it (annular) but also has the ability to close metal rams on the pipe if the annular preventer fails or if the wellhead pressure closely approached the working pressure of the annular preventer. The blind ram allows the well to be shut-in on open hole. The outlets on the casing should be utilized for circulating, etc., only if there is no spool and the outlets on the rams can't be used. The preferable kill, choke or circulating valves should be between the rams.

Testing

The BOPE should be tested on every installation or every 21 days, whichever is less.

Maintenance

During normal operations, the annular or ram elements should be replaced whenever they fail a pressure test or when visual inspection indicates the elements are damaged in any manner. The rams should be shopped immediately if they fail an operator test. In addition, the rams should also be shopped every three years for teardown and inspection.

Class 4 wells

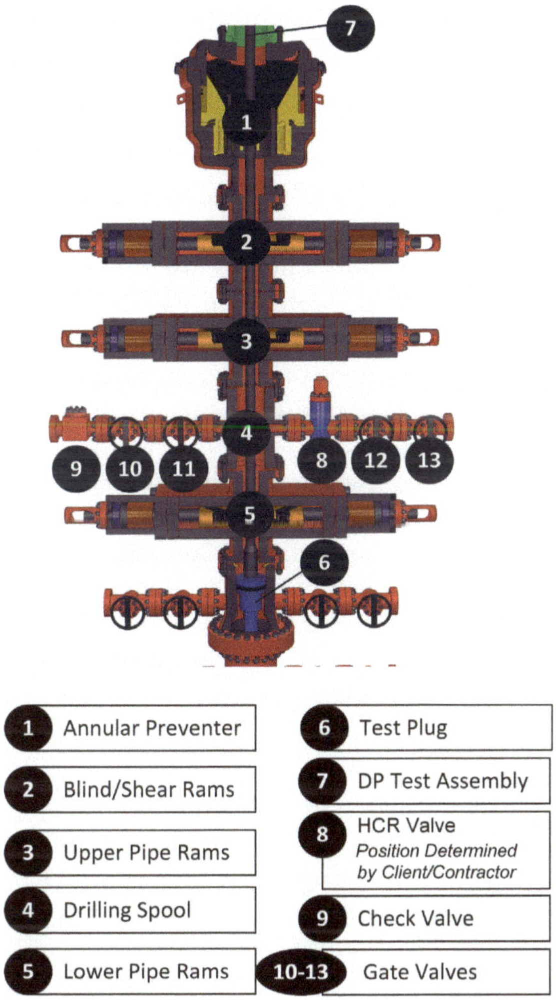

Fig. 5.67 Class 4 BOP stack.

The Class 4 preventer stack is designed for operations where the anticipated surface pressure is greater than 5000 psi and the drill string is nontapered (unless variable bore rams are used). The stack is composed of a single hydraulically operated annular preventer on top, then a blind ram preventer, a single upper pipe ram preventer, a drilling spool, and a single lower pipe ram preventer on the bottom (Fig. 5.67).

The choke and kill lines are installed onto the drilling spool, and have a minimum internal diameter of 3". All side outlets on the preventers or drilling spool must be flanged, studded, or clamped. An emergency kill line may be installed on the wellhead. A double ram preventer may be used for the blind rams and upper pipe rams in all instances if a drilling spool is being used. If this stack is used in conjunction with a tapered drill string, a set of variable bore pipe rams should be installed in the upper pipe ram preventer and large pipe rams should be installed in the lower pipe ram preventer.

Testing

The BOPE should be tested on every installation or every 21 days, whichever is less.

Maintenance

During normal operations, the annular or ram elements should be replaced whenever they fail a pressure test or when visual inspection indicates the elements are damaged in any manner. The rams should be shopped immediately if they fail an operator test. In addition, the rams should also be shopped every three years for teardown and inspection.

Class 5 wells

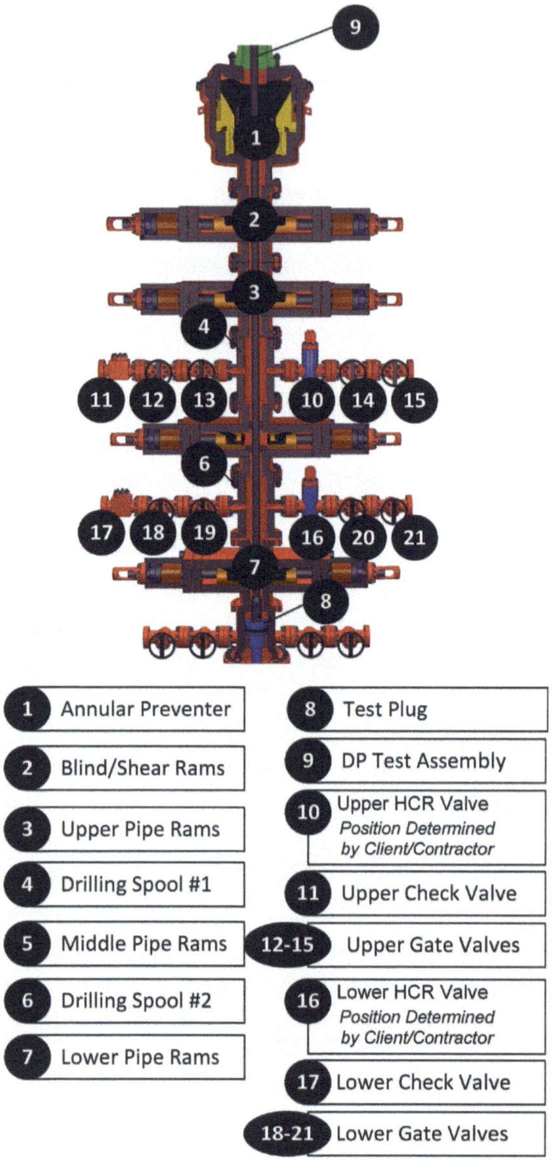

Fig. 5.68 Class 5 BOP stack.

The Class 5 preventer stack is designed for High-pressure/High Temperature operations where the anticipated surface pressure is greater than 10,000 psi and the drill string is nontapered (unless variable bore rams are used). The stack is composed of a single hydraulically operated annular preventer on top, then a series of four preventers with drilling spools located between a set of two preventers (Fig. 5.68).

The choke and kill lines are installed onto each drilling spool, and have a minimum internal diameter of 4-1/16". All side outlets on the preventers or drilling spool must be flanged, studded, or clamped. An emergency kill line may be installed on the wellhead.

A double ram preventer may be used for the upper pipe rams and blind/shear rams. If this stack is used in conjunction with a tapered drillstring, a set of "variable bore" pipe rams should be installed in the upper pipe ram preventer and large pipe rams should be installed in the lower pipe ram preventer.

Testing

The BOPE should be tested on every installation or every 21 days, whichever is less.

Maintenance

During normal operations, the annular or ram elements should be replaced whenever they fail a pressure test or when visual inspection indicates the elements are damaged in any manner. The rams should be shopped immediately if they fail an operator test. In addition, the rams should also be shopped every three years for teardown and inspection.

Generalized BOP testing procedures

Generalized testing procedures are herewith provided for Class 1, 2, 3, and 4 BOPE configurations. All regulations and company guidelines should be followed.

Safety Notes: All Rig personnel should be cleared from the floor and instructed to stand back in a safe area when the BOP stack pressure testing is being performed. Extreme care should be taken when pressure testing modified flanges (those flanges with 2000 and 3000 psi bolt holes to mate up with either 2000 or 3000 psi wellheads) as this alters the original equipment manufacturers specifications and therefore its pressure rating.

Class 1 Testing procedures with cup tester

For large diameter casing, appropriate test plugs may not be available or even manufactured. In these cases, cup testers are used

Testing

This generalized procedure assumes there is not sufficient room for a spool between and the wellhead with BOPE or only an adapter spool is in use (Fig. 5.69).

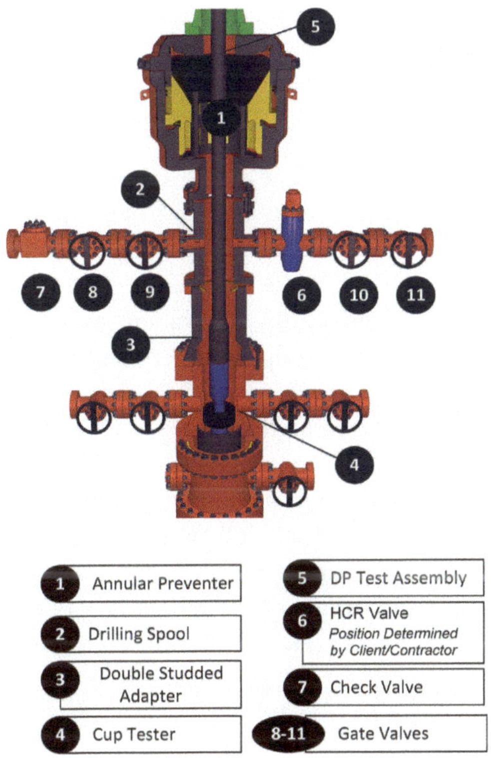

Fig. 5.69 Class 1 BOP stack with cup tester.

Procedure

1. Run cup tester to 15–30 feet below casing head.
2. Shut annular preventer with 1200 psi closing pressure.
3. Install kill/pump line to one casing valve and close casing valve on opposite side.
4. Pressure test to 250 psi for 5 min and inspect for integrity. Shut kill line valve, release pressure, and inspect for leaks.
5. Reopen kill line valve and pressure test to a minimum of 1000 psi, 70% of rated working pressure of BOP, 70% of casing burst, or casing head working pressure, whichever is less. Hold for 5 min and inspect for integrity.
6. Shut kill line valve, release pressure, and inspect kill line connection for leaks.
7. Release pressure, open annular, and retrieve cup tester.

FREQUENCY: ON EVERY INSTALLATION OR EVERY TWENTY-ONE DAYS, WHICHEVER IS LESS.
USE APPROPRIATE TEST PUMP
DO NOT UTILIZED RIG PUMP TO TEST BOP AS MOST WILL NOT PROVIDE SUFFICIENT PRESSURE.

Class 2: Testing procedure with test plug

If the outlets on the rams cannot be operated or a spool cannot be installed, a cup tester should be utilized. Casing valves must be used for circulating kill line and choke lines in this case (Fig. 5.70).

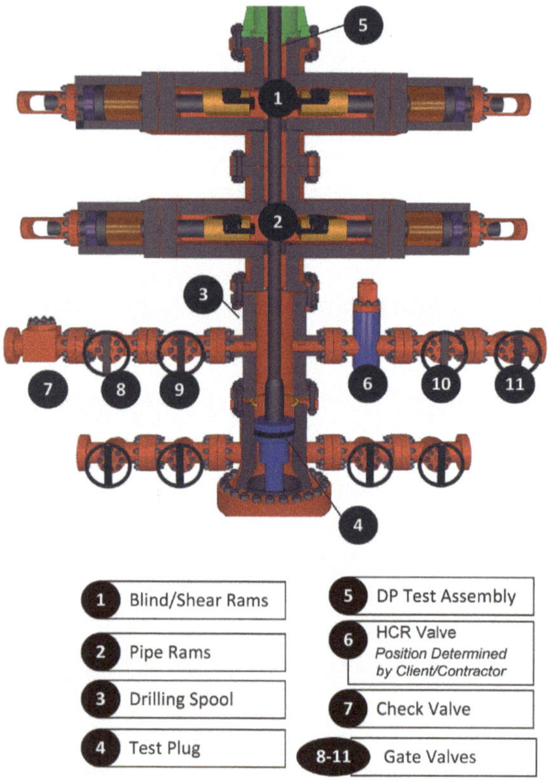

Fig. 5.70 Class 2 BOP stack with test plug.

Testing

A test plug should be used if the outlets on bottom ram can be operated or, if there is room for a spool between the rams and the casing head, the spool should be operated.

Isolation choice in order of preference:

- Between rams
- Spool
- Under Blind rams
- Casing head

Procedure

1. Run and set test plug in casing head.
2. Close pipe rams.
3. Close all valves except kill pump-in line. Leave casing valves open to check for test plug leaks.
4. Pressure test to 250 psi for 5 min and inspect for integrity. Shut kill line valve, release pressure, and inspect for leaks.
5. Reopen kill line valve and test to a minimum pressured of BOP rating, 70% of casing burst or wellhead working pressure, whichever is less. Hold for 5 min and inspect for integrity.
6. Shut kill line valve, release pressure, and inspect for leaks.

7. Release pressure, open pipe rams, and blackout of test plug.
8. Shut blind rams.
9. Pressure test to 250 psi for 5 min and inspect for integrity.
10. Pressure test to rated working pressure of BOP, 70% of casing burst or casing head working pressure, whichever is less. Hold for 5 min and inspect for integrity.
11. Release pressure, open blind rams.
12. Sting into test plug and retrieve same.

FREQUENCY: ON EVERY INSTALLATION ON EVERY TWENTY-ONE DAYS, WHICHEVER IS LESS. DO NOT UTILIZE RIG PUMP TO TEST BOP AS MOST WILL NOT PROVIDE SUFFICIENT PRESSURE.

Class 3: Testing procedure with test plug

If the outlets on the rams cannot be operated or a spool cannot be installed, a cup tester should be utilized. Casing valves must be used for circulating kill line and choke lines in this case.

Testing

A test plug should be used if the outlets on the rams can be operated or if there is room for a spool.

Isolation choice in order of preference:

- Spool between annular and rams.
- Between pipe and blind rams.
- Spool below rams.
- Below pipe rams.
- Casing head.

Procedure

See Fig. 5.71.

1. Run and set test plug in casing head.
2. Shut annular with 1200 psi closing pressure.
3. Close all valves except kill/pump-in line. Leave casing valves open to check for test plug leaks.
4. Pressure test to 250 psi for 5 min and inspect for integrity. Shut kill line valve, release pressure, and inspect for leaks.
5. Reopen kill line valve and test to a minimum 70% of rated working pressure of the BOP, casing head working pressure, whichever is less. Hold for 5 min and inspect for integrity.
6. Shut kill line valve, release pressure, and inspect for leaks.
7. Release pressure, open annular preventer, and shut-in pipe rams.

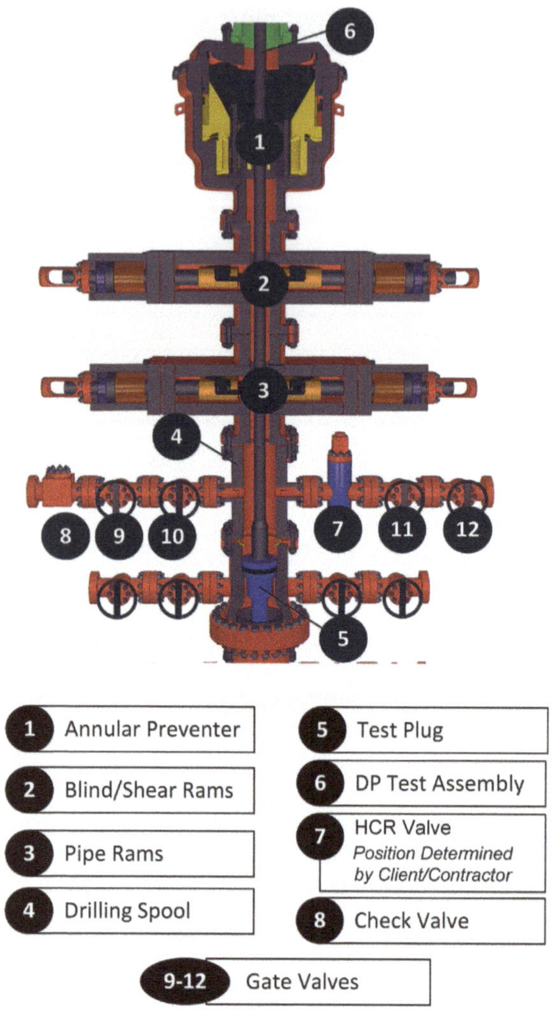

Fig. 5.71 Class 3 BOP stack with test plug.

8. Pressure test to 250 psi for 5 min and inspect for integrity.
9. Pressure test to working pressure of BOP or wellhead, whichever is less. Hold for 5 min and inspect for integrity.
10. Release pressure, open pipe rams, and back out test plug.
11. Shut blind rams and pressure test to 250 psi for 5 min and inspect for integrity.
12. Pressure test to working pressure of BOP or wellhead, whichever is less. Hold for 5 min and inspect for integrity.
13. Release pressure, open blind rams. Sting into test plug and retrieve same.

FREQUENCY: ON EVERY INSTALLATION ON EVERY TWENTY-ONE DAYS, WHICHEVER IS LESS. DO NOT UTILIZE RIG PUMP TO TEST BOP AS MOST WILL NOT PROVIDE SUFFICIENT PRESSURE.

Class 4: Testing procedure

If the outlets on the rams cannot be operated or a spool cannot be installed, a cup tester should be utilized. Casing valves must be used for circulating kill line and choke lines in this case.

Testing

A test plug should be used if the outlets on the rams can be operated or if there is room for a spool.

Isolation choice in order of preference:

- Spool between annular and rams.
- Between pipe and blind rams.
- Spool below rams.
- Below pipe rams.
- Casing head.

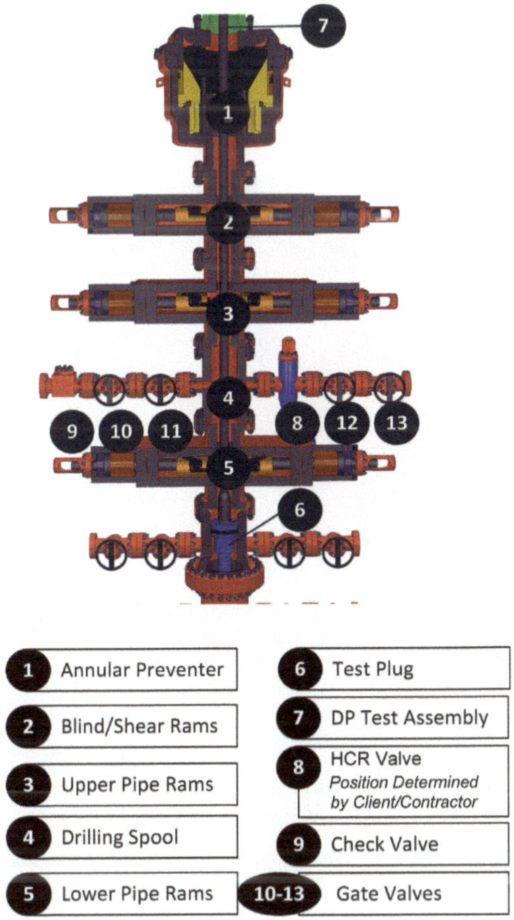

Fig. 5.72 Class 4 BOP stack with test plug.

Procedure

See Fig. 5.72.

1. Run and set test plug in casing head.
2. Shut annular with 1200 psi closing pressure.
3. Close all valves except kill/pump-in line. Leave casing valves open to check for test plug leaks.
4. Pressure test to 250 psi for 5 min and inspect for integrity. Shut kill line valve, release pressure, and inspect for leaks.
5. Reopen kill line valve and test to a minimum 70% of rated working pressure of the BOP, casing head working pressure, whichever is less. Hold for 5 min and inspect for integrity.
6. Shut kill line valve, release pressure, and inspect for leaks.
7. Release pressure, open annular preventer, and shut-in pipe rams.
8. Pressure test to 250 psi for 5 min and inspect for integrity.
9. Pressure test to working pressure of BOP or wellhead, whichever is less. Hold for 5 min and inspect for integrity.
10. Release pressure, open pipe rams, and back out test plug.
11. Shut blind rams and pressure test to 250 psi for 5 min and inspect for integrity.

12. Pressure test to working pressure of BOP or wellhead, whichever is less. Hold for 5 min and inspect for integrity.
13. Release pressure, open blind rams, and sting into test plug and retrieve same.

FREQUENCY: ON EVERY INSTALLATION ON EVERY TWENTY-ONE DAYS, WHICHEVER IS LESS.
DO NOT UTILIZE RIG PUMP TO TEST BOP AS MOST WILL NOT PROVIDE SUFFICIENT PRESSURE.

Choke manifold for class 1 and class 2 wells

A choke manifold should be readily available on location (Fig. 5.73).

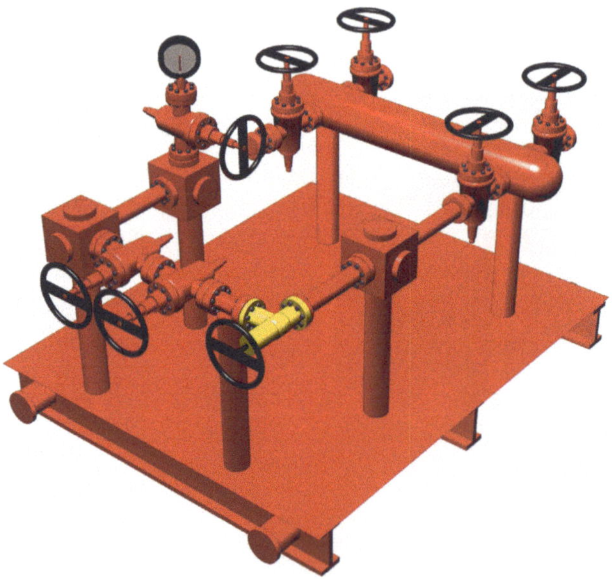

Fig. 5.73 Class 1 or 2 choke manifold.

All choke lines should be constructed of steel, installed in a straight line to the pit and properly anchored.

If a choke assembly is not in use, a steel circulating line, properly anchored should be utilized whenever there is a potential for a kick, whether or not the fracture gradient is below normal gradients.

Choke manifold for class 3 wells

A choke manifold will be rigged up on every Class 3 well (Fig. 5.74).

Fig. 5.74 Class 3 choke manifold.

All choke lines should be constructed of steel, installed in a straight line to the pit and properly anchored.

Choke manifold for class 4 wells

A choke manifold will be rigged up on every Class 4 well (Fig. 5.75).

Fig. 5.75 Class 4 choke manifold.

All choke lines should be constructed of steel, installed in a straight line to the pit and properly anchored.H_2S trim on all valves and lines.

Choke manifold for class 5 HPHT wells

All choke lines should be constructed of steel, installed in a straight line to the pit and properly anchored. H_2S trim on all valves and lines. Class 5 HPHT choke manifold is equal to Class 4 choke manifold (Fig. 5.76).

Fig. 5.76 Class 5 choke manifold.

BOP test equipment

Mandrel test plug

While testing the preventers, it is generally not desirable to expose the casing or open hole sections to the test pressures used to test the preventers. Some type of plug must be set in the bottom of the preventers to prevent this occurrence. The plugs most commonly used are the wellhead mandrel test plug, cup-type plug, and a combination mandrel cup-type test plug.

The mandrel test plug is designed to seat in the wellhead and each will generally seal in only one type of head. The plug is lowered into the head with a special test joint to test the pipe rams and annular preventer. The test joint (modified drillpipe) is removed with only the test plug resting in the wellhead in order to test the blind/shear rams (Fig. 5.77).

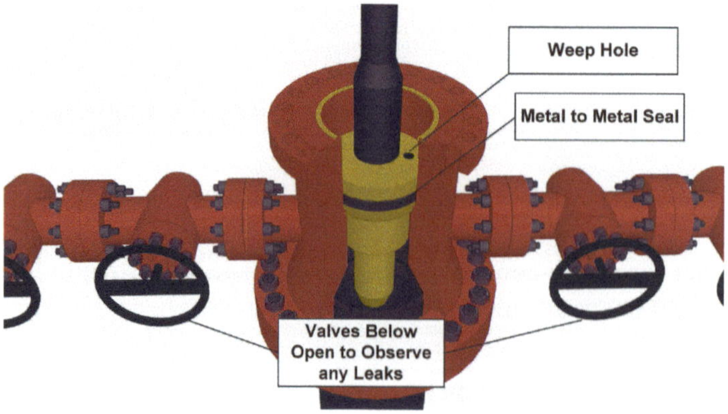

Fig. 5.77 Metal-to-metal sealing mandrel test plug.

If the mandrel test plug is to be used, care must be taken to ensure the plug used is designed for the existing wellhead. Wellheads with the same basic dimensions oftentimes require different plugs.

Care should also be taken to ensure the plug contains a "weep-hole" and is free of debris. If the "weep-hole" does not exist in the test plug, the test joint must have one.

Cup-tester

Cup-type testers are more universal as they are designed to create an effective seal in the casing and not in the wellhead. The government recommends "using a cup tester positioned 30–90' below casing head to test the casing with generally the most wear." Although the cup can be placed at any point in the casing, pressure testing specialists recommend positioning it opposite the slips setting area of the bowl in the casing spool or head (Fig. 5.78).

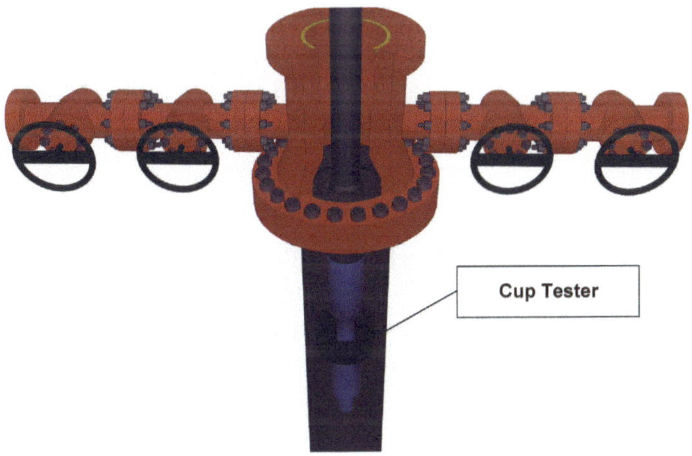

Fig. 5.78 Cup-type test plug.

Since the cup is not supported by the wellhead, the force (cup cross-sectional area (in) x pressure (psi)) created by the pressure test must be supported by the test joint. This will often require the use of drillpipe for testing because drillpipe/tubing yield strength might be exceeded.

THE CUP TYPE TESTER CANNOT BE USED TO TEST THE BLIND RAMS

Combination test plug

Combination plugs are available offering advantages of both the cup and the mandrel type test plugs. The combination test plug is supported by the casing head and allows testing of the blind rams while the cup tester creates and effective pressure seal for testing blind/ shear rams (Fig. 5.79).

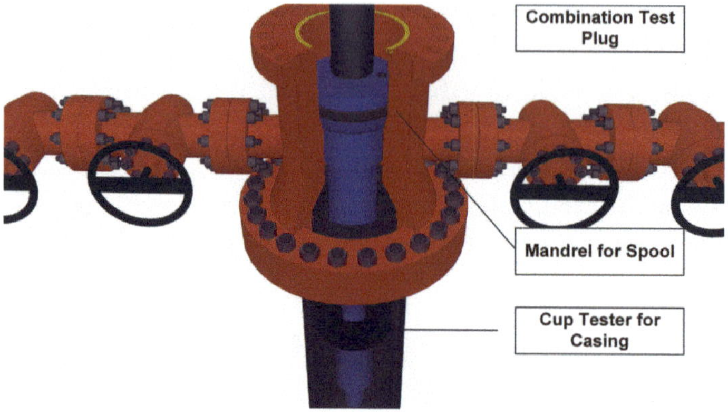

Fig. 5.79 Combination test plug (mandrel top with cup tester on bottom).

Test joint

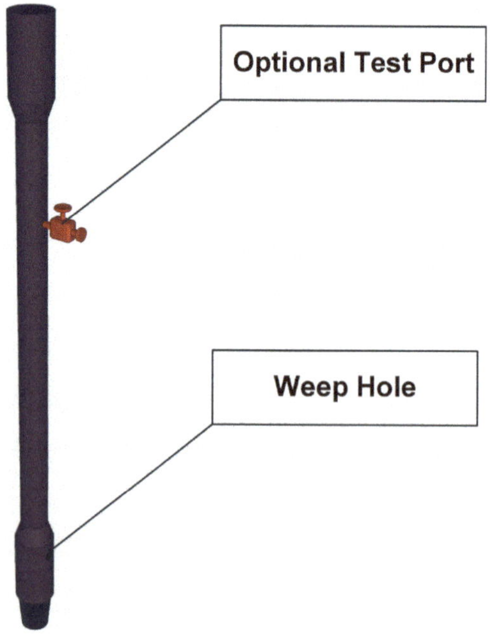

Fig. 5.80 Test joint with weep hole and test port.

The test joint should be made of pipe of sufficient weight and grade to safely withstand tensile yield, collapse, or internal pressures that will be placed on it during testing operations. The test joint should have a tapped or welded connection below the box end connection equipped with a valve, gauge, and fitting have a working pressure at least equal to the rated word working pressure of the preventer stack (Fig. 5.80). Weep holes may be drilled in the pin end of the test joint or may be installed in the test plug.

Considerations for BOP testing

Blowout prevention equipment is classified as **emergency** equipment and must be maintained in its proper working condition at all times. The Person-in-Charge must be an active participant in each test. Several maintenance tips are provided below:

Considerations before testing

1. Check accumulator, manifold, and annular pressure gauges for correct reading. Check fluid level in reservoir for proper testing level. **Maximum pressure test is 70% of casing burst rating.**
2. Check for proper control lines installation and ensure the lines will not be damaged by dropped tools.
3. Check proper position of preventer controls in either open or closed position (not blocked or neutral position). Ensure no leaks are evident.
4. Check the preventer stack is stabilized or guyed so vibrations are minimized.
5. All preventers should be tested on initial rig up or every 21 days (maximum). If blind rams cannot be tested on locations due to configuration limitations, they should be stump tested every three months.
6. Kill and choke lines should be flushed daily. Clear water must always be used.
7. Avoid circulating wet or noncured (green) cement through the preventer stack or choke manifold whenever possible. Always thoroughly flush with water any piece of blowout prevention equipment that has come in contact with wet or noncured (green) cement.
8. Ensure the Rig unit is centered over the well to reduce tubular and blowout prevention equipment contact and abrasion.
9. Do not use the kill line as a fill up line during trips.
10. Position ram preventer housing over casing head valves to provide an effective shield against damage from falling objects.

Considerations in performing testing

The objective of BOPE testing is to eliminate all leaks and to determine that the equipment will properly close and hold pressure as designed.

1. All tests are to be performed using clear water.
2. The initial "low-pressure" test of each piece of blowout prevention equipment must be conducted at a pressure at 250 psi.

IF THE BOPE DOES NOT PASS THE LOW-PRESSURE TEST, DO NOT TEST TO HIGH-PRESSURE AND THEN DROP BACK DOWN AND RETEST AT THE LOW-PRESSURE.
IF THE BOPE DOES NOT PASS THE LOW-PRESSURE TEST, IT DOES NOT PASS.

3. The "high-pressure" test of the annular preventer should be conducted at 70% of its rated working pressure.
4. The subsequent "high-pressure" test is conducted at the rated working pressure of the equipment for all:
   (a) Ram-type preventers
   (b) Choke manifolds and valves
   (c) Kill lines and valves
   (d) Upper and lower Kelly cocks, IBOP, FOSV.

5. "*BOPE should be tested in accordance to local laws or as follows:*
   (a) *When installed.*
   (b) *Following the disconnection or repair of any wellbore pressure seal in the wellhead or BOP stack assembly.*
   (c) *Blind rams are the only exception; they must be only when installed.*
   (d) *Every 21 days on extended jobs.*"

6. All valves not current being tested should be placed in the open position, according to local laws. "*All valves located downstream of the valve being tested must be placed in the open position.*"
7. All pressure tests must be held for a minimum duration of 5 min with no observable pressure decline.
8. If applicable, record all pressure tests on a recording chart. Ensure and label which piece of equipment is being tested.
9. The results of all pressure test, actuation, and inspection must be recorded on daily report.

Summary test procedure for a class 4 stack

The following steps represent a generic testing of a land Class IV BOP stack with a choke manifold with two (2) hydraulic chokes and two (2) manual chokes. These procedures are offered for suggested practices used in testing equipment. Actual field equipment and testing procedures will vary.

BOP tests are distinguished between the Initial Test (BOP stack first installed on well) and Subsequent Tests (all tests done after the initial test). On initial tests the rams have to be tested on ram locks with the closing pressure bled off at commissioning and annually.

See tables 1 and 2 at end of section for pressure test details.

Summary test procedure

Step #1: Function and flow testing
Step #2: Fill stack with water

Low-pressure test and high-pressure test

Step #3: Low and high-pressure testing of blind/shear rams on new completion or closed system.
Step #4: Low and high-pressure testing of annular preventer on new completion or closed system.

Low pressure test

Step #5: Testing buffer chamber and gate valves downstream of chokes
Step #6: Testing choke manifold valves downstream of chokes and pressure gauge
Step #7: Testing chokes and gate valves upstream of chokes
Step #8: Testing gate valves upstream of manual chokes
Step #9: Testing hydraulic chokes and gate valves upstream of chokes
Step #10: Testing casing pressure gauge valve
Step #11: Testing choke manifold inlet valve
Step #12: Testing upper pipe rams and drilling spool valves
Step #13: Testing upper pipe rams and drilling spool valves
Step #14: Testing upper pipe rams, hcr and drilling spool valves
Step #15: Testing lower pipe rams

High pressure test

Step #16: Testing buffer chamber and gate valves downstream of chokes
Step #17: Testing choke manifold valves downstream of chokes and pressure gauge
Step #18: Testing chokes and gate valves upstream of chokes
Step #19: Testing manual chokes and gate valves upstream of manual chokes
Step #20: Testing hydraulic chokes and gate valves upstream of hydraulic chokes
Step #21: Testing casing pressure gauge valve
Step #22: Testing choke manifold inlet valve
Step #23: Testing upper pipe rams and drilling spool valves
Step #24: Testing upper pipe rams and drilling spool
Step #25: Testing upper pipe rams, HCR and drilling spool valves
Step #26: Testing lower pipe rams

Low-pressure test and high-pressure test

Step #27: Test Surface Equipment

Step-by-step test procedure for a class 4 stack (with a test plug)

As each rig's equipment component complement varies and requires a customized BOP testing procedure. The following general BOP testing procedure was developed for a Class 4 Stack with a Class 4 Choke manifold. This set of generalized procedures was developed to provide an overview in the number of steps necessary to complete a thorough BOP test procedure (Figs. 5.81 and 5.82).

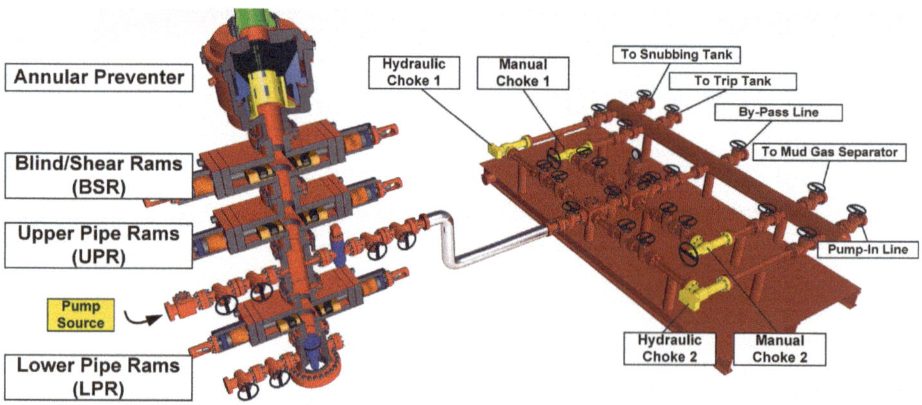

Fig. 5.81 BOP and manifold diagram with labels (note pump source connection).

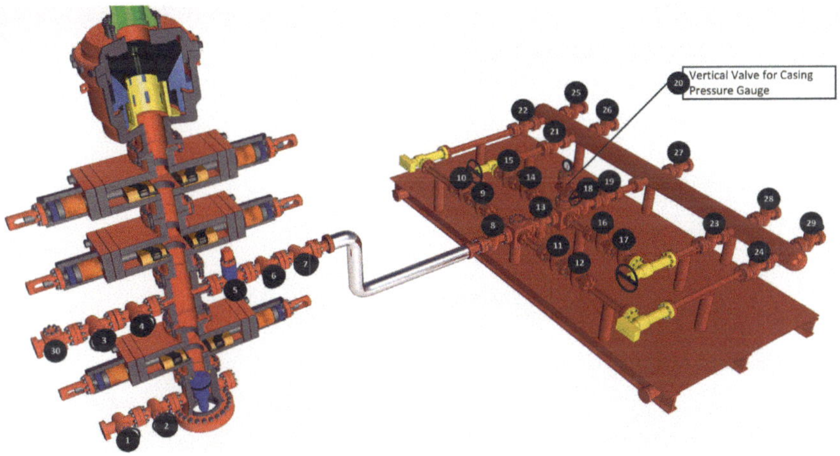

Fig. 5.82 BOP and manifold gate valve labels (note vertical valve 20 connecting casing pressure gauge).

INDIVIDUAL BOP TESTING PROCEDURE WILL VARY

Step #1: Function and flow testing

Before applying test pressure to the preventer, perform the following:

a. Close and open all preventers. Do not close pipe rams or annular preventer on open hole.

b. Pump through the kill line, flowline, (mud gas separator-if applicable), and choke lines with **water** to ensure none are plugged.

c. Record and retain all test records

Step #2: Fill stack with water

Drain the mud from the BOP stack and fill with **clear water.**

Step #3: Low and high-pressure testing of blind/shear rams on new completion or closed system

See Fig. 5.83.

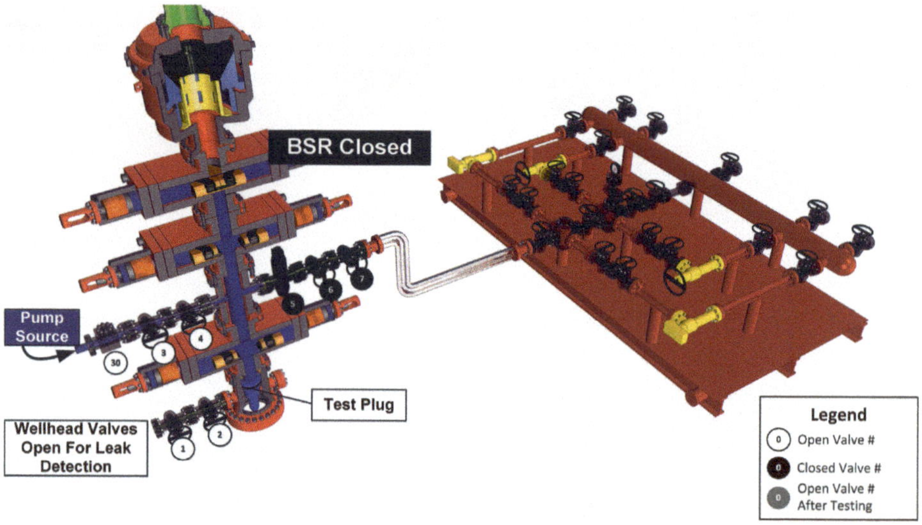

Fig. 5.83 Low pressure test of blind/shear rams (note: low pressure = *blue*).

(a) Connect pressure source to kill line. Open kill line valves #3 and #4.

(b) If applicable, the kill line Check Valve (valve #30) should be made inoperable (only temporarily) by removing the top cover, removing the inner workings, and replacing the top cover.

(c) Open **all** valves and chokes in choke manifold.

(d) Close the choke live valves #5 (HCR) and wellhead gate valves #6 and #7. The HCR should be operated from separate panels and BOP Closing unit.

(e) Close Blind Ram.

(f) Pump into the well through kill line monitoring and recording the test pressure at the test pump.

Perform low-pressure test

- Conduct the low-pressure test of 250–350 psi.
- Hold pressure for 5 min or more.
- Verify the Blind Ram does not leak by observing test pressure and performing a visual inspection of preventer housing and element.

Perform high-pressure test

See Fig. 5.84.

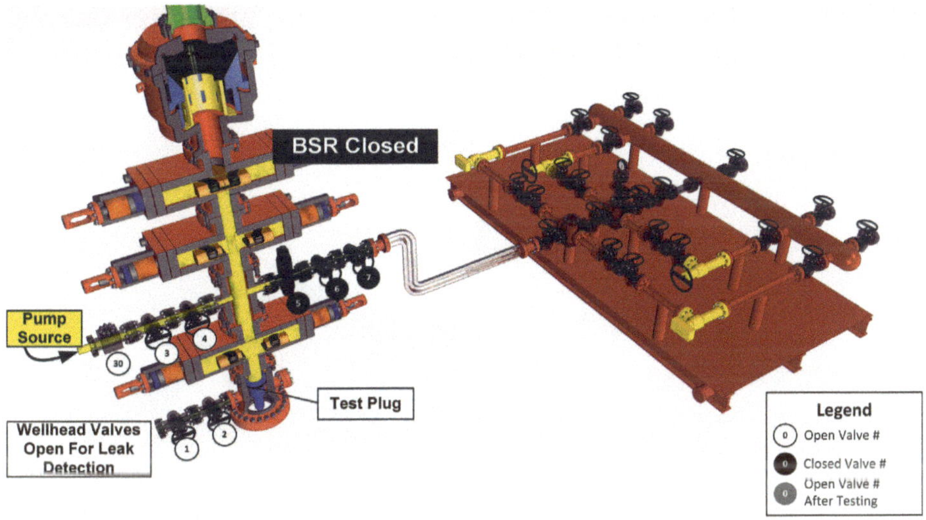

Fig. 5.84 High pressure test of blind/shear rams (Note: high pressure = *yellow*)

- Conduct the High-pressure test (**test plug may be required**) to the lesser of (1) or (2)
  - Initial Test
    1. Rated working pressure (RWP) of ram
    2. Rated working pressure (RWP) of wellhead
  - Subsequent Test

    Casing test pressure
- Hold pressure for 5 min or more.
- Verify the Blind Ram does not leak by observing test pressure and performing a visual inspection of the preventer housing and element.

(g) Bleed pressure from system, open valve 5, and close valves #3 and #4. Open Blind Ram.

(h) Proceed to next step.

Step #4: Low and high-pressure testing of annular preventer on new completion of closed loop system

See Fig. 5.85.

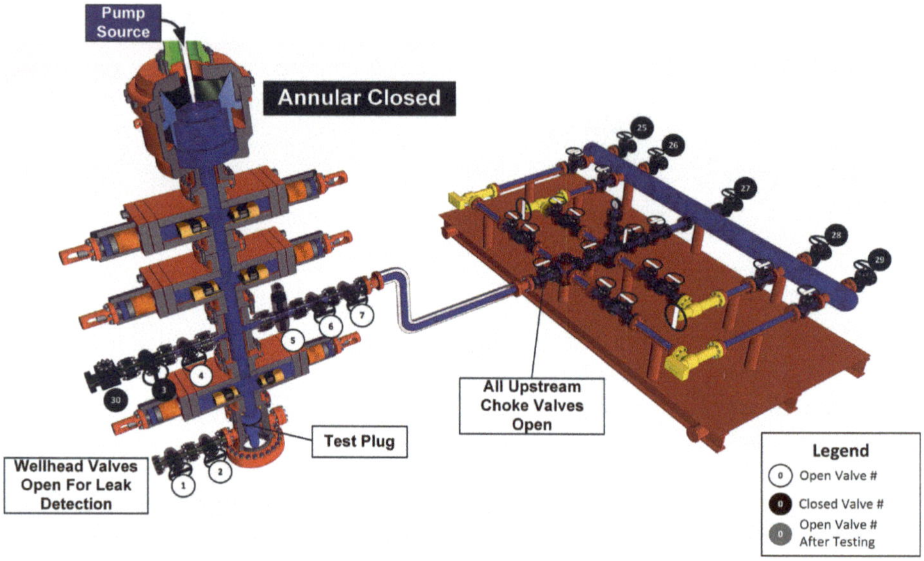

Fig. 5.85 Low-pressure testing of annular preventer and choke manifold.

(a) Connect the pressure source to test joint at Rig floor. Pick up and test joint so the lower tooljoint is at least three feet **below** lowest ram preventer.

(b) Close the check valve #30 and kill line valve #3.

(c) Close outermost buffer chamber gate valves #25, #26, #27, #28, and #29. Open remaining valves and chokes in the choke manifold.

(d) Close Annular Preventer on test joint.

(e) Pump into well through the test joint recording the test pressure at the test pump.

Perform low-pressure test

- Conduct the low-pressure test of 250–350 psi.
- Hold pressure for 5 min or more.
- Verify the Annular Preventer does not leak by observing test pressure and performing a visual inspection of preventer housing and element.

Step #4A: Low and high-pressure testing of annular preventer on new completion of closed loop system

Perform high-pressure test

See Fig. 5.86.

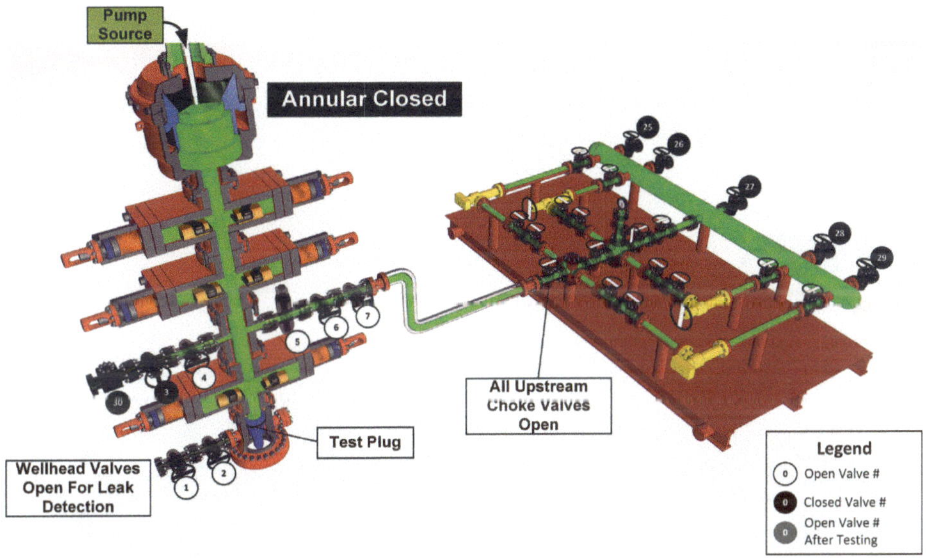

Fig. 5.86 High-pressure testing of annular preventer and choke manifold (Note: limited high pressure = green).

- Conduct the high-pressure test at the lesser of (1), (2), or (3).
  - Initial

    If there is a change out of a component, and elastomer or a gasket

(1) Rated Working Pressure (RWP) of preventer

If there is NO change out of a component, and elastomer or a gasket

Lesser of 1 or 2

(1) 70% Rated Working Pressure (RWP) of annular preventer

(2) Maximum Allowable Surface Pressure (MASP)

- Subsequent (lesser of 1 or 2)
  - **(1)** 70% RWP of Annular Preventer.
  - **(2)** Maximum Allowable Surface Pressure (MASP) of Hole Section

- Hold pressure for 5 min or more.
- Verify the Annular Preventer does not leak by observing test pressure and performing a visual inspection of the preventer housing and element.

- **Buffer chamber test pressure may be limited and isolated depending on design, service, and construction.**

Note: Verify the accuracy of the gauge installed downstream of choke manifold (Valve #20) by observing test pressure.

(f) Bleed all pressure from the system, Open Annular Preventer, and proceed to next step.

Step #5: Testing buffer chamber and gate valves downstream of chokes, low-pressure test only

See Fig. 5.87.

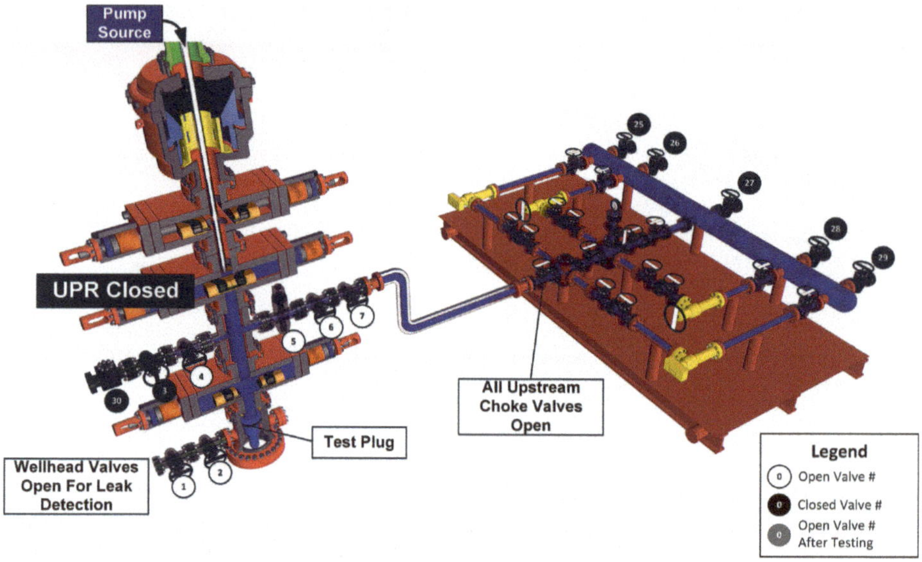

Fig. 5.87 Low pressure testing buffer chamber and gate valves downstream of chokes.

(a) From preceding step, ensure Check Valve #30 and Gate Valve #3 are closed. Ensure outermost choke manifold Gate valves #25, #26, #27, #28, and #29 are closed.

(b) Close Upper Pipe Ram on test joint.

(c) Pump into well through the test joint recording the test pressure at the test pump.

Perform low-pressure test

- Conduct the low-pressure test of 250–350 psi.
- Hold pressure for 5 min or more.
- Verify the Upper Pipe Ram does not leak by observing test pressure and performing a visual inspection of preventer housing and element.

(d) without bleeding pressure, proceed to next step.

Step #6: Testing choke manifold valves downstream of chokes and pressure gauge, low-pressure test only

See Fig. 5.88.

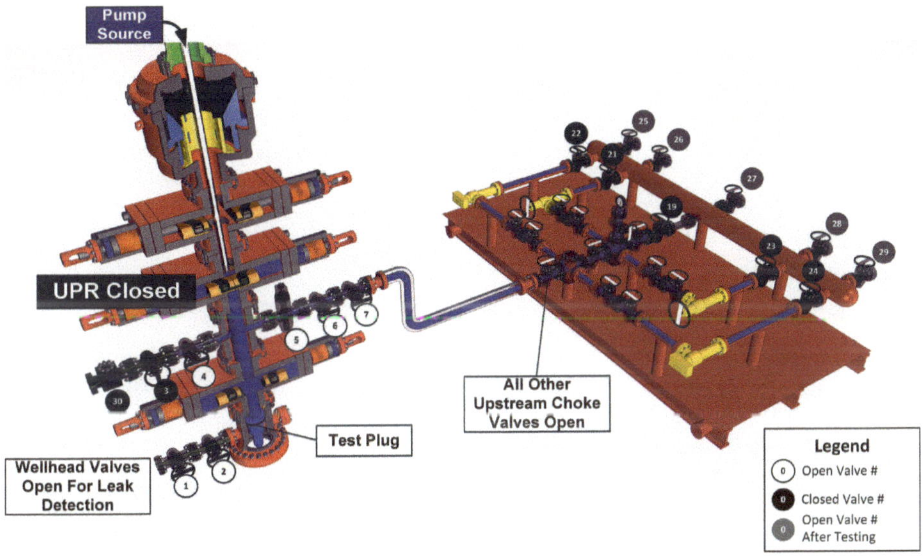

Fig. 5.88 Testing gate valves downstream of chokes.

While maintaining a low-pressure 250–350 psi, perform the following test on Gate valves Downstream of Chokes:

(a) Close gate valves #22, #21, #19, #23, and #24 upstream of the buffer chamber. Open Gate valves #25, #26, #27, #28, and #29 downstream of buffer chamber.

(b) **Operate all choke(s) in choke manifold. Chokes are not considered to be positive sealing, so leakage may be observed.**

Note: Continued flowing (trickle) indicates malfunctioning valve that must be repaired.

(c) Perform low-pressure test

- Conduct the low-pressure test of 250–350 psi.
- Hold pressure for 5 min or more.
- Verify the valves do not leak performing a visual inspection of gate valve housing and stem.

(d) without bleeding pressure, proceed to next step.

Step #7: Testing chokes and gate valves upstream of chokes, low pressure test only

See Fig. 5.89.

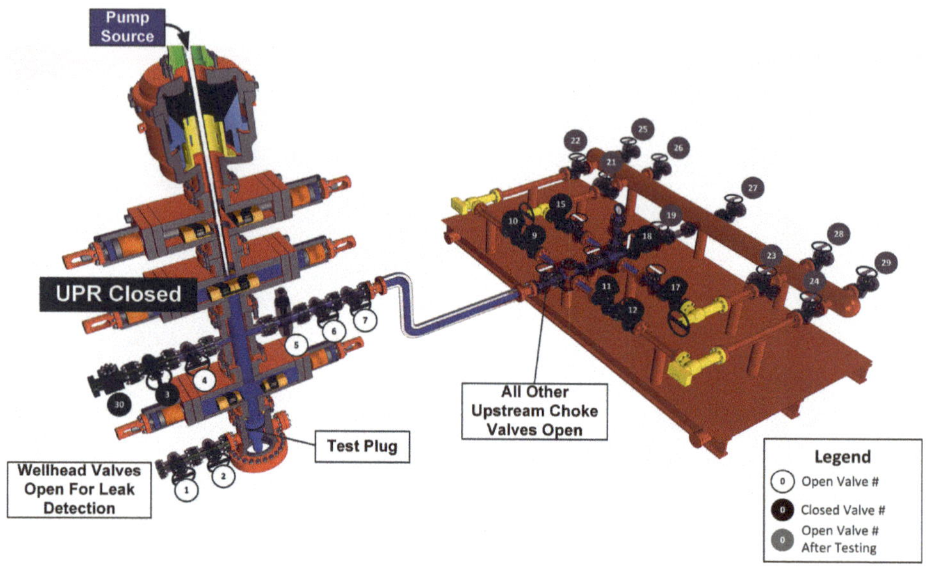

Fig. 5.89 Testing manual chokes and gate valves upstream of chokes.

While maintaining a low-pressure 250–350 psi, perform the following test on Manual Chokes and Valves Upstream of Chokes.

(a) Close Gate valves #9, #10, #11, #12, #15, #17, and #18.Open Gate valves upstream of buffer chamber #21, #22, #19, #23, and #24.

(b) **Operate manual choke(s) in choke manifold. Chokes are not considered to be positive sealing, so leakage may be observed.**

Note: Continued flowing (trickle) indicates malfunctioning valve that must be repaired.

(c) Perform Low-Pressure Test

- Conduct the low-pressure test of 250–350 psi.
- Hold pressure for 5 min or more.
- Verify the valves do not leak performing a visual inspection of gate valve housing and stem.

(d) without bleeding pressure, proceed to next step.

Step #8: Testing gate valves upstream of manual chokes, low-pressure test only

See Fig. 5.90.

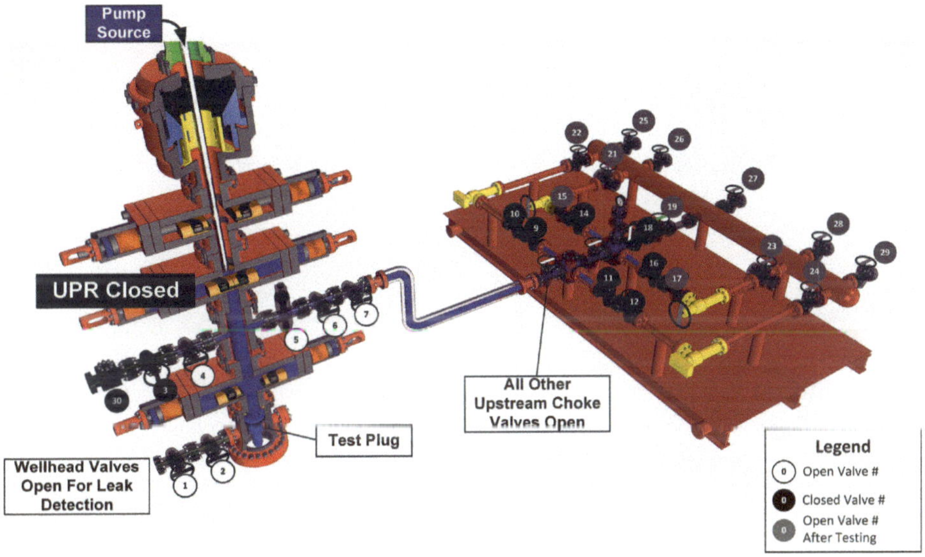

Fig. 5.90 Testing gate valves upstream of manual chokes.

While maintaining a low-pressure 250–350 psi, perform the following test on Gate valves Upstream of Manual Chokes.

(a) Close Gate valves #14 and #16.Open Gate valves #15 and #17.

Note: Continued flowing (trickle) indicates malfunctioning valve that must be repaired.

(b) Perform low-pressure test
- Conduct the low-pressure test of 250–350 psi.
- Hold pressure for 5 min or more.
- Verify the valves do not leak performing a visual inspection of gate valve housing and stem.

(c) without bleeding pressure, proceed to next step.

Step #9: Testing hydraulic chokes and gate valves upstream of chokes, low-pressure test only

See Fig. 5.91.

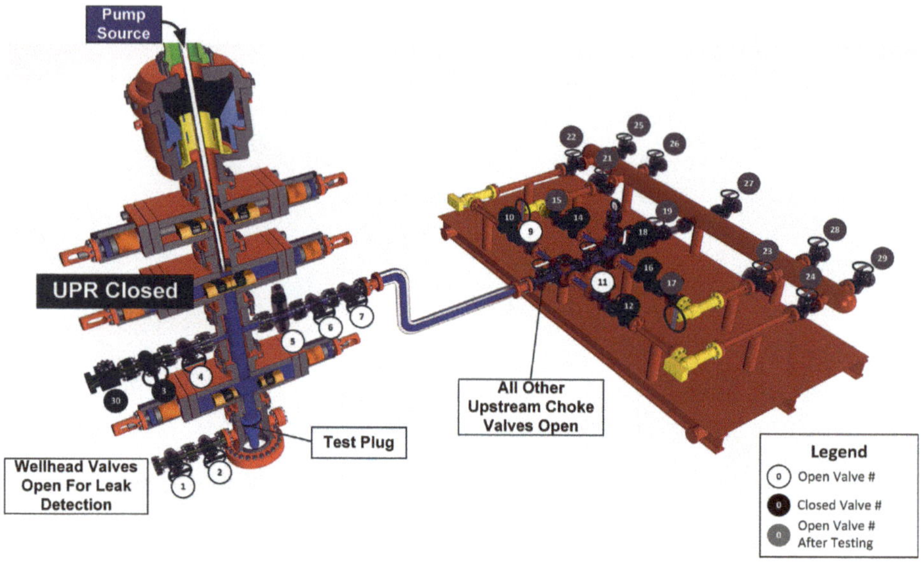

Fig. 5.91 Testing, hydraulic chokes and gate valves upstream of hydraulic choke.

While maintaining a low-pressure 250–350 psi, perform the following test on Hydraulic Chokes and Gate valves Upstream of Chokes:

(a) Close Gate valves #10, #12, #14, #16, and #18.Open Gate valves #9, #11.

(b) Operate hydraulic choke(s) in choke manifold. Hydraulic chokes are not considered to be positive sealing, so leakage may be observed.

Note: Continued flowing (trickle) indicates malfunctioning valve, which must be repaired.

(c) Perform low-pressure test

- Conduct the low-pressure test of 250–350 psi.
- Hold pressure for 5 min or more.
- Verify the valves do not leak performing a visual inspection of gate valve housing and stem.

(d) without bleeding pressure, proceed to next step.

Step #10: Testing casing pressure gauge valve, low-pressure test only

See Fig. 5.92.

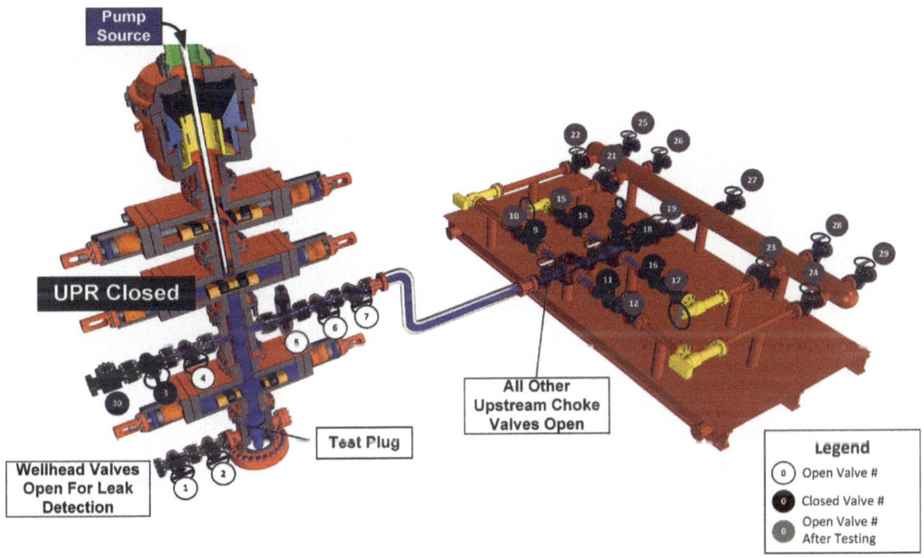

Fig. 5.92 Testing casing pressure gauge valve.

While maintaining a low-pressure 250–350 psi, perform the following test on Casing Pressure Gauge Valve.

(a) Close Gate valves #9, #11, #14, #16, #18, and #20 (Vertical Valve to isolate Casing Pressure Gauge.Open Gate valves #8 and #13).

Note: Continued flowing (trickle) indicates malfunctioning valve, which must be repaired.

(b) Perform Low-Pressure Test
- Conduct the low-pressure test of 250–350 psi.
- Hold pressure for 5 min or more.
- Verify the valves do not leak performing a visual inspection of gate valve housing and stem.

(c) without bleeding pressure, proceed to next step.

(d) ***Not shown***: Once complete, Close #13 and Open #20, #18, #16, #14, #11, and #9. Repeat High Pressure test on #13. After completion of low-pressure test, Open #13.

Step #11: Testing choke manifold inlet valve, low-pressure test only

See Fig. 5.93.

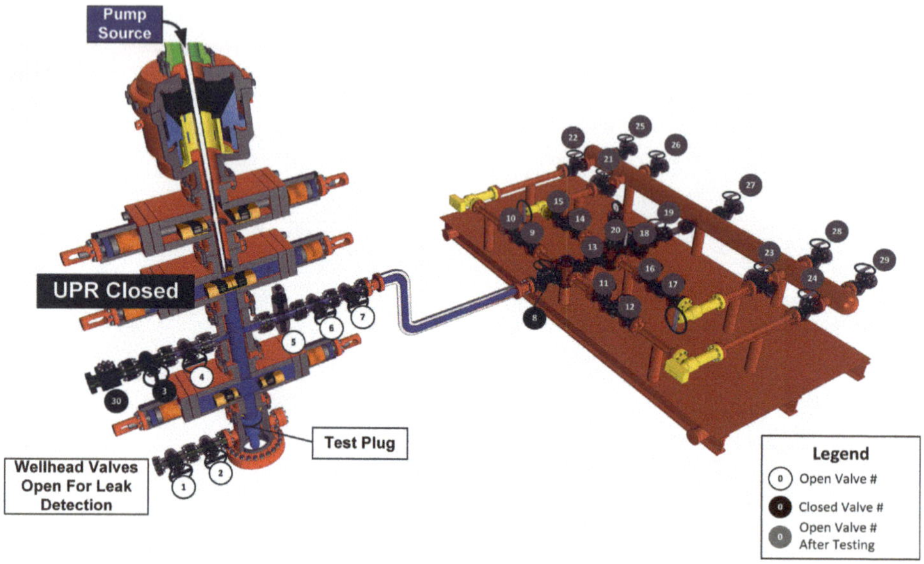

Fig. 5.93 Testing choke manifold inlet valve.

While maintaining a low-pressure 250–350 psi, perform the following test on Choke Manifold Valve

(a) Close gate valves #8. Open all other choke manifold gate valves on choke manifold.

Note: Continued flowing (trickle) indicates malfunctioning valve, which must be repaired.

(b) Perform low-pressure test
- Conduct the low-pressure test of 250–350 psi.
- Hold pressure for 5 min or more.
- Verify the valves do not leak performing a visual inspection of gate valve housing and stem.

(c) Without bleeding pressure, proceed to next step.

Step #12: Testing upper pipe rams and drilling spool valves, low-pressure test only

See Fig. 5.94.

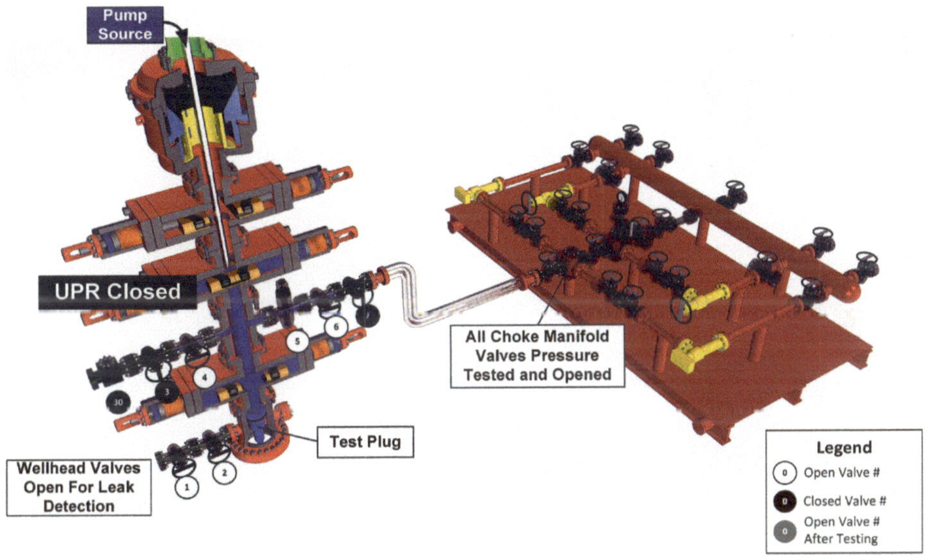

Fig. 5.94 Testing upper pipe rams and drilling spool valves.

While maintaining a low-pressure 250–350 psi, perform the following test Drilling Spool Valves:

(a) Close Gate valves #7.Open all other choke manifold gate valves on choke manifold.

Note: Continued flowing (trickle) indicates malfunctioning valve, which must be repaired.

(b) Perform Low-Pressure Test

- Conduct the low-pressure test of 250–350 psi.
- Hold pressure for 5 min or more.
- Verify the valves do not leak performing a visual inspection of gate valve housing and stem.

(c) without bleeding pressure, proceed to next step.

Step #13: Testing upper pipe rams and drilling spool valves, low-pressure test only

See Fig. 5.95.

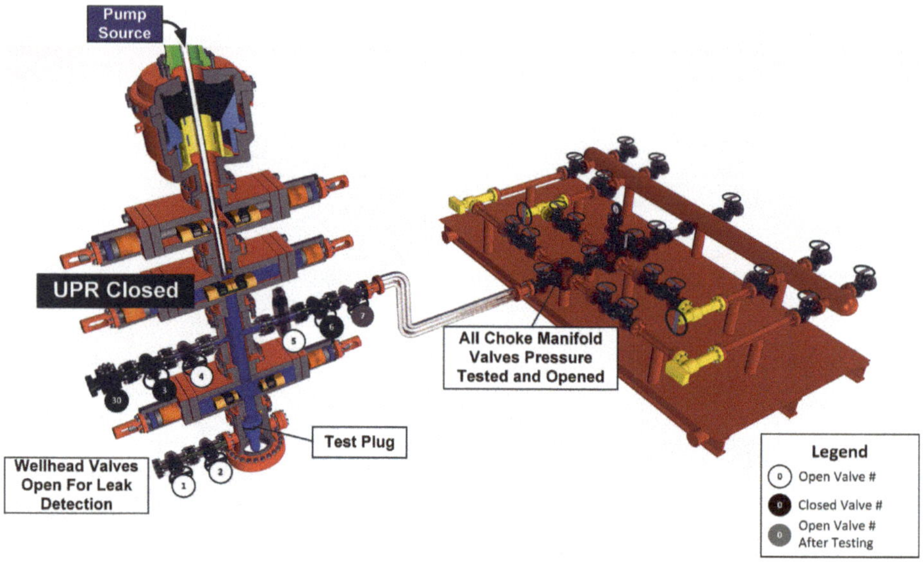

Fig. 5.95 Testing upper pipe rams and drilling spool valves.

While maintaining a low-pressure 250–350psi, perform the following test Drilling Spool Valves.

(a) Close Gate valves #3 and #6. Open Check Valve #30 and Gate Valve #7.

Note: Continued flowing (trickle) indicates malfunctioning valve, which must be repaired.

(b) Perform low-pressure test
- Conduct the low-pressure test of 250–350psi.
- Hold pressure for 5min or more.
- Verify the valves do not leak performing a visual inspection of gate valve housing and stem.

(c) without bleeding pressure, proceed to next step.

Step #14: Testing upper pipe rams, HCR, and drilling spool valves, low-pressure test only

See Fig. 5.96.

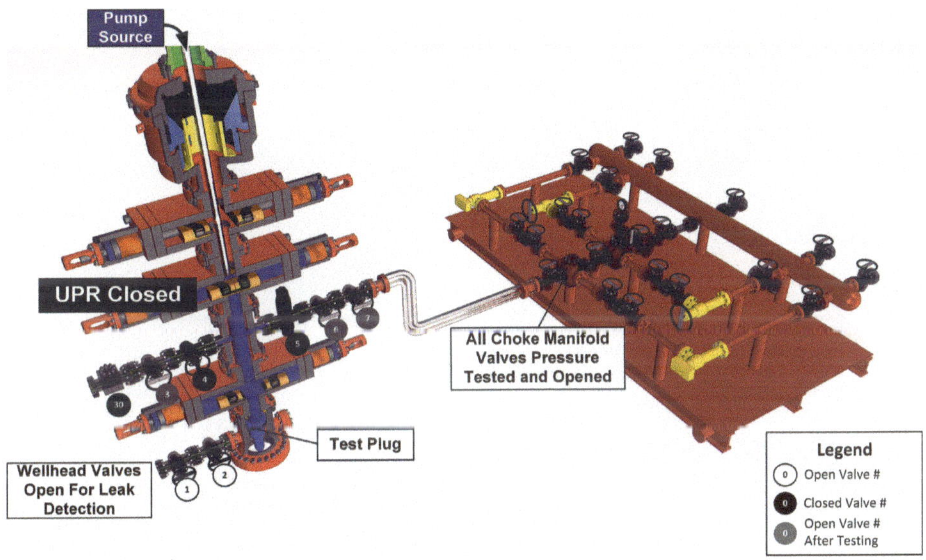

Fig. 5.96 Testing upper pipe rams, HCR and drilling spool valves.

While maintaining a low-pressure 250–350 psi, perform the following test HCR and Drilling Spool Valves.

(a) Close Gate valves #3 and #6. Open Check Valve #30 and Gate Valve #3, #6, and #7.

Note: Continued flowing (trickle) indicates malfunctioning valve, which must be repaired.

(b) Perform Low-Pressure Test
- Conduct the low-pressure test of 250–350 psi.
- Hold pressure for 5 min or more.
- Verify the valves do not leak performing a visual inspection of gate valve housing and stem.

(c) without bleeding pressure, proceed to next step.

Step #15: Testing lower pipe rams, low-pressure test only

See Fig. 5.97.

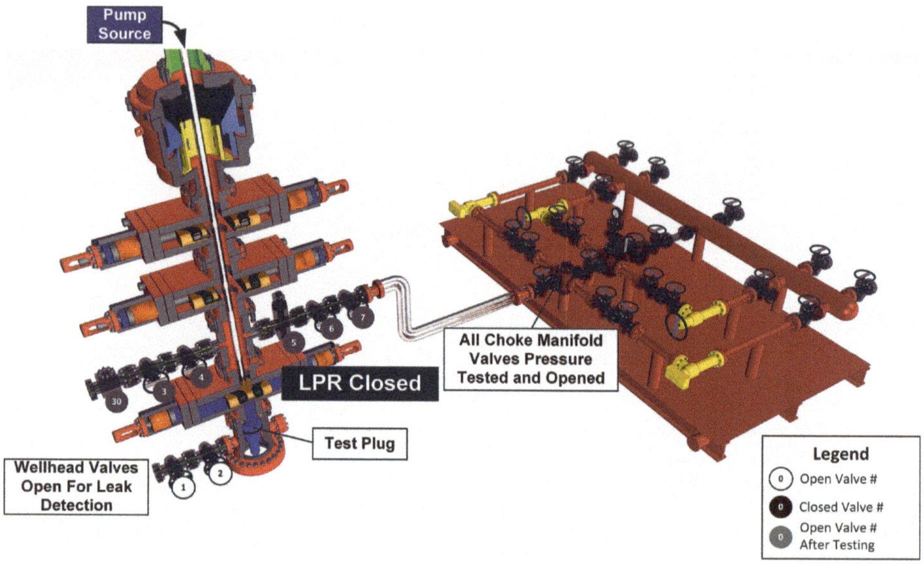

Fig. 5.97 Testing lower pipe rams.

(a) Open all valves above Lower Pipe Ram.
(b) Open Upper Pipe Rams, Close Lower Pipe Rams.
(c) **Note:** Continued flowing (trickle) indicates malfunctioning valve, which must be repaired.
(d) Perform low-pressure test
- Conduct the low-pressure test of 250–350 psi.
- Hold pressure for 5 min or more.
- Verify the valves do not leak performing a visual inspection of gate valve housing and stem.

(e) Bleed off pressure
(f) Open Lower Pipe Rams. Close Upper Pipe Rams. Proceed to next step.

Step #16: Testing buffer chamber and gate valves downstream of chokes, high-pressure test only

See Fig. 5.98.

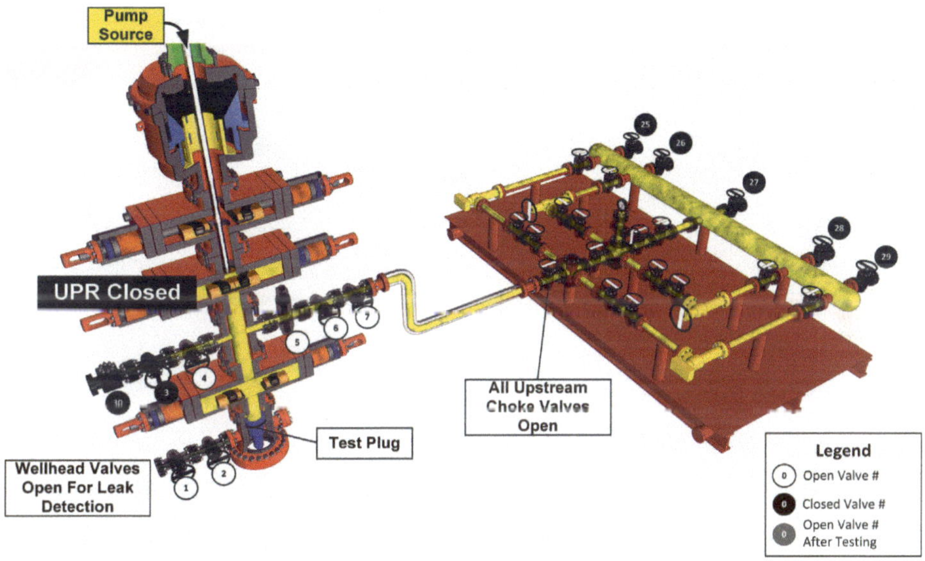

Fig. 5.98 High pressure testing buffer chamber and gate valves downstream of chokes.

(a) From preceding step, ensure Check Valve #30 and Gate Valve #3 are closed. Ensure outermost choke manifold Gate valves #25, #26, #27, #28, and #29 are closed.

(b) Close Upper Pipe Ram on test joint.

(c) Pump into well through the test joint recording the test pressure at the test pump.

- Conduct the high-pressure test at the lesser of (1), (2), or (3).

  Initial

  1. Maximum Allowable Surface Pressure (MASP) for well program
  2. Rated Working Pressure (RWP) of downstream valve(s)
  3. Rated Working Pressure (RWP) of downstream line(s)

  Subsequent lesser of (1), (2), or (3)

  1. Maximum Allowable Surface Pressure (MASP) for hole section
  2. Rated Working Pressure (RWP) of downstream valve(s)
  3. Rated Working Pressure (RWP) of downstream line(s)

- Hold pressure for 5 min or more.
- Verify the rams do not leak performing a visual inspection of ram preventer housing and element.

(d) without bleeding pressure, proceed to next step.

Step #17: Testing choke manifold valves downstream of chokes and pressure gauge, high-pressure test only

See Fig. 5.99.

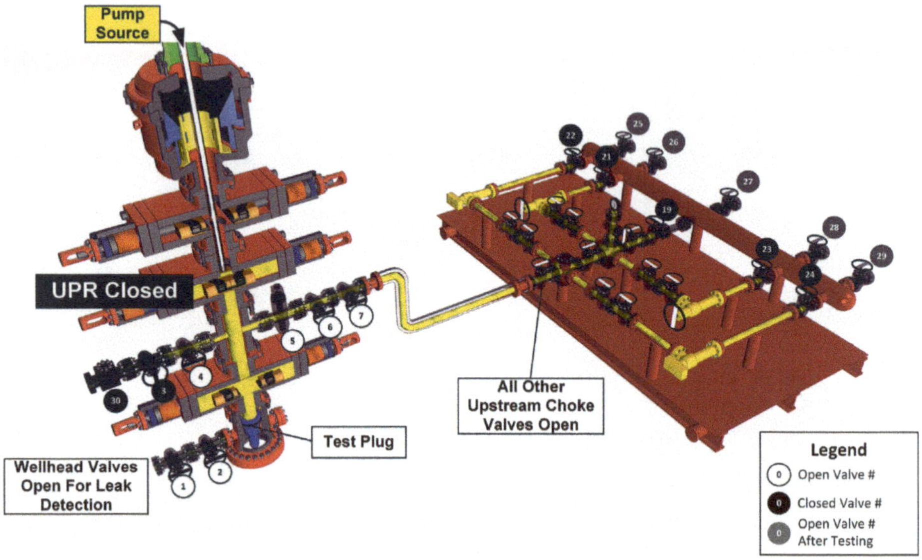

Fig. 5.99 Testing choke manifold valves downstream of chokes and pressure gauge.

(a) Close Gate valves #22, #21, #19, #23, and #24 upstream of the buffer chamber. Open Gate valves #25, #26, #27, #28, and #29 downstream of buffer chamber.

(b) Operate all choke(s) in choke manifold. Chokes are not considered to be positive sealing, so leakage may be observed.

Note: Continued flowing (trickle) indicates malfunctioning valve that must be repaired.

(c) Perform <u>High-Pressure Test</u>

- Conduct the high-pressure test at the lesser of (1), (2), or (3).

  Initial

  1. Maximum Allowable Surface Pressure (MASP) for well program
  2. Rated Working Pressure (RWP) of downstream valve(s)
  3. Rated Working Pressure (RWP) of downstream line(s)

  Subsequent lesser of (1), (2), or (3)

  1. Maximum Allowable Surface Pressure (MASP) for hole section
  2. Rated Working Pressure (RWP) of downstream valve(s)
  3. Rated Working Pressure (RWP) of downstream line(s)

- Hold pressure for 5 min or more.
- Verify the rams do not leak performing a visual inspection of ram preventer housing and element.

(d) without bleeding pressure, proceed to next step.

Step #18: Testing manual chokes and gate valves upstream of chokes, high-pressure test only

See Fig. 5.100.

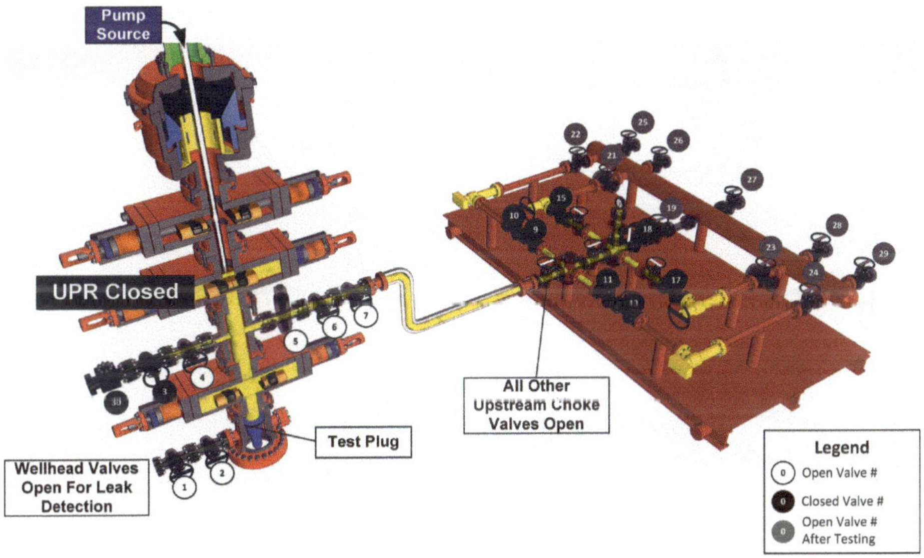

Fig. 5.100 Testing manual chokes and gate valves upstream of chokes.

(a) Close Gate valves #9, #10, #11, #12, #15, #17, and #18.Open Gate valves upstream of buffer chamber #21, #22, #19, #23, and #24.

(b) **Operate manual choke(s) in choke manifold. Chokes are not considered to be positive sealing, so leakage may be observed.**

Note: Continued flowing (trickle) indicates malfunctioning valve that must be repaired.

(c) Perform high-pressure test

- Conduct the high-pressure test at the lesser of (1) or (2).

  Initial

  1. Rated Working Pressure (RWP) of ram preventer
  2. Rated Working Pressure (RWP) of wellhead

  Subsequent

  1. Maximum Allowable Surface Pressure (MASP) for hole section

- Hold pressure for 5 min or more.
- Verify the rams do not leak performing a visual inspection of ram preventer housing and element.

(d) without bleeding pressure, proceed to next step.

Step #19: Testing gate valves upstream of manual chokes, high-pressure test only

See Fig. 5.101.

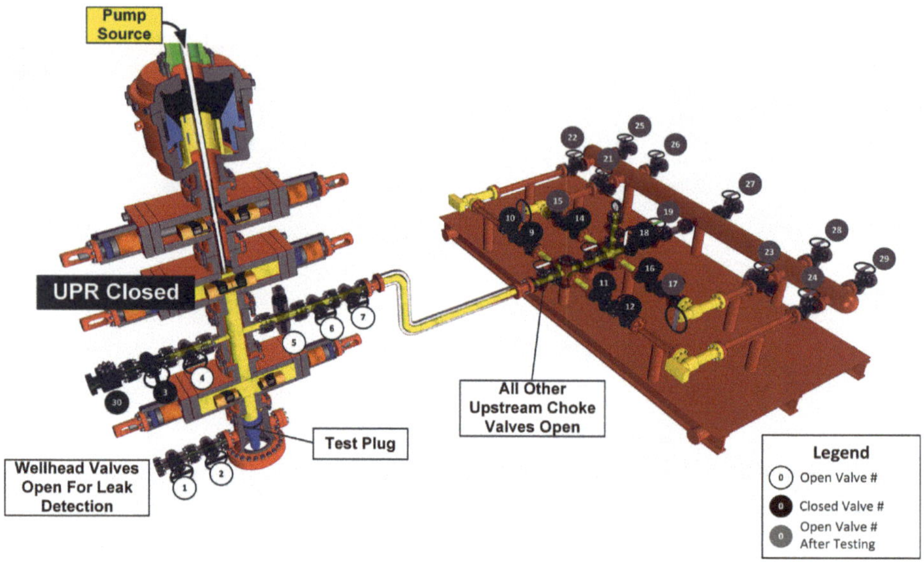

Fig. 5.101 Testing gate valves upstream of manual chokes.

(a) Close Gate valves #14 and #16.Open Gate valves #15 and #17.

Note: Continued flowing (trickle) indicates malfunctioning valve that must be repaired.

(b) Perform high-pressure test

- Conduct the high-pressure test at the lesser of (1) or (2).
  Initial
  1. Rated Working Pressure (RWP) of ram preventers
  2. Rated Working Pressure (RWP) of wellhead
  Subsequent
  1. Maximum Allowable Surface Pressure (MASP) of hole section
- Hold pressure for 5 min or more.
- Verify the rams do not leak performing a visual inspection of ram preventer housing and element.

(c) Without bleeding pressure, proceed to next step.

Step #20: Testing hydraulic chokes and gate valves upstream of chokes, high-pressure test only

See Fig. 5.102.

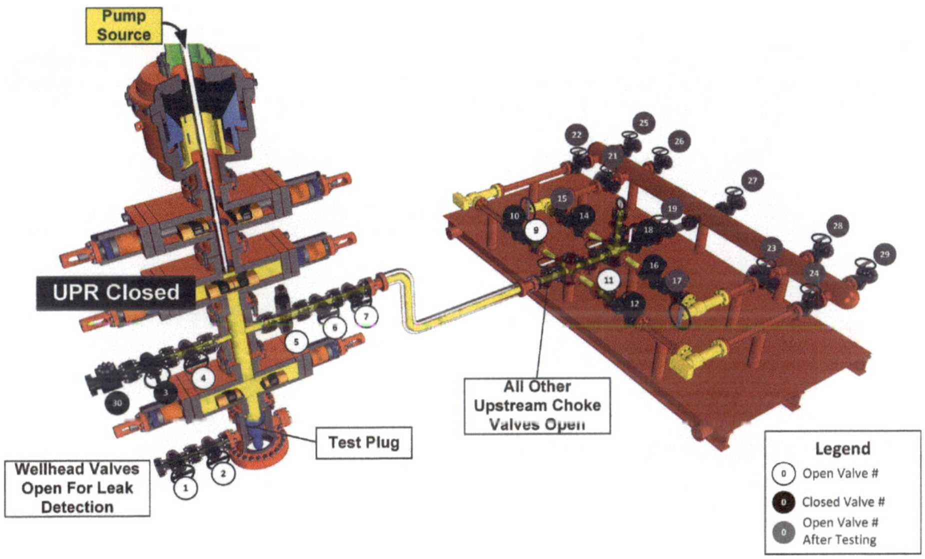

Fig. 5.102 Testing hydraulic chokes and gate valves upstream of hydraulic choke.

(a) Close Gate valves #10, #12, #14, and #18. Open Gate valves #9, #11. Test valves upstream of hydraulic chokes.

(b) Operate hydraulic choke(s) in choke manifold. Hydraulic chokes are not considered to be positive sealing, so leakage may be observed.

Note: Continued flowing (trickle) indicates malfunctioning valve, which must be repaired.

(c) Perform high-pressure test

- Conduct the high-pressure test at the lesser of (1) or (2).

Initial

1. Rated Working Pressure (RWP) of ram preventer
2. Rated Working Pressure (RWP) of wellhead

Subsequent lesser

1. Maximum Allowable Surface Pressure (MASP) for hole section

- Hold pressure for 5 min or more.
- Verify the rams do not leak performing a visual inspection of ram preventer housing and element.

(d) without bleeding pressure, proceed to next step.

Step #21: Testing casing pressure gauge valve, high-pressure test only

See Fig. 5.103.

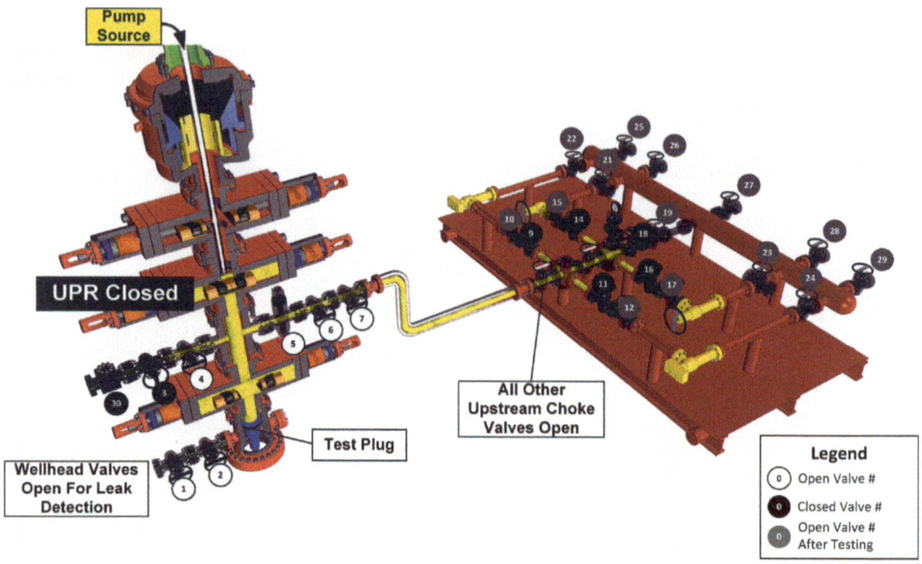

Fig. 5.103 Testing casing pressure gauge valve.

(a) Close Gate valves #9, #11, #14, #16, #18, and #20 (Vertical Valve to isolate Casing Pressure Gauge.Open Gate valves #8 and #13.

Note: Continued flowing (trickle) indicates malfunctioning valve, which must be repaired.

(b) Perform high-pressure test

- Conduct the high-pressure test at the lesser of (1) or (2).

Initial

1. Rated Working Pressure (RWP) of ram preventer
2. Rated Working Pressure (RWP) of wellhead

Subsequent lesser

1. Maximum Allowable Surface Pressure (MASP) for hole section

- Hold pressure for 5 min or more.
- Verify the rams do not leak performing a visual inspection of ram preventer housing and element.

(c) Without bleeding pressure, proceed to next step.

(d) Not shown: Once complete, Close #13 and Open #20, #18, #16, #14, #11, and #9. Repeat high-pressure test on #13. After completion of high-pressure test, Open #13.

Step #22: Testing choke manifold inlet valve, high-pressure test only

See Fig. 5.104.

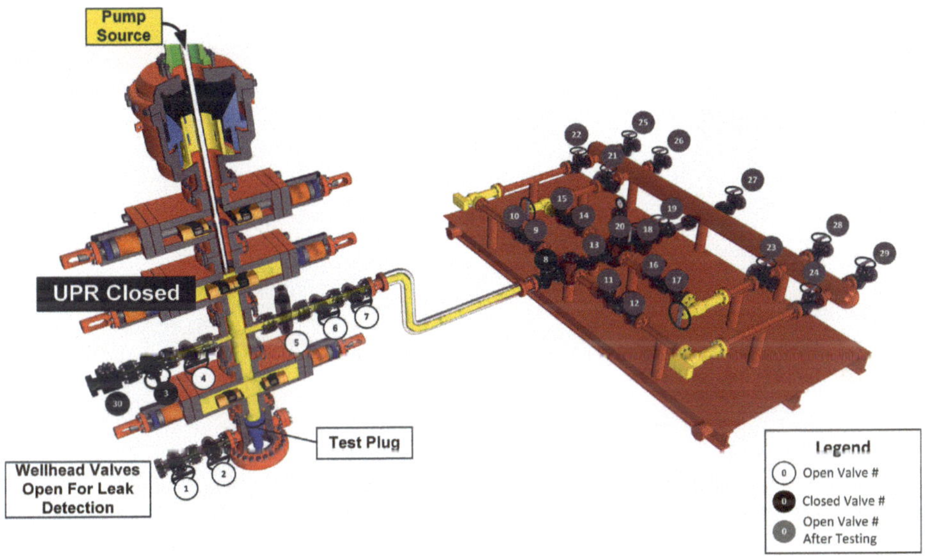

Fig. 5.104 Testing choke manifold inlet valve.

(a) Close Gate valves #8. Open all other choke manifold gate valves.

Note: Continued flowing (trickle) indicates malfunctioning valve, which must be repaired.

(b) Perform high-pressure test

- Conduct the high-pressure test at the lesser of (1) or (2).

  Initial

  1. Rated Working Pressure (RWP) of ram preventer
  2. Rated Working Pressure (RWP) of wellhead

  Subsequent lesser

  1. Maximum Allowable Surface Pressure (MASP) of hole section
- Hold pressure for 5 min or more.
- Verify the rams do not leak performing a visual inspection of ram preventer housing and element.

(c) without bleeding pressure, proceed to next step.

Step #23: Testing upper pipe rams and drilling spool valves, high-pressure test only

See Fig. 5.105.

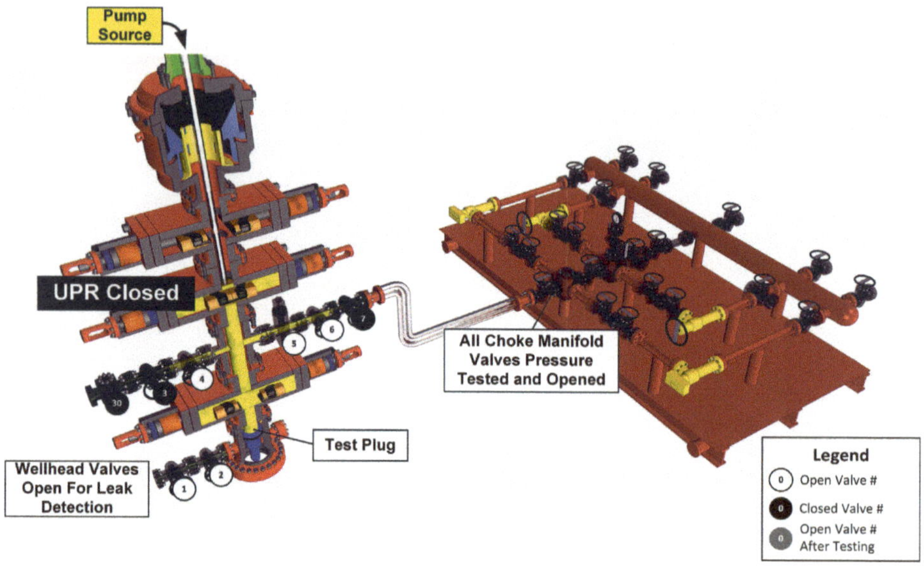

Fig. 5.105 Testing upper pipe rams and outer drilling spool valves.

(a) Close gate valves #7.Open all other choke manifold gate valves.

Note: Continued flowing (trickle) indicates malfunctioning valve, which must be repaired.

(b) Perform High-Pressure Test

- Conduct the high-pressure test at the lesser of (1) or (2).

  Initial

  1. Rated Working Pressure (RWP) of ram preventer
  2. Rated Working Pressure (RWP) of wellhead

  Subsequent lesser

  1. Maximum Allowable Surface Pressure (MASP) of Hole Section
- Hold pressure for 5 min or more.
- Verify the rams do not leak performing a visual inspection of ram preventer housing and element.

(c) Without bleeding pressure, proceed to next step.

Step #24: Testing upper pipe rams and drilling spool valves, high-pressure test only

See Fig. 5.106.

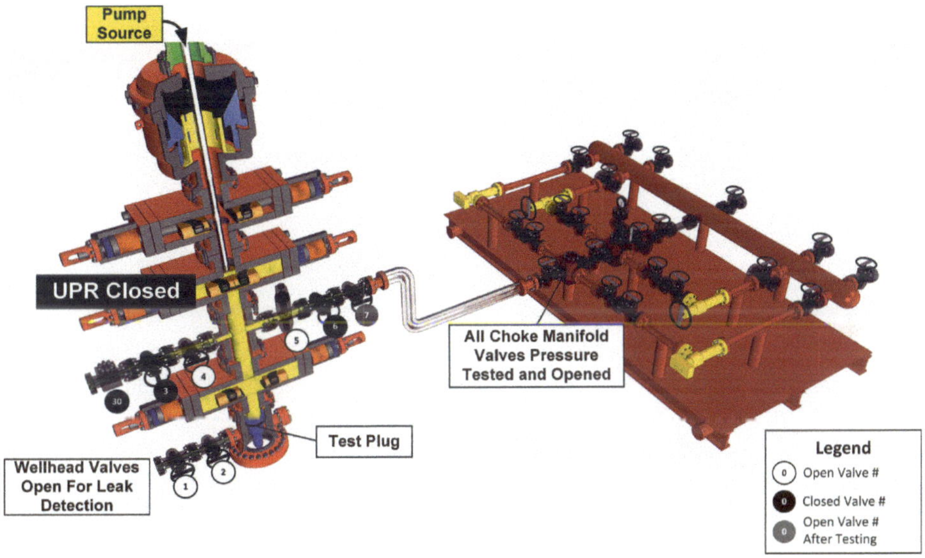

Fig. 5.106 Testing upper pipe rams and inner drilling spool valves.

(a) Close gate valves #3, and #6. Open Check Valve #30 and Gate Valve #5, and #7. **Note:** Continued flowing (trickle) indicates malfunctioning valve, which must be repaired.

(b) Perform <u>High-Pressure Test</u>

- Conduct the high-pressure test at the lesser of (1) or (2).

  Initial

  1. Rated Working Pressure (RWP) of ram preventer
  2. Rated Working Pressure (RWP) of wellhead

  Subsequent lesser

  1. Maximum Allowable Surface Pressure (MASP) of Hole Section
- Hold pressure for 5 min or more.
- Verify the rams do not leak performing a visual inspection of ram preventer housing and element.

(c) Without bleeding pressure, proceed to next step.

Step #25: Testing upper pipe rams, HCR, and drilling spool valves, high-pressure test only

See Fig. 5.107.

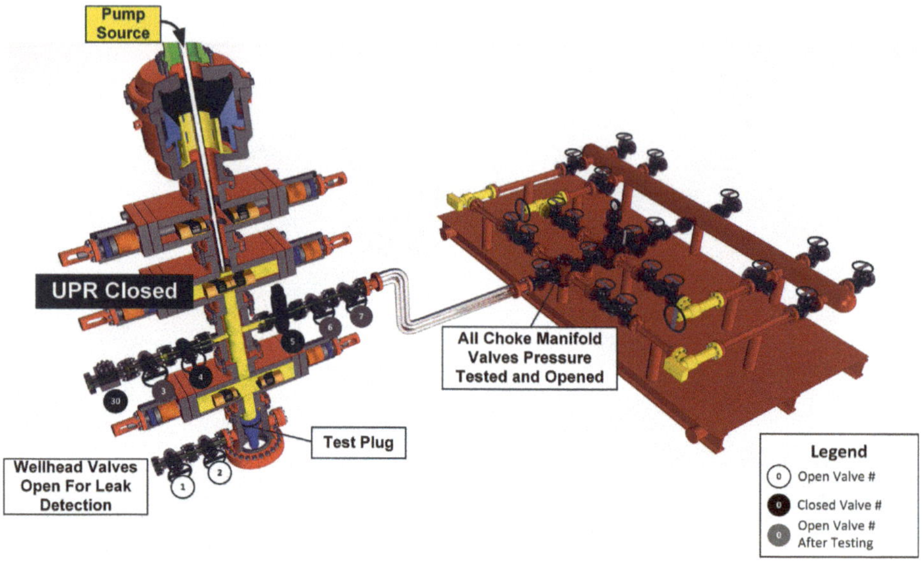

Fig. 5.107 Testing upper pipe rams, HCR.

(a) Close HCR Valve #5 and #6. Open Check Valve #30 and Gate valves #6 and #7. **Note:** Continued flowing (trickle) indicates malfunctioning valve, which must be repaired.

(b) Perform high-pressure test

- Conduct the high-pressure test at the lesser of (1) or (2).

  Initial

  1. Rated Working Pressure (RWP) of ram preventer
  2. Rated Working Pressure (RWP) of wellhead

  Subsequent lesser

  1. Maximum Allowable Surface Pressure (MASP) of Hole Section
- Hold pressure for 5 min or more.
- Verify the rams do not leak performing a visual inspection of ram preventer housing and element.

(c) Without bleeding pressure, proceed to next step.

Step #26: Testing lower pipe rams, high-pressure test only

See Fig. 5.108.

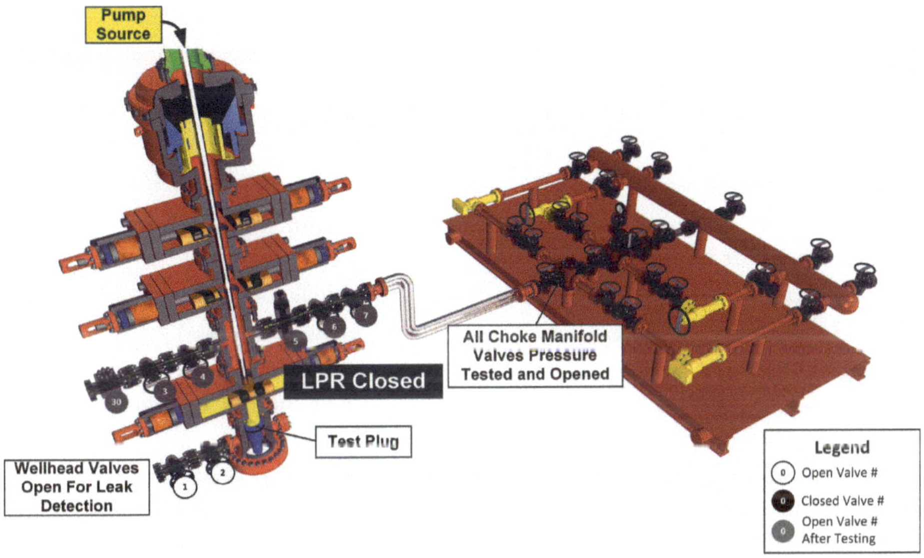

Fig. 5.108 Testing lower pipe rams.

(a) Open all valves above Lower Pipe Ram.

(b) Open Upper Pipe Rams, Close Lower Pipe Rams.

Note: Continued flowing (trickle) indicates malfunctioning valve, which must be repaired.

(c) Perform High-Pressure Test

- Conduct the high-pressure test at the lesser of (1) or (2).

  Initial

  1. Rated Working Pressure (RWP) of ram preventer
  2. Rated Working Pressure (RWP) of wellhead

  Subsequent lesser

  1. Maximum Allowable Surface Pressure (MASP) of Hole Section

- Hold pressure for 5 min or more.
- Verify the rams do not leak performing a visual inspection of ram preventer housing and element.

(d) Bleed off pressure

(e) Open Lower Pipe Rams. Close Upper Pipe Rams. Proceed to next step.

Step #27: Test surface equipment

(a) Install appropriate test sub on bottom of Top Drive.

(b) Pump into Top Drive recording the test pressure at the test pump.

Perform low-pressure test

- Conduct the low-pressure test of 250–350 psi.
- Hold pressure for 5 min or more.
- Verify the valve(s) do not leak performing a visual inspection of Top Drive, hose, and standpipe.

Perform high-pressure test

- Maximum Allowable Surface Pressure (MASP) for the well program
- Hold pressure for 5 min or more.
- Verify the Top Drive, hose, and standpipe do not leak.

(c) Bleed off pressure.

(d) Install full opening safety valve and inside BOP on the Top Drive

(e) Pump into Top Drive recording the test pressure at the test pump.

Perform low-pressure test

- Conduct the low-pressure test of 250–350 psi.
- Hold pressure for 5 min or more.
- Verify the valve(s) do not leak performing a visual inspection of full opening safety valve and inside BOP.

Perform high-pressure test

- Maximum Allowable Surface Pressure (MASP) for the well program
- Hold pressure for 5 min or more.
- Verify the full opening safety valve and inside BOP do not leak.

(f) Bleed off pressure.

This concludes the testing of BOPE

See Tables 5.22 and 5.23 for test pressure requirements.

Surface BOP initial pressure test

See Table 5.22.

Table 5.22 Suggested equipment standards (reference applicable standards)

		High pressure test	
Tested item	**Low pressure Test psi (MPa)**	**Change out of component, elastomer, or ring gasket**	**No change out of component, elastomer or ring gasket**
Annular Preventer	250–350 (1.72–2.41)	Rated Working Pressure (RWP) of annular preventer	Lesser of 1. 70% of RWP 2. Maximum Allowable Surface Pressure (MASP) of Hole Section
Fixed Preventers Variable Bore Rams Blind or Blind/Shear Rams	250–350 (1.72–2.41)	Lesser of 1. RWP of Ram 2. RWP of wellhead	Initial Test Pressure
Choke and Kill Line and BOP side outlet valves below rams(both sides)	250–350 (1.72–2.41)	Lesser of 1. RWP of side outlet valve 2. RWP of wellhead	Initial Test Pressure
Choke Manifold upstream of choke(s)	250–350 (1.72–2.41)	Lesser of 1. RWP of Ram 2. RWP of wellhead	Initial Test Pressure
Choke Manifold downstream of choke(s)	250–350 (1.72–2.41)	Lesser of 1. MASP 2.RWP of downstream valve(s), line(s)	
Kelly, Kelly valves, drillpipe safety valves, IBOPs	250–350 (1.72–2.41)	MASP for well program	

All low- and high-pressure tests must be stabilized and with no leaks for a minimum of 5 min
Annular and VBRs to be tested on smallest and largest (OD) pipe used
Adjustable chokes are not required to be full sealing devices. Pressure testing against a closed choke is not required.

Surface BOP subsequent pressure tests

See Table 5.23.

Table 5.23 Suggested equipment standards (reference applicable standards)

Tested item	Low pressure test psi (MPa)	High pressure test psi (MPa)
Annular Preventer	250–350 (1.72–2.41)	Lesser of 1. 70% of RWP 2. MASP of Hole Section
BOP side outlet valves above pipe ram preventers (wellbore side)	250–350 (1.72–2.41)	Lesser of 1. 70% of RWP 2. MASP of Hole Section
BOP side outlet valves above pipe ram preventers (nonwellbore side)	250–350 (1.72–2.41)	MASP for hole section
Fixed Preventers Variable Bore Rams	250–350 (1.72–2.41)	MASP of Hole Section
Blind and blind/shear rams	250–350 (1.72–2.41)	Casing test pressure
Choke and kill line and BOP side outlet valves below pipe ram preventers (both sides)	250–350 (1.72–2.41)	MASP of Hole Section
Choke Manifold upstream of choke(s)	250–350 (1.72–2.41)	MASP of hole section
Choke Manifold downstream of choke(s)	250–350 (1.72–2.41)	Lesser of 1. Rated Working Pressure (RWP) of choke(s) valves 2. Rated Working Pressure (RWP) of downstream line(s) 3. MASP of hole section
Kelly, Kelly valves, and Safety Valves	250–350 (1.72–2.41)	MASP of hole section

All low- and high-pressure tests must be stabilized and with no leaks for a minimum of 5 min
Annular and VBRs to be tested on smallest and largest (OD) pipe used
Adjustable chokes are not required to be full sealing devices. Pressure testing against a closed choke is not required

CHAPTER SIX

Subsea well control

Introduction

Well control in a deepwater environment is more complicated than well control on land or fixed bottom rigs. A three-step process is needed after a well is shut-in on an influx. The well has to be circulated and killed with the correct mud weight, the riser has to be circulated with the kill mud and any gas remaining below the shut-in preventer has to be cleared in a safe manner prior to opening the closed BOP.

The proper shut-in procedures, kill procedures, and stack gas clearing procedures are discussed. Fingerprinting, measuring choke line friction pressures and reviewing shut-in procedures are all done prior to drilling. Emergency disconnect procedures are included along with several dual gradient drilling methods. A discussion of factors which affect deepwater well control is included.

This section is intended to provide guidance on handling a kick on a deepwater well. Each rig is unique and the rig specific procedure which are in place need to be reviewed and discussed prior to starting drilling operations.

Commentary

The well control methods employed in subsea well control are not unlike those used for surface location well control (i.e., Driller's Method, Wait and Weight Method, etc.). But there are many additional elements which contribute to the increased complexity of drilling wells using a subsea stack as opposed to wells using a surface BOP stack. The subsea stack and control systems for the stack are larger, more complicated, and are remote distances from the rig. The only ways to manipulate valves or rams on a subsea stack are from the surface using the control system or using a remote operating vehicle (ROV). Drilling in deepwater comes with additional potential problems such as Overpressured Water Flows, hydrates, near freezing cold water temperatures at the mudline, low kick tolerances, and rig station-keeping.

Subsea well control on deepwater rigs has many of the same possible problems as land-based drilling but at remote distances. Additional problems include longer reaction times due to long choke and kill lines, larger volume BOPs, potential for formation of hydrates inside BOPs stack and choke lines, possible trapped stack gas after a kill, and increased probabilities of gaseous influxes entering the riser. If the rig has to disconnect due to

Universal Well Control
https://doi.org/10.1016/B978-0-323-90584-8.00005-8

weather, the BOPs may have trapped pressure to deal with after reconnection of the LMRP to the stack.

Influx

Most deepwater wells penetrate formations with little or no gas due to the overburden pressures of deepwater. If formation gases do exist, due to pressure and temperature, they will remain in solution until circulated near surface. This means when deepwater influxes occur, they will normally present themselves as fluid kicks and generate the false perspective that the influxes do not contain expandable gases.

Rig crews have to be vigilant because if the influx is not recognized quickly due to being in solution, very large size kicks may occur. These large kicks may lead to lost return problems. In order to ensure personnel safety, every kick taken in deepwater should be considered to be extremely dangerous as they may well contain gases. These gases will normally remain in solution until the kick is within the choke line and may easily displace fluid in the choke line as the gases break out of solution. When gas fills the choke line, the bottom-hole pressure can rapidly drop allowing an additional kick to enter the wellbore and resulting surface pressures may also rapidly change.

Overpressured shallow water flows

Deepwater wells represent exploration within frontier areas where formation waters may easily become trapped within shallow formations due to rapid deposition. These zones are frequently associated with salt beds or intrusions. If these trapped waters occur deeper than conductor casing seats, resultant water flows most likely will occur during the drilling or cementing of the surface casing stage. Once water flows occur, liquids and solids may enter the annulus causing difficulties in controlling the flows.

These water flows can be persistent and long lasting. Enormous amounts of time and money can be expended on mitigating a single water flow event. An understanding of the detailed stratigraphy below the conductor casing can help avoid drilling through shallow water flow zones.

Implementing improved drilling practices will ensure drilling gauge holes as possible for the conductor casing sections and will help assure a competent cement job can be performed thus lessening the risk of a water flow reaching the seabed.

Remote

Most deepwater locations may be considered "remote" areas presenting problems with scheduling delivery of kill weight muds, removal of excess volumes of displaced mud during circulations and any needed additional personnel. All efforts must be available to schedule and mobilize needed transportation to ensure proper timing of deliveries.

Remote locations for explorations wells may not have any offset drilling data for improved well planning. The multidisciplinary planning team must use all available geological and seismic data to anticipate shallow water flows, hydrate zones, pressure regimes, environmental, and surface conditions. The availability of subsea capping equipment including BOPs and vessels must be included within preplanning activities.

Communications

Due to the large number of people involved with drilling and workover operations on a floating vessel, effective communications may be hampered because of vast distances between shore bases, offices, boats trucks, helicopters, and multiple levels of rig personnel. The PIC must strive to keep "everyone" involved with well control and ensure all personnel are properly informed of preplanning, on-going processes, and future operations.

Other communications should include a Drilling the Well on Paper exercise, Well Control Bridging documents, Tour Handover notes, Prejob safety meetings, and many other daily briefings. The better informed everyone is, the less chance for mistakes.

Environmental conditions

Heavy sea state and strong currents can be problematic for subsea drilling operations. The wellhead must first be installed vertically so that the subsea BOP stack can be installed properly. In most areas, the deeper the water, the less current there is on bottom. The BOP stack must be landed on the wellhead connector. If the connector is damaged while landing the stack and cannot be repaired, a new well would have to be started.

The current will affect landing the BOP stack onto the wellhead connector by pushing or bowing the riser in the direction of the current. Even if the current is low near the seafloor, mid column currents can bow the riser and push the stack away from the wellhead. Bowing of the riser can be accommodated by lowering the stack down "up-current" and letting the current push the BOP stack over the wellhead. Lowering of the riser may require a combination of rig movements during lowering operations. The effect of rig heave in heavy seas is lessened by using the heave compensator to land the stack. The biggest obstacle to landing the stack without damage to the wellhead connector is the pitch and roll of the rig. Each situation must be evaluated and the environmental conditions weighed prior to landing the stack.

A storm may necessitate the rig having to disconnect and move off the well to prevent damage to the wellhead, riser, or rig. If normal operations are underway and time permits, a planned disconnect can be performed. If this occurs during a well control situation, properly closing in the well to disconnect is only half of the problem. Once the LMRP is reconnected, opening the proper BOPs under safe conditions may be complicated depending on how the well was shut-in. The top of the drillstring may have to be milled off to provide a smooth top for an overshot connection.

Equipment

Deepwater drilling rigs contain equipment not found on land or bottom-supported rigs. The wellheads, BOP stack, BOP closing unit, riser system, ROVs, and the dynamic positioning system are some of the different systems. It is important that company supervisory personnel to learn as much as possible about these different systems on a deepwater rig. Many of these systems can affect well control decisions. Rig personnel are trained on the use of their equipment and need to understand the limitations of the equipment in support of well control operations.

Deepwater vessels

Today's rig fleets have been developed over many years with drillship construction greatly outpacing semi–submersible construction. Therefore, the primary subsea drilling vessel is the Drillship. Semi–Submersibles are discussed but are mainly used for conventional shelf drilling. Occasionally, the Semi–Subsea drilling operations are conducted from floating vessels such as well intervention services.

Drillships

Fig. 6.1 Example of drillship.

As the primary means of drilling subsea wells, the drillship represents a custom built-for-purpose ship which utilizes thrusters (in lieu of propellers) and provides gantry systems for deployment of Subsea BOPs and other subsea equipment. These ships are also comprised of multiple cranes to assist with loading/unloading of material along with a large complement of crew quarters. These vessels incorporate one or more specialized derricks with motion compensation as well as a large "moon pool" where subsea equipment can be raised and lowered into the sea (Fig. 6.1).

A large complement of worldwide drillships incorporates dual function-compensated derricks. These unique systems allow for simultaneous operations to occur such as running casing while drilling operations are on-going. Today's drillships are increasing the complements for subsea BOPs from one to two stacks. By having redundant stacks, if one stack is damaged or is not operating correctly, the stack can be removed to the surface for service and testing. The second identical stack can then be installed.

The classes of drillships normally follow their design-era, maximum water depth rating along with overall drilling capability and equipment specifications. Unfortunately, no formal classification system has been defined by the drilling industry.

Semisubmersibles

Fig. 6.2 Example of semisubmersible.

The Semi–Submersible drilling vessel consists of a pontoon style system, where the pontoons are located below the water line offering the ballast to keep the vessel afloat. These vessels may be outfitted for moored service (anchor and cabling for less than 5000′ of water depth) or include a series of thrusters for self-propelled movement. In deepwater applications, most Semisubmersibles use thrusters for active station-keeping. Those outfitted with anchors must include a larger complement of anchor handling vessels to perform moves from one well to another (Fig. 6.2).

Semi–Submersibles offer a greater envelop for stability, as their ballast pontoons are located below water. This increase level of stability may be needed in areas affected by adverse weather. The increased stability of the Semisubmersible vessel is largely favored for field development.

Although they possess a greater stability, variable deck loads are smaller when compared to drillships. With a large equipment complement located above the water line, variable deck loads are smaller when compared to those of the drillships. This means more coordination efforts are needed to manage equipment and fluids during all phases of drilling activities.

Water depth

Operating water depth or maximum water depth rating is probably the key component for selection of the drilling vessel. This rating incorporates many of the design characteristics needed to drill subsea wells, specifically the marine riser tensioning system. The riser tensioning system is designed to support the weight of the riser and provide sufficient tension of the riser while in position. Additional buoyancy systems can be added to the riser to improve overall performance within the operational water depth range.

General Water Classifications. Since the drilling industry does not have one worldwide standard for defining water depth ratings, the following table represents a set of general descriptions.

- Shallow water—0′ to 400′: Shelf properties within Continental Shelf used for commercial fishing. Bottom-supported vessels (i.e., jack-ups, rig-assisted floating barges, platform rigs). For shallow lagoons and near shore wells, inland barges and posted barges are used.
- Midwater—400′ to 5000′: Transition properties beyond Continental Shelf used for commercial fishing. Serviced by moored and nonmoored Semisubmersibles and Drillships.
- Deepwater—5000′ to 7500′: Deepwater beyond transition properties mainly used for mineral extraction. Serviced by thruster-driven Semisubmersibles and Drillships.
- Ultra-Deepwater—>7500′: Located beyond deepwater, mainly used for mineral extraction. Serviced by latest generation Drillships outfitted for 10,000 to 12,000′ operating water depths.
- Technical Limit—~12,000′ and greater: Current technical limit due for conventional drilling practices due to mechanical limits of the number of casing strings which can be deployed within current wellheads.

Rig station-keeping

Rig Station-Keeping is the methodology and procedures necessary to keep the rig centered over the well. The actual performance of rig station-keeping will be dependent upon the vessel, water depth, and equipment complement of the vessel.

For drillships and semisubmersibles outfitted with thrusters, thruster maneuvering is the primary means to provide station-keeping. Using a computer-controlled system, called dynamic positioning systems (DPS), the thrusters are automatically adjusted in speed, thrust, and sometimes pitch to center the vessel over the well. This system is interlinked to positioning satellites which provide accurate global position surveys (GPS) along with sonic beacons placed on the seafloor. The DPS uses each of these key data points to calculate exact positioning with great accuracy. Under normal conditions, the drillship's bow is normally pointed into the wind which allows for easier positioning movements.

Semisubmersibles, if anchored, use a multipoint spread mooring system which arranges the anchors in a specific pattern to stabilize the vessel over the well. Using a winch system for tightening and loosening cables attached to the anchors, these vessels are manipulated into a centered position. Unfortunately, as the water depth increases, the ability to provide station-keeping reaches a technical limit around 5000′.

Subsea drilling considerations

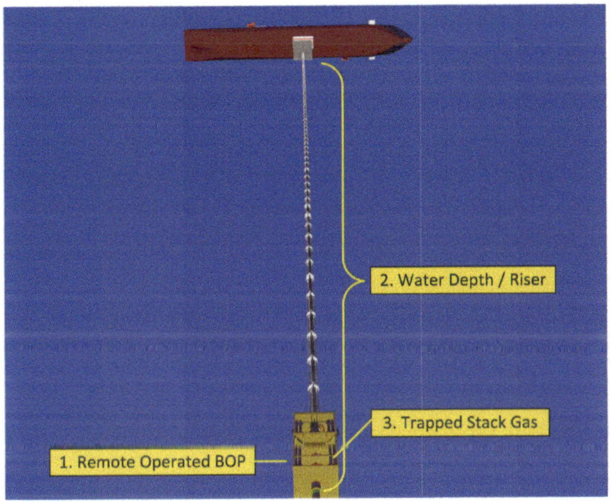

Fig. 6.3 Subsea well control contains three sequenced events.

Although well control practices do not differ substantially between onshore and floating operations, subsea well control is complicated by remote operation of the BOP stack located underwater. These subsea well control operations include more complex mechanical systems, movement of the floating vessel, installation of an extended riser used to convey drilling fluids to the surface, and the need for additional surface systems used in close proximity to crew quarters (i.e., MGS, Diverting Lines, etc.) (Fig. 6.3).

Subsea well control consists of three steps, each of which has to be completed in sequence BEFORE the well is dead.

Step 1: Kill well

Using Remote-Operated BOPs, the well is shut-in and killed using routine or nonroutine well control methods.

Step 2: Fill riser

The Riser is filled with KMW.

Step 3: Clear trapped stack gas

Finally, trapped stack gas is cleared, BEFORE rams are opened to the riser to perform a flow check.

All well control methods previously discussed are used in subsea operations. The way in which these well control methods are used becomes more complex due to operational needs of more complex mechanical equipment. The following discussions and procedures have been developed as general overview for these systems.

As each rig is custom designed and outfitted with various manufacturer's equipment, individual rig operating systems must be studied independently to improve understanding of how their specific equipment operates.

Kick detection and fingerprinting

Accurately determining a positive indicator of a well flowing such as (1) Increase in Flow and (2) Increase in Pit Gains becomes difficult to identify. This is primarily due to vessel motion. As the vessel pitches, heaves and rolls, flowing returns will increase/decrease due to this vessel motion. Volume within the pits will appear to vary due to the sloshing of fluids within the pits. The floating vessels are outfitted with computerized vessel Pit Volume Totalizers which use sonic pulses placed on opposite sides of each pit, tracking volume movement and normalizing reading between points to develop accurate develop accurate "trend lines" or the normal range of mud contained within the pits. Unless the sea state is flat, these variances make it difficult to determine small kicks.

The motion also impacts flow detection due to compression and decompression of the riser slip joints. Surging flow as well as decreasing circulation can occur from sea state. The surge/decrease will affect the ability to discern small kick volumes. Once again, Automated Flow Devices must be properly calibrated to establish trend lines between surges and decreased flow, as well as Peak/Bottom of the amplitude of each flow wave. With all three established, small kicks should be able to be discerned.

Procedures to improve kick detection

1. Frequent Flow Checks: The Driller can conduct frequent flow checks, especially while drilling wildcat wells.
2. Trip Tanks: All flow checks to be performed using the trip tank to assess possibility of a small well flow.
3. Calibrated PVT: Acoustic (sonic) sensors used to define fluid deviations within pits. If placed in opposite corners of the pit will yield an average fluid level and mitigate motion-induced sloshing.
4. Well Flowing: If well flows, immediately shut-in well.

Fingerprinting guidelines

Fingerprinting data are collected to be able to identify an actual influx by comparing real-time data to previously fingerprinted data quickly and correctly. This is necessary so the well can be shut-in quickly if there is an influx. The data recorded during a given operation become the fingerprint for the next time where the same operation is performed. By establishing trends without an influx and comparing the trend with subsequent checks can greatly assist in identifying the presence of an influx.

The recording of fingerprint data is normally the responsibility of the driller, mud logging company, and MWD engineer. Fingerprinting data are recorded and continually updated while drilling the well. Graphical easy-to-read fingerprint plots will be available within the Driller's cabin for instant reference. Copies are also available to rig supervisory personnel and client representatives.

The most accurate fingerprint data are recorded prior to drilling out of the shoe using properly conditioned mud with the bit near casing or liner shoe. If the well is fingerprinted in open-hole, the results may be influenced by fluid loss to the formation, entrained gas, cuttings load, ballooning, etc. **Tables and Rig-Specific procedures should be written and implemented for each fingerprint test.**

Gauge and PVT synchronization

The recording of fingerprint data begins after conducting BOP testing and prior to drilling out casing or liner. To ensure data are accurately recorded, a variety of pressure gauges are observed for fingerprinting including driller's pressure gauges, choke manifold gauges, choke panel gauges, mud logging gauges, and cement pump gauges at various circulating rates. All pressure gauges and pit volume instrumentation shall be calibrated (Tables 6.1 and 6.2).

Table 6.1 Generic pressure sensor calibration.

Generic Pressure Sensor Calibration						
Date	**Time**	**Gauge**	**Test Pressure**	**Driller Sensor**	**Mud Logger Sensor**	**ML HPLV Unit Sensor**
		SPP				
		Choke				
		Kill				
Test Pressure is the pressure read at the HPLV Unit gauge. A range of pressures is required for each gauge.						

Table 6.2 Generic pit volume calibration.

Generic Pit Volume Calibration							
			Flow Line Temp	Agitator		Pit Volume	
				On	Off	Driller Sensor	Mud Logger Sensor
		Active 1					
		Slug					
		Reserve 1					
		Desander					
		Sand trap					
		Degasser					
		Desilter					
		Trip Tank					
Pit volumes sensors are to be checked weekly.							

Trip sheet—displacement measurement

On the trip in the hole for drilling out the casing shoe, measure and record the actual displacement for drillpipe. It is important that the pipe is full and void of air at the beginning and at the end of the measurement for each section of drillpipe. Use the factors established on subsequent trips.

Drain back volume with generic procedure

Drain back is defined as the volume of mud observed as a rise in pit level when circulation stops and surface equipment "drains back" into the active pits or to the trip tank. It is usually recorded and plotted every time the pumps are switched off throughout the remainder of the well. While POOH or RIH, obtain fingerprinting data for start/top of ancillary equipment. These records are used to determine the effect on mud volumes by starting or stopping various pieces of equipment such as degasser, MGS, centrifuge, or charging pumps. Independently, start and stop each piece of equipment and record volume changes. Develop rig-specific procedures.

- Stop pipe movement and circulation (from the normal drilling flow rate expected).
- Record the active pit levels at 1-min intervals.
- Conduct a second test at the flow rate that is planned to be used when pumping out of hole.
- Plot the data on a chart of "Volume against time." This information is plotted so that the driller can use it to compare later flow backs.
- Repeat this test taking returns to the trip tank and also for the switching on and off of individual pieces of solids control equipment.

Each connection will then have a drain back fingerprint (on the active) and a trip tank fingerprint. These volume fingerprints along with the LWD pressure fingerprint provide the complete picture of a connection.

This data shall be recorded every minute throughout the connection in a table similar to the generic table (Table 6.3). The data shall also be plotted for easy comparison to other connections (Table 6.3).

Table 6.3 Generic drain back responses.

Generic Drain Back Responses			
Stop Circulation			
Return Flow vs Pump Rate			
Pump Off (min)	Pump Rate (gpm)	Pump Rate (gpm)	Pump Rate (gpm)
0			
1			
2			
10			
Drain Back			
Drain Back Volumes			
Pump Off (min)	Return Flow	Trip Tank	Active Pit
0			
1			
2			
10			
Mud Mixing and Treating Equipment			
Equipment Drain Back Volumes			
Minutes	Degasser	Mixing Pump	Centrifuge

In case of a flow back or ballooning event, it is essential that detailed fingerprinting data are recorded and plotted. Both volume and downhole pressure data should be fingerprinted in order to build up a detailed picture of the event. This will be important to identify any ballooning from well flows or further ballooning events.

The pit volume increase will vary with depth and temperature and shall be regularly recorded when opportunity arises. The data shall be recorded in a table.

Mud compressibility

Before drilling out casing with clean out assembly, perform mud compressibility fingerprinting data when conducting casing test. To determine the Compressibility Factor, divide the volume pumped for the test by the test pressure times the hole volume.

Well control simulation

Perform well control circulating exercise. Crews will practice holding casing pressure constant while individual kill pump(s) is brought up to kill rate speed. From the choke panel, choke(s) will be manipulated to lower or raise casing pressure by 100 psi. As pressures are achieved, note the response delays between DP and CP. Kill pumps are to be shut down while maintaining constant SICP.

Stripping drill

Perform a five (5) stand stripping drill. Close uppermost annular ram with reduced pressure and strip in hole to practice holding casing pressure constant by bleeding fluid back to trip tank. Verify the fluid bled back equals the capacity of DP stripped in hole. Bleed off pressure and record trapped pressure volume.

Drillstring rotational effect

Obtain fingerprinting data for rotational effects by circulating at drilling pump rate and record pump pressure, equivalent circulating density (ECD), and pressure while drilling (PWD). Rotate drillstring at several different speeds and record pump pressure, ECD, and PWD data. Repeat test with planned pump rates and rotating speeds for the upcoming section.

Generic procedure

- Record standpipe pressures while rotating drillpipe at 50, 100, and 150 rpm for each of three flow rates with the bit at bottom.
- The rpm shall be increased in stages allowing sufficient time for the parameters to stabilize.
- Record data in a table and plot the results similar to Table 6.4: Example Generic Rotational Effects on ECD.

Table 6.4 Example generic rotational effects on ECD.

Example Generic Rotational Effects on ECD					
Pump Rate – No String Rotation					
Depth	Pressure	1200	1000	900	800
	SPP				
	PWD				
Pump Rate – 50 rpm					
Depth	Pressure	1200	1000	900	800
	SPP				
	PWD				
Pump Rate – 100 rpm					
Depth	Pressure	1200	1000	900	800
	SPP				
	PWD				
Pump Rate – 100 rpm					
Depth	Pressure	1200	1000	900	800
	SPP				
	PWD				

PWD data shall also be recorded to determine the effects on ECD. Do not reciprocate the drillstring while performing this test.

This test shall be repeated after any significant change in mud weight or mud parameters.

Trapped pressure effect

Pressure can be trapped on a well if it is shut-in with a BOP immediately after stopping circulation. The higher the flowrate and faster the BOP closes the higher the chance for trapping pressure. To simulate trapped pressures, pump(s) is shut down and well is closed-in well immediately. Record the time from pump shut off to well shut-in. Record SICP (trapped pressure). Bleed off pressure. Trapped pressures illustrate the magnitude of potential trapped pressure when shutting in the well.

Thermal pressure effect

Thermal pressure effects resulting from the changing temperature profile in the well can contribute or reduce trapped pressure. The thermal expansion test procedure assesses the effects of well temperature on shut-in pressures. The shut-in fingerprint provides valuable data for determining the stabilized shut-in pressures following a well kick. After the well has been circulated and is up to circulating pressure, shut the well in and apply 100–300 psi. Record the shut-in pressures to determine the value of thermal expansion pressure (Table 6.5).

Table 6.5 Example generic thermal pressure effects.

Generic Thermal Pressure Effects						
	Annular Pressure			Drill Pipe Pressure		
Pump Rate Before Shut-in						
Minutes						
1						
2						
30						

Flow checks

Mud volume can increase or decrease in the trip tank during a flow check due to temperature changes in the well. This is dependent on the temperature profile of the well. Both are valid and both can mask kick detection during a flow check.

All flow checks should be monitored and recorded to ensure the driller has recent history fingerprinting information available. Compare each flow check with previous flow checks. Display the current flow check against the last one on the mud loggers drill floor display. **If flow is not decreasing over time, then it can be assumed an influx is occurring and the well should immediately be shut-in.**

Shut-in well

Kick detection may be possible only by shutting-in the well and fingerprinting the pressure build up. A baseline shut-in fingerprint should be recorded prior to drilling ahead. This pressure build up data can then be used for comparison for the well shut-in to determine if there is a kick in the hole.

The following example procedure shall be used as a guideline for the flow checking:

- With the well is shut-in, compare a plot of the annulus pressure build up over 30 min to the current fingerprint baseline.
- If an influx is considered likely, initiate kill procedures.
- If no influx is suspected, open the BOP after performing a flow check through the choke and confirming it is negative (Table 6.6).

Table 6.6 Example generic well shut-in effects.

Generic Well Shut-in Effects								
Status Prior to Shut-in					Shut-in Pressure			
Mud Weight		Mud Temperature		Pit Gain				
In	Out	In	Out		Minutes	Drill Floor Choke	Mud Loggers Choke	PWD
					1			
					2			
					30			

Slugging

Slugging fingerprint records volume changes as the slug settles in the drillstring. The slug volume and displacement volume should be determined and verified by returns to the trip tank.

- Record Active, Trip Tank, and slug pit volumes both on the Driller's and Logger's sensors.
- Line up mud pump to slug pit. Pump required volume of slug.

- Line up to Active pit and reset counters.
- Pump ~100 strokes to place slug in string.
- Switch over to Trip Tank and initiate trip tank pump.
- Open bleed off and allow slug to settle in pipe. Monitor trip tank volume against time until trip tank volume stabilizes.
  - Note: the gain in the trip tank from the slug settling and the time for the slug to settle.
- Record the Trip Tank, Slug, and Active Pit volumes.
- SCRs should be taken at regular intervals to establish trends.

Response due to weather

The heave, pitch, and roll of the rig will impact instrument readings. These effects on the mud pit volumes and return flow meter caused by the prevailing weather conditions are to be recorded. The impact on mudflow and volumes resulting from crane movements, helicopter landings, etc. is also be recorded within the fingerprint table (Table 6.7).

Table 6.7 Example generic weather response table due to weather.

Generic Weather Response Table due to Weather					
Heave					
Pitch					
Roll					
Observation Point	Mud Pits	Flow meter	Crane	Ballast Rig	Other
Driller					
Mud logger					

Large hole volumes

Knowing subsea drilling operations center around the size and depths of hole to be drilled in order to run protective casing, hole sizes are normally larger resulting in larger volumes of fluid needed. Surface holes are normally drilled with weighted water-based muds using seawater wherever possible. As the wells are drilled deeper, hole parameters will change to higher pressures and temperatures. For the most part, water-based drilling fluids are switched over to synthetic oil-based (SOBM) muds. Managing two types of fluid at the same time with limited pit volumes can be challenging. To assist, support vessels offer space for temporary storage.

Larger volumes of mud require more processing equipment, replenishment stock, mixing, and storage capabilities. Managing surface volumes of mud together with support vessels requires great communications. Since subsea operations are normally conducted in isolated areas at great distances from shore base support operations, fluid management is further challenged.

Large kicks and increases in kill mud densities may take much longer when compared to onshore operations due to surface capabilities and long delivery times by support vessels.

Variable deck loads

Most floating vessels used today attempt to perform various functions at the same time. For those vessels outfitted with multiple derricks, understanding large variable deck loading may impact abilities to quickly observe if the well is kicking. If the decks are outfitted with risers and Subsea BOPs, the overall vessel draft will increase. If multiple tasks are being performed such as drilling and running casing, deck load will change and may impact fluid movement in pits and in flowlines. Therefore, base line information on pits and flowlines is needed to ensure changes to variable loads won't cause an influx to be masked by other operation.

Remote operations

Since Subsea BOPs are installed on the ocean seabed, the distance between the Subsea BOPs and surface vessel can be in the thousands of feet. Although normally, fluid is considered to be incompressible, this discussion is only relevant over short distances. When choke line and kill line lengths exceed thousands of feet, this assumption is no longer valid, and fluid will have to be compressed in order to have sufficient hydrostatic pressure to perform functions.

In order to activate Subsea BOPs remotely, more sophisticated blowout preventer closing units are used. These BOP Closing Units employ accumulator banks and 4-way valve pressure operation. More fluid is needed, so Subsea BOP Closing systems have larger accumulator bottles for more volume and larger OD pressure lines. System pressures are also increased to allow fluid to flow faster for improved closing times. Today's Subsea Rigs utilize the latest multiplexer (MUX) systems where hydraulic controls are replaced by fiber optic cables run from surface to subsea, utilizing electricity to open solenoids and direct control fluid through regulators, subplate mounted valves, and shuttle valves to function the appropriate subsea BOP and gate valves.

Another consideration is the large amounts of fluid needed to open and close preventers, compared to onshore BOPs. Therefore, BOP operations are somewhat slower due to large compressible volumes, longer distances, and larger equipment requirements. In order to offset these challenges, Subsea BOPs have closure assist accumulator bottles within the Subsea BOP structure. Furthermore, functions occur with fiber optic cables, downhole computer-driven solenoids, and surface supplied hydraulic fluid.

With these upgrades, function times are greatly reduced. Overall, closure times remain somewhat longer when compared to onshore operations.

Temperature extremes

When deep-seawater depths approach thousands of feet, water temperatures can decrease to about 1 °F above freezing (33 °F/1 °C). This cold environment can play havoc on the properties of drilling fluids. As formations are drilled deeper, for the most part, these fluids encounter greater pressures and temperatures. When weighted drilling muds are circulated into the riser, these fluids quickly cool causing gelation problems. If the fluid become static for a period of time, the gelation problems can be much worse.

During Well Control operations where these fluids are circulated within a much smaller choke line, the cooling effects occur over a shorter period of time and can lead to creation of hydrates (crystalline structures of frozen gas). They can actually plug these lines. Therefore, mechanical interfacing with antifreeze chemicals may need to take place to ensure choke lines are kept free of plugging.

Communication within a larger workforce organization

To offset individual responsibilities, the organizational structure on a floating vessel is extremely large (when compared to onshore operations) requiring many individual experts' knowledge to keep systems running, tested, and operating. The major system experts include

1. Captain and support crew (operating vessels thrusters for station-keeping)
2. Subsea Engineer (responsible for BOP stack component installation, testing, and operation)
3. Mud Engineer (responsible for larger quantity and variety of mud equipment, storage, and operation)

Effective communication between all parties becomes paramount to success when dealing with a well control event.

Subsea well planning, design, and construction

Drilling a subsea well, when compared to an onshore well, is more complex in design and execution. The core difference will be the number of casing strings needed and the wellhead necessary to land the casing string. The time of construction may easily exceed 180 days, costing companies upward of $100MM/well, depending on spread costs. Due to the high cost of daily operations, design, planning, and development of contingent plans for each step in the construction are greatly scrutinized.

Additional considerations for subsea well planning include:

- Water depth and Vessel Selection: Planning begins with understanding the specifications for a selected vessel and the water depth in which the vessel will be operating. Included in this analysis is a detailed study of the yearly wind, waves, and environment.
- Downhole Geology: The downhole geology, including studies of formations which may flow or fracture will drive the number of casing strings needed to safety drill the well. Contingency, or back-up casing strings are also incorporated.
- Subsea Wellhead: The appropriate subsea wellhead is selected based upon both the drilling and production phases. The subsea wellhead will be the basis of developing a detailed drilling prognosis for each hole section.
- Subsea production tree type: The selection of subsea production trees range from vertical trees to horizontal trees. The type of production selected will drive the completion sequence and the BOP/LMRP configurations during completion operations.
- BOP stack requirements: With pressure containment defined by casing design and fluid weights, the chosen vessel will be outfitted with at least 5 BOPs in their stack. Each BOP provides critical functions necessary to shear a variety of tubulars planned to be run into the well during operations. These BOPs must permit rapid release of the LMRP for emergency and planned disconnects. Due to the complexities of these BOP systems, current efforts are underway to outfit vessels with two complete sets of BOPs contained within their own surface testing facility.
- Station-keeping requirements: Station-keeping due to environmental constraints may impact vessel selection. In harsh weather environments with high sea states, vessel station-keeping criteria will be critical. Multipoint spread mooring may be preferable in isolated locations and not for field development.

Subsea drilling sequence

The following is a generic drilling sequence used for subsea operations. As a general sequence, the following procedure has been developed to improve understanding of operations necessary to drill a subsea well. This general sequence will deviate from every subsea drilling prognosis, as they are developed for a specific set of equipment complements and equipment ratings for the specific vessel.

1. After performing a subsea survey outlining potential seabed anomalies and shallow flows, the vessel is moved to location. The vessel and site are prepared for drilling operations.
2. Jet structural casing in place (30″ or 36″) and cement in place. The key for success is to install the structural casing as straight as possible (no deviation or lean in this phase).

Mud systems will be simple seawater mix with gel with returns at the seafloor. Install low-pressure wellhead housing (LPWHH) to the uppermost casing, jet into place where the low-pressure wellhead housing is located at sufficient distance above seabed.

   a. If needed, a permanent guide base (structural mud mat) can be attached to the low-pressure wellhead housing to provide support for a soft.seafloor.

   b. Soak or cement structural casing. Soaking will allow time for formation to collapse around the structural casing.

3. Release drill ahead tool and drill surface hole. Drill the surface hole section with seawater/gel mud and returns at seafloor. Displace hole with weighted mud and POH with drilling assembly and Drill Ahead Tool (DAT).

   a. For shallow hole hazards, an intermediate casing may be used to thwart shallow flow, requiring drilling and installation of one additional casing string (28″ or 26″).

   b. Shallow flow zones will be drilled riserless, using pump and dump mud system with returns at seafloor.

4. Run, set, and cement surface casing (22″ or 20″) with the high-pressure wellhead housing (HPWHH) and extension on drillpipe. Returns are taken at the seafloor via ROV-activated ball valves.

5. Rig up, run, and install Drilling Riser and Subsea BOP stack. The BOP stack is latched onto the high-pressure wellhead housing.

   a. With riser in place, drilling returns are now routed to the vessel for the remainder of drilling sequences.

6. Drill each subsequent hole section, run, set, cement, and test each casing string until TD.

Subsea wellheads

Wellhead systems must include fatigue and failure modeling due to variety of compression, burst, and tensional loading forced acting upon these systems. Analysis of these systems should include bending loads and emergency stresses imparted during emergency drive-off. Well control analysis and stress analysis due to BOP stack weight are also included. Well control and production analysis to include hang-off weights imparted on the BOP stack. Once installed, only the outer casing string can be pressure monitored, as the unitized design of the wellhead prevents access to all other casing annuli.

Subsea casing design

As water depths increase, the formation fracture pressure limits of the well decrease. This translates to a narrowing of operating mud weights (difference between formation fracture pressures and pore pressures).

Step 1: Calculate total depth (See Table 6.8)

Table 6.8 Total depth example calculations.

Water depth	Air gap	Measured depth	Total depth
500 ft	50 ft	1100 ft	1650 ft
6000 ft	50 ft	1100 ft	7150 ft
11,000 ft	50 ft	1100 ft	12,150 ft

Step 2: Calculate fracture pressure (See Table 6.9)

Table 6.9 Fracture pressure example calculations.

Water depth	Water grad	Water press	MD	Fracture PG	Form Frac pressure	Frac pressure
500 ft	0.465 psi/ft	233 psi	1100 ft	0.7 psi/f	770 psi	1003 psi
6000 ft	0.465 psi/ft	2790 psi	1100 ft	0.7 psi/f	770 psi	3560 psi
11,000 ft	0.465 psi/ft	5115 psi	1100 ft	0.7 psi/f	770 psi	5885 psi

Step 3: Calculate fracture pressure in mud weight equivalent (MWE) (See Table 6.10)

Table 6.10 Fracture pressure as fracture mud weight equivalent.

Frac pressure	Total depth	Frac gradient	Frac MWE
1003 psi	1650 ft	0.61 psi/ft	**11.7 ppg**
3560 psi	7150 ft	0.49 psi/ft	**9.4 ppg**
5885 psi	12,150 ft	0.48 psi/ft	**9.2 ppg**

.

AS HORIZONS APPROACH 10,000 FT, THE FRACTURE MUD WEIGHT EQUIVALENT BECOMES LESS AND LESS DUE TO THE AMOUNT OF OCEAN HYDROSTATIC PRESSURE ACTING ON THE FORMATIONS

Considerations used in subsea casing design are as follows:

1. The water depth and decreasing formation fracture pressure limits will drive the casing selection needed to reach TD (size and number of casing strings).
2. Shallow flows may require an additional casing string.
3. Salt zone penetration. When salt zones are penetrated, the plastic deformation of the salt requires installation of additional/heavy-walled casing strings. Salt zones

will present challenges in drilling to assure salt does not flow (plastic deformation) during drilling and installing casing.

4. For produced fields with highly depleted zones, additional casing may be needed to isolate troublesome formations.
5. Annular Pressure Build up (APB) will need to be studied. APB is the pressure build up as a result of wellbore temperature changes during drilling and production phase. Fluids within a closed annulus will attempt to expand imparting forces against wellhead, casing, and formation. To overcome these issues, fluids left.in the annulus may consist of fluid which can be compressed or allow the casing hanger to vent pressures.

After installation of the high-pressure wellhead housing, BOP stack, and riser, the maximum casing size which can be run will be limited by the drift diameter of the drill-through equipment. The high-pressure wellhead housing can hang-off 3 to 4 casing strings, while well design limits may require up to seven casing strings. To provide coverage for additional casing strings within a fixed number of hung-off casing, designers employ drilling liners. Using specialized underreamers and bicenter bits, holes beneath hung-off casing shoes are enlarged allowing tight clearance drilling liners to be installed and set. Designs may include expandable casing to be installed (Figs. 6.4 and 6.5).

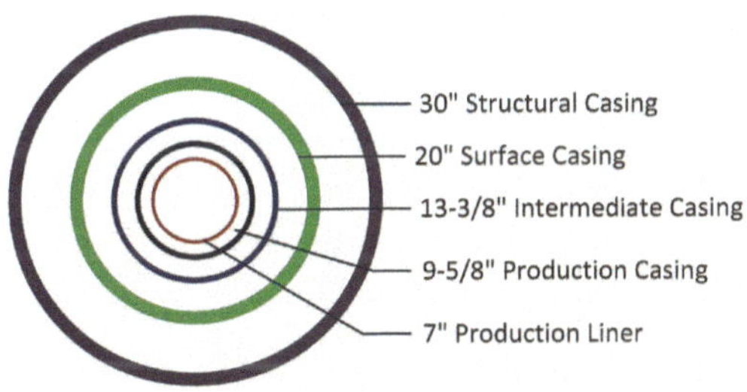

Fig. 6.4 Normal clearance casing string.

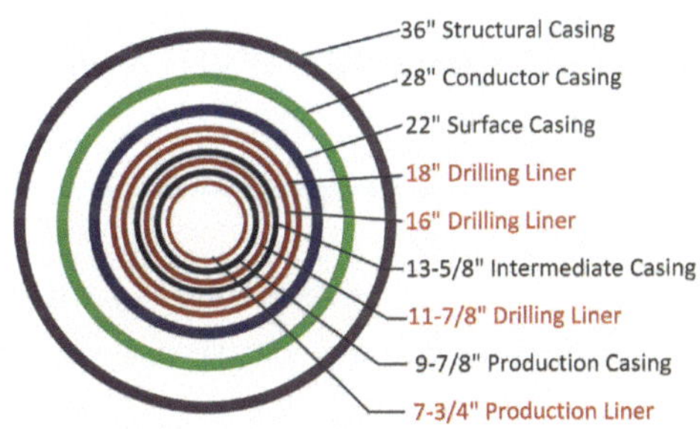

Fig. 6.5 Tight clearance casing string.

Riserless drilling and shallow well flows

Riserless drilling is the process in which fluids are pumped down the drillpipe, through the bit with returns vented at the seafloor. These returns will cause a cuttings pile to form on the seafloor near the wellbore. Shallow flows may occur at any time.

Structural casing string section

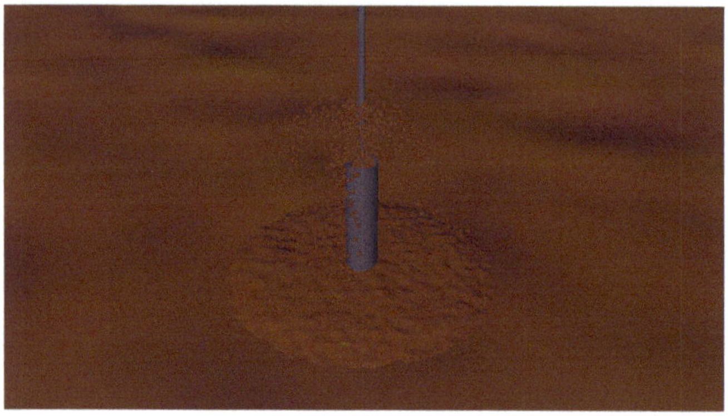
Fig. 6.6 Riserless jetting of structural casing.

The first hole section is normally installed using jetting technique where a bit is suspended on drillpipe just beyond the open-ended shoe joint of structural casing. Using a drill ahead tool (DAT) latched into the low-pressure wellhead, this system is affixed to the top of the structural casing string. Seawater/gel mud is pumped at a high velocity down the drillpipe, the structural casing is lowered into the seabed. The high-rate velocity seawater/gel mud "blows" the soft.formations and creates an annulus behind the structural casing, called jetting (Fig. 6.6).

The structural casing is jetted to an appropriate planned depth (~100–300 ft. below mudline). When the structural casing is set at the predetermined depth, the low-pressure wellhead (LPWH) is positioned several feet above the mudline and above the debris field.

STRUCTURAL CASING IS THE FIRST CASING STRING JETTED IN PLACE.

A 2ND CASING STRING, CALLED THE CONDUCTOR CASING, MAY BE USED TO OFFSET SHALLOW FLOWS.

If jetted, structural casing is not cemented in place. Sufficient time is allocated to ensure soft formations will slough into the annulus forming "skin friction." The formation sloughing occurs during a soaking period where the formation to casing bond forms "skin friction." The skin friction force must be greater enough to support the weight of the structural casing string plus anticipated drilling loads. After the soaking period, the drill ahead tool along with the drillpipe is released and the surface hole drilling commences. Once again, the returns for this hole section will be taken on the seafloor.

Once drilling operations resume, the suspended drillstring is released from the inside of the casing, using the drill ahead tool, to permit drilling the surface hole section. If the mudline formations are too firm to allow jetting, a large diameter pilot hole is drilled to TD with returns at the seafloor. A densified pill is spotted within the hole section and casing is installed and cemented.

Surface casing string section

As with the structural casing, the surface casing string is drilled riserless with returns at the seafloor. The surface casing is made-up and run with the high-pressure wellhead attached to the top section of the casing. After placing the surface casing in the drilled hole section, the high-pressure wellhead housing is hung-off in the low-pressure wellhead housing and cemented in place. After cement has cured, the subsea BOP stack, LMRP, and riser are run, landed, and connected to the high-pressure wellhead housing. After installation of the BOP stack and riser, conventional well control practices can be implemented.

Riserless tophole drilling fluids

Structural and Surface casing holes are drilled with a seawater/gel sweep drilling fluid with returns to the seafloor. These systems are designed to be environmentally safe as large hole sizes will require large volumes of fluid. To offset potential shallow flows, the surface mud system contains several hole volumes (normally 1–1/2 to 2) of kill weight fluid mixed, stored, and readied for emergency displacement.

After drilling the surface hole, a weighted mud pill is placed for entire hole section (referred to as pad mud) to ensure hole remains open and free of cuttings. If a shallow flow formation in encountered, the kill mud will be added to the hole. The density of the kill mud along with the seawater overburden pressure should be sufficient to curtail flow.

Shallow hazards

Sonar, side-scanning sonar, high-resolution shallow seismic surveys as well as visual inspection of seabed are conducted before moving a vessel onto location. These surveys are used to identify potential shallow flow formations or gas hydrates. As discussed previously, gas hydrates are formed under pressure and temperature extremes where gases form a crystalline structure much like ice. Natural deposits of these hydrates can be found in ultra-deepwater and shallow sediments below the mudline. Formations containing hydrates as referred to as in-situ gas hydrates (in place gas hydrates).

A hazardous situation can develop after drilling out the surface casing and encountering gas hydrate formations. As the gas hydrate formations are drilled, the heated drilling mud may cause the hydrates to transform into light-weight gas. If gas enters the riser, it may expand and push mud out of the hole. The surveys are helpful in determining potential hydrate areas and allow designers to avoid this hazard.

Shallow water flow (SWF)

Shallow water flows (SWF) can occur when a shallow water bearing formation becomes charged and the subsequent pressure exceeds the hydrostatic pressure of the drilling fluid. The shallow water formation must consist of a trap formed with high-density formation above the SWF forming a trap. When penetrated, the formation pressure can exceeds drilling hydrostatic pressure. When taking returns to the seabed, the hydrostatic pressure of the wellbore consists of seawater overburden pressure plus the hydrostatic pressure of drilling fluid and dynamic friction pressure of fluid flowing up the annulus.

Shallow water flows occur in deeper waters (>500′) at depths ranging from 200 to 2000′ below the mudline. Water flows in structural casing holes are extremely rare due the unconsolidated formations found immediately below the mudline. Shallow well flows occur most frequently in the surface hole section of the wellbore. The rates of these flows range from barely detectable to formation carrying flows with visible sand clouds around seabed returns. For small shallow well flows, surface casing section is drilled with spotting of a kill mud weight pill. When BHA is pulled, the kill pill hydrostatic pressure exceeds of the kick zone.

Concerns of high-rate shallow well flows

1. Time and cost to remediate any SWF event, including running protective casing string (Conductor Casing).
2. Formation debris build up around low-pressure wellhead housing, obstructing future access.
3. Subsidence of seafloor in and around wellbore and low-pressure wellhead housing. If significant, can buckle structural casing.
4. Loss of Wellbore Integrity. Hole collapses due to flow of unconsolidated formations which may fill the entire well.
5. Inability to Reenter Well. With collapsed wellbore, inability exists to reenter and perform kill operations.

Drilling procedure for suspected shallow well flow formations

1. Pilot Hole: Drill a smaller pilot hole for surface casing section (typically 8–1/2″ or 9–7/8″) followed by enlargement of the hole section by drilling with bicenter bit or hole opener. The smaller pilot hole allows for faster displacement of kill mud weights needed to thwart Shallow Well Flows. The smaller hole size also increases the chance of wellbore bridging.

2. Keep Drillstring BHA Simple: To decrease chances of sticking the BHA within a shallow well flow, consider keeping the complexity of the BHA to a minimum. One consideration is to use Pressure While Drilling tools to assist in real-time detection of shallow flows.
3. Use Conductor Casing: Run, install, and cement a second string of casing across the SWF to enhance strength of casing while isolating zones with flow potential.
4. Visually Monitor Drilling: Using a subsea camera or ROV, provide visual monitor of the seabed for quick reaction. Because of riserless operations, mud and cutting discharges will decrease visibility. When well flow occurs, the visibility is further decreased. The ROV can be used to scan for the presence of gas bubbles in and around wellbore.

Riser margin

Riser margins are developed in the event of a riser disconnect. Riser margin is the additional mud weight increase in drilling mud required to balance formation pore pressure if the riser hydrostatic pressure is replaced with seawater hydrostatic pressure. It may be possible to drill with the higher mud weight so that even if an accidental riser disconnect occurs, the mud column will overbalance the formation.

Decreasing formation strengths in deepwater wells generally prohibit drilling with a riser margin as the open hole cannot withstand increases in mud weights.

Calculation of riser margin

Calculate the bottom-hole hydrostatic pressure reduction due to a riser disconnect point.
Note: Riser length and water depth are not the same

$$\mathbf{BHP\,Reduction_{psi}} = \left(\mathbf{Riser\,Length_{TVDft}} \times \mathbf{Mud\,Weight_{ppg}} \times \mathbf{0.052}\right) - \left(\mathbf{Water\,Depth_{TVDft}} \times \mathbf{Seawater\,Density_{ppg}} \times \mathbf{0.052}\right) \tag{6.1}$$

Riser margin calculation

$$\mathbf{Riser\,Margin_{ppg}} = \frac{\left(\mathbf{BHP\,Reduction_{psi}}\right)}{\left(\mathbf{Distance\,from\,Disconnect\,Depth\,to\,Well\,TVD_{ft}} \times \mathbf{0.052}\right)} \tag{6.2}$$

Calculate mud weight needed to maintain current BHP after riser disconnect

$$\mathbf{Mud\,Weight\,Needed_{ppg}} = \mathbf{Riser\,Margin_{ppg}} + \mathbf{OMW_{ppg}} \tag{6.3}$$

Choke line friction

Understanding choke line friction pressures (CLFP) is essential to subsea well control. Circulating through the choke lines can induce high back pressures on the well due to the size and length of the lines. The additional back pressure induced can overpressure the well and cause lost return problems. To compensate for the additional friction pressure incurred by circulating through the choke line(s), this additional pressure must be reduced from the circulating casing pressure during pump ramp up to kill speed. Conversely, it must be added back in during pump slowdown.

CLFP must be measured prior to taking a kick. The methods discussed later for measuring CLFP are for cased-hole and open-hole scenarios.

Cased-hole CLF discussion

Choke line friction (CLF) pressures are measured by pumping fluids down the drillpipe and back-up through the choke line. These pressures are additive to the bottom-hole pressure. These friction pressures, if not accounted for, will increase the BHP.

UNDER NO CIRCUMSTANCES SHOULD CASED HOLE CLF BE MEASURED WITH OPEN HOLE

To accurately measure cased-hole CLF, first the circulating drillpipe pressures are recorded for a normal circulation (down drillstring, up annulus, and riser). Next, the wellbore is isolated by closing the uppermost annular preventer and opening the appropriate choke line. Circulation is established by pumping down the drillstring, up the annulus, up the choke line, and through choke manifold. The difference in these two measured pressures is the choke line friction pressure.

$$\begin{aligned}\mathbf{HP_{DS} + Circ\,DP_{DS} - Friction_{drillstring}} &= \mathbf{BHP_{Dynamic}} \\ &= \mathbf{HP_{Ann} + Circ\,CP_{Ann} + Friction_{Ann}} \quad (6.4)\end{aligned}$$

where

HP_{DS}	=	**Hydrostatic pressure at the bottom of the drillstring (psi)**
Circ DP_{DS}	=	Surface pressure at the top of the drillstring, i.e., the standpipe pressure, equal to the pump pressure less the surface piping pressure losses (psi)
$Friction_{DS}$	=	Sum of the friction pressure losses within the drillstring, including the bit pressure loss. This pressure is read on the standpipe manifold (pump pressure minus surface piping pressure losses). Since the well is being forward circulated, a minus sign is used, meaning the drillstring/bit friction pressure losses do not add to the bottom-hole pressure.

$BHP_{Dynamic}$	=	Dynamic bottom-hole pressure while circulating (psi)
HP_{Ann}	=	Hydrostatic pressure at the bottom of the annulus (psi)
Circ CP_{Ann}	=	Surface pressure at the top of the annulus (psi)
$Friction_{Ann}$	=	Sum of the friction pressures in the annulus. A plus sign is used because the annulus pressure acts on the bottom of the well (psi)

Annulus friction pressure term ($Friction_{Ann}$) is the sum of two friction components:

1. Friction due to flow through the drillstring × cased-hole/open-hole annulus
2. Friction due to flow through the choke line/choke/choke manifold.

The cased-hole CLF measures the choke line friction for the range of slow pump rates and current mud weight. Since length of the choke line remains fixed, the CLF for a given slow pump rate and mud weight will remain constant throughout drilling of the well. For every 0.2 ppg change in mud weight or at each casing point, the CLF must be remeasured.

FOR EVERY 0.2 PPG CHANGE IN MUD WEIGHT OR AT EACH CASING POINT, THE CLF MUST BE REMEASURED

Cased-hole CLF procedure

A generic cased-hole CLF measurement procedure is as follows:

1. At various SPRs, Obtain Circulating Drillpipe Pressures Up the Riser.
   - a. Position the bit near the float collar of the surface casing shoe track, circulate, and condition mud.
   - b. Before drilling out shoe track, circulate down the drillstring and up the riser annulus to surface.
   - c. Record Circulating DP Pressure at various slow pump rates (P_{riser}).
2. Space Out, Shut Down, and Shut-in the Well. Rig Up to Circulate Through the Choke:
   - a. Verify the tooljoint space-out in the subsea BOP stack, then close the upper annular BOP.
   - b. Open the inner and outer hydraulic valves on the uppermost choke and kill line side outlets beneath the closed BOP.
   - c. Open valves on choke manifold to selected choke and open choke. Route returns to mud pit.
3. At the same SPRs, Obtain Circulating Drillpipe Pressures Up the Riser and through Choke
   - a. Circulate down the drillstring and up the annulus through the choke line/choke/choke manifold, along the same flow path planned for any well kill.
   - b. Record Circulating Drillpipe Pressure at the same SPR used before ($P_{choke\ line}$).

4. Calculate Choke Line Friction (CLF) For Each Slow Pump Rate:
   a. To calculate, use CLF= $P_{\text{choke line}} - P_{\text{riser}}$ for each SPR
5. Calculate Incremental Change In Choke Line Friction (Δ CLF) Between SPR
   a. To calculate, use $\Delta\ CLF_2 = CLF_{SPR2} - CLF_{SPR\ 1}$ and use $\Delta\ CLF_3 = CLF_{SPR3} - CLF_{SPR\ 2}$
6. Develop table of SPR, CLF and Δ CLF

As pumps are brought up to speed holding casing pressure constant, at the first selected rate, the circulating CP is dropped by the $CLF_{SPR\ 1}$. If the pump rate is increased to the second chosen rate holding casing pressure constant. At the second selected rate, the circulating DP pressure is dropped by $\Delta\ CLF_2$.

If the pump rate is increased to a third chosen rate while holding casing pressure constant, the circulating DP is dropped by $\Delta\ CLF_3$.

Open-hole CLF discussion

If mud weight increases by 0.2 ppg while drilling the open-hole section, the CLF should be remeasured. In order to keep the drillstring from sticking, this method is a one-step method and employs reverse circulation. The mud pump flow is routed to the choke manifold and pumped through a fully opened choke and down the choke line, back into the Subsea BOP stack and up the annulus (drillstring × riser annulus).

Since the bottom hole is isolated, the friction pressure generated by pumping down the choke line does not affect the wellbore pressure. The casing shoe and bottom-hole pressure will only be exposed to the friction pressure of the riser × drillstring annulus.

$$\mathbf{HP_{Riser} + Circ\ CP_{Riser} + Friction_{Riser} = BHP_{Dynamic@subsea\ wellhead} = HP_{CL} + Circ\ CP_{CL} - Friction_{CL}} \tag{6.5}$$

where

HP_{Riser}	=	Surface pressure at the top of the riser, which is equivalent to atmospheric pressure since the riser is open to the atmosphere (psi)
$Circ\ CP_{Riser}$	=	Surface pressure from top of the hole to the flowline (psi)
$Friction_{Riser}$	=	Friction pressure loss across the drillstring × riser annulus. Since the well is being reverse circulated down the choke line, a plus sign is used for the friction pressure loss in the riser × drillstring annulus, meaning this friction pressure is added to the bottom-hole pressure. However, this pressure loss is minimal at the pump rates used on a well kill, due to the relatively large flow area, and can therefore be ignored
$BHP_{Dynamic\ @\ subsea\ wellhead}$	=	Dynamic bottom-hole pressure at the subsea wellhead (psi)
HP_{CL}	=	hydrostatic pressure of the choke line, at the subsea BOP stack (psi)

Circ CP_{CL}	=	Surface pressure at the top of the choke line, as read off the casing pressure gauge (psi)
$Friction_{CL}$	=	Friction pressure loss in the choke line. A minus sign is used for reverse circulation, meaning the choke line friction pressure has been consumed in circulating the mud through the choke line from surface to the subsea BOP stack, and thus it does not add to the bottom-hole pressure

Open-hole CLF procedure

The Open-Hole Procedure for measuring CLF is as follows:

1. Rig Up to Reverse Circulate Down the Choke Line and Take Returns Up the Riser Annulus:
   a. Rig up to pump through choke manifold and through a fully opened choke. The fluid will be pumped down the choke line, entering the annulus through Subsea BOP side outlet and into the drillstring × riser annulus.
   b. The HCRs (two hydraulic spring-assist closure valves) on the selected BOP side outlet must be opened to permit circulation into the wellbore annulus.
2. Obtain Slow Pump Rates Through the Choke Manifold/Choke/Choke Line:
   a. Perform reverse circulating by pumping at predetermined slow pump rates. Record Circulating Casing Pressures at the selected slow pump rate. Since the drillstring × riser annulus friction pressure can be negligible, this friction pressure is not recorded.
   b. Record Casing Pressure gauge ($P_{\text{choke line}}$) is a direct measure of the choke line friction at the given SCR.

Managing choke line friction pressure

A static kill line can be used to manage choke line friction pressure and keep bottom-hole pressure constant as the pumps are ramped up to kill speed or reduced. This method is very simple. If a static kill line is not available, the pumps can be staged up or down using the CLF data obtained at different pump rates.

Static kill line compensation method

Using a static kill line to compensate for back pressures imposed by choke line friction is straightforward. The static kill line is a kill line not being used for circulating the well. It has to be opened under the closed BOP and will be used to read shut-in surface pressure. It must be closed at the surface to provide a static line. This arrangement prevents any flow through the kill line and will allow monitoring of annulus pressures at the Subsea Wellhead.

As pumps are sped up to SPRs, the choke operator only needs to keep the static kill line surface pressure constant. This is accomplished by choke manipulations. As the rates are slow, anticipate lag times during choke manipulations.

Many rigs are equipped with a gauge on the BOP stack so that pressure can be read below the choke line.

WELLS WITH NARROW MUD WEIGHT RANGES MAY NOT ALLOW COMPENSATION OF CLF AT HIGHER SPR

THESE WELLS MAY REQUIRE UTILIZATION OF MULTIPLE OPEN CHOKES ALONG WITH SMALL SPR OR TAKING RETURNS SIMULTANEOUSLY UP CHOKE AND KILL LINES.

As wells are drilled deeper, the operating mud weight range (between pore pressure and fracture pressures) become narrow. Choke line friction pressures will increase accordingly. Therefore, the ability of compensate at higher SPR may not be possible. This may require opening multiple chokes or taking the flow returns simultaneously up both the choke and kill lines.

Subsea well control

The basic principles of all well control methods are the same for subsea operation as they are for onshore operations. However, due to complexities of the well control systems as well as long distances between the vessel and subsea BOP stack, noticeable supplementary practices must be employed.

List of concerns during subsea well control

- Diverter: Upon completion of the kill and following displacement of KMW into riser, the diverter is closed and the overboard divert lines are opened. With fluids directed overboard, the kill rams are opened and the remaining trapped stack gas is allowed to enter the riser and expand to the surface. During this time, the upwind overboard divert line is closed.
- Trapped Stack Gas: A proper procedure is needed to clear pressurized gases from BOP Stack after the conclusion of a successful well control method. The choke and kill lines are circulated with low-density fluid and pressures are bled off. Once they are displaced with KMW, the riser is also displaced with KMW, before opening BOPs. Failure to account for Trapped Stack Gas can void the riser and allow another kick in the well.

- Lower Formation Fracture Pressures: Generally speaking, due to increasing water depths, fracture pressures are lowered due to the effect of water depth.
- Remote Subsea BOP Stack: Due to remoteness below sea level, closing and opening reaction times become longer. Complex system operation can also affect closure times. Emergency systems must be provided to ensure ROV intervention can be accomplished if surface systems become compromised.
- Subsea BOP Hydraulic Control Fluid: In order to facilitate closure of BOPs, when closing fluid is routed to the closing side of the BOP piston, the responding opening chamber fluid is vented into the ocean. This means the hydraulic fluid is not routed back to the surface and recovered into the reservoir tank.
- BOP Closing Unit Complexity: As the need of larger volumes of hydraulic fluid is required to efficiently operate the Subsea BOP Stack, the complexity of the surface BOP Closing Unit increases as well as the flow paths to the Subsea BOP Stack.
- Higher Choke and Kill Line Friction Pressures: Due to the length, internal diameter and fluid properties, choke lines exert added back pressure during well control circulations.
- Emergency Station-Keeping: Motion and location movement of the vessel can impart significant stresses on riser, subsea BOP stack, and wellhead equipment.
- Hydrate Plugging: Gas hydrate plugging can become problematic in deeper waters requiring remediation during well control.
- Riser Gas Clearance: Gas within the riser can lead to catastrophic events, even complex voiding of the rise. This can occur from BOP leaking during a well control event, failure to address trapped stack gas after well control event, loss of barriers, etc.
- Unplanned Riser Disconnect: Can lead to loss of well control due to loss of rig station, equipment malfunction, or human-induced error.
- Choke Line Length: Depending on the overall length of the choke line, the lag time needed to operate equipment can become excessive complicating well control methods which support constant bottom-hole pressures.
- Water Depth/Test Fluid Relationship: As the water depth increases, the test fluid density will impact the maximum BOP surface pressure.

Diverter

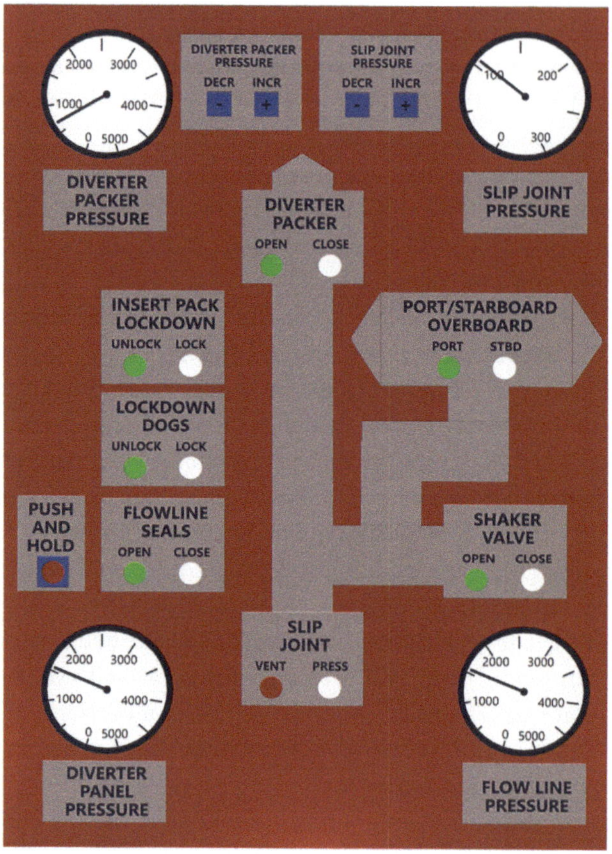

Fig. 6.7 Subsea diverter panel.

During floating operations using a subsea BOP stack, the diverter is primarily used to clear trapped stack gas or for any influx which enters the riser. Unlike diverters on land, floating operational diverters are NOT used to direct uncontrolled flow of formation fluids away from the floating rig, as risers are not designed to be voided of fluid (Fig. 6.7).

The diverter is located below the rig floor and is connected to overboard divert valves. The diverter panel, as shown, is used to close the diverter packer and regulate the slip joint to accommodate vertical movement while the influx is circulated from the riser. The diverter packing lockdown dogs, lock, and flowline seals are energized. As gas exits the diverter lines, the shaker valves are closed and the flow is directed to the overboard divert lines (port and starboard). The final step is to close the upwind divert valve to ensure gas is not blown back onto the rig.

Trapped stack gas

Trapped Stack Gas Procedure is the proper procedure needed to clear pressurized gases from BOP Stack after the conclusion of a successful well control method. The choke and kill lines are circulated with low-density fluid and pressures are bled off. Once they are displaced with KMW, the riser is also displaced with KMW, before opening BOPs. Failure to account for Trapped Stack Gas can lead to loss of riser integrity and allow another kick in the well.

As the riser provides a conduit for drilling mud to the surface by connecting to the subsea BOP stack, the riser is not a high-pressure well control device. Riser integrity can be compromised through loss of mechanical support, BOP Stack leaks, unintentional disconnect, and loss of drilling mud due to riser leak (displaced with seawater). In each of these cases, the hydrostatic pressure exerted by a column of homogenous fluid is lowered allowing the BHP to drop below the pore pressure and allowing an influx into the well (riser margin).

Lower formation fracture pressures

As water depth increases to ultra-deepwater depths, the drilling margin between pore pressures and facture pressures decrease. This margin can narrow to less than 1 ppg mud weight. Such operations require diligent flow checks and early kick detection.

Remote subsea BOP stack

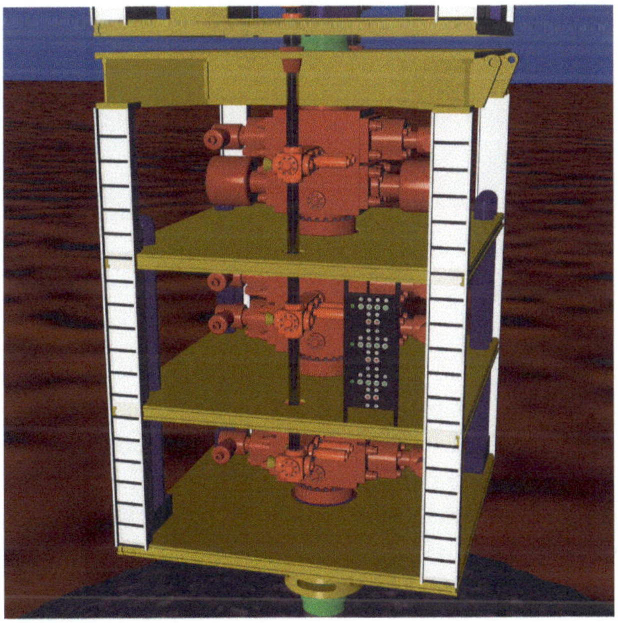

Fig. 6.8 Six ram subsea BOP stack.

A Subsea BOP Stack is extremely tall and heavy. Due to its complexity, a structure is needed to house all the components and their redundant systems. These BOPs are designed to operate within freezing waters for extremely long periods of time without maintenance. While on surface, these systems are thoroughly tested within a testing facility. Access to the BOP Stack, once the stack is underwater, can only be accomplished with the ROV (Fig. 6.8).

The Subsea BOP Stack consists of BOP with large bore sizes (18–3/4″) needed to accommodate multiple sizes of casing strings. With larger bores, closing piston diameters increase requiring more fluid to operate. The BOP Hydraulic Closing Units must have the ability to provide fluid to close BOPs and operate from surface. Therefore, hydraulic control fluid must be able to be provided from surface to stack, requiring even larger volumes.

BOP closing unit complexity

The BOP Closing Unit routes pressurized hydraulic control fluid from surface down to the operating chamber of a subsea BOP to execute a desired function (open or close). The opposing chamber of the piston is vented to the sea. Therefore, new control fluid must be mixed to replace the volumes lost to the sea (Fig. 6.9).

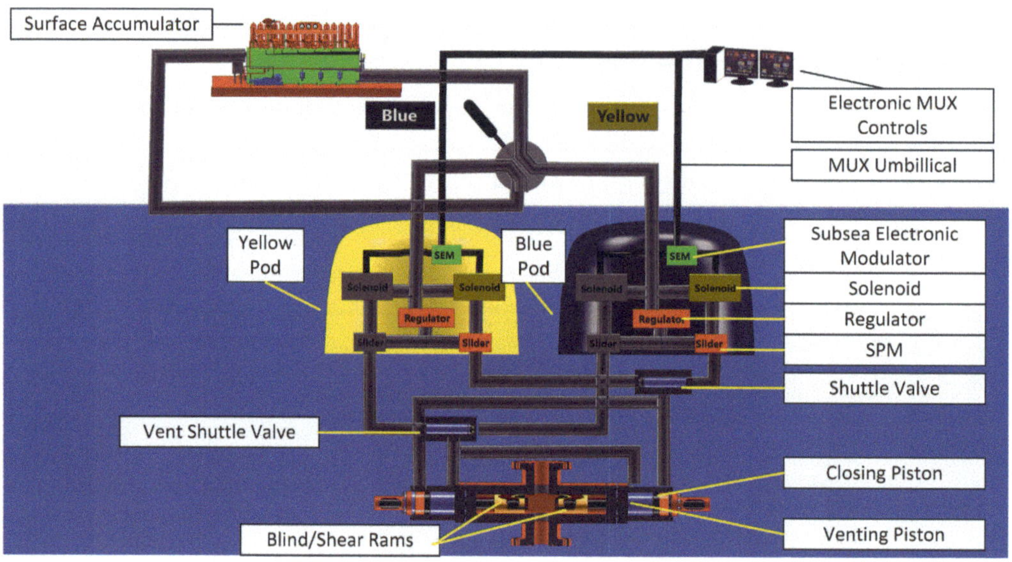

Fig. 6.9 *Yellow/blue* pod accumulator flow diagram (see equipment section for more examples).

For redundancy, supplemental volumes of pressurized hydraulic fluid are available within the BOP Stack and LMRP for critical stack functions. This system is in place in case hydraulic connections to the surface are severed for any reason and critical stack operation is needed. Unlike surface accumulator bottles, Subsea BOP Stack and LMRP accumulator bladders are precharged at the surface with pressures which account for water depth, temperature effects, and cooling effects of bladder inert gas expansion during discharge.

1. Water depth: Subsea Accumulator bladders are to be precharged to the sum of the hydrostatic pressure of the control fluid column at the bottle depth plus the accumulator unit's surface operating pressure. When functioned, the pressurized hydraulic control fluid is routed to the operating piston chamber, while the opposing piston chamber fluid is vented to the sea. Thus, the pressure in the vented chamber of the BOP is equal to the hydrostatic pressure of the seawater at the BOP depth.

2. Subsea Temperature: The Subsea Accumulator bladders are located within accumulator bottles positioned on the seafloor at temperatures of 33 °F/1°C. The seawater cools the accumulator bottle and bladder, causing the inert gases of the bladder to cool, become denser and reducing the precharge pressure.
3. Cooling Effects of Sudden Bladder Inert Gas Expansion: When the BOPs are functioned, the bladders within the Subsea Accumulators will rapidly expand driving pressurized hydraulic fluid to the operating pistons. As the inert gases within the bladder expand rapidly, the gases cool and become denser, resulting in reductions to the precharge pressure.

The BOP Stack contains redundant choke and kill lines to provide multiple means of circulating fluids to the surface. These redundant circulation points require hydraulic fluid to operate. On a Subsea Stack, multiple choke and kill lines can be operated interchangeably. (Of particular note, the kill lines won't have check valves.) This allows maximum flexibility when circulating influxes to surface, clearance of stack gas and displacing the riser.

Choke and kill line lengths

Choke and kill lines on Subsea BOP stacks at the mudline have excessive lengths (water depth plus air gap). As water depths increase, the lengths of these lines will also increase. This results in ever increasing frictional pressures caused when fluids are circulated through them. As an additive pressure transmitted downhole to the BHP, these friction pressures can significantly increase BHP to the point they might fracture formations during a well control event.

Emergency station-keeping

Motion and location movement of the vessel can impart significant stresses on riser, subsea BOP stack, and wellhead equipment.

Hydrate plugging

Gas hydrate plugging can become problematic in deeper waters requiring remediation during well control.

Riser gas clearance

Gas within the riser can lead to catastrophic events, even complex voiding of the rise. This can occur from BOP leaking during a well control event, failure to address trapped stack gas after well control event, loss of barriers, etc.

Unplanned riser disconnect

Can lead to loss of well control due to loss of rig station, equipment malfunction, or human-induced error.

Water depth/test fluid relationship

As the water depth increases, the test fluid density will impact the maximum BOP surface pressure.

Well shut-in methods

Hard shut-in

In order to minimize the volume of the influx, a hard shut-in is used in subsea applications. To minimize problems during the hard-shut-in procedures, the following must be considered:

DO NOT CLOSE ON TOOLJOINT. In order to facilitate the fastest closure, the top annular preventer is usually closed during kicks. To minimize problems, the Driller must always know the location of the drillstring tooljoints within the Subsea BOP Stack at all times. An up-to-date tooljoint space-out calculation worksheet must be constantly updated with latest draft.and tidal fluctuations.

To minimize space-out problems, the following general hard shut-in process is normally used.

1. Close upper most annular preventer for initial shut-in.
2. Perform tooljoint space-out calculation.
3. Verify tooljoint space-out by pulling up until the Annular Preventer shows tooljoint deflections in pressure.
4. The appropriate pipe ram is closed and the drillstring is lowered and hung-off.
5. The uppermost annular preventer is reopened after pressure is bled off above the ram preventer.

Shut-in procedures must be developed and refined for each particular operation being performed such as during drilling, tripping or other nonroutine well operations. As each vessel represents a customized complement of rig equipment, these procedures will be rig-specific. The procedures should be reviewed individually during each prejob safety meeting and posted on rig floor. Drills must be conducted to ensure sufficient competency is shown by each rig crew member.

General subsea shut-in procedure for tripping

The following is a basic shut-in procedure for Tripping. As a generic procedure, the following thoughts are offered in order to review individual rig procedures to ensure they encompass the following topics. This procedure applies to tripping or running casing as long as either the drillstring, casing landing string or the casing/liner are across the BOP stack.

1. Driller sounds the alarm.
2. The tooljoint is positioned above the rig floor and pipe is set in the slips.
3. Floor hands install the FOSV in open position, fully make it up and then close the FOSV.
4. The drillstring is spaced out to ensure that a tooljoint is not across a ram BOP.
5. A flow check is done by the driller to see if the well is flowing.
   a. If positive shut-in well. If unsure, observe well for flow and shut-in if the check is positive or resume operations if negative.
6. Driller closes in the well by using the Upper Annular on the LMRP (determined by drilling contractor procedure).
7. Driller confirms that closing unit volumes and pressures are correct.
8. Driller ensures that the well is properly shut-in and that there are no leaks including into the riser.
   a. If there is no flow from the riser, read the SICP per step 10.
   b. If there is flow from the riser, the BOP could be leaking or there is gas above the BOP stack. Appropriate steps to close an additional BOP should be taken.
9. The Driller should confirm the proper lineup of the choke manifold valves, MGS, and pump lines.
10. Read the SICP:
    a. Record the SICP from both lines.
    b. The Driller will keep a log of pressures and volumes.
11. Notify supervisory personnel of well shut-in
12. Record the kick parameters
    a. The driller will record the mud pit volume gain.
    b. The driller will determine the SIDPP by bumping the drillpipe float.
13. Monitor the riser on the trip tank.
14. Prepare kill sheet.
15. Carry out well control operations which may include killing off bottom, stripping, or volumetric gas migration.

General shut-in procedure for drilling

The following is a basic shut-in procedure for Drilling. As a generic procedure, the following thoughts are offered in order to review individual rig procedures to ensure they encompass the following topics.

1. Driller sounds the alarm.
2. The drillstring is picked up and a tooljoint is spaced out above rig floor.
3. The riser boost pump and mud pumps are shut down.
4. A Flow Check is done by the driller to see if the well is flowing.
   a. If positive shut-in well. If unsure, observe well for flow and shut-in if the check is positive or resume operations if negative.
5. Stop Rotation of drillstring.
6. Driller closes in the well by using the Upper Annular on the LMRP (determined by drilling contractor procedure)
7. Driller confirms that closing unit volumes and pressures are correct.
8. Driller ensures that the well is properly shut-in and that there are no leaks including into the riser.
   a. If there is no flow from the riser, read the SICP per step 10.
   b. If there is flow from the riser, the BOP could be leaking or there is gas above the BOP stack. Appropriate steps to close an additional BOP should be taken.
9. The Driller should confirm the proper lineup of the choke manifold valves, MGS, and pump lines.
10. Read the SICP:
    a. Record the SICP from both lines.
    b. The Driller will keep a log of pressures and volumes.
11. Notify supervisory personnel of well shut-in.
12. Record the kick parameters
    a. The driller will record the mud pit volume gain.
    b. The driller will determine the SIDPP by bumping the drillpipe float.
13. Monitor the riser on the trip tank.
14. Prepare kill sheet.
15. Carry out well control operations which may be the Driller's Method or the Wait and Weight Method.

General subsea shut-in procedure when out of the hole or the bit above the subsea BOPs

The following is a generic shut-in procedure with the bit above the subsea BOPs. As a generic procedure, the following thoughts are offered in order to review individual rig procedures to ensure they encompass the following topics.

1. Driller sounds the alarm.
2. A Flow Check is done by the driller to see if the well is flowing.
    a. If positive, shut-in well. If unsure, observe well for flow and shut-in if the check is positive or resume operations if negative.
3. The well is shut-in using the Blind/Shear rams. Avoid having any pipe across the blind/shear rams.
4. If there is any pipe in the hole, the Floor Hands will install the FOSV in open position, fully make it up and then close the FOSV.
5. Driller confirms that closing unit volumes and pressures are correct.
6. Driller ensures that the well is properly shut-in and that there are no leaks including into the riser.
    a. If there is no flow from the riser, read the SICP per step 8.
    b. If there is flow from the riser, the BOP could be leaking or there is gas above the BOP stack. Appropriate steps to increase the closing pressure on the BOP should be taken.
7. The Driller should confirm the proper lineup of the choke manifold valves, MGS, and pump lines.
8. Read the SICP:
    a. Record the SICP from both lines.
    b. The Driller will keep a log of pressures and volumes.
9. Notify supervisory personnel of well shut-in
10. Record the kick parameters
    a. The driller will record the mud pit volume gain.
    b. The driller will determine the SIDPP by bumping the drillpipe float.
11. Monitor the riser on the trip tank.
12. Prepare kill sheet.
13. Carry out well control operations which may include bullheading, stripping, or volumetric gas migration.

Well kill preparations

Once surface casing has been set, mud systems are swapped out to synthetic oil-based muds (SOBM). In deeper hole sections, the probability of influx migration remains low. But, in surface sections where water-based muds are used, migration may be a problem.

If an influx enters the wellbore, the hydrocarbon constituents of the kick are soluble in the SOBM. During a well control event, the time to establish stabilized SICP and SIDPP due to more complex procedures will be longer when compared to onshore operations. For SOBM, a lower probability exits of gas migration and although SICP is continually monitored, observations should show little or no gas migration.

For water-based muds, the probability of active gas migration is higher. Therefore, SICP must be constantly monitored during shut-in procedures and as kill preparation are made to define gas migration. If gas migration occurs, rise rate should be recorded and its effects on SICP. If SICP grows rapidly, action to address these increases may become immediate and the use of the Volumetric Method can be used to keep BHP constant.

With SICP, SIDPP, and kick volume, well kill sheet calculations are performed. Mud Engineers are informed of KMW and develops a plan to address anticipated volume gains due to gas expansion from the kill circulation. Kicks in deepwater may not show much in the way of volume gains due to the gas staying in solution until it enters the choke line or at times until it exits the choke.

A prejob safety meeting is to occur before kill circulation is started where each rig crew member's individual responsibility is reviewed. The meeting should be free flowing and only conclude when all questions have been answered. When in doubt, notify Person-in-Charge, if leaks are observed, shut down and shut-in.

Drill crew duties

1. Record the stabilized SICP and SIDPP pressures. Verify and confirm the volume gain.
2. Perform hang-off of drillstring in preselected BOP (if applicable).
3. Notify Person-in-Charge of well shut-in and secured.
4. Monitor the SICP and SIDPP throughout shut-in procedures and during kill preparations for increases.
5. Ensure all valves for kill are in the proper position.
6. Ensure rig crews perform individual responsibilities including all preparations for the well kill.
7. Review and ensure emergency disconnect plan is readied, in case of an emergency.

Gas in riser after BOP shut-in

After well has been shut-in, flow may be noticed from the riser. This can mean the BOP is leaking. Any potential gas entrance into the riser can easily become catastrophic. If flow is detected after shut-in, additional BOP is to be closed to isolate the leak. Then, the riser flow must be mitigated before kill circulation begins.

The following is a generic shut-in procedure for gas in riser mitigation. As a generic procedure, the following thoughts are offered in order to review individual rig procedures to ensure they encompass the following topics.

1. For safety considerations, personnel located near the moonpool and rig floor must be reduced to minimum number of qualified personnel throughout operations involving gas in the riser.
2. Open Diverter Lines and Close the Riser Diverter (located below the rig floor bushings).
   (a) Close upwind diverter valve.
   (b) Account for the number of barrels of mud diverted overboard. Diverting is essential for the safety of all personnel.
3. Increase Pressure on the Riser Slip Joint Seal
4. Circulating the Riser to prevent riser collapse
   (a) If the hydrostatic pressure within the riser decreases due to gas expansion, riser circulation can be used to keep the riser full of mud. Use slow pump rates to allow gas to expand.
   (b) If the rig has been outfitted with a riser gas handling system, actuate the Annular BOP atop the Subsea Diverter and vent mud and gas underwater while bleeding off mud and gas through the bleed spool and into the manifold choke.
5. Once riser has been displaced with KMW, proceed to applicable Well Kill Method.

Step 1: Kill well

The Driller's Method and the Wait and Weight Method are typically the preferred kill methods. Regardless of the selected well kill method, subsea procedures will be more complex. The following represents a set of generic considerations which should be made for any subsea well kill method.

1. Manage choke line friction
   (a) During a subsea well kill circulation, the pumps are gradually increased to an appropriate slow pump rate (SPR). Even at these slower pump rates, the extended length of the choke line will add significant friction back pressure to the BHP. In the course of operations, these choke line friction pressures which have been measured must be accounted for by reducing circulating casing pressure by the measured amount.
   (b) During the circulation, if the pump needs to be swapped out or slowed, the process of slowing down would be to account for the loss of friction pressure at lower speed by adding the SPR pressure of the lower rate, then slowing the pumps.
2. Low formation fracture pressures
   (a) As previous stated, for deepwater operations, the margin between pore pressures and formation fracture pressures is narrow. To avoid fracturing the formation during well kill circulation, low pump rates are selected to minimize pressures

downhole and allow time for choke manipulations at the surface. The circulation rate may have to start as low as ½ BPM.

3. Gas in Choke Line.
   (a) As gas enters the choke line and displaces mud, the friction pressure will drop as gas has less circulating friction compared to fluid. The hydrostatic pressure in the choke line will also be reduced as the gas displaces fluid. The greatest change will be the hydrostatic pressure decrease by several orders of magnitude. For a gas-filled choke line, the following equation can be used to estimate the reduction or change **(CL ΔHP)** in choke line hydrostatic pressure for gas vs mud:
4. Rapid pressure/volume changes:
   (a) Rate of increase in circulating casing pressure
   (b) Rapid increase in mud pit gain

$$\mathbf{CL\,\Delta HP_{psi} = \left(Mud\,Weight_{ppg} - Gas\,Weight_{ppg}\right) \times 0.052_{psi/ft/ppg} \times Choke\,Line\,Length_{TVDft}} \tag{6.6}$$

Example given data:

- Mud weight in choke line: 12 ppg
- Gas density = 2 ppg
- Choke line length: 6000 ft. TVD
- Choke line is filled with gas

The estimated CL ΔHP under this scenario is:

$$\mathbf{Est.CL\,\Delta HP_{psi} = [(12ppg) - (2.0\,ppg)] \times 0.052 \times 6000\,ft\,TVD = 3120\,psi} \tag{6.7}$$

The circulating casing pressure would have to be increased by 3120 psi while maintain constant slow pump rate in order to keep BHP constant. These changes in circulating casing pressure will occur rapidly as gas enters the smaller capacity of the choke line versus the larger capacity of casing. For each barrel of mud, the fluid will occupy much greater vertical distance in the choke line when compared to casing.

Choke operator must react properly on the choke to ensure constant circulating bottom-hole pressure is maintained. As choke line lengths are extremely long, each manipulation will take at least 2 s per 1000′ to observe on the drillpipe gauge. If the gas has to be compressed in order to maintain pressure, this lag time can easily increase.

If these changes in choke line friction are not properly managed, the BHP pressure may drop below the pore pressure and allow additional influxes into the wellbore or overpressure the well. Therefore, gas arrival into the choke line must be anticipated in order to maintain constant BHP.

After gas has exited the choke line, the added friction and hydrostatic pressures for mud in the choke line must be alleviated by reducing the circulating casing pressure by the same amount which had previously been added (our example was 3120 psi, so 3120 psi is taken off by opening the choke).

Rig crews should be on alert to react to the following noticeable changes:

1. Rate of decrease/increase in circulating casing pressure
2. Rapid decrease/increase in mud pit gain
3. If choke is fully open and pressure on the casing is rising too high, there are several actions that can be taken.
4. Reduce pump rate
5. Open a second choke
6. Take returns up the kill line in addition to the choke line.

TIME BETWEEN CHOKE MANIPULATION AND RESPONSE ON DRILLPIPE GAUGE IS 2 SECONDS PER 1000 FT.

THIS LAG TIME MAY INCREASE IF GAS HAS TO BE COMPRESSED IN ORDER TO KEEP BHP CONSTANT.

Subsea complications

Physical and chemical compositions of the drilling fluid and influx may introduce unique complications in subsea well control operations. The following is several generic considerations of these physical and chemical complications.

Gas solubility in the SOBM/WBM

The solubility of influx gases in various drilling muds is based on composition of the fluid phase of the mud and the introduced gas. Solubility increases with increasing pressure (deeper holes) and higher specific gravities of the mud but it decreases with increasing temperature. For the variety of Water-Based Muds (WBM), gas solubility is minimal as water represents the entire fluid phase of the mud.

Synthetic Oil-Based Muds (SOBM) are designed in various ratios of oil/water and provide improved drilling rates, wellbore stability, improved characteristics for rheology, and reduce the risk of gas hydrate formation. The downside of using this type of mud is the issue of gas solubility.

Solubility means the gases in a compressive state enter the compressible oil phase of the drilling fluid. At high pressures, these gases act as a liquid meaning very little expansion. Soluble gas translates into no increases on initial shut-in casing pressures. Subsequently, volume gains will be slight and may be un-noticeable until the mud containing the gas reaches a pressure (bubble point) at which it comes out of solution and no longer act as a liquid but as a gas which can rapidly expand.

Challenges of well control with influxes in SOBM/NAF

As discussed previously, both an increase in flow and an increase in pit gain are difficult to detect for soluble gas influxes. These difficulties in detection may yield larger and harder to kill influxes. Larger influxes will generate higher pressures while circulating and higher density kill muds.

As the soluble influx is circulated from the wellbore and bubble point is reached, the gas will rapidly expand as it breaks out of the oil phase. The rapid expansion may impact the choke operator's ability to keep BHP constant and fluid, followed by gas, rapidly escapes from the wellbore. For subsea systems, the bubble point may take place within the riser, further complicating well control.

Higher downhole pressures compress the mud but higher temperatures increase volume and decrease density; therefore, the density of mud at the surface is not the same as the density at the bottom of the hole. Static or circulating densities must be measured or calculated along the depth of the well to obtain the equivalent static density and equivalent circulating density.

Precautions used to maintain well control with solubilized influxes

1. Early recognition of kicks (increase in flow, increase in pit gain) is key.
2. Realistic pit drills to keep crews sharp.
3. Be aware activities such as mud transfers, starting and stopping mud cleaning equipment, crane lifts, or boat activity can affect pit levels. Driller should be notified of these events.
4. Pressure gauges and Pit level indicators must be properly installed (opposite corners of the pit for pit level indicators), maintained, serviced, and calibrated.
5. Mud compressibility: The synthetic oil-based fluid makes up the majority of the fluid in the drilling mud with remaining fraction consisting of emulsified brine and solids needed to control rheology. The compressibility of the SOBM is dependent on pressure and temperature. Ensure SOBM are controlled within proper specifications.
   - **(a)** Compressibility will also lengthen the time required for flowing annulus fluids to stop after pumps are shut down and ECD is removed
   - **(b)** Mud compressibility translates to the need of a longer duration to achieve stabilized shut-in pressures during kicks.
   - **(c)** Time between choke manipulation and observing changes on the drillpipe will lengthen due to compressibility.

Gel strengths of the mud

When pumping is stopped and SOBM muds become static, high gel strengths or viscosities can develop especially at low temperatures affecting the mud's pumpability.

Accurate pressure readings may be impacted by gelled mud in the choke and kill lines. The increase in gel strength means it will take more effort to start movement of the fluid once pumping starts.

To offset problems due to gelation, fresh mud can be pumped down the choke and kill lines, prior to obtaining a SICP. This is done by circulating fresh mud in place by pumping down choke and kill lines and taking returns up the riser above a closed ram. After proper pressure readings are taken, the kill procedure can then be initiated.

Each rig or company should have their own procedure to flush the choke and kill lines.

WHEN DISPLACING FRESH MUD FROM SURFACE THROUGH THE KILL LINES AND UP THE CHOKE LINES, THE ACTIVE BOP (WHICH INITIALLY WAS USED FOR WELL SHUT-IN) SHOULD REMAIN CLOSED.

Weak casing shoes

Kick tolerance is a function of the strength of the casing shoe, weakest formation and the kick parameters. For subsea formations, the overburden gradient is of seawater (0.465 psi/ft) plus the rock overburden gradient. For the same depth of rock drilled, the combined gradient will be less as the water depth increases yielding potential lost circulation zones. If a subsea wellbore is compromised due to formation fracturing, conventional well control practices may no longer be applicable. Losses must be cured before well control can be regained.

Rig crews must be vigilant is observing operations which might affect formation fracturing such as:

- Drillpipe pressure gradually or suddenly decreases while circulating at a constant kill rate.
- Loss of fluid downhole reflected on pit level losses.
- Manipulation of the choke is unable to control circulating drillpipe pressure.

Gas hydrates

Hydrates are a solid mixture of gas and water which looks like ice; however, they can form at temperatures above freezing. During well control operations, the combination of low temperatures, high pressure, water, and gas can contribute to hydrate formation. Hydrate formation is more likely in a water-based mud than SOBM. These hydrates can form on the inside and outside of the wellbore and can be penetrated during drilling operations as hydrate formations are drilled.

Free gas hydrates are gas hydrates formed during well control activities when gas reaches the subsea choke line with correct pressure and temperature needed to form hydrates. As free gas hydrates are formed, a portion of the pressurized gas will become trapped below the closed BOP and within the wellhead and BOP ports and lines isolated by closed choke and kill line valves. If hydrates form inside the wellbore, they may plug off the annulus as well as the choke and kill lines.

To stop hydrates from forming, disruption of conditions that lead to hydrates is key. Actions to help reduce the formation are:

1. No gas in the hole
2. Lightest mud weight practical
3. Composition of drilling fluid such as the addition of salts and glycol
4. Circulate to keep mud warm

For external hydrate development on wellbore, BOPs, and choke and kill lines, a source gas must be available. This would include gas bearing seabed formations near the Subsea BOP Stack. A gas diversion system may be deployed to ensure gas percolating near the wellbore is directed away from the Subsea BOP stack. Additional ROV accessible injection ports for injecting antifreezing agents may be employed to thwart hydrates.

Hydrate formation on Subsea BOP Stack may prevent release of the BOP/LMRP from the high-pressure wellhead housing and the LMRP from the Subsea BOPS Stack.

Step 2: Fill riser with KMW

Although the well is killed and pressures are now zero, two more steps remain before a successful well kill operation is concluded. After the well has been secured, the next step will be to fill the riser with KMW. This is a simple precautionary step which ensures sufficient hydrostatic pressure exists within the riser which would prevent the well from flowing in case of a BOP malfunction. The following graphical illustrations demonstrate one method for filling the riser. Actual steps used will be determined by individual rig equipment complements and may vary.

The illustration Fig. 6.10 demonstrates shut-in conditions after the well has been shut-in on hang-off rams. The potential for trapped gas exists between the bottom of the operating ram in the closed position down to the circulating port below the ram. This volume may represent up to several bbls. of gas trapped at a pressure equal to the mud hydrostatic in the kill line above the circulating port (Fig. 6.10).

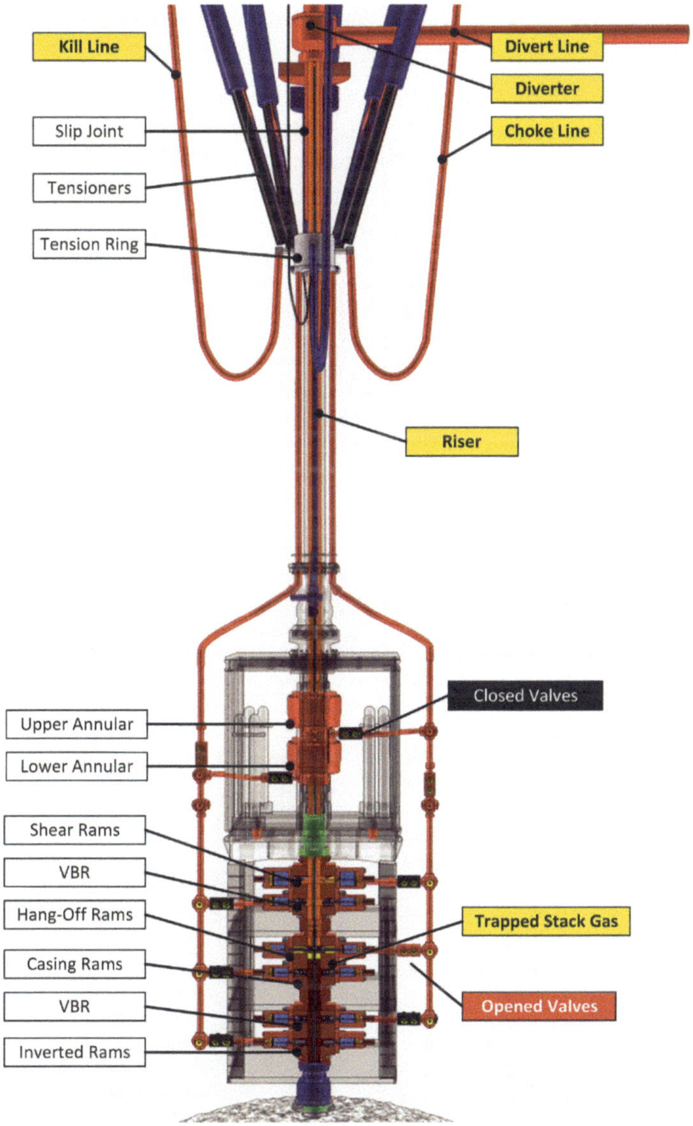

Fig. 6.10 Shut-in conditions with trapped stack gas below hang-off rams.

If the rams were opened, the trapped gas would enter the riser and if allowed to expand uncontrollably, could displace a majority of the fluid in the riser! It is imperative, that as much of the trapped gas as possible be removed and pressure reduced in a controlled manner to ensure gas expansion will not displace fluids from the riser.

The three illustrations below represent using both the kill line and the booster line to displace the riser with kill mud weight. Once KMW has been circulated throughout the riser, upper choke and kill lines may also be displaced with KMW. At conclusion, all Kill Line and Booster lines may be isolated or closed (Figs. 6.11–6.13).

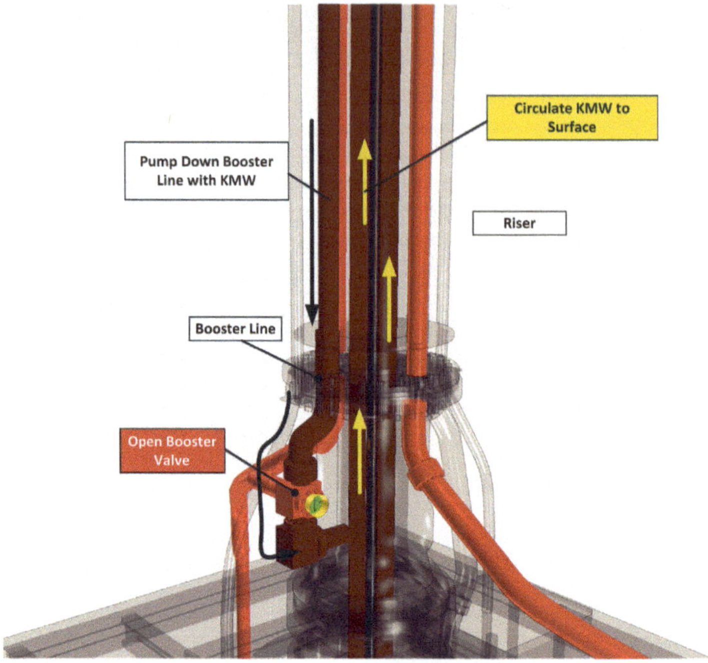

Fig. 6.11 Displace riser with KMW using booster line (Part 1 of 3).

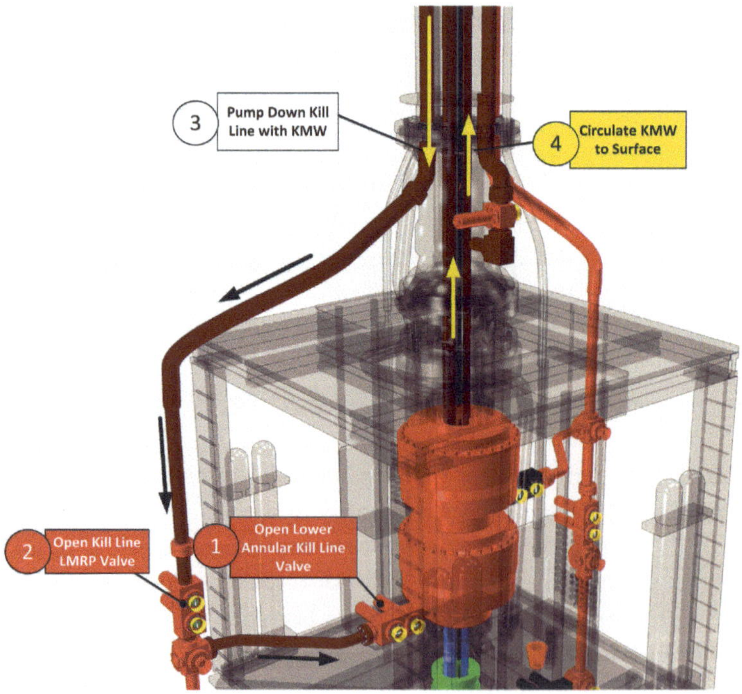

Fig. 6.12 Displace LMRP with KMW using kill line (Part 2 of 3).

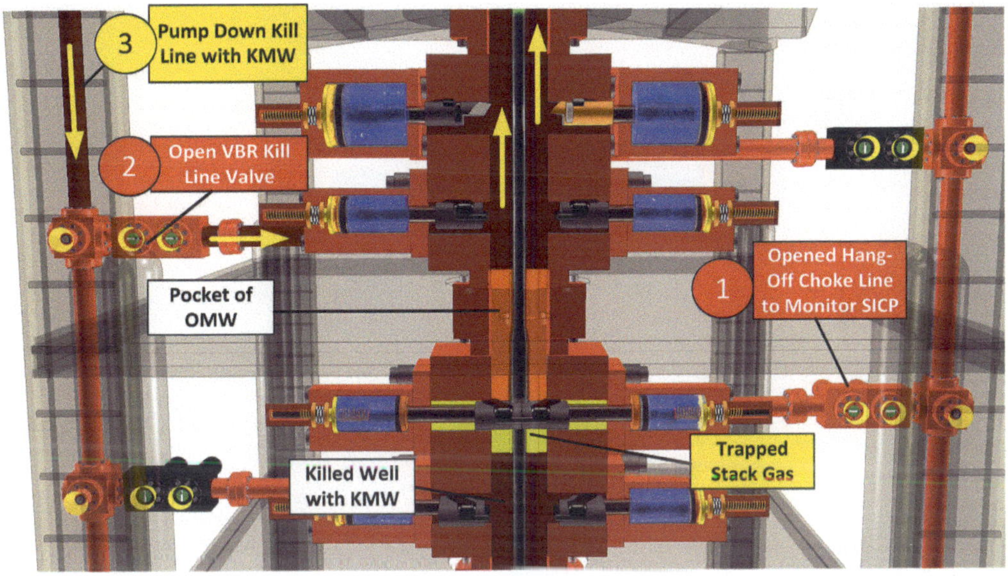

Fig. 6.13 Displace rams cavities above hang-off rams with kmw using kill line (part 3 of 3).

Step 3: Clear trapped stack gas

Trapped Stack Gas refers to the volume of gas trapped between the shut-in ram and the circulation port used below the ram. Due to the large size of the BOPs, this volume can range up to several barrels. If the BOPs were opened after a kill circulation, the trapped gas would enter the riser and could expand to a thousand or more barrels depending on pressure, cause potential loss of the riser, well, and injuring personnel. To prevent such a catastrophic event from occurring, the stack gas is allowed to expand up the choke line as much as possible in a controlled manner (Figs. 6.14–6.21).

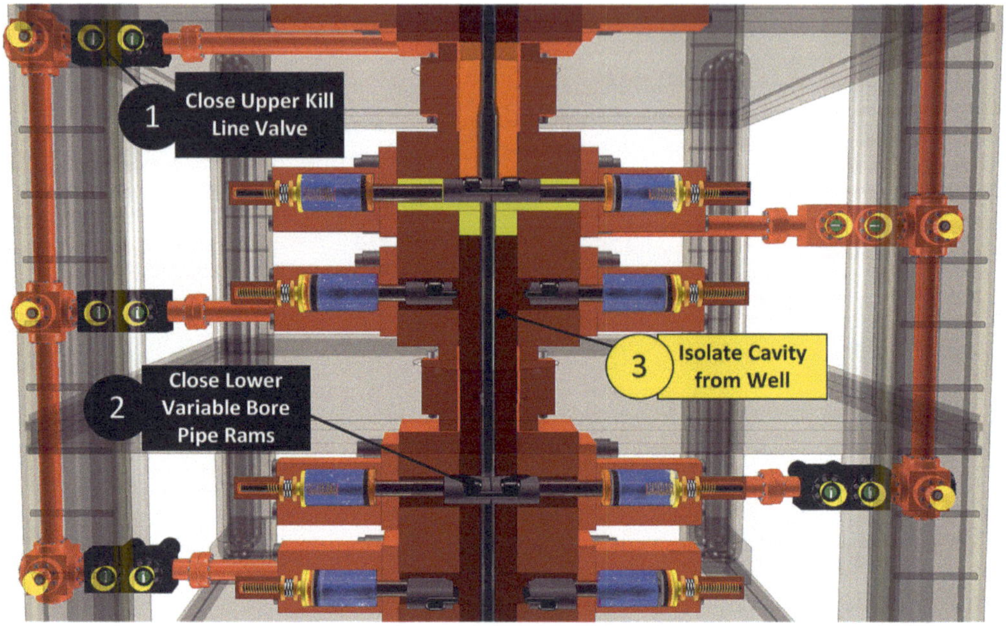

Fig. 6.14 Isolate wellbore by closing lower variable bore pipe rams.

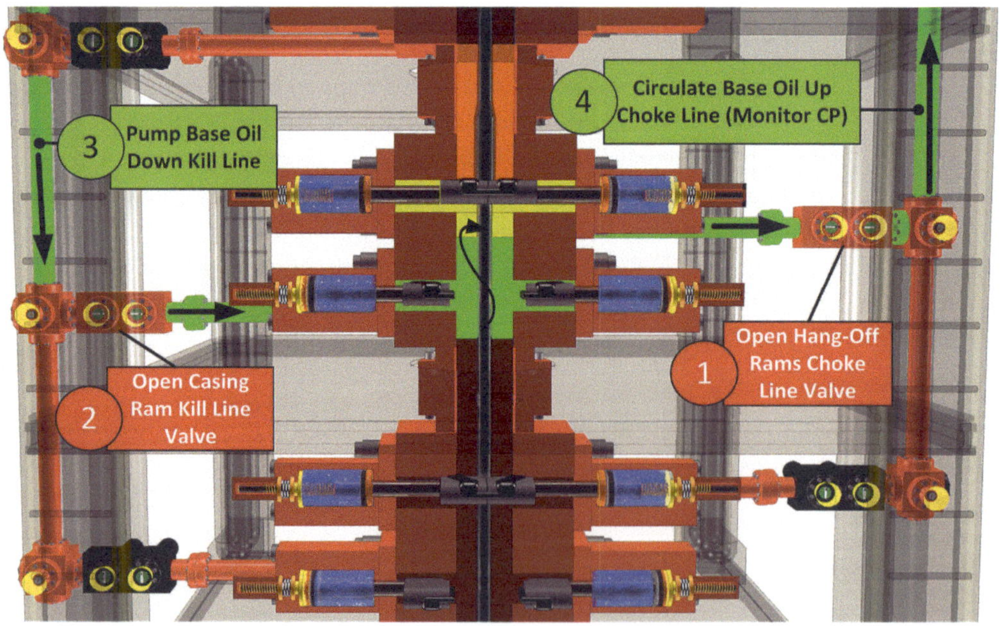

Fig. 6.15 Displace hang-off ram cavity with base oil.

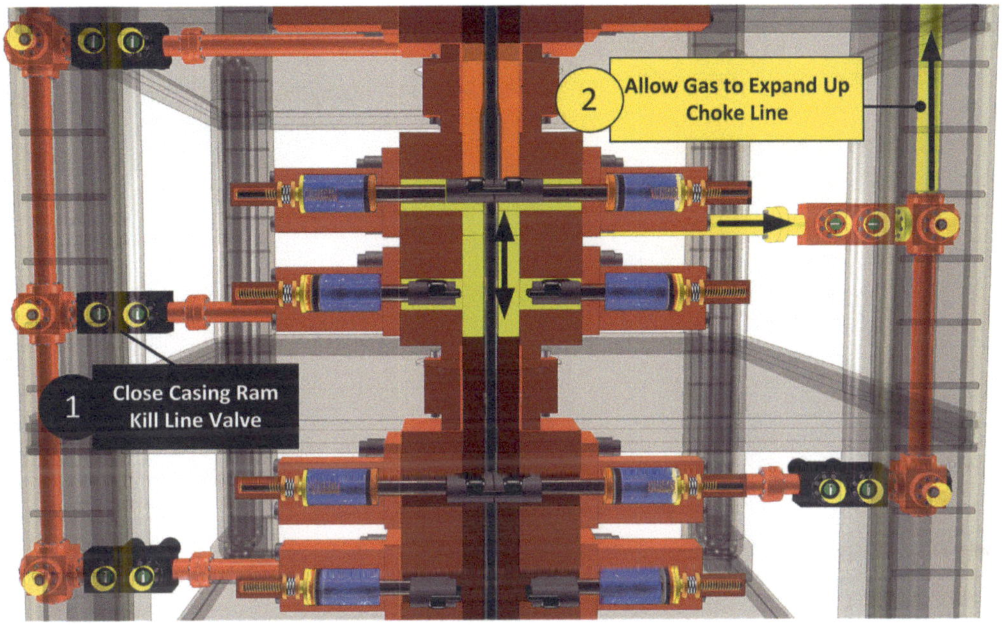

Fig. 6.16 Allow trapped gas to expand up choke line and displace base oil.

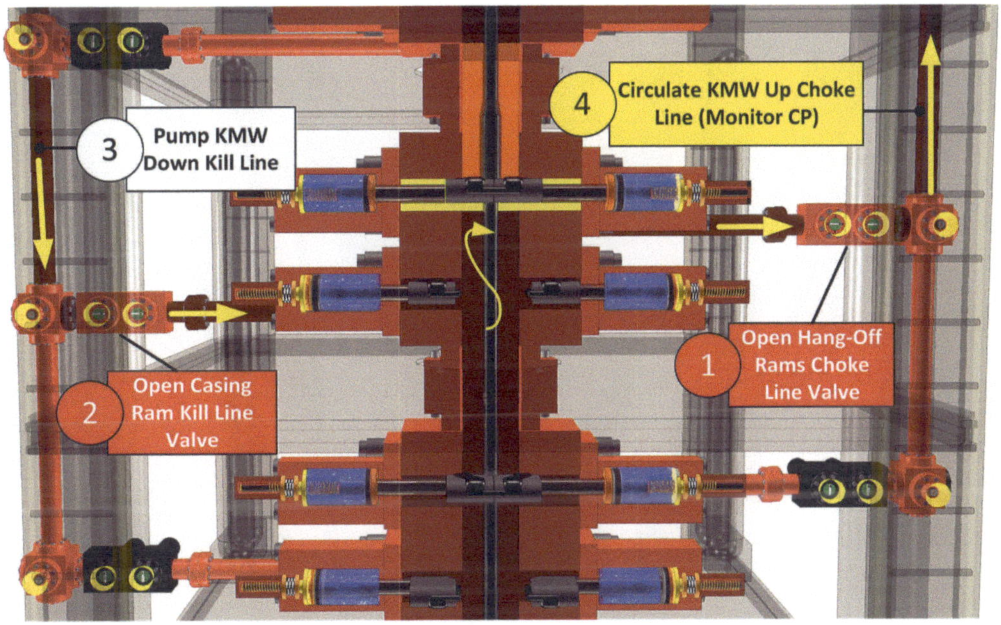

Fig. 6.17 Flush hang-off ram cavity with KMW.

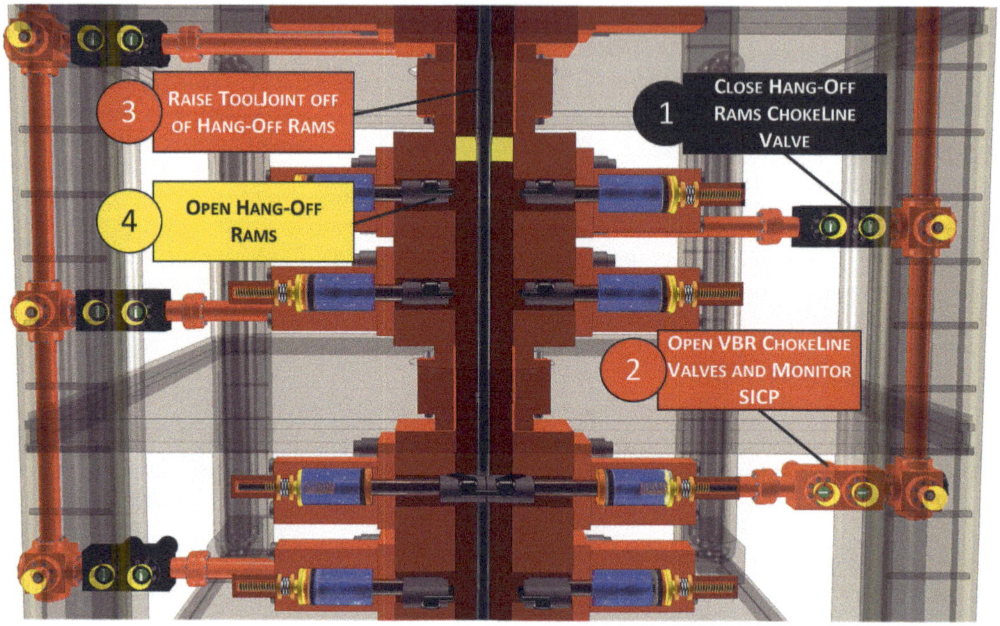

Fig. 6.18 Open hang-off rams using procedures outlined in the following two illustrations.

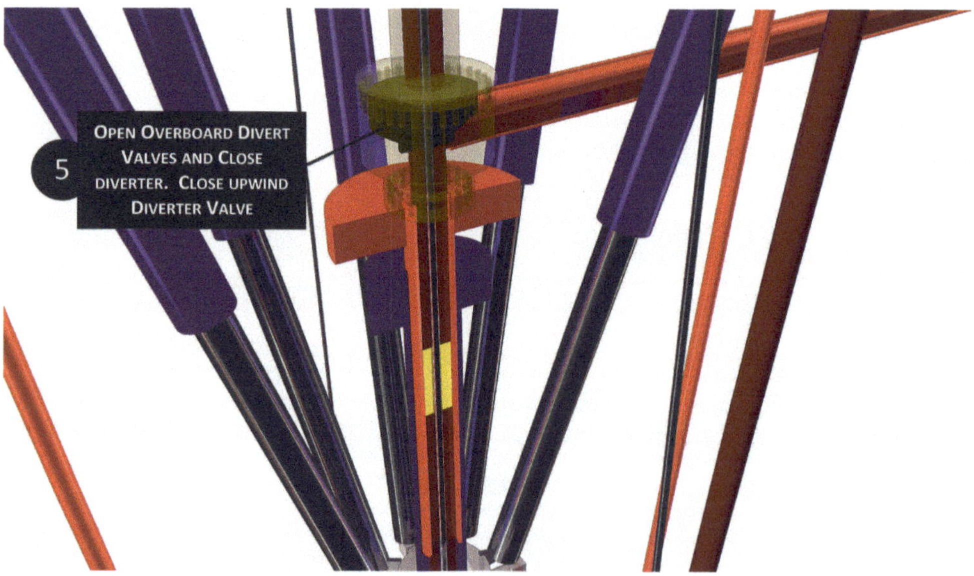

Fig. 6.19 Use diverter as part of the hang-off ram opening procedure.

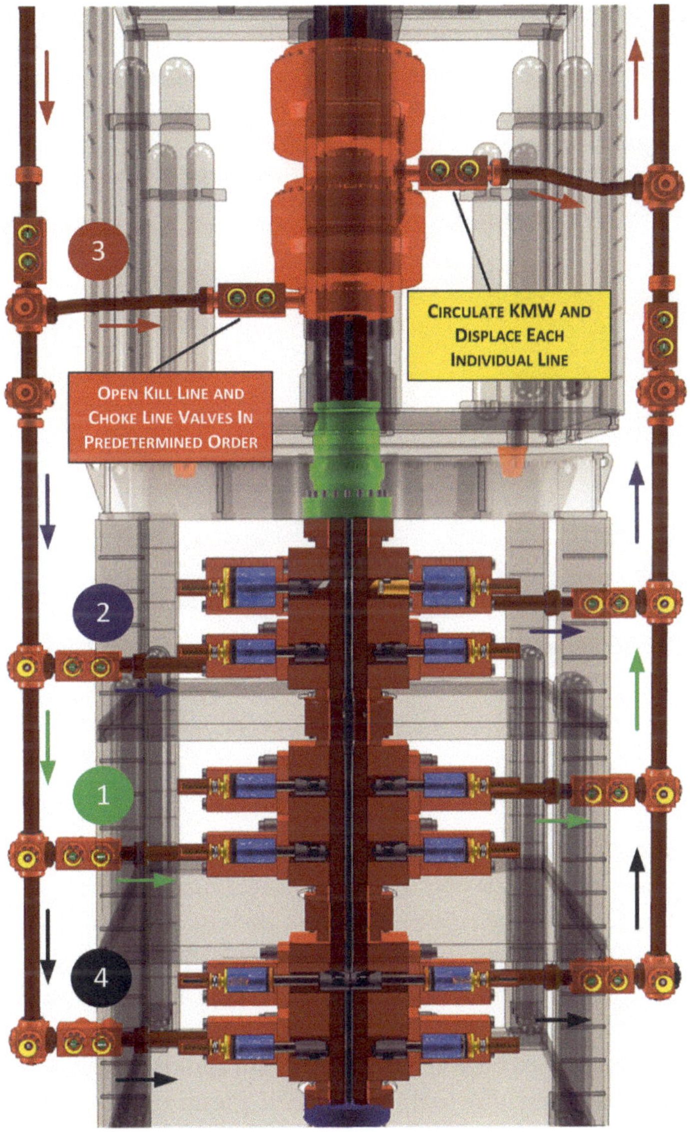

Fig. 6.20 Flush all circulating lines with KMW. Isolate same.

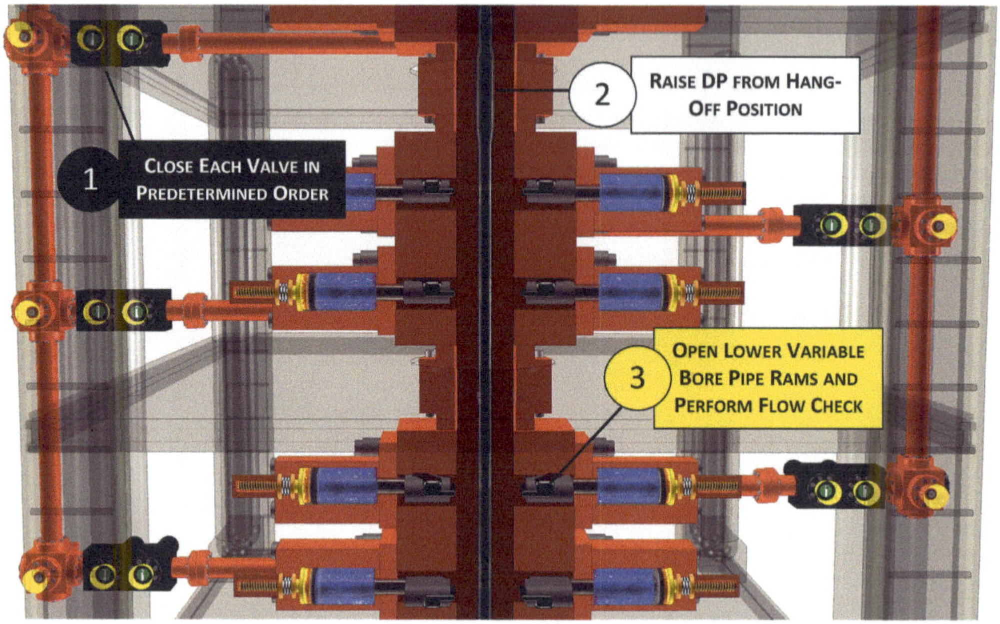

Fig. 6.21 Isolate by closing all valves. Open diverter and lower variable bore pipe rams. Perform flow check.

Pressure of trapped stack gas after kill

Assume

1. The gas volume trapped below the closed ram is 1 barrel.
2. The kill mud density is 11.5 ppg.
3. The TVD distance from the rig floor RKB to the circulating side outlet used on the kill is 7500' TVD.
4. The well is dead at this point, with kill mud circulated around (0 psi on the casing and drillpipe).
   (a) The stack gas trapped between the closed ram and the circulating side outlet is at a pressure of 4485 psi, due to the hydrostatic pressure of the kill mud acting on it.

$$\textbf{Gas Pressure}_{\text{psi}} = \textbf{KMW}_{\text{ppg}} \times \textbf{TVD of closed BOP} \times \textbf{0.052}_{\text{psi/ft/ppg}} \quad (6.8)$$

$$\textbf{Gas Pressure}_{\text{psi}} = \textbf{11.5 ppg} \times \textbf{7500 ft} \times \textbf{0.052} = \textbf{4485 psi} \quad (6.9)$$

A1 bbl volume of pressurized trapped stack gas at 7500 ft. will expand to 305 bbls if it enters the riser and migrates to the surface. This gas expansion can be calculated using Boyle's Gas Law for an Ideal Gas, as follows:

$$\mathbf{V_2\,(bbl)} = \frac{\mathbf{(P_1\,psi)(V_1\,bbl)}}{\mathbf{P_2\,psi}} \quad (6.10)$$

$$V_2\,(\mathrm{bbl}) = \frac{(4,485\,\mathrm{psi})(1\,\mathrm{bbl})}{14.7\,\mathrm{psi}} = 305\,\mathrm{bbls} \tag{6.11}$$

If allowed to expand inside the riser, the gas volume could easily result in a catastrophic event. Therefore, a procedure to sweep the trapped gas volume from the well in a controlled manner must be implemented prior to opening the shut-in BOP.

A generic stack clearance procedure

1. Close the Lowermost VBR BOP to isolate the well below this point.
2. Displace the choke line to base oil or water whichever is appropriate to reduce the hydrostatic pressure in the choke line.
   (a) Route to circulate down the choke line, through the choke line side outlet valves located immediately beneath the closed rams, through the kill line side outlet valves located immediately beneath the lower VBR rams, and up the kill line.
   (b) Circulate the choke line to base oil or fresh water, if applicable. Pump only enough fluid to reach the choke line side outlet at the subsea BOP stack. If the mud weight in the kill line is less than the mud weight in the choke line, hold sufficient back pressure at surface on the kill line side while displacing the choke line to base oil (or fresh water) to prevent the choke line fluids from u-tubing into the kill line. Normally, the mud weight in both choke and kill lines will be the same at the end of circulating out a kick.
   (c) If mud weights differ, hold back pressure as calculated below on the kill line side while circulating base oil into place in the choke line to maintain pressure within the trapped stack gas.

$$\mathrm{BP_{psi}} = \left(11.5_{\mathrm{ppg\ mud\ inchokeline}} - 11.0_{\mathrm{ppg\ inkillline}}\right) \times 7500_{\mathrm{ft\ TVD\ depth\ of\ the\ side\ outlet}} \times 0.052_{\mathrm{psi/ft/ppg}} = 195\,\mathrm{psi} \tag{6.12}$$

3. Close kill line side outlet valves.
   (a) Isolate the kill line from the wellbore by closing the kill line side outlet valves. Then shut-in the choke line at the surface.
4. Close choke line at the surface.
5. The trapped stack gas pressure is equivalent to a column of KMW acting against a column base-oil (~7 ppg) in choke line. The pressure has been reduced after displacing to base oil in the choke line

$$\mathrm{TrappedStack\ Gas\ Pressure_{psi}} = (11.5\,\mathrm{ppg} - 7.0\,\mathrm{ppg}) \times 7500'\mathrm{TVD} \times 0.052 = 1755\,\mathrm{psi} \tag{6.13}$$

6. Evacuate trapped stack gas through the choke line
   (a) Route to evacuate as much trapped stack gas in a controlled manner by taking returns from the choke line and directing them to the MGS.
   (b) The trapped gas will flow through the choke line to the MGS under controlled conditions. The remaining pressure would be a column of gas remaining in the choke line.
   (c) The final gas pressure can be calculated as follows:

$$\textbf{Final trapped stack gas pressure}_{psi} = 2.0\,\text{ppg} \times 7500\,\text{ft TVD} \times 0.052_{psi/ft/ppg} = 780\,\text{psi} \quad (6.14)$$

(d) The pressure differential acting upon the ram or element is as follows:

$$\Delta P_{AgainstAnnular\,psi} = 11.5\,\text{ppg}_{Riser\,KMW} - 2\,\text{ppg}_{chokeline\,gas\,density}) \times 7500\,\text{ft TVD} \times 0.052_{psi/ft/ppg} = 3705\text{psi} \quad (6.15)$$

(e) The pressure differential acting against the ram or element increases with water depth. Since preventers are designed to hold pressure from below, they would be prone to failure with this much differential pressure.
   (i) The maximum differential pressure rating acting from above is dependent upon the make and model of annular in use. Therefore, the BOP manufacturer should be consulted to determine this limit.

7. Once expansion has taken place, reroute and sweep the trapped stack gas through the choke line with an open choke using the kill mud in the riser.
   (a) Use the trip tank circulating on the riser to keep the riser full of KWM as the riser kill mud is u-tubing into the choke line.
8. Displace all of the remaining choke and kill line side outlet lines to KMW prior to resuming operations.
   (a) Circulate all ports, choke and kill lines containing original MW with KMW.
9. When the well, riser, and all flow pathways in the well are full of KMW, the Well is DEAD.

Mud gas separator (MGS)

Loading and emergency bypass

As part of the well control equipment spread, a floating vessel MGS includes additional instruments to monitor the MGS internal pressure for comparison to the mud leg hydrostatic pressure (providing back pressure). If the pressure monitor on the gas portion of the MGS approaches the same value the mud leg, blow through may occur. If blow through is imminent, the circulating rate can be slowed while keeping casing pressure constant or the flow should be routed overboard through the diverter line. Blow through puts personnel at risk as gas will be routed to the shakers area. Continual monitoring of the MGS should occur during all portions of well control.

Ballooning

Ballooning is typically more pronounced when SOBM is used due to its compressibility versus a water-based mud (WBM). With reduced margins between the pore pressure and fracture pressure normally found on a subsea drilling operation, the potential for micro-fractures increases. These fractures are needed for ballooning. Accurate hole volume losses are needed to verify ballooning as flow back should diminish overtime and return volume of losses.

Fingerprinting should identify normal flow back volumes and allow a determination of whether the well is flowing, ballooning or just the normal flow back volume. Accurate fingerprinting charts will help assess it the well is ballooning or kicking.

Abnormal well flow back

Any abnormal well flow back should be treated as a kick, until it is proven otherwise. The following is a suggested procedure to help evaluate the cause of abnormal well flow back after the pumps have been stopped:

1. Evaluate the well upon detection of an abnormal flow back:
   a. Verify if the flow continues after allowing for the volume of surface line drainage and residual well flow, or the rate of flow back is above expectations.
   b. If the flow continues shut-in the well on a suspected kick.
2. Record and monitor the shut-in pressures and pit volume gain:
   a. Record the shut-in drillpipe pressure (SIDPP), the shut-in casing pressure (SICP), and the pit volume gain.
   b. Monitor the shut-in pressures for any changes.
3. Assess the flow back:
   a. Flow back fingerprinting can be useful in assessing whether it is normal flow back or an abnormal flow back.
   b. Further diagnostics such as bleeding off a small volume of mud and observing the pressure response may be needed to determine the cause of the flow back.

Nonshearable tubular across the BOP stack

If a nonshearable tubular is across the subsea BOP stack when a kick occurs, the annular BOP is normally used to close in the well. The annular preventer is sized to provide a pressure seal for the shut-in and subsequent well kill. Without a means of hanging off the drillstring, vessel heave can damage the annular element and compromise the well kill operation and safety of the rig.

The rig should have a contingency plan in place to accommodate the potential failure of the annular BOP. The force of pressure against the bottom of the workstring is calculated to ensure the upward force is not greater than the weight of the pipe. This typically is not a problem in deepwater and ultra-deepwater. If the annular preventer fails, rig personnel must be prepared to quickly implement appropriate contingency plan.

These plans will be rig and operation specific. One option is to strip the nonshearable tubular below the BOPs to position drillpipe across the BOP stack in order to hang-off the workstring.

Water depth and test fluid density

Each BOP is rated to a maximum working pressure (e.g., 5 K, 10 K, 15 K, etc.). This is the maximum internal pressure a BOP should be exposed to. This internal pressure is not a differential pressure rating and should be considered an absolute limit. The internal pressure rating applies to such all BOP component parts which isolate well pressure.

The effect of hydrostatic pressure on a subsea BOPs must be determined, as the hydrostatic pressure from rig floor to seabed may greatly impact test pressures. The hydrostatic pressure of the test fluid acting within the BOPs must be accounted for in determining the maximum allowable surface test pressure to avoid exceeding the rated working pressure for each BOP.

Use of a differential pressure rating is not suggested as the seawater hydrostatic pressure may not act to offset the internal test pressure, thereby creating the possibility of exceeding the rated working pressure of the tested equipment.

DO NOT USE DIFFERENTIAL PRESSURE WHEN DETERMINING TEST PRESSURE

Loss of rig station-keeping

A dynamic positioning (DP) system or a mooring system keeps a floating drilling rig over a well during operations. A DP vessel can move off station involuntarily due to operator error, system malfunction, and/or heavy weather. In the case of heavy weather, limits of thruster capability can be reached, and the vessel gradually loses its position. System malfunctions resulting in unacceptable loss of position can be put into two categories: Drive-Off and Drift-Off. Drive-Off occurs when, due to either position reference system or thruster control malfunction, the controller commands the thrusters to full power in the wrong direction.

A Drift-Off situation occurs when power is lost to the thrusters in a power loss situation and the vessel is left to drift. Whatever the reason for a loss of position, serious riser and BOP stack damage can occur, leading to a significant amount of downtime or possibly a well control event. Therefore, the DP computer is often programmed to issue alarms to warn the DP operator and the driller when certain predetermined limits of offset from the wellhead have been exceeded.

A **green light** designates the station-keeping is within normal operating limits.

A **yellow light** warning given when a preselected limiting criterion is reached and requires the Driller to position and hang-off the drillpipe in preparation for securing the well with the blind shear rams. Since the consequences of not having the drillstring in the correct position when the Emergency Disconnect Sequence (EDS) is actuated could be catastrophe, the yellow watch circle should be set as small as possible to provide maximum time for the Driller to correctly position and hang-off the drillstring.

The yellow light setting (generally 2%–3% of water depth) is based on past experience and water depth. This alarm should be set such its limit is not frequently exceeded, but allows the drilling crew sufficient time to complete the hang-off procedure (approximately 45–60 seconds) prior to receiving a disconnect signal.

When a **red light** disconnect alarm is indicated, the operating procedures typically require the driller to activate the emergency disconnect sequence. The criteria for setting the **red light** alarm (generally 4%–6% of water depth) is based on the rig's ability to safely disconnect the riser prior to reaching the limiting value of any one of several critical parameters such as flex (ball) joint angle, slip joint stroke, tensioner stroke, or riser/casing/wellhead stresses. The time required to complete the emergency disconnect sequence (EDS) varies depending on the type of BOP control system. A typical multiplex BOP control system requires approximately 45–60 seconds to complete the EDS. Note: The Driller is empowered to initiate this EDS without any approval. If the well has not been secured and the riser disconnected by the time the Point of Disconnect (POD) is reached, mechanical damage to the drilling equipment or wellhead is likely.

Certain vessels employ a **blue light** to indicate loss of system capability (i.e., thruster down for maintenance). This condition applies when the rig is able to maintain position in an "ADVISORY" operating condition, but an equipment, system or component failure which when combined with another similar failure, could result in a change of operational status.

A detailed Drive-Off and Drift-Off analysis is conducted by skilled analysts to calculate time and distance traveled by the vessel under various power settings and environmental conditions. This is done to help establish yellow and red light alarms at specific vessel offsets from the wellhead.

Disconnecting from the Well

An unplanned disconnect from the well may be required during the course of operations. An unplanned disconnect can either be implemented manually, if time and circumstances permit, or automatically in an emergency. In an unplanned disconnect, minimal if any disconnect preparations can be completed. Loss of well control or station-keeping failure are two primary reasons for unplanned disconnects.

A generic disconnect procedure is described in the following section for an unplanned disconnect. Any procedure will need to be well and rig specific. Emergency disconnect procedures should be reviewed prior to drilling.

Unplanned manual disconnect

The following generic procedure is for an unplanned manual disconnect. There is ample time to hang-off the drillstring on a set of rams prior to shearing the drillstring.

Space out the drillstring.

1. Space out the drillstring and ensure the tooljoint is above the hang-off rams.
2. Close the hang-off pipe ram with normal operating pressure (1500 psi).
3. Lower the drillstring onto the hang-off pipe ram. Take a slight overpull of 10,000 lbs. to keep the drillstring in slight tension above rams. Lock the hang-off pipe ram.
4. Close the lowermost pipe ram with normal operating pressure (1500 psi), and lock same.
5. Use the blind/shear rams utilizing full operating pressure (typically 3000 psi) to shear the drillstring, and lock same.
6. Disconnect and pull the LMRP/riser using normal procedures.
   a. Note: Caution must be taken at the surface with the sheared stub of the recovered drillpipe. High strength drillpipe can shatter when recovered.

Note: During actual conditions when the rig is offset from the wellbore, the tooljoint location will be different than the calculated location. Since offset conditions will increase the distance to the BOP stack, the tooljoint location will always be closer to the rotary during an emergency disconnect of the BOPs. The Driller must understand the concept of the tooljoint location during various scenarios and conditions to ensure correct positioning of the tooljoint for hang-off.

For such emergencies where no time is available to manually hang-off and shear, an Emergency Disconnect System (EDS) will rapidly secure and disconnect from the well.

Emergency disconnect system (EDS)

Emergency Disconnect Sequence (EDS) system means a safety system which is designed to be manually activated to disconnect the LMRP, in the event of an emergency situation. Once the EDS button is manually pushed, the functioning of the EDS is automatic. The minimum functions of an EDS should at least include closing the blind/shear rams and then disconnecting the LMRP. The specific functions are unique to each rig.

There will typically be a shearable sequence, a casing shear sequence and nonshearable sequence on the BOP control system. The shearable sequence is used when thin-wall tubulars such as drillpipe which can be severed by the blind/shear rams are across the BOP stack. The casing shear sequence is used when thick-wall tubulars are across the BOP stack, such as casing, certain drill collars, landing strings, or high-grade drillpipe. The nonshearable sequence is used when a thick-walled drill collars which are not shearable are across the BOP stack. In the nonshearable mode, the nonshearable tubulars must be removed from the BOP stack. An acoustic system or ROV can then be used to shut the well in.

Autoshear

Autoshear system is safety system which is designed to automatically shut-in the wellbore, in the event of the LMRP disconnect or parted riser. The autoshear system uses hydraulic fluid from a stack mounted subsea accumulator bottles to function the blind/shear rams and secure the well.

Deadman

Deadman system means a safety system which is designed to automatically shut-in the wellbore in the event of a simultaneous absence of hydraulic supply and signal transmission capacity in both subsea control pods. If communication is only lost to one subsea control pod, the Deadman won't activate since the other pod can be used.

Dual gradient drilling

Dual Gradient Drilling (DGD): Creation of multiple pressure gradients within select sections of the annulus to manage the annular pressure profile. These methods are used to reduce the ECD (equivalent circulating density) on the hole which may allow for fewer casing strings, reduction in mud volumes, and reduction in lost circulation. Methods include use of pumps, fluids of varying densities, or combination of these.

There is one primary method for pre-BOP deployment and there are three primary types for post-BOP deployment in use or in development.

Pre-BOP Deployment.

1. Riserless mudline pumping (RMP)

Post-BOP Deployment

1. Controlled annular mud level (CAML or CML) or Low Riser Return System
2. Subsea MudLift Drilling
3. Dilution, or Continuous annular pressure management (CAPM)

Kick detection measures and well control methods are unique to each rig and each system. These procedures are not described in this document. A description is provided below for the reader.

Riserless mudline pumping (RMP)

Riserless mudline pumping employs a subsea pump to return drilling fluid from the subsea wellhead to the rig. The wellhead collection point is through open top Suction Control Module. Variable speed pump automatically compensates for operations changes.

- Pump control based on differential pressure/fluid level at Suction Control Module. Operator can drill and circulate the hole without running out of fluid.
- Increase casing shoe depth and eliminate one or more casing or liner.

Controlled annular mudline level (CAML/CML) or low riser return system

A pump is suspended from the rig to a fixed depth in the seawater column. It is attached to the riser where the mud is withdrawn from the wellbore and returned to the rig through an external line. This allows the fluid level inside the drilling riser to be adjusted, so bottom-hole pressure can be managed. BHP can be reduced by changing the fluid level in a riser. The riser is equipped with pressure sensors to measure the riser fluid level thus permitting real-time BHP calculations. The system does not use an RCD. Changes in flow are monitored so that an influx results in an increase in the riser fluid level or by an increase in the subsea pump speed.

Subsea mudlift drilling

A seawater-driven positive displacement pump (MLP) is located above the BOP/LMRP. It withdraws the mud from the well and pumps it back to the surface through a line attached to the drilling riser. The riser is filled with a seawater-density fluid, or the mud/riser fluid interface can be maintained at any depth.

A Subsea Rotating Device (SRD) sits above the MudLift.Pump which can be used to rapidly change the pressure profile in the well.

Continuous annular pressure management

Continuous Annular Pressure Management (CAPM) is a method for injecting diluted drilling fluid into the riser at or near the seabed. This reduces the hydrostatic pressure at the mudline and thus at the casing seat. All pumping is done from the rig.

Drilling fluid generally consists of two components: a base fluid (water, oil, or synthetic fluid) and a weighting material (usually barite or hematite). Mud returns to the rig by moving up the marine riser and goes through the shale shaker to remove cuttings. A degasser, a desander, and/or centrifuge may be used prior to the mud going back to the suction pit. The centrifuge is designed to separate the weighting material from the base fluid. The light base fluid typically weighs between 8 and 9 ppg, and the heavily barite laden fluid is typically 16 ppg. The lighter fluid is normally reinjected into the riser annulus at or near the seabed and reduces the hydrostatic pressure on the formation. The heavy fluid is returned to the pits.

CHAPTER SEVEN

Reference: Subsea equipment

Commentary

As a planned event, subsea well control becomes more critical due to long lead times for manufacturing equipment such as subsea wellhead systems. These systems must be selected based upon the best subsurface pressure information available at the time of ordering. During the manufacturing process, more subsurface data may become available altering the design.

The overall well construction time line may approach 9 months due to nonproductive time and weather windows. In terms of well control, this means hole sections must include barriers which may provide pressure barriers over several weeks. One example would be an approaching hurricane or storm, which may require disconnection from the subsea stack. Reentry into open hole environments may lead to more complex well control kill situations.

Equipment operating and technical specifications must be closely followed to ensure uninterrupted service over a wide range of operating parameters. Manufacturer specifications must be studied and understood, to ensure proper operation during any subsequent well control event. This equipment includes 100% redundancy, which provides sufficient back-up operation during emergency events.

Subsea stack and riser

The following illustration demonstrates how the Subsea BOP is installed and is connected to the floating vessel (Drillship). Major components begin with wellhead, connector, six-ram BOP, LMRP with dual annular preventers and connector, Flex Joint, Riser, Riser Tension Ring/Slip Joint, and Diverter (Fig. 7.1).

Universal Well Control
https://doi.org/10.1016/B978-0-323-90584-8.00003-4

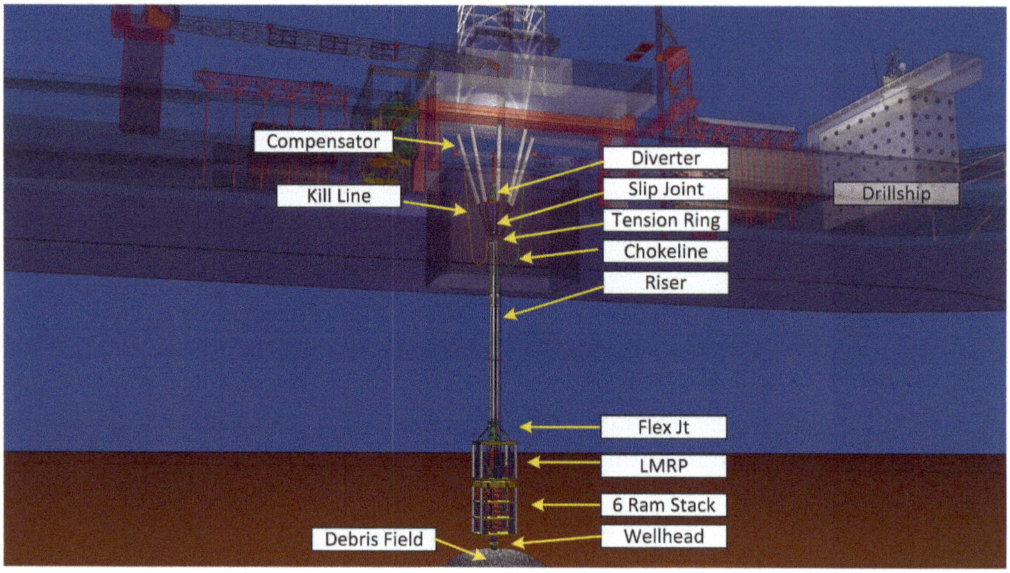

Fig. 7.1 Subsea stack and riser

Subsea wellhead

Subsea wellheads are designed as unitized housing allowing installation, isolation, and sealing of a series of casing strings. Each casing string is run with a landing joint with a specific profile and sealing ring which allows the landing joint to land, transfer weight, and energize the seal ring. Each subsequent casing string will be installed above the first casing hanger in much the same manner, allowing the casing string to be landed, provide weight transference, and energize isolation pressure seal (Figs. 7.2 and 7.3).

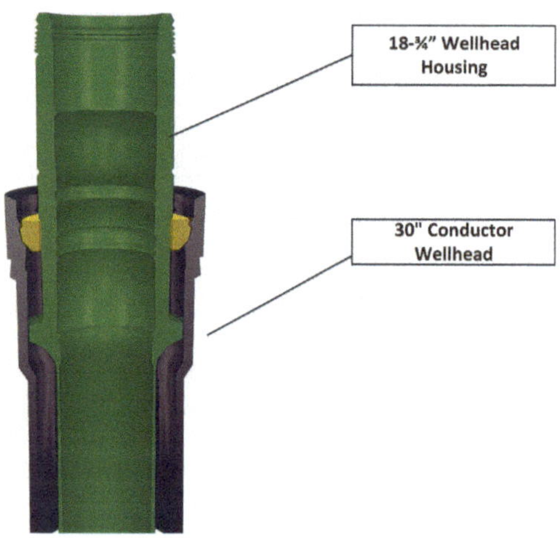

Fig. 7.2 18–3/4″ Wellhead housing in 30″ conductor wellhead.

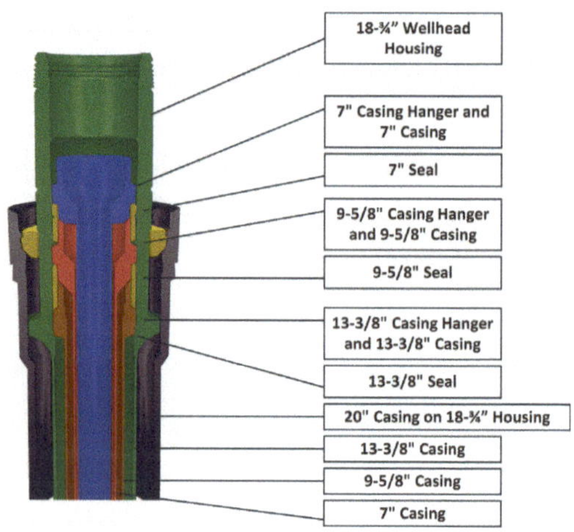

Fig. 7.3 7″, 9–5/8″, 13–3/8″, 20″, and 30″ installation.

To install the wellhead, the surface hole is drilled with returns taken to the seafloor. The surface hole is normally drilled to a depth that exceeds the length of structural casing by a combined length of at least structural casing plus subsea wellhead and rat hole of at least two additional lengths of casing for those wells in which casing strings hang off in the high-pressure wellhead housing. After drilling, the surface hole is sufficiently cleaned with sweeps to ensure minimum fines within wellbore. The structural casing with wellhead is installed. The installation angle is to be kept at 0 degrees, as each subsequent casing string and BOP will be installed at **0-degree inclination**. A top down view of normal clearance and tight clearance casing designs shows the versatility of the hanger system (Figs. 7.4 and 7.5).

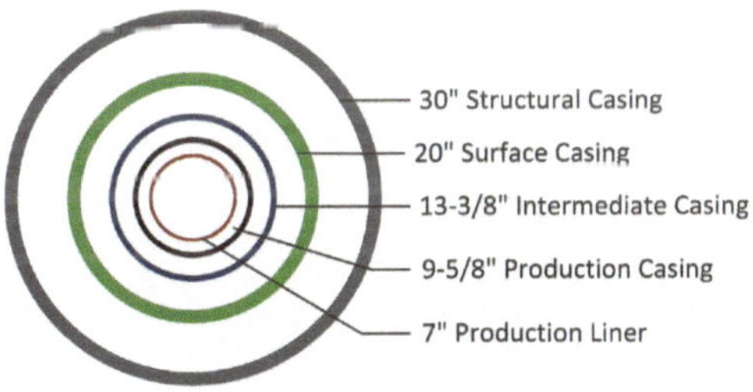

Fig. 7.4 Normal clearance casing strings.

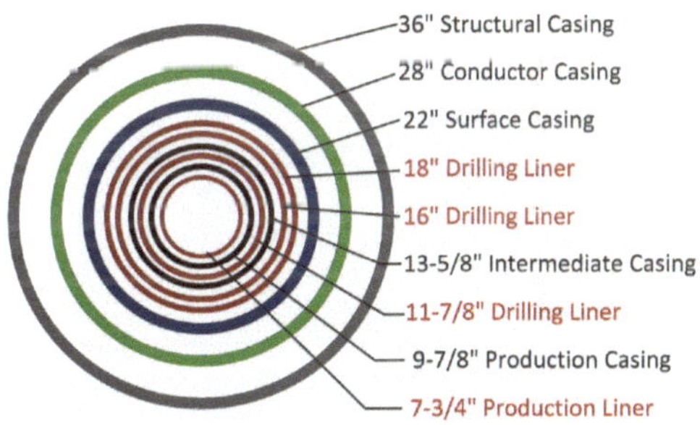

Fig. 7.5 Tight clearance casing strings.

- Conductor casing is installed to ensure subsea wellhead connector protrudes at least 10 ft. above seabed. Installation distance will be determined by the height and diameter of surface hole debris pile.
- Debris pile may be raked by ROV to ensure wellhead assembly can be accessed.

Big bore subsea wellhead systems

As the subsea industry continues to expand and explore in deeper water depths, requirements wellhead components have changed. Due to the excessive water depths, shallow formations are generally weaker while the probabilities of shallow "flows" increase. Big bore wellhead systems have been designed to accommodate more strings of casing which will ensure higher-pressure formations can be penetrated.

Conductor and intermediate casing strings can be redesigned to provide isolation barriers for the potential of multiple shallow flows. Conventional subsea wellhead assemblies could not accommodate the needs of additional casing strings and heavier, thicker-walled casing strings.

Subsea wellhead and LMRP connectors

Piston Force or Collet Lever connectors represent the majority of wellhead and LMRP connector groups (Figs. 7.6 and 7.7). Piston Force connectors swallow the connector pin. The length of the swallow will limit the angle of disconnect. The Collet Style connector uses a segmented collet that moves the segments to grasp the connector pin (Figs. 7.8 and 7.9). Without much distance needed to swallow, the Collet Style connectors are preferred when connecting the LMRP to Subsea BOP stack.

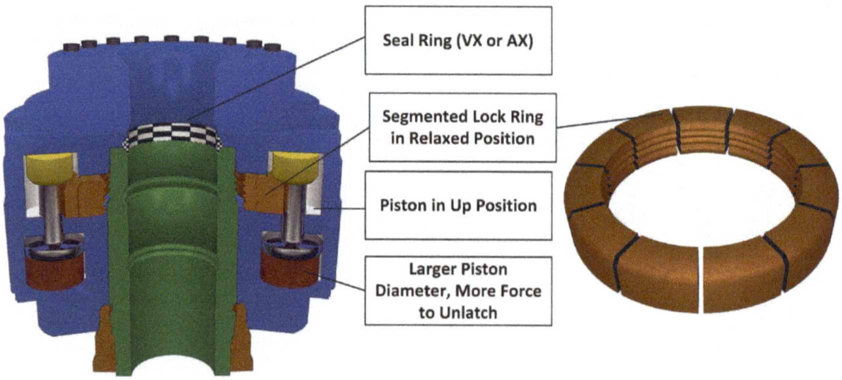

Fig. 7.6 Unlatched wellhead connector in running position.

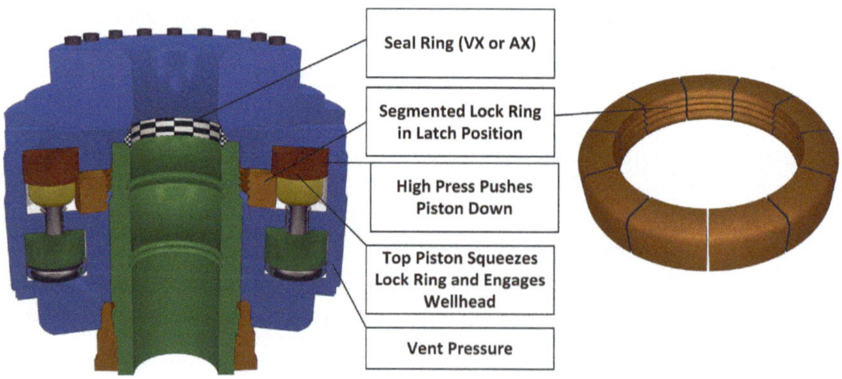

Fig. 7.7 Wellhead connector in latched positions.

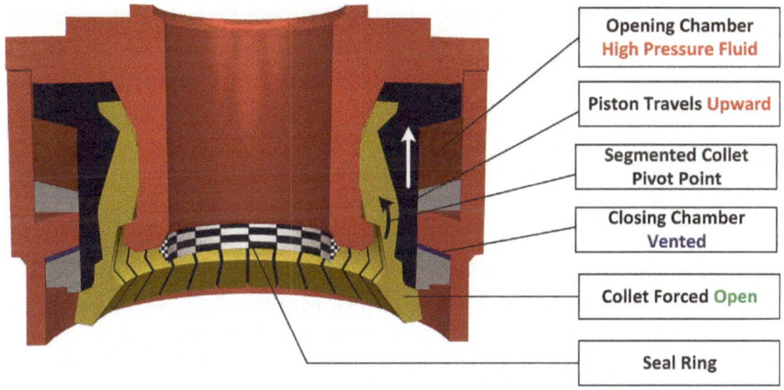

Fig. 7.8 Collect wellhead connector in unlatched position.

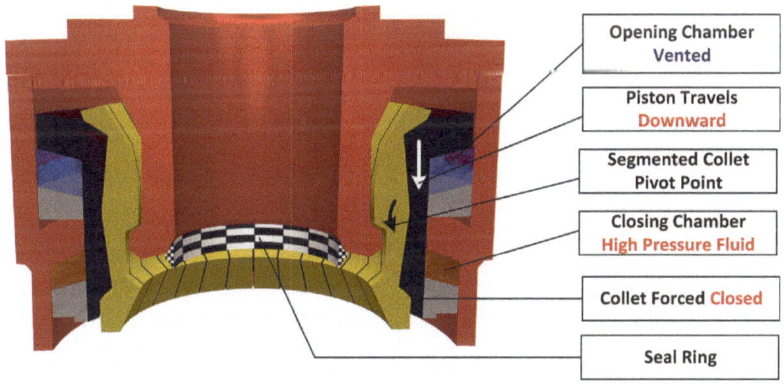

Fig. 7.9 Collect wellhead connector in latched position.

Piston force connectors

Piston Force connectors incorporate a "swallow" feature in which the wellhead connector is lowered over the wellhead pin connector. The wellhead connector protrudes within the connector until it impacts an internal lip guide. Eleven pistons are energized by pumping high-pressure fluid into the top portion of the piston. The piston is driven downward impacting the segmented split-locking dogs. The segmented split-locking dogs are "locked" into the opposing grooves on the pin connector.

Conversely, the opening force is applied by pumping high-pressure fluid to the opening chamber located below the piston. The larger diameter closing piston provides a much greater opening force. As the piston is driven upward, the closing ring becomes relaxed and opens. Once unlatched, the BOP stack must be raised until the connector clears the pin connector.

PRIMARY UNLATCH ENERGY USES ONE HALF OF THE PISTONS.
SECONDARY UNLATCH ENERGY USES REMAINING PISTIONS TO MAXIMIZE OPENING FORCE.
PISTON FORCE CONNECTORS MUST BE RAISED ONCE UNLATCHING TAKES PLACE
OPERATES UP TO A MAXIMUM OF 12° ANGLE RELEASE

Collect style connectors

Collet Style connectors utilize a design where a segmented collect engulfs the wellhead or LMRP connector. This type of connector does not require shallowing a long portion of the wellhead or LMRP connector. The minimum distance along with multitude of segments allows emergency disconnect to take place rapidly and at high angles (up to 30 degrees).

COLLET STYLE CONNECTORS ARE PREFERRED FOR LMRP TO SUBSEA STACK CONNECTIONS
DUE TO HIGH ANGLE RELEASE AND SMALL PENETRATION DISTANCE

Subsea BOP stacks (five ram and six ram)

The following illustrations demonstrate the complexities of a five-ram and six-ram stack. Today's deepwater rigs are designed to contain a complement of up to two separate Rams Stacks and LMRP. Older deepwater rigs may have only a complement of one-ram stack. Due to overall weight and size, these deepwater rigs are designed with a series of gantry cranes and hoists capable of lifting and moving subsea BOPs in various positions over water. Deployment techniques will vary with individual rig equipment complement and rig design (Figs. 7.10–7.12).

Fig. 7.10 Five-ram subsea BOP stack.

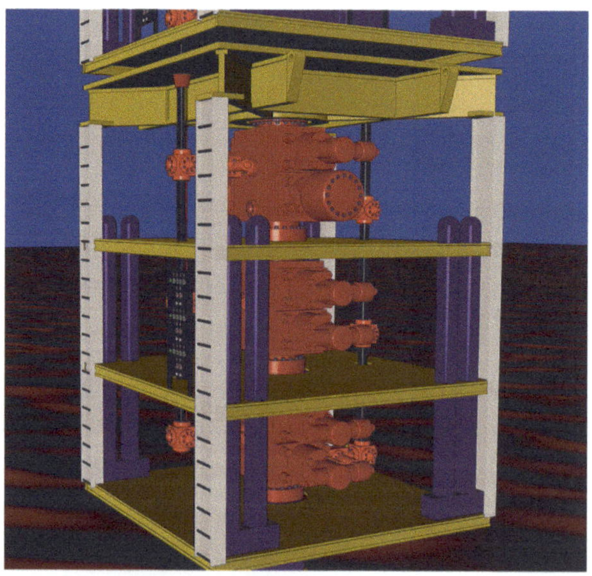

Fig. 7.11 Six-ram subsea BOP stack.

Fig. 7.12 Five-ram subsea BOP Stack and LMRP on deck of drillship prior to deployment.

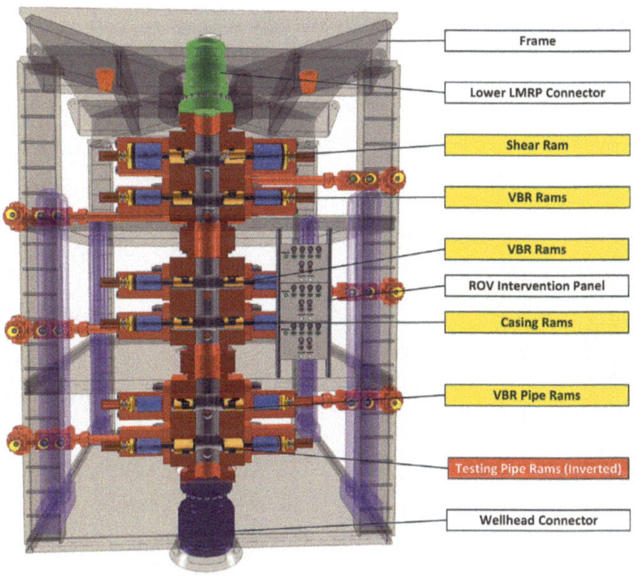

Fig. 7.13 Six-ram subsea BOP stack cut-a-way drawing.

Six-ram subsea BOP stacks were developed to offer the greatest number of options for well control and testing. The bottom ram is represented by an inverted pipe ram that allows testing of all components above this ram while providing isolation of the wellbore. This innovation eliminates the need for pulling completely out of the hole to pick up and install a test plug. With the inverted testing pipe rams, normal operations will allow for pulling the bit into the lowermost casing shoe, close the test rams, and test the BOP stack (Fig. 7.13).

THE INVERTED BLIND RAM IS <u>NOT CONSIDERED</u> TO BE A BLOWOUT PREVENTER

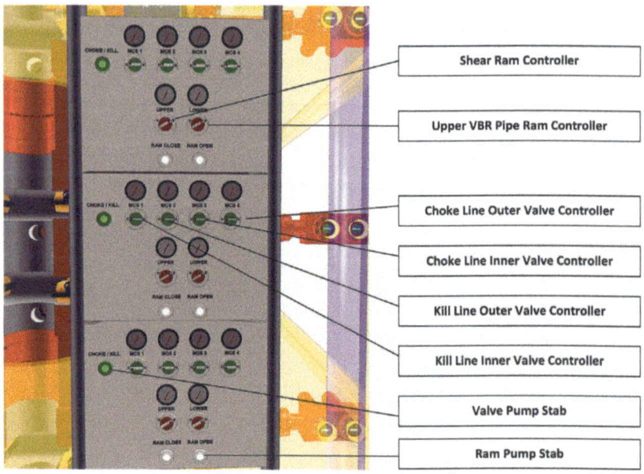

Fig. 7.14 Six-ram stack ROV intervention panel.

The ROV Intervention panel is affixed to the subsea BOP frame at a location where emergency operations can be initiated and allow access to each ram or valve. Each panel contains a stab tap that allows the ROV to install emergency pumping services from ROV pump or hot line. The ROV pump source fluid is seawater; therefore, in emergency operations, high-pressure seawater is used to function the appropriate ram or valve (Fig. 7.14).

ONLY SELECT BOPE (RAMS, CONNECTORS, ETC.) WILL HAVE ROV INTERVENTION CAPABILITIES

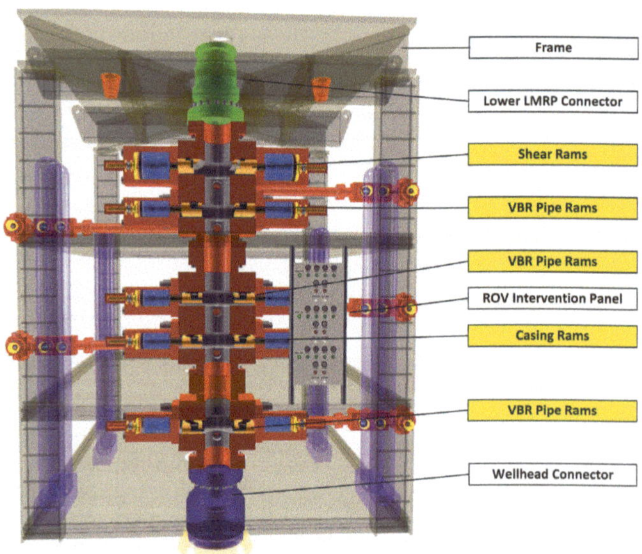

Fig. 7.15 Five-ram subsea BOP stack cut-a-way drawing.

After installation of pumping stab, the ROV can manipulate the ram or valve using the Mast Control Station (MCS) either "open" or "closed." The MCS is a dedicated system that controls and retrieves data from subsea equipment on the ocean floor.

The five-ram stack is offered on vessels with less operating draft. Unlike the six-ram stack's inverted blind ram placed on bottom, this feature is eliminated on the five-ram stack (Fig. 7.15).

The latest generation floating drillships and semi-submersibles use five- to six-ram subsea BOP stacks. The stacks are maintained within an antifall structure that provides protection from falling debris during a catastrophic event. Independent BOP operation is provided via the 100% redundant Blue Pod/Yellow Pod control fluid operation. Operating pressurized hydraulic fluid (1500 psi) and emergency by-pass pressurized hydraulic fluid (5000/3000 psi) are provided from each pod and are directed to the appropriate closing or opening piston of each ram.

Below each ram, an integrated circulation port is provided along with appropriate hydraulic remote operating gate valve to direct wellbore fluids to the appropriate choke or kill line. These ports allow for circulation into/out of the BOP stack and for potential stack gas clearance.

Subsea stacks employ automatic locking systems on rams which remain locked in the closed position, even if closing pressure is lost or wellbore pressures breach internal sealing assemblies. Hydraulic opening pressure is required to unlock the automatic locking device. One type of automatic locking device is the clutch-driven automatic locking device (Figs. 7.16–7.18).

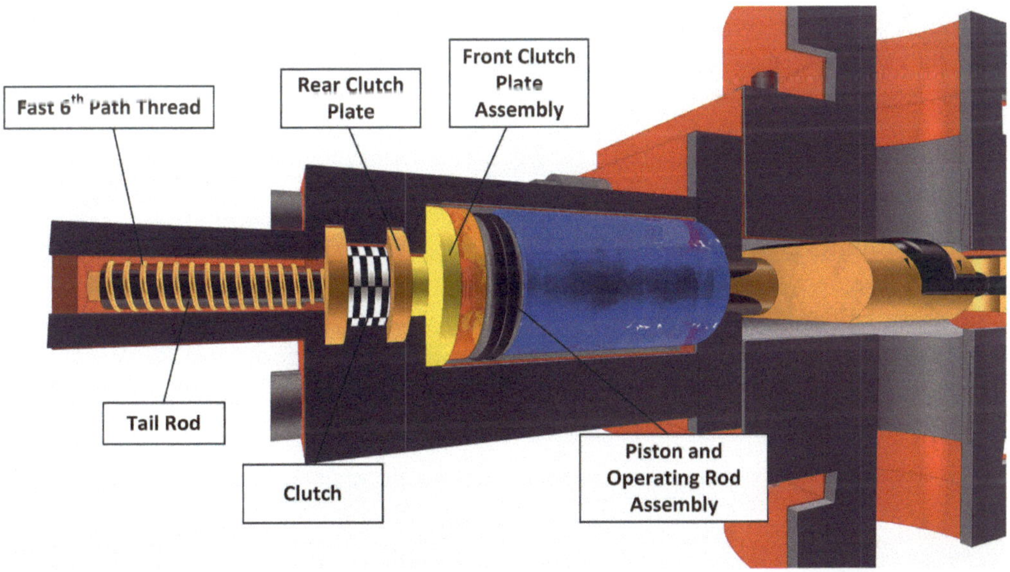

Fig. 7.16 Clutch-driven automatic locking device diagram.

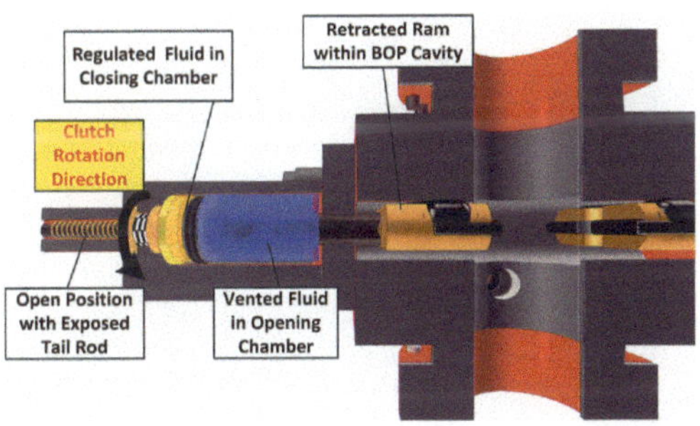

Fig. 7.17 Clutch-driven automatic locking device at beginning of closing operation.

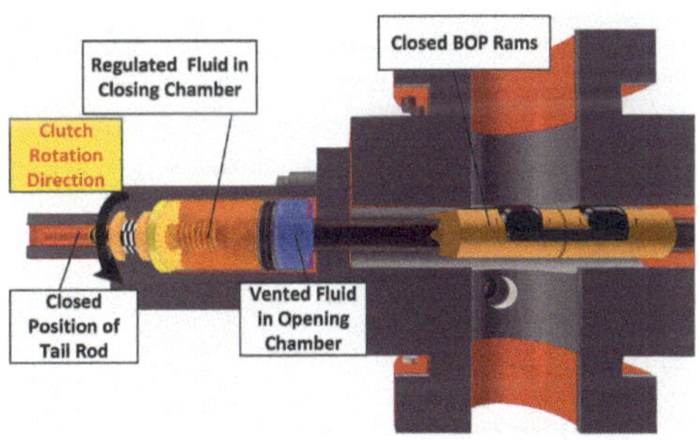

Fig. 7.18 Clutch-driven automatic locking device at the end of closing operation.

The clutch-driven automatic locking device contains a three-part clutch mechanism servicing a threaded tail rod attached to the closing ram piston. The clutch is sandwiched between two clutch plates used to allow the clutch to rotate as the piston is closed. The external clutch plate allows the clutch to rotate clockwise as the piston ram travels into the closed position. The external clutch plate ensures proper clockwise locking rotation (cannot travel counterclockwise). As the clutch is rotated, the tail ram rod is "automatically locked" at any closed travel position. This feature locks the rams in the closed position regardless of ram packer wear.

To open, automatic locking systems require applications of hydraulic opening pressure. For the clutch-driven automatic locking device, hydraulic opening pressure shifts a transfer ring to disengage the rear clutch plate from the front clutch plate. The hydraulic opening pressure ensures these clutch plates remain separated during the opening sequence. The opening pressure moves the ram and closing ram piston into the open position, while the front clutch plate and clutch "freely" rotate counterclockwise. The tail rod attached to the closing ram is allowed to extend freely, allowing the piston ram to freely travel to the open position.

Lower marine riser package (LMRP)

Consisting of a separate antifall structure, the LMRP contains BOP connector, two annular preventers, circulating ports below each preventer, landing receptors for yellow/ blue pods, kidney plates for installation of yellow/blue pods, pressure assist chambers with pressurized fluids, flex joint (with booster line connection), and ROV controls. To provide access for recovery of an individual pod, the upper antifall cage deck has access hatchway panels called kidney plates (Fig. 7.19).

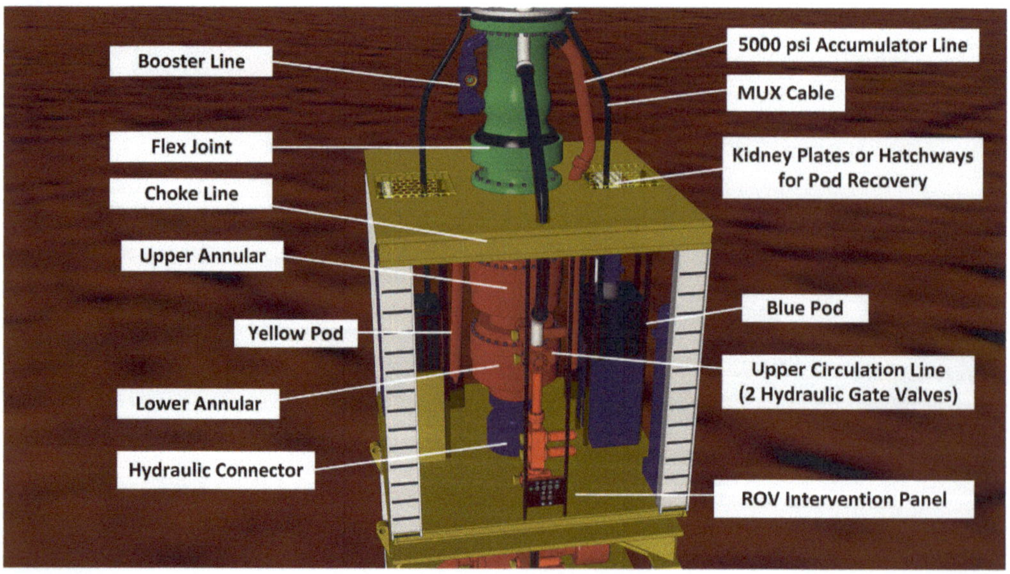

Fig. 7.19 Lower marine riser package (LMRP).

NOT ALL PODS HAVE RECOVERY FEATURES

The most common type of LMRP hydraulic connectors are Cameron connectors. These hydraulic connectors do not "swallow" protruding connector and provide increased disconnect angles in the event of an emergency (Fig. 7.20).

Fig. 7.20 Recovery of LMRP.

Remote-operated vehicles (ROV)

Remote-operated vehicles are robotic self-propelled systems which allow interfacing with subsea remote-control stations on LMRP and BOPs. Using robotic arms, these vehicles can be used to direct and operate emergency systems, deploy pumping services, and provide means to deploy hydrate inhibition services. Lighting and camera rigs allow for visual remote operation. Due to the wide range of designs and service, recovery and deployment systems will vary. Shown below is one type of cage recovery systems.

As ROVs are custom designed, features will vary between manufacturer and application. Design features include:

- Designing subsea BOPE to include intervention of ROV such as operation of Blind/Shear Rams.
- ROVs can be used along with hot lines to provide power and pumping abilities.
- ROVs normally have capabilities to be piped into selected BOPE.

- Pumping abilities may include injecting seawater into BOP cavities during emergency operations to close rams.
- Pumping abilities may be volume limited. Must provide sufficient volume to operate shuttle valves and close Shear Rams with 3000 psi.
- Pumping abilities may include operating at ranges of 3000 and 1500 psi (Figs. 7.21–7.23).

Fig. 7.21 Remote-operated vehicle (ROV).

Fig. 7.22 ROV in deployment cage.

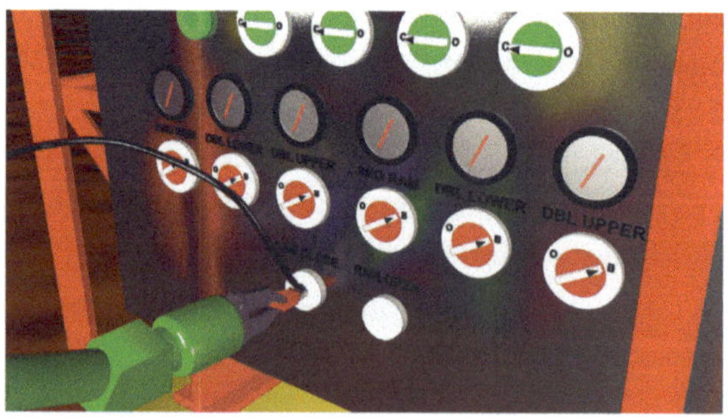

Fig. 7.23 ROV rotational arm with hot stab.

MUX BOP closing systems

Multiplex (MUX) BOP Closing Systems represent the latest generation of control systems for Subsea BOPE. These systems utilize computer-controlled, electronically operated solenoid/SPM valves (which share a single pair of wire bundles) to assist in directing pressurized hydraulic control fluid to the appropriated Subsea Blowout Preventer. These systems offer the fastest, most reliable closing times available (Fig. 7.24).

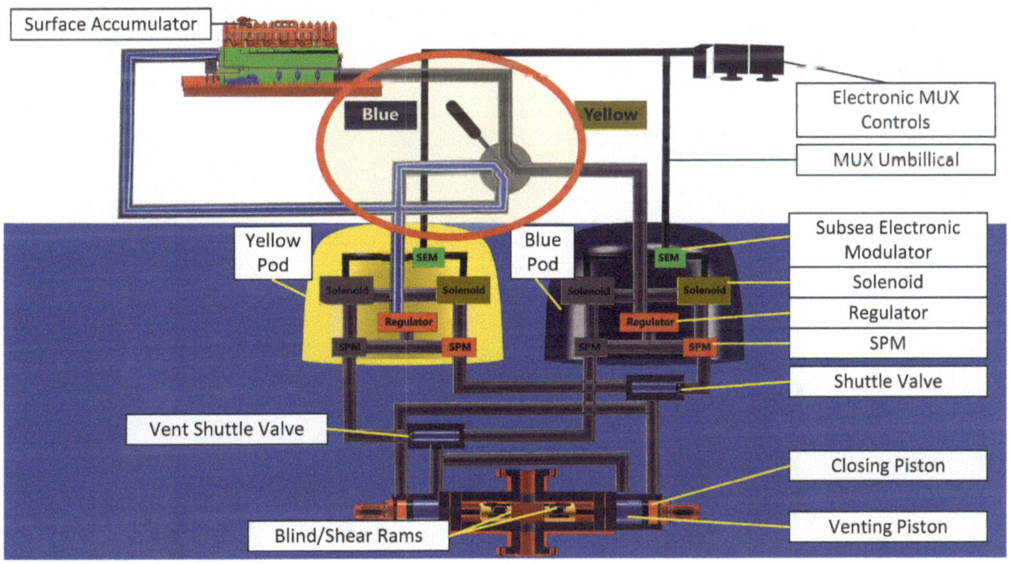

Fig. 7.24 Pods with umbilical control system.

Designed for 100% redundancy, each pod contains exact copies of equipment and service. Pilot fluid is provided to both pods along with electronic signals. This system allows switching operations from one pod to another, in case the first pod does not function correctly. The entire fluid system is designed to accommodate 5000 psi pressure thresholds. Pilot fluid and shearing system pressures are regulated to 3000 psi. The remaining BOPE is serviced by system pressures regulated to 1500 psi operating pressures (Figs. 7.25–7.28).

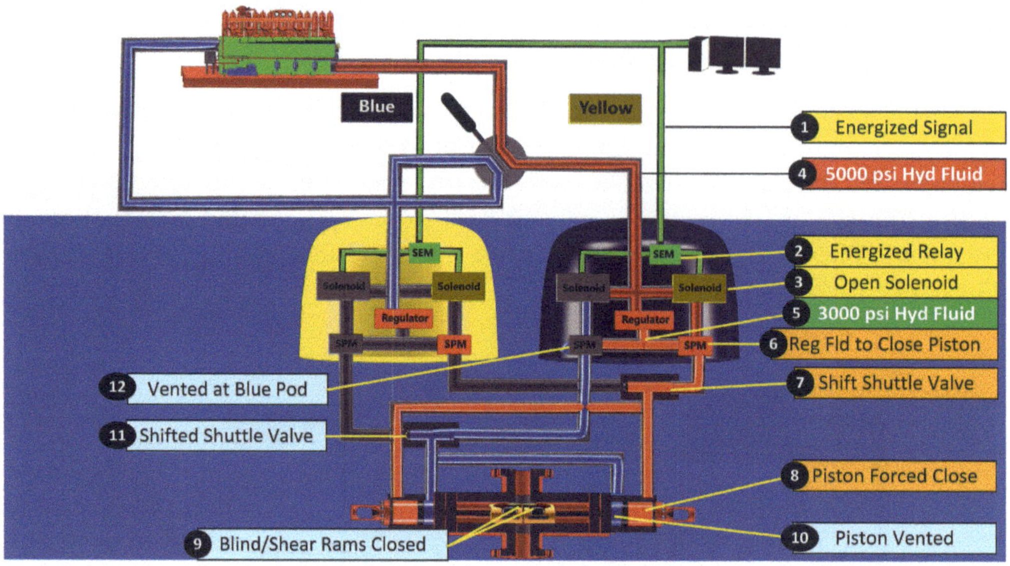

Fig. 7.25 Blue pod closing sequence for blind/Shear rams using 3000 psi regulated closing pressure.

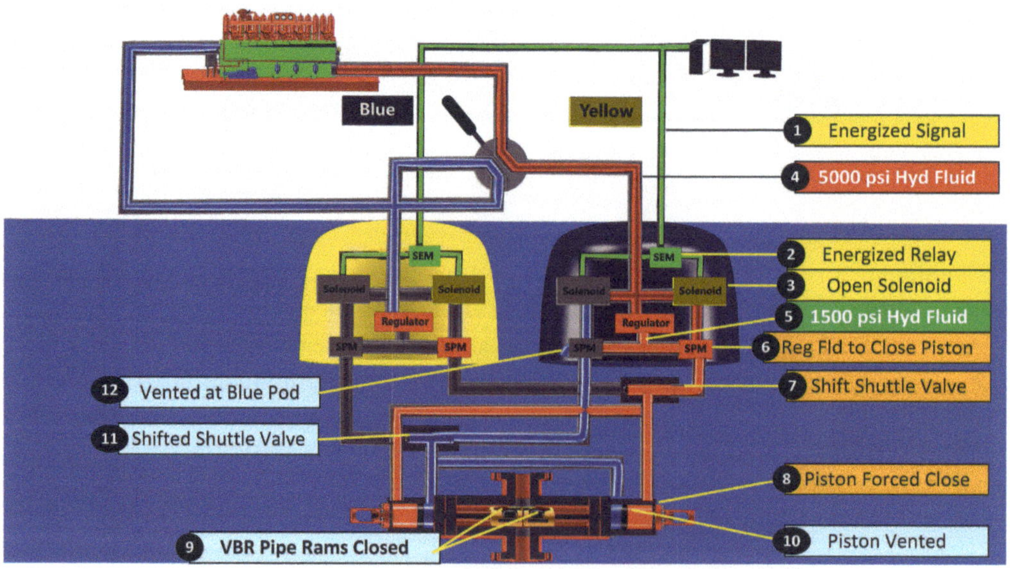

Fig. 7.26 Blue pod closing sequence for VBR pipe rams using 1500 psi regulated closing pressure.

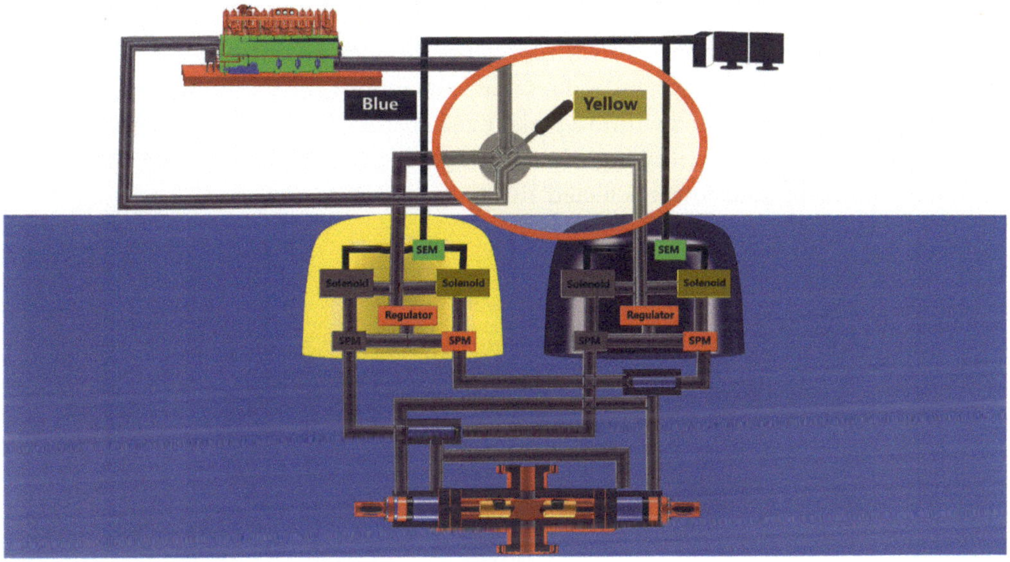

Fig. 7.27 Shift to yellow pod with selector valve.

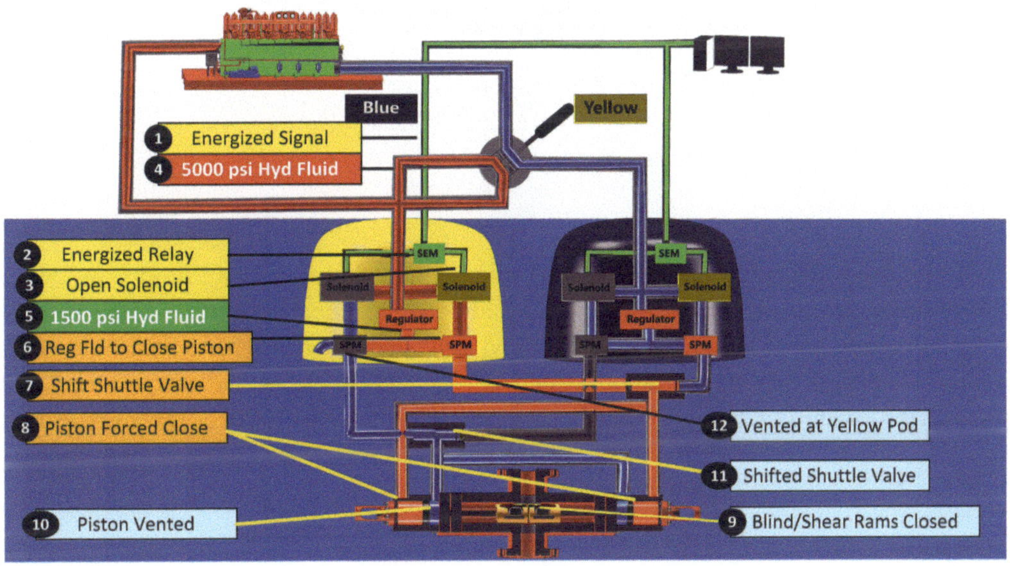

Fig. 7.28 Yellow pod sequence to close blind/shear rams with 3000 psi regulated closing pressure.

Subsea electronics module (SEM)

The subsea electronics module (SEM) represents the subsea controller, receiving electronic signals from the surface and "directing" operations on the Subsea BOP. Using a processing computer housed within a pressurized chamber with electronic back-up battery, the SEM directs electrical signals to an appropriate subsea solenoid that directs hydraulic operating fluid to the selected BOPE (Fig. 7.29).

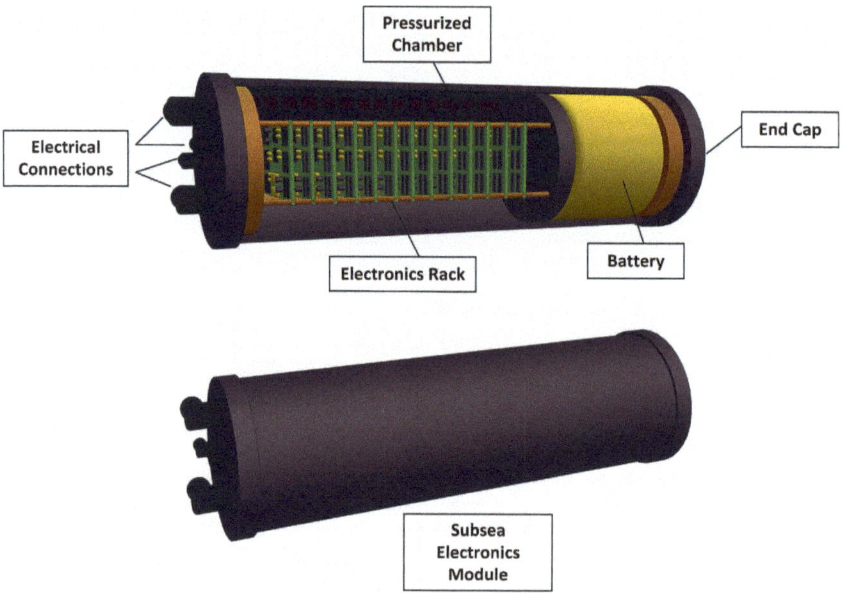

Fig. 7.29 Subsea electronics module.

The subsea electronic module (SEM) oversees direct control of "fluid flow" with full redundancy of all operations on valves and communication with surface control systems. When a pod or corresponding SEM fails, the other SEM is used to independently operate the BOP or valve. When energized, the SEM will direct regulated pressurized operating fluid (1500 psi) or emergency bypass fluid to appropriate ram or valve.

Solenoid

Depending on application, solenoids may be used in the subsea pod to provide positive fluid direction and flow. Each solenoid is positioned within each pod, providing 100% redundancy for fluid flow. When not energized, the solenoid provides a positive seal preventing fluid to flow to BOP equipment. When the electrical circuit is energized, a steel rod is raised inside the electromagnet and pulls the sealing gate upward. Once opened, pressurized pilot fluid is allowed to flow to appropriate BOP equipment (Figs. 7.30 and 7.31).

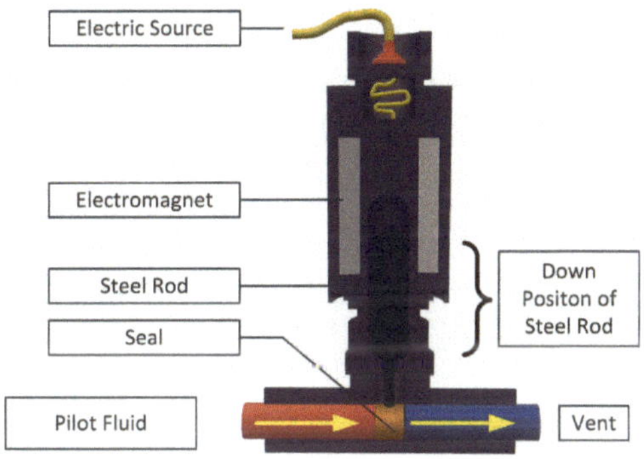

Fig. 7.30 Solenoid with no pressure.

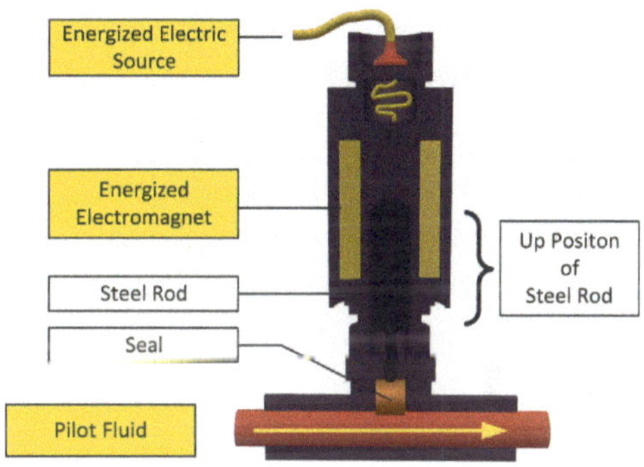

Fig. 7.31 Solenoid with pressure.

Subplate mounted valves (SPM)

Subplate Mounted (SPM) valves possess two chambers that assist in directing regulated hydraulic pressure. The upper chamber houses a spring/piston combination. The lower chamber contains a sliding sleeve which connected by a shaft to the upper chamber piston. With no pressure applied to upper chamber, the piston is forced downward due to the spring force. The downward motion forces the adjoined sliding sleeve downward, sealing off regulated hydraulic pressurized fluid and venting the returning fluids from the BOPE to the sea (Fig. 7.32 and 7.33).

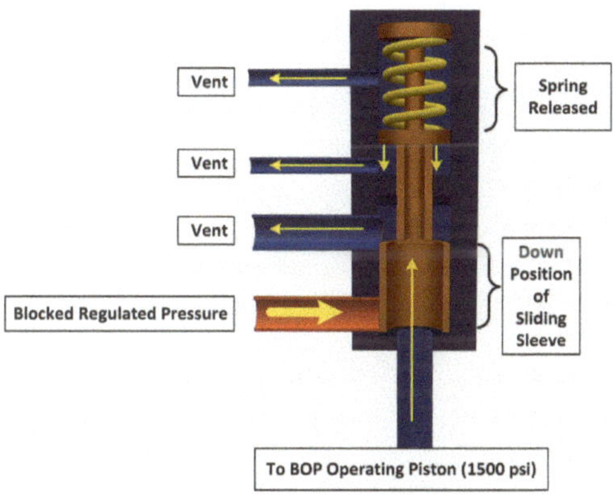

Fig. 7.32 Subsea sliding valve with no pressure.

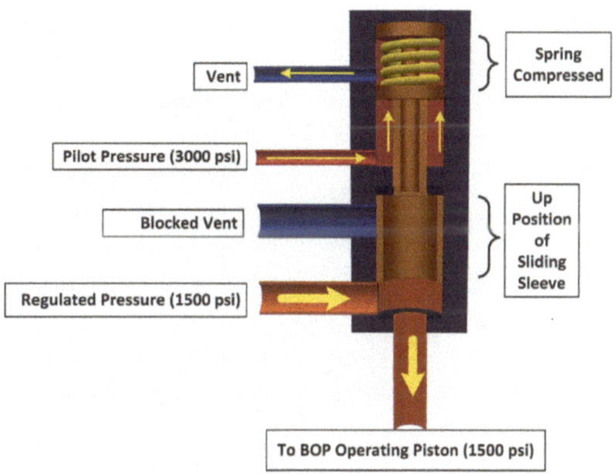

Fig. 7.33 Subsea sliding valve with pressure.

With emergency by-pass hydraulic pressurized fluid (3000 psi) is applied to the upper chamber, the upper piston is forced upward compressing the spring. The attached sliding sleeve is forced upward, sealing off the vent pressure and providing means for the regulated operating pressure (1500 psi) to be directed to proper ram or valve.

Subsea regulator

The subsea regulator provides means of reducing pressure from the maximum 5000 psi operating pressure to a regulated emergency by-pass pressurized hydraulic fluid (3000 psi) or regulated operating pressurized hydraulic fluid (1500 psi). The regulator also allows means of venting pressurized fluid. For system control, the operating regulator provides means of directing pressurized pilot hydraulic fluid and pressurized readback hydraulic fluid to showcase proper operation (Figs. 7.34 and 7.35).

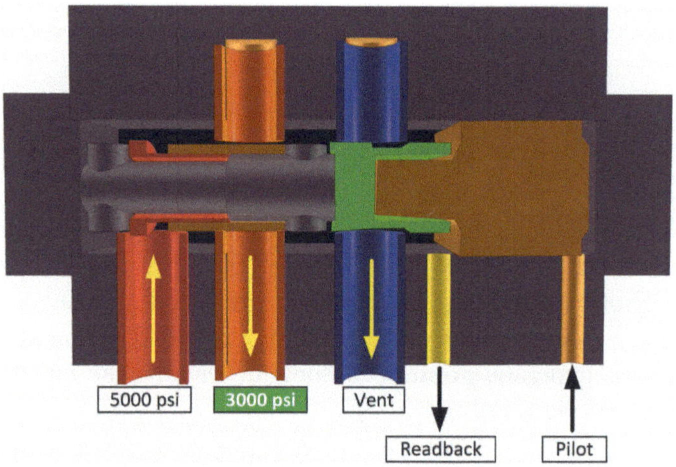

Fig. 7.34 Subsea regulator from 5000 to 1500 psi for subsea BOPE.

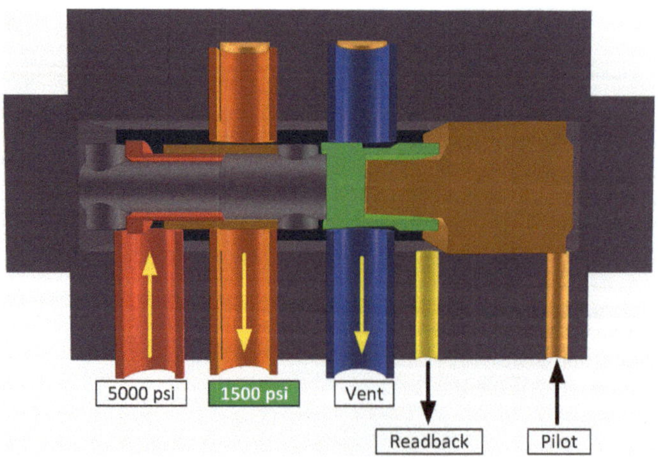

Fig. 7.35 Subsea regulator from 5000 to 3000 psi for subsea shear rams.

Shuttle valve

Shuttle valves are designed to direct flow of fluid from two sources. A simple sealing element is designed and installed to move freely within the shuttle valve. When pressurized fluid enters the shuttle valve chamber (shown in orange at 1500 psi), the pressure of this fluid forces the shuttle sealing element to the opposite end of the chamber. When landed, the sealing element provides a shut off of the chamber from the second flow source, but allows flow through the middle opening (shown as the lower flow path). The shuttle valve prevents backflow from one flow source to the other flow source (Figs. 7.36–7.41).

Fig. 7.36 Shuttle valve on subsea BOP.

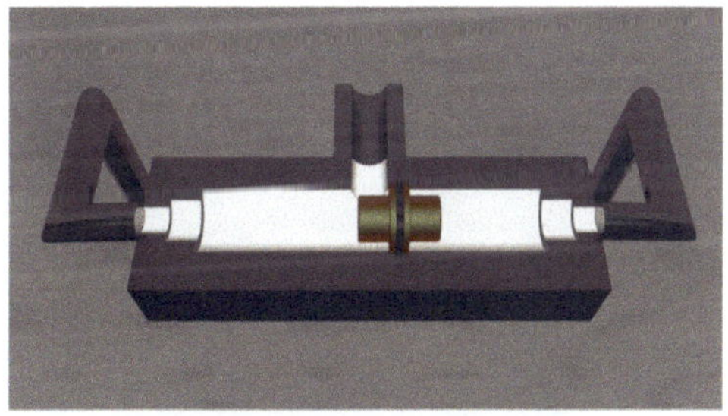

Fig. 7.37 Shuttle valve cut-a-way.

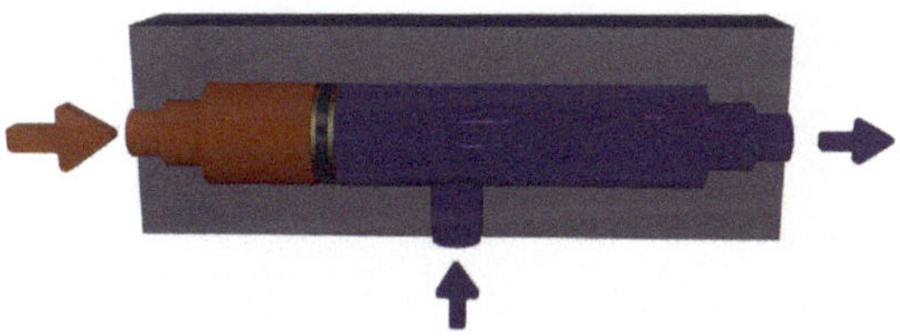

Fig. 7.38 Regulated hydraulic pressure from yellow pod (beginning)P.

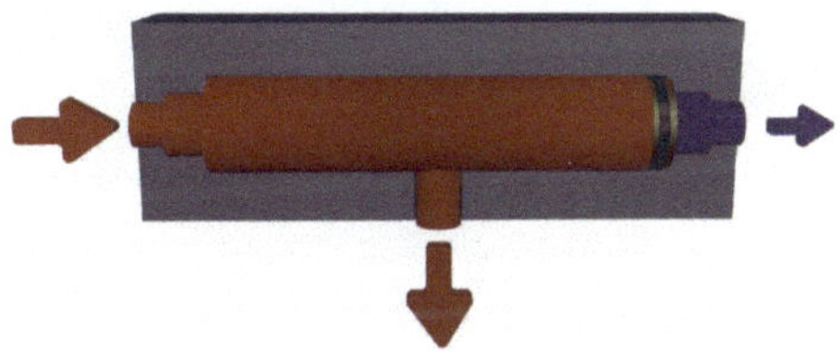

Fig. 7.39 Regulated hydraulic pressure from yellow pod (shuttle valve isolation).

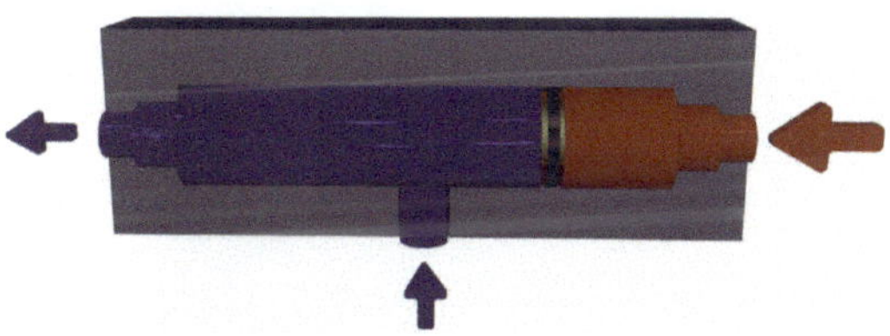

Fig. 7.40 Regulated hydraulic pressure from blue pod (beginning)P.

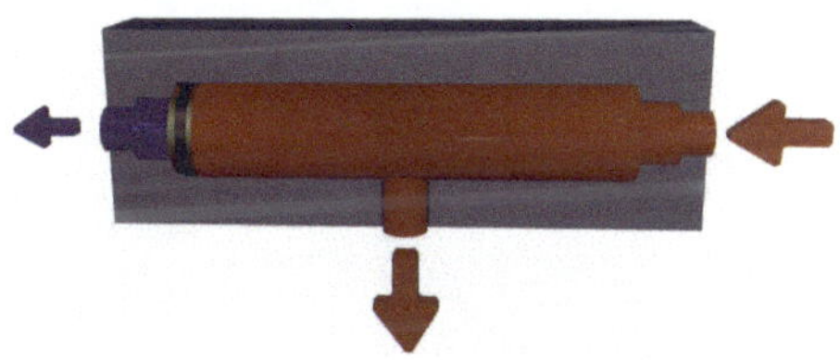

Fig. 7.41 Regulated hydraulic pressure from blue pod (shuttle valve isolation).

Risers

Top tension risers (TTR) commonly referred to as "risers" connect a subsea well to the floating vessel. As part of the drilling vessel's closed-loop circulating systems, the drilling riser provides the pathway of circulating fluid from the vessel to the wellbore and from the wellbore back to the surface. The riser system is a complex vertical pipeline that incorporates specialized equipment needed to offset dynamic loading requirements of a floating vessel along with maximum forces for tensional, compressional, and burst loading. Individual drilling riser lengths are normally limited to road transportation nonpermitted lengths (~35–50 ft.) (Figs. 7.42 and 7.43).

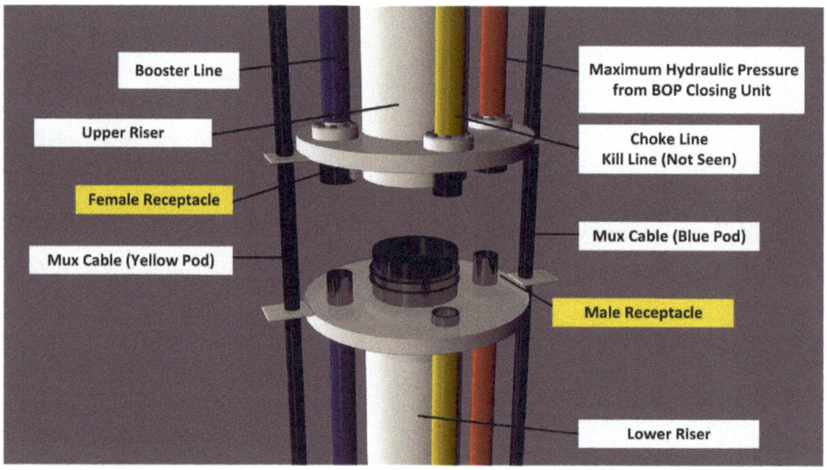

Fig. 7.42 Riser connection assembly illustration.

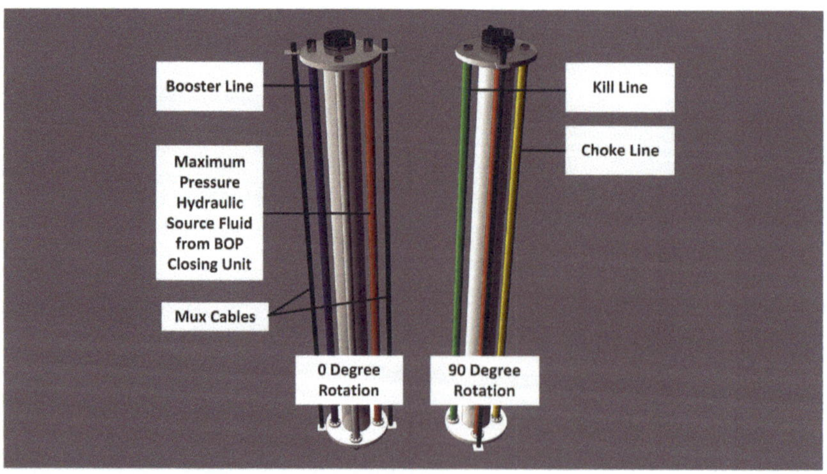

Fig. 7.43 Color-coded riser for illustrative purposes only.

The high-pressure choke lines, kill lines, booster lines, and MUX cable guides are affixed externally to the riser terminating with riser guide plates located at either end of the riser. The guide plates contain a series of guides, male and female receptacles which allow the riser and external systems to be easily connected and provide needed pressure seals. The MUX control cables are deployed independently via a spool reel and are attached to the guide plate after connection seating.

Today's risers are designed for ease of handling, installation, testing, and long life. Risers are "tensioned" to maintain structure stability. The floating vessel provides a nearly constant tensional force needed to adequately stabilize the riser and support the buoyed weight of this vertical pipeline. The tensional requirements reflect the weight of the riser assembly, needed buoyancy compensation, forces of wave and current, hydrostatic pressure of drilling fluids, and most important—allowances for equipment failure. The main torsional force applied to risers is ocean currents. Current impact is normally limited to near surface, unless the subsea location is near the outflow of a river. Buoyancy modules are normally added to riser assemblies near the ocean surface to reduce overall tension force needed for the system. Female connections are used on top with male connections on bottom to prevent debris from falling down into the connection during assembly and possibly resulting in a leak.

Booster lines

Booster lines are one of the external systems within the riser assembly and normally terminate at the flex joint on top of the LMRP. The booster line provides a supplementary fluid supply line to assist circulation of drill cuttings up the drilling riser. As drilling fluids are circulated from the wellbore and BOP stack, they will enter the larger diameter riser. As the fluids enter the larger area, the "return" velocity becomes less due to the larger area, and cuttings removal efficiency decreases. With reduced removal efficiency, cuttings may fall downhole and create solids loading within the drill fluid near the subsea BOPs (Fig. 7.44).

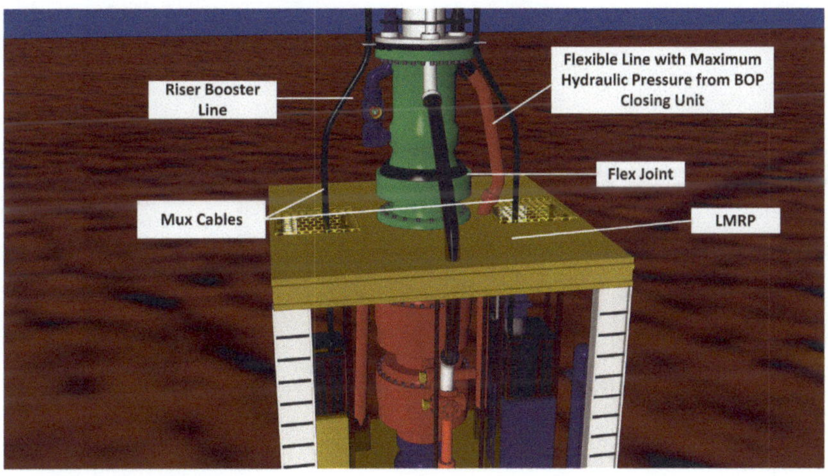

Fig. 7.44 Booster lines (*blue*) and maximum hydraulic source fluid (*red*).

By providing a secondary means of fluid flow, the cuttings removal efficiency within the riser is maintained equal to the efficiency within the open hole section, ensuring cuttings are transported to surface, separated, and removed by mechanical means.

Booster lines are also used during well control situations. After successfully killing the well using an appropriate well control method, the riser fluid density is raised to the appropriate KMW density before stack gas clearance procedures are performed.

Choke and kill lines

Choke and Kill Lines are installed from the floating vessel to the Subsea BOP stack via external riser system assemblies. These vertical lines consist of the largest internal diameter possible (4–1/16″ to 6″) and are routed from the riser assembly plate to the LMRP using a flexible stainless-steel wrapped line. The flexible choke and kill lines provide movement of the Flex Joint without impairing capabilities to flow. Using hard piping, the choke and kill lines are routed to flow ports below each preventer. Normally, at least two hydraulically operated gate valves are used for each of these connections.

The choke and kill lines provide durable, high-pressure systems for directing fluid flow in normal drilling operations as well as hazardous well control operations. By providing multiple flow paths, these lines are also an integral portion for testing equipment while minimizing trapped stack gas. As shown below, the LMRP annular preventers contain means of flushing these preventers using assigned choke and kill lines (Figs. 7.45 and 7.46).

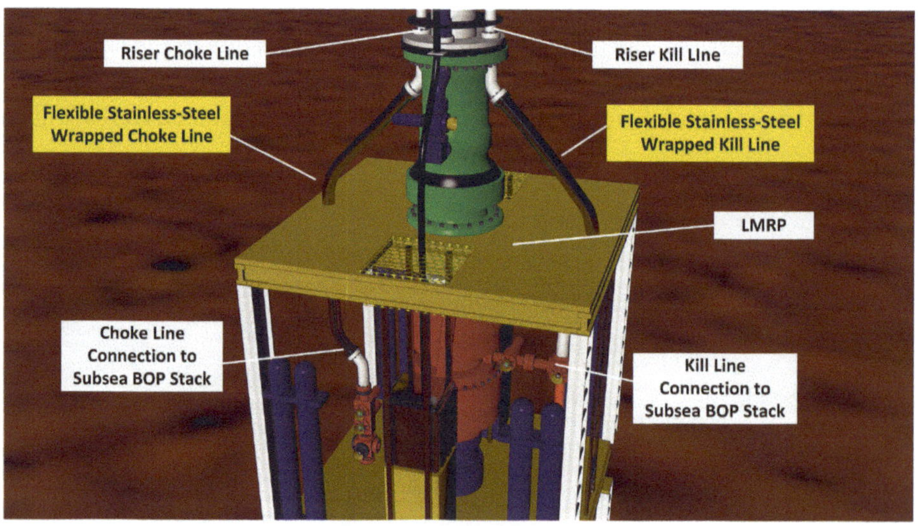

Fig. 7.45 Choke and kill lines from riser and routed around flex joint to LMRP.

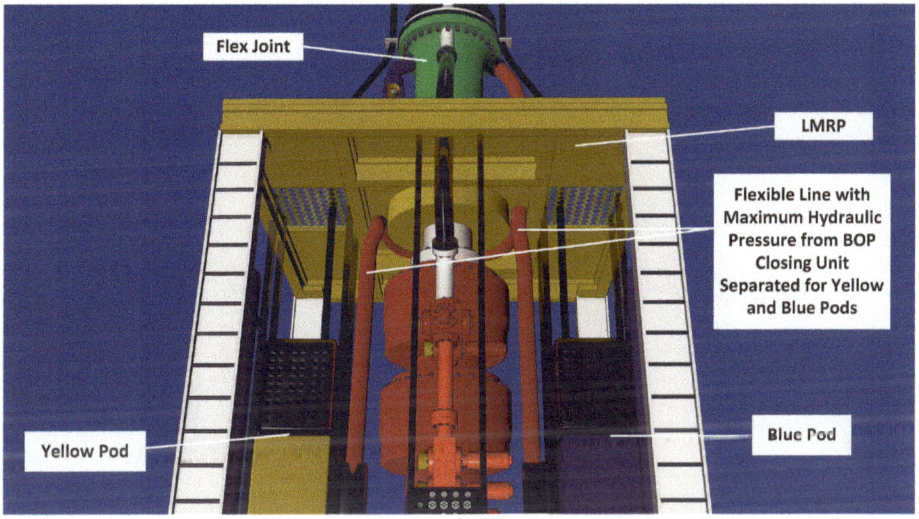

Fig. 7.46 Separation of maximum hydraulic source fluid to *yellow* and *blue* pods.

Flex joints

Flex joints provide means to allow some angular movement of the riser, due to vessel movement or environmental forces. Consisting of a steel and elastomer assembly, the flex joint provides multiple low-pressure sealing surfaces to ensure riser fluid integrity. Flex joints minimize wear-and-tear fatigue on the riser. They also provide means for installation of the booster line (Figs. 7.47–7.49).

Fig. 7.47 Flex joints (3 degree port).

Fig. 7.48 Flex joints (3 degree starboard).

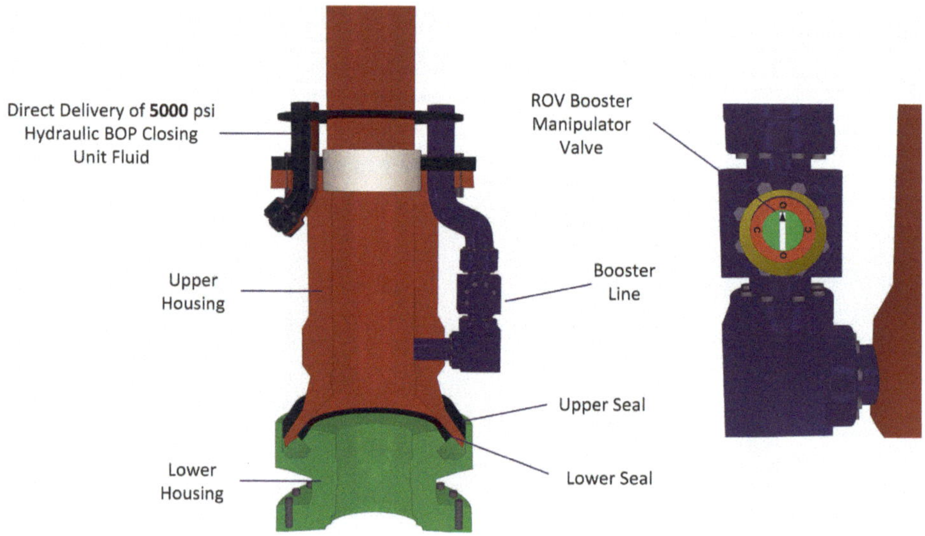

Fig. 7.49 Flex joint assembly illustration.

Slip Joints and tension rings

Tension Rings provide means of tensional support of the entire riser assembly and transference of this weight to the floating vessel by affixed tensioner cabling systems. The slip point provides means of compensation for wave-induced vertical motion and movement. The slip joint contains an inner and outer barrel with a low-pressure internal sealing element. Sealing element pressure is lowered to allow inner barrel to freely travel and provide sufficient seal to prevent leakage of drilling fluids.

In well control applications, the sealing element pressure must be increased to prevent leakage of potential formation influxes (i.e., gas) from causing catastrophic failure of the riser system. The sealing pressure is normally routed to a separate panel, which must be actuated during a well control operation (Figs. 7.50 and 7.51).

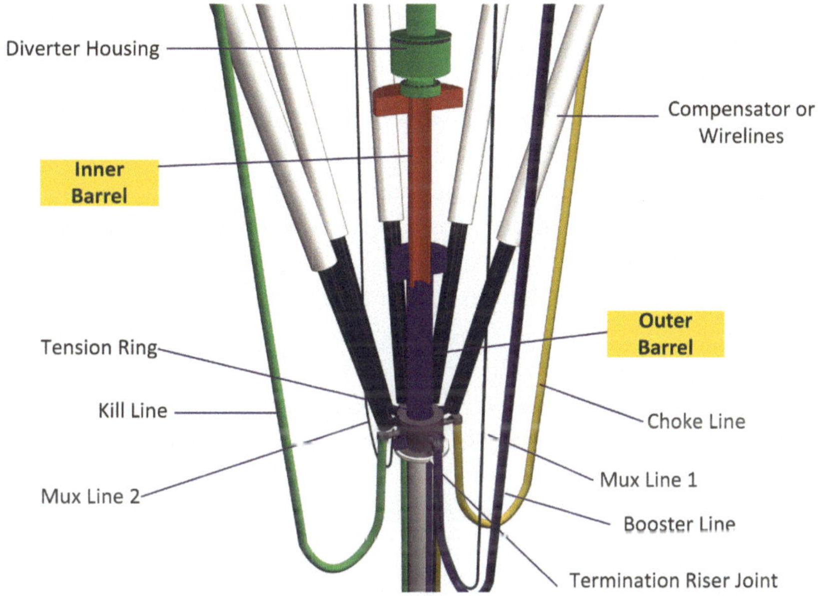

Fig. 7.50 Slip joint in the up position.

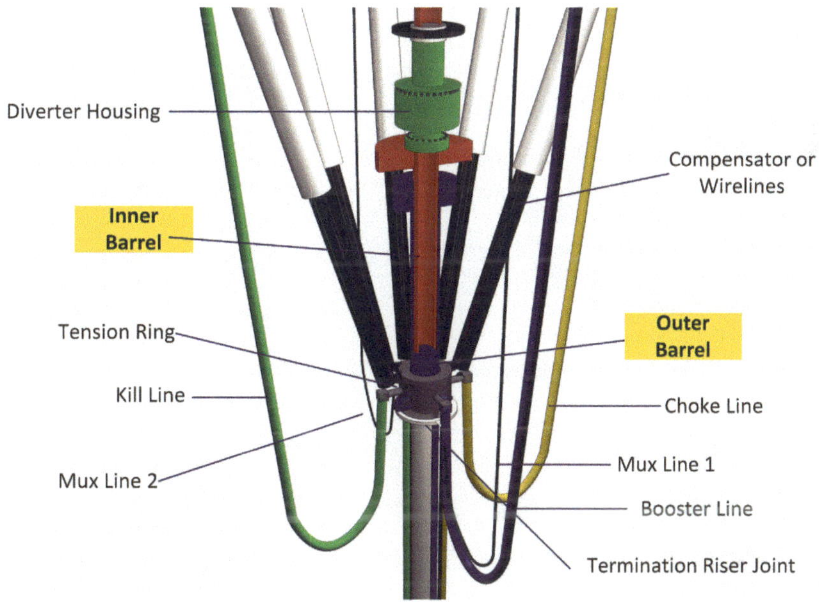

Fig. 7.51 Slip joint in down position.

DURING WELL CONTROL OPERATIONS, ENSURE SLIP JOINT PRESSURE IS INCREASED ACCORDING TO EMERGENCY PROCEDURAL GUIDES

Diverter system

The diverter system includes the diverter, manifolding, valves (flow diverter), control panels, and operating systems. The diverter is normally installed below the rotary table and is affixed to the conduit with flex joint and/or slip joint. The diverter is not a well control pressure device, rather it is simply a device that diverts flow overboard and not onto the drill floor.

The diverter should be plumbed in such a fashion as to open the overboard divert valves before the diverter is closed. Once an open flow route is provided, the diverter is closed by either applying hydraulic pressure directly to the outside of the elastomer or forcing a piston upward using pressurized hydraulic fluid. As the piston is forced upward, the elastomer sealing element is forced inward and seals off the annulus or open hole.

In lieu of divert valves, a mass "flow selector" or "flow divider" may be used. This is a relatively simple system, whereas the overboard divert lines are never fully closed. The flow divider is always set to direct flow into the flowline downstream of the wind direction. With the divert line always opened, the diverter is simply closed and fluids (and/or influx) are directed downwind (Figs. 7.52–7.62).

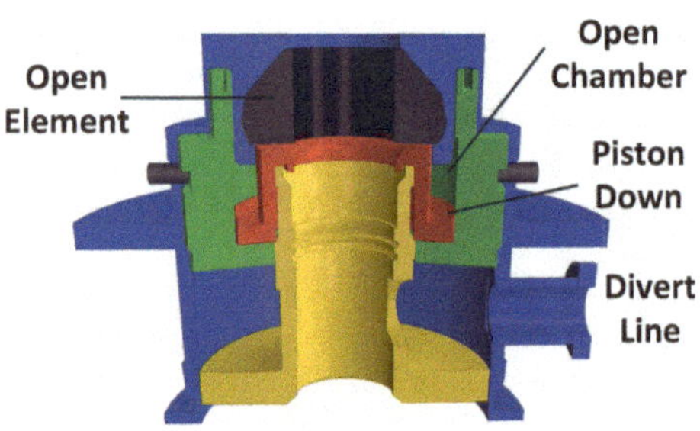

Fig. 7.52 Spherical diverter in open position.

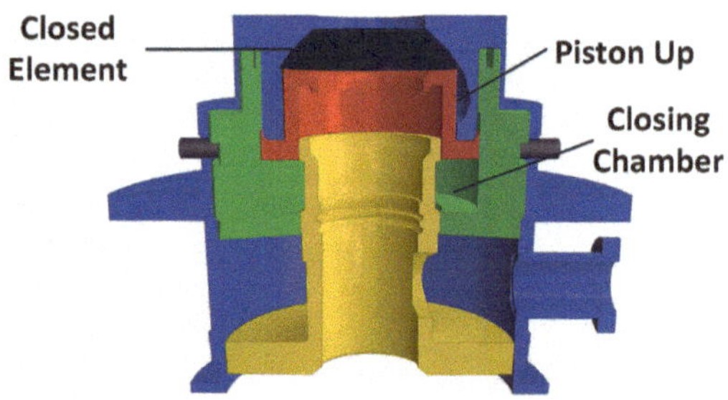

Fig. 7.53 Spherical diverter in closed position.

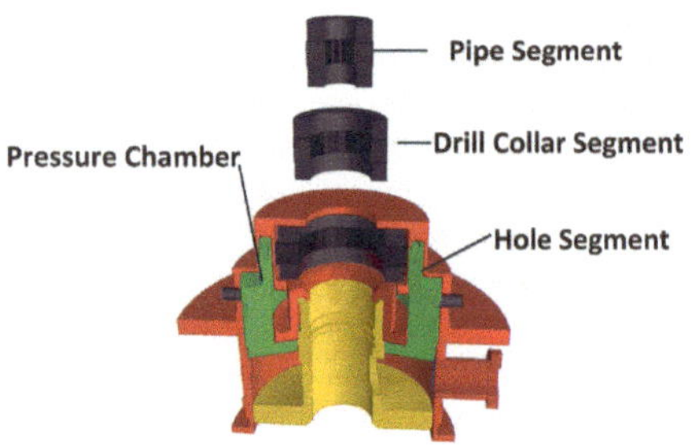

Fig. 7.54 Multiple segment diverter.

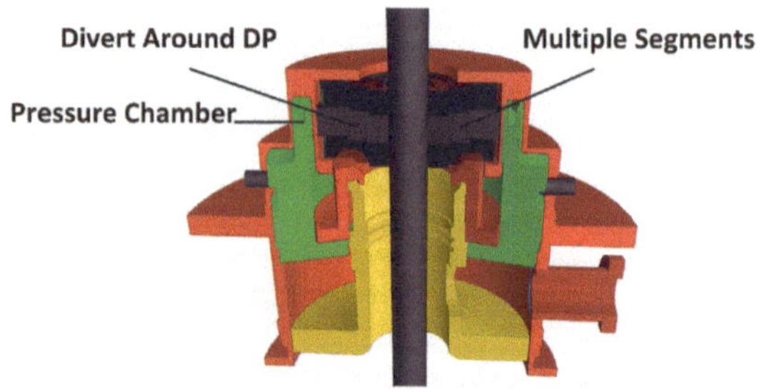

Fig. 7.55 Multiple segment diverter in closed position.

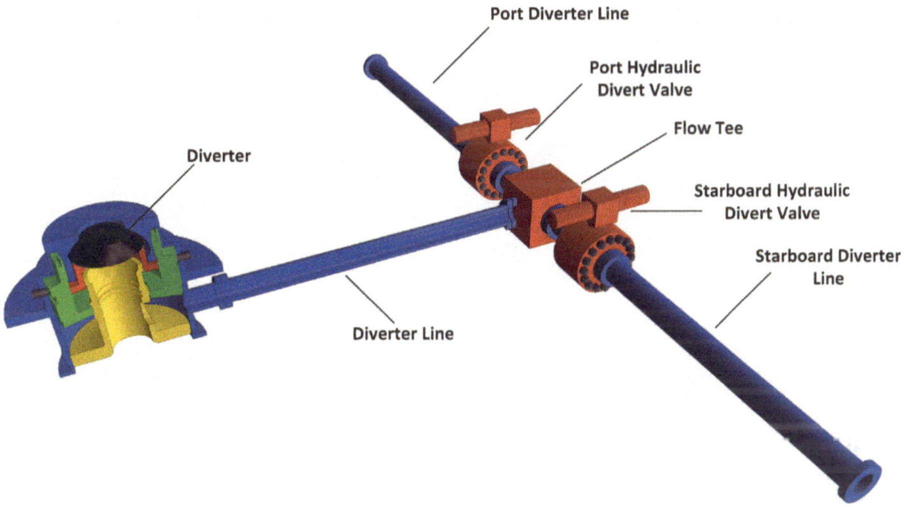

Fig. 7.56 Common starboard and port diverter plumbing arrangements (will vary per vessel).

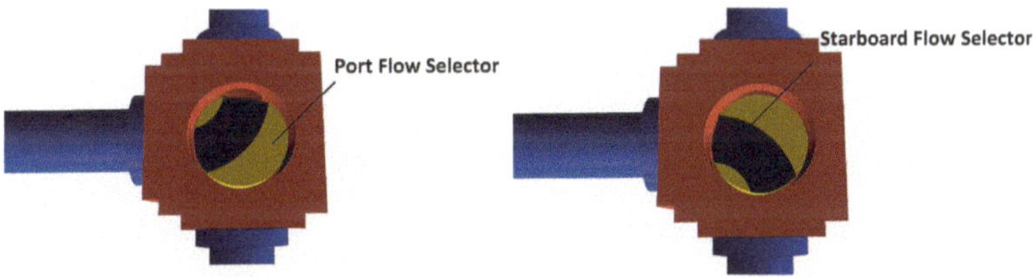

Fig. 7.57 Flow Selector cut-a-way diagram.

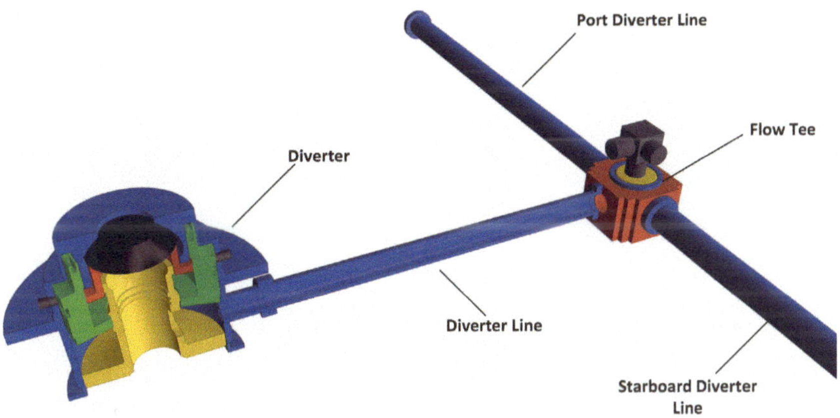

Fig. 7.58 Diverter with flow selector (no divert valves needed).

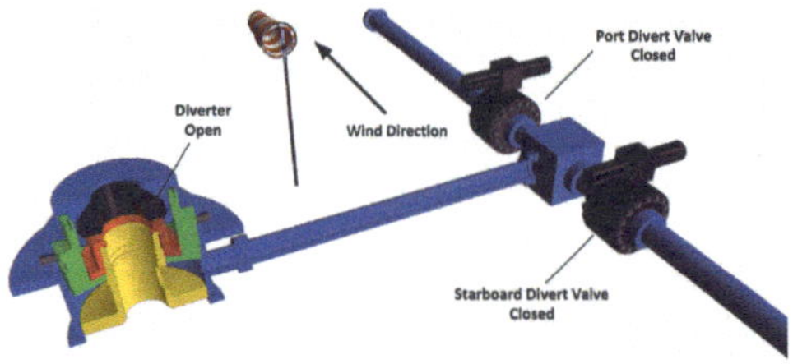

Fig. 7.59 While drilling, diverter open, valves closed.

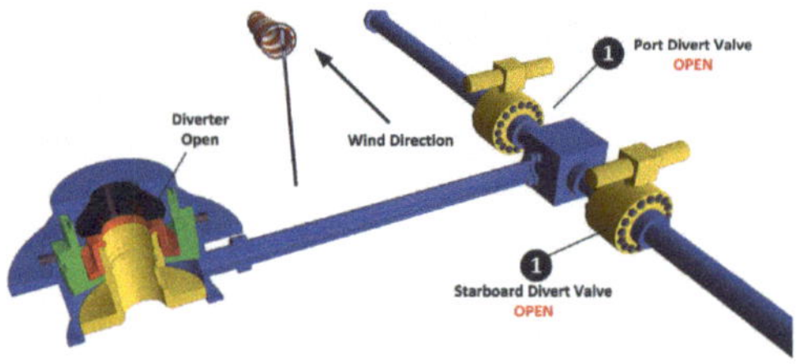

Fig. 7.60 Open both divert valves.

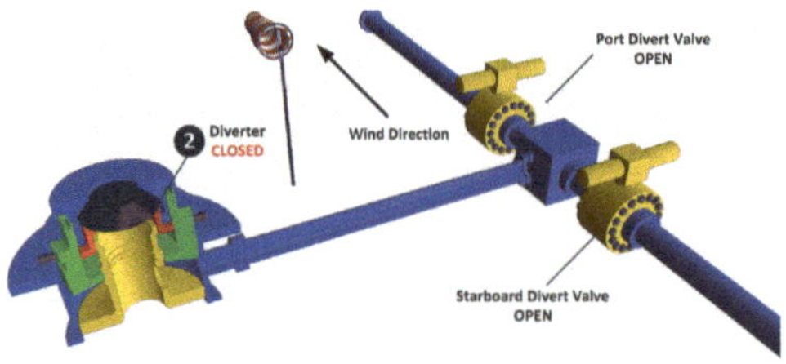

Fig. 7.61 With valves opened, diverter closed.

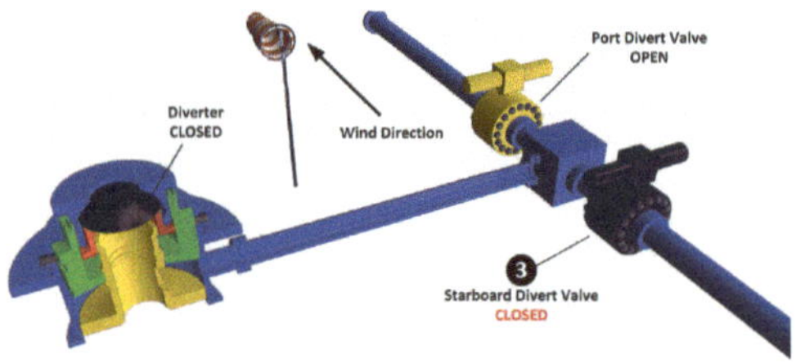

Fig. 7.62 Upwind divert valve closed.

SYSTEMS WITH DIVERT VALVES MUST BE PLUMBED IN SUCH AS FASHION AS BOTH DIVERT VAVLES ARE OPENED BEFORE THE DIVERTER IS CLOSED.
ONCE DIVERTER IS CLOSES, CLOSE APPROPRIATE UPWIND DIVERT VALVE TO DIRECT FLOW DOWNWIND

FOR SYSTEMS WITH FLOW DIVIDERS, ENSURE DIVIDER IS SET TO DIRECT FLOW DOWNWIND.

Abbreviations

ANN	annulus/annular
APB	annular pressure buildup
AZ	Azimuth
B/U	bottoms up
BAR	unit of barometric pressure (14.5 psi)
BBL	barrel (42 US gallons)
BBL/FT	barrels (US) per foot
BBL/MIN	barrels (US) per minute
BBL/STROKE	barrels (US) per stroke
BF	buoyancy factor
BHA	bottom hole assembly
BHP	bottom hole pressure
BML	below mud line
BOP	blowout preventer
BPV	back-pressure valve
BSR	blind shear ram
C&K	choke and kill
CFM	cubic feet per minute
CLFP	choke line friction pressure
CSG	casing
DAT	drill ahead tool
DC	drill collars
DP	drill pipe
DP	dynamic positioning
ECD	equivalent circulating density
EMW	equivalent mud weight
FG	fracture gradient
FIT	formation integrity test
FT	foot or feet
FT/HR	feet per hour
FT/MIN	feet per minute
FT-LBS	foot-pounds (torque generated by line pull per foot of lever)
GAL/STROKE	gallons (US) per stroke
GPM	gallons per minute
GPS	global positioning system
HCR	high closing ratio. Referring to the automatic choke valve
HP	horsepower
HWDP	heavyweight drill pipe
IADC	international association of drilling contractors
ID	inside diameter
IN	inches
KOP	kickoff point
LB	pounds (weight)
LCM	lost circulation material

LEL	lower explosive limit
LMRP	lower marine riser package
LOT	leak-off test
MASP	maximum allowable surface pressure
MCFPD	thousand cubic feet per day
MD	measured depth
MMscfd	million standard cubic feet per day
MSL	mean sea level
MSV	maintenance supply vessel
MUX	multiplex
MWD	measurement while drilling
NAF	nonaqueous fluid
OBM	oil-based mud
OD	outside diameter
P&A	plug and abandon
PDC	poly-crystalline diamond compact (bit cutters)
PIC	person in charge
POB	personnel on board
PPE	personal protective equipment
PPG	pounds per gallon
PPM	parts per million
PR	public relations
PSI	pounds per square inch
PSI/FT	pounds per square inch per foot
PSI/HR	pounds per square inch per hour
PWD	pressure while drilling
RES	reserve pit/reserve mud
RKB	rotary kelly bushing
ROP	rate of penetration
ROV	remotely operated vehicle
SCBA	self-contained breathing apparatus
S.G.	specific gravity
SICP	shut-in casing pressure
SIDPP	shut-in drill pipe pressure
SOBM	synthetic oil-based mud
SPM	strokes per minute
SPP	standpipe pressure
SPR	slow pump rate
SWF	shallow water flow
TBG	tubing
TD	total depth (of section or hole)
TQ	torque
TVD	true vertical depth
UGBO	underground blowout
VDL	variable deck load
WOB	weight on bit
WP	working pressure
WT	weight

Glossary

A

Abnormal pressure Pore pressure in excess of or less than pressure resulting from a vertical salt water column of normal salinity for the geographic region.

Accumulator A vessel containing both hydraulic fluid and gas stored under pressure as a source of fluid power to operate opening and closing of blowout preventer rams and annular preventer elements. Accumulators supply energy for connectors and valves remotely controlled.

Accumulator bank isolator valve A valve located upstream of the accumulator bottles which prevents flow of fluids into or out of the accumulators.

Accumulator relief valve An automatic device located in the accumulator piping that opens when the preset pressure limit has been reached to protect the accumulators.

Accumulator unit The assembly of pumps, valves, lines, and accumulators, which open and close the blowout preventer equipment.

Air breather A device permitting air movement between the atmosphere and the component in which it is installed.

Air pressure switch bypass valve A device which either directs air flow to the shutoff switch or allows the air pumps to run.

Air pump suction valve The opening and closing device which allows fluid from the reservoir into the fluid end of the pump.

Air regulator The adjusting device to vary the amount of air pressure to be discharged down the piping lines. A device to allow downstream pressure to be either increased or decreased as needed.

Air supply valve Opens or shuts off air flow to run the air pumps.

Ambient temperature The temperature of all the surrounding atmosphere within a given area.

American Petroleum Institute A trade organization founded to lead standardization for oil field drilling and producing equipment.

Annular blowout preventer An emergency closure device, normally installed above the ram preventers which forms a seal in the space between the pipe and wellbore or on the wellbore itself. Comprised of large elastomer doughnut style element, the element is designed to close on a variety of tubulars, collars, kellys, etc.

Annular regulator A device located on the accumulator unit to control the amount of closure pressure to the annular preventer.

Annulus The space between two concentric objects such as between the wellbore and casing or between casing and tubulars.

Annulus friction pressure Circulating pressure loss in the annulus between tubulars or open hole.

B

Back off To unscrew the drillstring downhole.

Back pressure (casing pressure, choke pressure) The surface pressure on the casing side of the drill pipe/casing annulus. In reverse circulation, it is the surface pressure at the top of the drillstring.

Baffle A partition plate inside the MGS to help liberate gas of the mud.

Ball up A mass of drill cuttings on drill bits and downhole tools. A bit with such material attached to it is termed a balled-up bit.

Barite Ground barium sulfate which is added to mud to increase density. It has a specific gravity of 4.37 and a size range of 4-74 microns (μm).
Barite plug A heavy slurry of settled barite placed in the wellbore to seal off a pressured zone.
Barite slurry A mixture of barium sulfate, chemicals, and water mixed up to 22 pounds per gallon (lb/gal).
Bell nipple (mud riser, flow nipple) A large piece of pipe, connected to the top of the blowout preventer/annular or marine riser with a side outlet to the shale shaker.
Bent sub A short bent drill collar run above a downhole motor to deflect the downhole motor from vertical to drill a directional hole.
Bentonite A montmorillonite clay used as a major component of drilling mud.
BHA Bottom Hole Assembly. The lower portion of the drillstring which is not DP.
Bleeding Controlled release of fluids from a pressurized system to reduce the pressure.
Blind rams Rams which seal against each other to effectively close and isolate the hole. They do not close on pipe.
Blind/shear rams Blind rams with a built-in cutting edge which will shear tubulars and seal the hole. Used primarily in subsea systems.
Blowout An uncontrolled and unwanted flow of gas, oil, or other well fluids from the well to the atmosphere.
Blowout preventer The equipment installed on the wellhead to enable the well to be sealed to prevent the escape of wellbore fluids and pressure. It is a secondary barrier for well control.
Blowout preventer closing unit Same as accumulator unit.
Blowout preventer drill A training drill conducted to determine the rig crews are competent and familiar with correct practices to safely shut-in a well in the event of a kick. Also referred to as a kick drill.
Blowout preventer stack Blowout preventers along with all the associated spools, valves, and nipples connected to the top of the wellhead.
Blowout preventer test tool A tool to allow pressure testing of the blowout preventers' stack and accessory equipment.
Borehole See wellbore.
Bottom hole pressure Pressure exerted by a column of fluid contained in the wellbore or the formation pressure at the depth of interest.
Bottoms up The volume of displacing mud from total depth to surface. Normally, this displacement is used to coordinate timing of cutting samplings and ensure they represent formations at total depth.
Broaching Venting of fluids to the surface or to the seabed through channels outside of the casing.
Bullheading Pumping into a closed-in well without returns.
Buoyancy The upward force acting on an object submerged in a fluid. The buoyancy force is equal to the weight of fluid displaced by the object.
Bushing A removable-lining or sleeve inserted or screwed into an opening to limit its size, resist wear, or serve as a guide.

C

Casing Large-diameter pipe run into an open hole and cemented in place. Casing is run to protect freshwater formations and isolate formations as the well is drilled deeper.
Casing head/spool The part of the wellhead to which the blowout preventer stack is connected.
Casing pressure See Back Pressure.
Casing seat test A test to determine the integrity of the formation immediately below the casing seat and the integrity of the casing cement job. For hard rock environments, the casing seat test is performed by bullheading fluid into the well at a low rate to a predetermined pressure limit (Formation Integrity Test). For sedimentary environments, fluids are bullheaded into well at low rate until such time the formations begin to accept fluids. When pumps are shut-off, formations close and squeeze fluid back into wellbore (Leak-Off Test). An equivalent mud weight can then be calculated.

Caving Sloughing off formation into the wellbore.
Cellar A pit in the ground within which the blowout preventers are installed on a land rig. It also collects drainage water and other fluids for disposal.
Centrifugal pump A pump which discharges fluid by using an impeller and centrifugal force.
Check valve A valve that permits flow in only one direction.
Choke A control device used to regulate the rate of flow of the drilling mud out of the hole when the blowout preventer is closed and a kick is being circulated out of the hole.
Choke line The high-pressure piping between blowout preventer outlets or wellhead outlets and the choke manifold.
Choke line valve The valve installed on the choke line and used to open or close flow to the choke manifold from the BOPs.
Choke manifold The valves, chokes, and piping system tied into the choke line which direct high-pressure fluids to control flow from the annulus.
Choke pressure See Back Pressure.
Christmas Tree (see Xmas Tree) A series of valves, flanges and housings on the casing head to direct and control the flow of formation fluids from the well.
Circulate To pump the drilling fluid through the whole of the active pumping system from the active tanks down the drillstring, up the annulus, over the solids removal system, and back to the active tanks. A reverse circulation path may be down the kill line and back up through the drillstring to the pits.
Circulating head A device attached to the top of drill pipe or tubing to allow pumping into the well without use of the Kelly or top drive.
Closing unit Accumulator unit.
Closing ratio The ratio of the wellhead pressure to the pressure required to close the blowout preventer.
Collapse pressure The pressure at which a tube will collapse and deform as a result of differential pressure acting from outside to inside.
Conductor pipe A relatively short string of large diameter pipe which is set and cemented at a shallow depth. It keeps the unconsolidated formations in the upper part of the hole open while drilling deeper.
Control manifold The system of valves and piping used to operate the various components of the blowout preventer stack.
Control panel, remote A panel containing a series of control that will operate the valves on the control manifold from a remote point.
Control pod Subsea valves and regulators which direct hydraulic fluid to operate subsea blowout preventer equipment.
Corrosion inhibitor Any substance which slows or prevents the chemical reactions of corrosion.
Crossover A short sub to enable two components of different thread types or sizes to be screwed together.
Cuttings The formation solids drilled up in the well and carried to the surface in the mud.
Cut drilling fluid Fluid which has been reduced in density as a result of contamination by less dense formation fluids or air.

D

Degasser A vessel used to separate entrained gases from the liquid phases of drilling fluids.
Density The mass or weight of a substance per unit volume. The density of a drilling mud may be stated as pounds per gallon (ppg), pounds per cubic foot (lb/ft^3), or kilograms per cubic meter (kg/m^3). Specific gravity and API gravity are other units of density.
Deviation The horizontal distance of departure of the wellbore from the vertical or the angle from which a bit has deviated from the vertical.
Deviated well A well which is directionally drilled away from vertical.

Differential pressure The difference between two fluid pressures; for example, the difference between the pressure in a reservoir and the hydrostatic pressure of the drilling fluid opposite the reservoir.

Differential sticking A condition in which the drill string becomes stuck against the wall of the wellbore. Differential pressure sticking occurs when there is a pressure differential from the mud to the formation and the drillstring is imbedded into the filter cake. Differential sticking occurs when the hydrostatic pressure of the fluid exceeds the formation pressure resulting in the drillstring becoming stuck against the formation.

Discharge check valve The valve which allows fluid flow out of the pump but not back into it.

Displacement 1. The volume of mud which is forced from the well and equal to the volume of tubulars when inserted or withdrawn from the wellbore. 2. The actual volume of steel that the tubulars occupy. 3. The act of replacing one fluid in the wellbore with another by circulation.

Diverter A directional flow device which opens a pathway to direct fluids away from rig, then closes and forms a seal around pipe and wellbore. It is attached to the wellhead or marine riser to close and direct any gas and fluid flow away from the rig.

Downhole motor A drilling tool which allows the bit to turn while the drill string does not turn. It is used most often with a bent motor housing or sub in directional drilling. It is also called mud motor.

Drilling break A sudden increase or decrease in the rate of drilling penetration. This may be due to formation changes or changes in bottom hole pressure.

Drilling fluid weight recorder An instrument which continuously measures drilling fluid density.

Drilling fluid The fluid used to circulate and carry cuttings to the surface, to support the wellbore, and to hold back formation fluids. It is usually called mud.

Drilling spool A component with ends either flanged or hubbed used to space out the BOP stack and provide side outlets for choke or kill lines. It must have an internal diameter at least equal to the bore of the blowout preventer.

Drill pipe safety valve A full-opening valve with threads to match the drill pipe in use and kept on the rig floor. This valve is used to close off the drill pipe to prevent flow from the drill pipe.

Drill stem test (DST) A short-term flow test conducted to determine formation flow rate and/or formation pressure prior to completing the well.

Drill string float A check valve which can be installed in the drill string that will allow fluid to be pumped into the well but will prevent flow from the well through the drill pipe.

Drive pipe A relatively short string of large diameter pipe driven or forced into the ground to function as structural or conductor pipe.

E

Elevation The altitude above a specific reference such as RKB (Rotary Kelly Bushings), MSL (Mean Sea Level), or ML (Mud Line).

Equivalent circulating density (ECD) The pressure on the bottom of the hole while circulating which includes the hydrostatic pressure of fluid, any drilled solids, and friction pressure losses in the annulus expressed in pounds per gallon (lb/gal) or pounds per cubic ft (lb/ft^3).

F

Fault A fracture in subsurface formations in which there has been significant displacement as a result of rock movement. Often, the formation on one side of the fault has moved (upward, downward, or laterally) relative to its original position.

Feed-in (influx, inflow) The flow of formation fluids into the wellbore.

Fill-up line A line usually connected into the bell nipple above the blowout preventers to allow the addition of drilling fluid to the hole.

Filter A device used to clean contaminants from either air flow or a fluid flow.

Filter cake The layer of mud solids that forms on the walls of the borehole opposite permeable formations; also called wall cake or mud cake. Filter cake can be measured using a standard filter press and filter. Known pressure is added to closed reservoir of mud and filter. Fluid is allowed to escape to a measured tube. Filter is removed and the cake is measured in fractions of inch or mm.

Final circulating pressure The initial circulating pressure adjusted for increase in mud density necessary to kill well at predetermined slow circulating rate. Once the KMW has reached the bit, FCP will be held constant until influx is circulated out and KMW reaches surface.

Fish An object left in the wellbore during drilling or workover operations which must be recovered before work can proceed. It may consist of a broken off part of the drilling assembly or production equipment such as packers or screens.

Fishing The procedure to recover lost or stuck equipment in the wellbore.

Flange A ring or collar normally fitted with holes for bolts and attached over the end of a tube or pipe to allow other like objects to be attached to it.

Flash set A premature thickening or setting of cement slurry, which makes it unpumpable.

Float A nonreturn or back-pressure valve located close to the bottom of the BHA to prevent the flow of gas, cuttings, or annular fluids from entering the drill string.

Float collar A short check valve inserted one or two joints above the bottom of the casing string. The float collar prevents drilling mud from entering the casing while it is being lowered and also prevents backflow of cement after a cementing operation.

Float shoe A short, rounded component attached to the bottom of the casing string. It contains a check valve and guides the casing toward the center of the hole.

Flow line A pipe through which drilling fluids travel from the bell nipple to the shale shakers and mud pits.

Flow meter A device which measures either flow rate, total flow, or a combination of both, which travels through pipe or tubing.

Flow rate The volume, mass, or weight of a fluid flowing through pipe or tubing, per unit of time.

Fluid A substance such as a liquid or gas that is capable of flowing and yields to any force tending to change its shape.

Fluid density The unit weight of fluid; e.g., pounds per gallon (lb/gal).

Formation A bed or strata composed substantially of the same kind of rock. Formations which are gas or oil bearing can potentially be hydrocarbon producing.

Formation breakdown Occurs when borehole pressure exceeds the matrix strength of the exposed formation creating fractures which allow it to accept whole fluid from the borehole.

Formation integrity The strength or fracture pressure of the formation.

Formation integrity test (FIT) Application of pressure from the surface on a fluid column in order to determine the strength of a zone to withstand a certain hydrostatic pressure.

Formation pressure (pore pressure) Pressure exerted by fluids within the pores of a reservoir

Flowline sensor A device to monitor rate of fluid flow from the annulus.

Fracture gradient (frac. gradient) The pressure gradient (psi/ft) required to induce fractures of the formation at a given depth.

Fracture zone A zone with naturally occurring fractures that can receive whole mud and cause a total loss of circulation.

Function Perform a specific activity such as operating the control valves of the accumulator system.

G

Gas cut mud Drilling mud that has entrained gas giving the mud a fluffy texture and lighter density.

Gate valve A valve which uses a sliding (fixed) gate or expanding gate to open or shut off flow.

Gauge An instrument for measuring fluid/gas pressure.

Gunk Plug A volume of gunk slurry placed in the wellbore.

Gunk Slurry A slang term to denote a mixture of diesel oil and bentonite typically used to help control mud losses.

H

H_2S An abbreviation for hydrogen sulfide.

Hard shut-in To shut-in a well by closing a blowout preventer with the choke and/or choke line valve closed.

HCR High Closing Ratio. The name given to the hydraulic gate which is between the BOP and the choke manifold.

Hydrostatic pressure The pressure at any point in the wellbore due to the density of the vertical column of fluid above that point.

I

Influx Flow into the wellbore of formation fluids.

Initial circulating pressure Shut-in drillpipe pressure plus slow circulating pressure. This is the initial pressure used when beginning a kill circulation.

Inside blowout preventer A device which can be installed in the drill string that acts as a check valve allowing drilling fluid to be circulated down the string but prevents back flow.

J

Jar A mechanical, mechanical/hydraulic or hydraulic device run downhole used to deliver a heavy downward or upward blow to the drill string if it becomes stuck.

K

Kelly cock (Kelly valve) A valve immediately above/below the Kelly/top drive which can be closed to prevent flow out of the drill string and keeps the mud column inside the Kelly or top drive when disconnecting from the drillstring.

Key seat A channel or groove cut lengthwise in the side of the hole of a well in which the drillstring can become stuck.

Kick An influx of formation fluids into the wellbore.

Kill fluid density The mud weight, e.g. (lb/gal), required to balance formation pressure.

Kill line A high-pressure line connecting the mud pump and the wellhead below a blowout preventer. This line allows drilling fluids to be pumped into the well or annulus underneath a closed blowout preventer.

Kill rate (slow circulating rate) A selected pump rate which is used to circulate the well to remove a kick; kill rate is usually less than the drilling pump rate.

L

Leak-off test (LOT) Application of surface pressure to the drilled hole to determine the pressure at which the exposed formation begins to fracture and accept mud.

Lost circulation (lost returns) The loss of whole drilling fluid to the wellbore.

Lost circulation material (LCM) Materials added into the drilling mud to mitigate or prevent seepage, partial or complete loss of mud to the formation.

Lost returns See Lost Circulation.

Lubrication Pumping a relatively small volume of fluid into a closed wellbore system with gas at the surface and waiting for the fluid to swap places with the gas. Gas can be bled off to reduce surface pressure by an amount equal to the hydrostatic pressure of the mud pumped.

Lubricator A section of wireline riser and stuffing box above wireline BOPs which houses the wireline tools for running and retrieving under pressure.

M

Make a trip To pull the drill string out of the wellbore and then to run the drill string back into the wellbore.

Manifold header A piping system which is able to divide a flow through several different outlets.

Manifold regulator A throttling device which reduces upstream pressure as fluid passes through and exits its chamber.

Manifold regulator bypass valve A valve which can route high-pressure fluid around the regulator allowing higher pressure to operate control valves.

Manifold relief valve A device located in the manifold piping that opens when the preset pressure limit has been reached to protect the manifold header.

Mud Drilling fluid used to provide hydrostatic pressure and prevent influxes from the formation. Other uses include keeping the drill bit cool, circulating cuttings to the surface, and preventing damage to the formation.

Mud cake see Filter Cake.

Mudflow indicator A device that continually measures and records the volume of mud returning from the annulus and flowing out of the mud-return line.

Mud pits The pits are steel storage tanks which provide for storage and use of the drilling mud. Mud is circulated from the well, and it is returned to the pits. Pits are used to settle out solids, treat mud properties, and provide suction to the mud pumps. These are also called mud tanks.

Mud pump A large reciprocating positive displacement pump used to circulate the mud on a drilling rig.

Mud-return line A pipe through which drilling fluids travel from the bell nipple to the shale shakers and mud pits.

Mud weight The density of a drilling fluid expressed as pounds per gallon (ppg), pounds per cubic foot (lbs/ft^3), or kilograms per cubic meter (kg/m^3).

MWD Measurement while drilling.

Mud gas separator A low-pressure vessel for removing free gas from the drilling fluid returns. It is also called a poorboy separator.

N

Needle valve Valve that incorporates a small port and a threaded needle shaped plunger. It can accurately control low flow rates.

Normal pressure Hydrostatic pressure of a vertical column of water from the surface to a subsurface formation. Normal pressure is the hydrostatic pressure within the pore spaces of a formation in a given area.

Nozzle A passageway through a bit that allows the drilling fluid to exit at high velocity and create hydraulic horsepower.

O

OD Outside diameter.

Open The portion of wellbore having no casing or a hole having no drill pipe or tubing suspended in it.

Outside diameter The distance from one outside edge to the other of pipe.

Overbalance The amount of hydrostatic pressure exerted by the fluid in the wellbore which is above the formation pressure.

Overburden The pressure on a formation due to the weight of the sediments above that formation. It is usually estimated at 1 psi/ft of depth.

P

Packer A sealing device that isolates and contains produced fluids and pressures within the tubing string. Packers are designed with flexible, elastomeric elements that expand after running it in the hole.

Packoff A blockage in the annulus or drillstring, usually due to cavings or cuttings, preventing circulation.

Packoff or stripper A device with an elastomer packing element that depends on hydraulic pressure to create a seal. It is used primarily to run or pull pipe or slickline under low or moderate pressures. This device is not dependable for service under high differential pressures.

Permeability The ability of a fluid to flow within the interconnected pore network of a porous formation. See porosity.

Pill A small volume of drilling fluid used for a specific purpose such as to stop loss of circulation or free stuck drillpipe. A heavy pill for lowering the mud level inside the drill string for tripping purposes is also called a Slug.

Pipe rams Blowout preventer ram blocks blocks which are designed to seal around a certain size pipe to close the annular space.

Pit volume indicator A device installed in the drilling fluid tank to register the fluid level in the tank.

Pit volume totalizer A device that combines all of the individual pit volume indicators.

Pore A small void within a rock which can contain water or hydrocarbons or both.

Porosity The percentage of pore space within a rock.

Pore pressure (formation pressure) Pressure exerted by the fluids within the pores of a formation.

Potable A liquid that is suitable for drinking.

Pressure gradient, normal The normal pressure divided by true vertical depth.

Pressure switch An automatic device to start and stop a pump when the present pressure limits are reached.

Pressure transmitter/transducer Device which sends a pressure signal to be converted and calibrated to register the equal pressure reading on a gauge.

Primary well control Maintaining a hydrostatic mud pressure equal to or greater than formation pressure.

Pump (air) A pump operated with an air pressure motor.

Pump (electric) A pump operated by an electric motor.

R

Rate of penetration The speed at which formations are drilled, usually expressed in feet (meters) per hour.

Ram The closing and sealing block on a blowout preventer. Blind rams form a seal on a hole that has no pipe; pipe rams seal around the pipe; shear rams cut through drillpipe and then form a seal.

Reciprocating pump A pump consisting of a piston that moves back and forth in a cylinder. The cylinder is equipped with inlet (suction) and outlet (discharge) valves. This is a positive displacement pump.

Regulator A valve for controlling the pressure of flowing gas or liquid to maintain a predetermined pressure.

Replacement The process whereby a volume of fluid equal to the volume of steel in tubulars and tools withdrawn from the wellbore is returned to the wellbore.

Reservoir 1. A container for storage of liquid. 2. A formation containing hydrocarbons.

Rotating Head A low-pressure sealing device which seals around the drill string above the top of the blow-out preventer stack and allows rotation under pressure.

S

Safety factor An incremental pressure added to shut-in pressures during several well control techniques to ensure hydrostatic pressure is maintained above reservoir pressure.

Set casing To run and cement casing at a certain depth in the wellbore.

Settling pit The mud pit into which mud flows and in which heavy solids are allowed to settle out.

Shale A finely-grained sedimentary rock composed of silt and clay-sized particles.

Shear ram The type of ram block that cuts or shears through pipe and forms a seal against well pressure. Shear rams are used in floating drilling rigs to provide a quick method of moving the rig away from the hole in an emergency.

Single One joint of drill pipe.

Soft Shut-in To shut-in a well by closing a blowout preventer with the choke and choke line valve open and then closing the choke. Theoretically least likely to damage to the formation and equipment, but is the slowest to secure the well.

Squeeze A secondary operation in which cement/mud pill is pumped under pressure to block off an uncemented zone behind casing, or plug perforations/channels, seal off a high pressure zone, or combat lost circulation.

Stabilizer A component placed in the BHA and used to control the deviation of the wellbore. More than one may be used to achieve the intended results.

Stack See blowout-preventer stack.

Stand Normally, two or three joints of pipe screwed together depending on the rig.

Standpipe A vertical pipe running up along the side of the derrick through which mud is pumped to the rotary hose.

Strata Layers of rock or soil which are distinct from those above or below them. An individual bed is a stratum.

Stringer A relatively thin layer of formation that is of a composition different from that of the surrounding strata. A stringer of sandstone, for example, may be imbedded in a shale formation.

String The entire length of casing, tubing, sucker rods, or drill pipe run into a hole.

Stuck pipe Drillpipe, collars, casing, or tubing that cannot be pulled free from the wellbore.

Sub A short, threaded piece of pipe used as a crossover screw together pipes of different sizes or threads.

Surface pipe (casing) The next string of casing (after the conductor pipe) which can range from several hundred feet down to several thousand feet. It is normally used to case off freshwater sands and other weakly consolidated formations.

Swab To reduce pressure in a wellbore by moving pipe, wireline tools, or rubber-cupped seals up the wellbore. If the pressure is reduced sufficiently, reservoir fluids may flow into the wellbore and toward the surface. Opposite term is Surge.

T

Torque The turning force applied to the drillstring causing it to rotate.

Total depth (TD) The maximum depth reached in a well measured along the well path.

Trajectory The directional path of the wellbore.

True vertical depth (TVD) The vertical depth of a well measured from the surface straight down to the bottom of the well.

U

Unconsolidated formation A section of rock which is not tightly bound and susceptible to mud losses while drilling

V

Vacuum degasser A piece of equipment designed to remove gas from drilling fluids.
Vug A cavity in a rock.

W

Wall cake See Filter Cake.
Wall sticking See Differential Sticking.
Washout Excessive wellbore diameter enlargement or a fluid-cut opening caused by fluid leakage.
Water-base mud Drilling mud in which the continuous phase is water.
Weighting material Nearly inert material that has a high specific gravity and is used to increase the density of drilling fluids or cement slurries without affecting the chemical properties of the mud.
Weight on bit (WOB) The weight put on the formation when the bit is resting on bottom.
Weight up To increase the weight or density of drilling fluid by adding weighting material.
Well A drilled hole in the earth constructed to bring hydrocarbons or water to the surface.
Wellbore A hole drilled by the bit and either cased or open (not cased).
Wellbore pressure Total pressure exerted in the wellbore by a column of fluid and/or back pressure imposed at the surface.
WOC Waiting on cement to set or harden.
Wireline A small-diameter metal line used in wireline operations; also called slick line.
Wireline operation The running of tools on wireline into the well for various purposes.

Z

Zone See Formation

Bibliography

IADC Drilling Manual, 2000. Cameron Ram BOPs – Operating Characteristics. ebook Version V.11/Chapter K/Section K.2/Table K 1-8-7. p. K-83, Technical Toolboxes, Inc.

Lapeyrouse, N.J., 1999. Formulas and Calculations for Drilling, Production and Workover. Gulf Professional Publishing Company, Houston, TX, pp. 18–21.

Index

Note: Page numbers followed by *f* indicate figures.

T

U